Real and Functional Analysis

MATHEMATICAL CONCEPTS AND METHODS IN SCIENCE AND ENGINEERING

Series Editor: Angelo Miele
Mechanical Engineering and Mathematical Sciences
Rice University, Houston, Texas

Volume 1 **INTRODUCTION TO VECTORS AND TENSORS**
Volume 1: Linear and Multilinear Algebra
Ray M. Bowen and C.-C. Wang

Volume 2 **INTRODUCTION TO VECTORS AND TENSORS**
Volume 2: Vector and Tensor Analysis
Ray M. Bowen and C.-C. Wang

Volume 3 **MULTICRITERIA DECISION MAKING AND DIFFERENTIAL GAMES**
Edited by George Leitmann

Volume 4 **ANALYTICAL DYNAMICS OF DISCRETE SYSTEMS**
Reinhardt M. Rosenberg

Volume 5 **TOPOLOGY AND MAPS**
Taqdir Husain

Volume 6 **REAL AND FUNCTIONAL ANALYSIS**
A. Mukherjea and K. Pothoven

Volume 7 **PRINCIPLES OF OPTIMAL CONTROL THEORY**
R. V. Gamkrelidze

Volume 8 **INTRODUCTION TO THE LAPLACE TRANSFORM**
Peter K. F. Kuhfittig

Volume 9 **MATHEMATICAL LOGIC**
An Introduction to Model Theory
A. H. Lightstone

Volume 10 **SINGULAR OPTIMAL CONTROLS**
R. Gabasov and F. M. Kirillova

Volume 11 **INTEGRAL TRANSFORMS IN SCIENCE AND ENGINEERING**
Kurt Bernardo Wolf

Real and Functional Analysis

A. Mukherjea and K. Pothoven
University of South Florida, Tampa

Plenum Press · New York and London

Library of Congress Cataloging in Publication Data

Mukherjea, A
Real and functional analysis.

(Mathematical concepts and methods in science and engineering; v. 6)
Bibliography: p.
Includes index.
1. Functions of real variables. 2. Functional analysis. I. Pothoven, K., joint author. II. Title.
QA331.5.M84 515'.8 77-14282
ISBN 0-306-31015-5

A Division of Plenum Publishing Corporation
227 West 17th Street, New York, N.Y. 10011

Printed in the United States of America

Preface

This book introduces two most important aspects of modern analysis: the theory of measure and integration and the theory of Banach and Hilbert spaces. It is designed to serve as a text for first-year graduate students who are already familiar with some analysis as given in a book similar to Apostol's *Mathematical Analysis*.†

This book treats in sufficient detail most relevant topics in the area of real and functional analysis that can be included in a book of this nature and size and at the level indicated above. It can serve as a text for a solid one-year course entitled "Measure and Integration Theory" or a comprehensive one-year course entitled "Banach Spaces, Hilbert Spaces, and Spectral Theory." For the latter alternative, the student is, of course, required to have some knowledge of measure and integration theory. The breadth of the book gives the instructor enough flexibility to choose what is best suited for his/her class. Specifically the following alternatives are available:

(a) A one-year course on "Measure and Integration" utilizing Chapters 1 (Sections 1.1–1.3 and 1.6), 2, 3, 4, portions of 5 (information on L_p spaces), and portions of 7 (left to the discretion of the teacher).
(b) A one-year course in "Functional Analysis" utilizing Chapters 1 (Sections 1.4–1.6), 5, 6, 7 (Sections 7.4 and 7.6), and the Appendix.

† T. M. Apostol, *Mathematical Analysis*, 2nd ed., Addison-Wesley (1974).

(c) A two-year course in "Real and Functional Analysis" utilizing Chapters 1–4 in the first year and Chapters 5 through the Appendix in the second year.

If time is a factor in covering the suggested material, certain portions of the text designated by (•) can be omitted.

The rudiments of point-set topology are presented in succinct form in Chapter 1. The material presented there is essential for analysis but is certainly not part of the main theme of the text. Thus the presentation of topology is deliberately concise and can be used as a reference for the essential facts of topology. Also included in Chapter 1 are various interesting extensions of the classical Banach Fixed Point Theorem that depend only on elementary concepts of metric spaces. Although the applicability of all these results has not yet been established, many of these results are recent and mathematically appealing.

Quite a few of the exercises in the text may prove to be difficult. Such problems are starred (⋆). They have been included not only to challenge the most serious student, but, more importantly, to give various extensions and variations of the material presented in the text.

Although it is not our intent to emphasize the applications of measure theory and functional analysis, sufficiently many illustrations and applications of most of the major theorems in the text have been included to demonstrate that this area of mathematics has been a standard subject of study because of its applicability in diverse disciplines of mathematics and science. Our intent is to present with clarity, sufficient motivation, and rigor, up-to-date and detailed information in the subject area of real and functional analysis. We regret that because of size limitations some important topics such as Banach algebras and semigroups of operators could not be included in this volume. The topics that are included reflect not only their importance in the context of what every graduate student of mathematics ought to know, but also our own preferences. Though most of the results are classical and well known, the novelty of the book lies in the arrangement and treatment of the subject matter along with the inclusion of much material so far available only in research papers.

It is with sincere thanks and appreciation that we acknowledge the help of various individuals whose efforts made the completion of the book possible. First, our gratitude to Betty Pothoven for assiduously typing a major portion of the original manuscript. We are also thankful to those friends and colleagues who read portions of the manuscript and gave many useful constructive criticisms, especially to K. Iseki, R. A. Johnson, E. B.

Saff, B. Schreiber, V. M. Sehgal, D. Snider, and J. Ward. We are also indebted to our able and kind students at the University of South Florida who studied from the manuscript and pointed out many needed improvements. Finally, our thanks to Plenum Publishing Company, especially to Ms. Betty Bruhns for her patience and efforts in the publication of this work.

A. Mukherjea
K. Pothoven

Contents

1

Preliminaries on Set Theory and Topology

One of the aims of this chapter is to introduce the reader to those parts of set theory and topology that are used frequently in the main theme of this text. While providing these preliminaries, for completeness we also consider here almost all basic concepts in point-set topology. The sections covering this material are presented in a manner somewhat different from that used for the other sections. Thus readers with very little or no background in topology may have to do a little (but not much) extra work while studying these sections.

We also consider fixed point theorems in Section 1.5. In recent years, the study of fixed point theorems has been found to be very fascinating and extremely useful in various problems in analysis, and it is now a subject in its own right. Here we have attempted to present a somewhat complete account of those fixed point theorems that depend only on very elementary concepts in metric spaces. A part of this account is given in the form of problems.

Finally, we present in this chapter two very important results in analysis that have been used frequently in the rest of this text: the Stone–Weierstrass Theorem and the Ascoli Theorem.

1.1. Basic Notions of Sets and Functions

It is not our intention to give a rigorous treatment of the theory of sets or even to give an elaborate discourse on this theory. Readers interested in an axiomatic treatment of the theory of sets should consult texts such as

Axiomatic Set Theory by P. Suppes[†] or *Naive Set Theory* by P. Halmos.[‡] It is our intention, rather, in this brief section to establish notation and terminology to be used in the text as well as to explicitly list needed relationships and theorems involving sets and functions, many of which the reader is probably already familiar with.

Sets will generally be denoted by capital letters as A, B, C, X, Y, or Z and elements of sets by small letters a, b, c, x, y, or z. The small Greek letters α, β, γ, ... will generally be used to represent real or complex numbers when used as scalars in vector spaces. Classes or families of sets will be denoted by capital script letters as $\mathscr{A}$, $\mathscr{B}$, $\mathscr{C}$, ...

As is standard, if $P(x)$ denotes some property about x, the notation

$$\{x: P(x)\}$$

will be used to mean "the set of all elements x (from some universal set) for which $P(x)$ is true."

In this text the following notations will be used:

$\varnothing$ is the empty set.
N or Z^+ is the set of natural numbers $\{1, 2, 3, \ldots\}$.
Z is the set of integers $\{\ldots, -3, -2, -1, 0, 1, 2, 3, \ldots\}$.
Q is the set of rational numbers.
R is the set of real numbers.

Definition 1.1. If A and B are sets, *A is a subset of B*, written $A \subset B$, if each element of A is an element of B. If in addition there is an element of B not in A we write $A \subsetneqq B$. Set A equals set B, written $A = B$, if $A \subset B$ and $B \subset A$.

The *union* $A \cup B$ of A and B is defined as

$$A \cup B \equiv \{x: x \in A \text{ or } x \in B\};$$

the *intersection* $A \cap B$ by

$$A \cap B \equiv \{x: x \in A \text{ and } x \in B\};$$

the *complement* of B in A by

$$A - B \equiv \{x: x \in A \text{ and } x \notin B\};$$

† P. Suppes, *Axiomatic Set Theory*, Dover, New York (1972).
‡ P. Halmos, *Naive Set Theory*, Springer-Verlag, New York (1970).

and the *symmetric difference* $A \triangle B$ by

$$A \triangle B \equiv (A - B) \cup (B - A).$$

If U is the universal set, the complement A^c of A is the set $U - A$.

If $\mathscr{A}$ is some family of sets, then

$$\bigcup_{A \in \mathscr{A}} A \equiv \{x\colon\ x \in A \text{ for some } A \text{ in } \mathscr{A}\},$$

and

$$\bigcap_{A \in \mathscr{A}} A \equiv \{x\colon\ x \in A \text{ for each } A \text{ in } \mathscr{A}\}.$$

If I is a nonempty set and for each $i \in I$ there is associated a set A_i (an indexed family of sets), then

$$\bigcup_{i \in I} A_i \equiv \{x\colon\ x \in A_i \text{ for some } i \in I\},$$

and

$$\bigcap_{i \in I} A_i \equiv \{x\colon\ x \in A_i \text{ for each } i \in I\}.$$

A collection of sets $\mathscr{A}$ is said to be *pairwise disjoint* if $A \cap B = \varnothing$ whenever A and B are in $\mathscr{A}$ and $A \neq B$. ∎

The following theorem is easy to verify.

Theorem 1.1. If A, B, and C are sets and $\mathscr{A}$ is a family of sets,

$$\left.\begin{array}{ll} \text{(i)} & B \cap \Big(\bigcup_{A \in \mathscr{A}} A\Big) = \bigcup_{A \in \mathscr{A}} (B \cap A) \\ \text{(ii)} & C \cup \Big(\bigcap_{A \in \mathscr{A}} A\Big) = \bigcap_{A \in \mathscr{A}} (C \cup A) \end{array}\right\} \quad \text{distributive laws}$$

$$\left.\begin{array}{ll} \text{(iii)} & B - \Big(\bigcup_{A \in \mathscr{A}} A\Big) = \bigcap_{A \in \mathscr{A}} (B - A) \\ \text{(iv)} & C - \Big(\bigcap_{A \in \mathscr{A}} A\Big) = \bigcup_{A \in \mathscr{A}} (C - A) \end{array}\right\} \quad \text{De Morgan's laws.}$$ ∎

Definition 1.2. Let X and Y be sets. The *Cartesian product* of X and Y is the set $X \times Y$ given by

$$X \times Y \equiv \{(x, y)\colon\ x \in X \text{ and } y \in Y\}.$$

Two elements (x, y) and (x', y') of $X \times Y$ are equal if and only if $x = x'$ and $y = y'$. [Note that if x and y are elements of $X \cap Y$, $(x, y) = (y, x)$ if and only if $x = y$.] ∎

Definition 1.3. Let X and Y be sets. A *relation* f *from* X *into* Y is any subset of $X \times Y$. The *domain* of a relation f is the set

$$D_f = \{x \in X: (x, y) \in f \text{ for some } y \in Y\}.$$

The *range* R_f of a relation f is the set

$$R_f = \{y \in Y: (x, y) \in f \text{ for some } x \in X\}.$$

If f is a relation from X into Y, the *inverse relation* f^{-1} is given by

$$f^{-1} = \{(y, x): (x, y) \in f\}.$$

(Note that $D_{f^{-1}} = R_f$ and $R_{f^{-1}} = D_f$.) If f and g are relations from X into Y and from Y into Z, respectively, the *composition* $g \circ f$ is the relation from X into Z given by

$$g \circ f = \{(x, z) \in X \times Z: (x, y) \in f \text{ and } (y, z) \in g \text{ for some } y \text{ in } Y\}.$$

(Note $g \circ f \neq \varnothing$ if and only if $R_f \cap D_g \neq \varnothing$.) If f and g are both relations from X into Y and $f \subset g$, then f is said to be a *restriction* of g or g is said to be an *extension* of f. ∎

As is customary, we write $y = f(x)$ if $(x, y) \in f$, where f is a relation. More generally we have the following definition.

Definition 1.4. If f is a relation from X into Y and A is a subset of X, the *image* of A under f is the set

$$f(A) \equiv \{y \in Y: y = f(x) \text{ for some } x \in A\}.$$

If B is a set in Y, the *inverse image* of B under f is the set

$$f^{-1}(B) = \{x \in X: y = f(x) \text{ for some } y \in B\}.$$ ∎

The reader may prove the following.

Proposition 1.1. Let f be a relation from X into Y and $\mathscr{A}$ a collection of subsets from X. Then

(i) $f(\bigcup_{A \in \mathscr{A}} A) = \bigcup_{A \in \mathscr{A}} f(A)$,

(ii) $f(\bigcap_{A \in \mathscr{A}} A) \subset \bigcap_{A \in \mathscr{A}} f(A)$. ∎

Definition 1.5. A *function f from X into Y*, denoted $f\colon X \to Y$, is a relation from X into Y such that

(i) $D_f = X$
(ii) if $(x, y) \in f$ and $(x, y') \in f$, then $y = y'$.

[Thus if $y = f(x)$ and $y' = f(x)$, then $y = y'$. We can thereby speak of y as *the image* of x under f, denoted by $f(x)$.] If $R_f = Y$ then f is called a *surjection* or an *onto* function from X into Y. If $f(x) = f(x')$ always implies $x = x'$, then f is called *injective* or a *one-to-one* function. If f is surjective and injective, it is called *bijective* or a *one-to-one correspondence*. A function is also called a *mapping* or *transformation*. ▮

Proposition 1.2. If $f\colon X \to Y$ and $\mathscr{B}$ is a collection of subsets of Y, then

(i) $f^{-1}\left(\bigcup_{B\in\mathscr{B}} B\right) = \bigcup_{B\in\mathscr{B}} f^{-1}(B)$,
(ii) $f^{-1}\left(\bigcap_{B\in\mathscr{B}} B\right) = \bigcap_{B\in\mathscr{B}} f^{-1}(B)$,
(iii) $f^{-1}(B^c) = [f^{-1}(B)]^c$,
(iv) $f(f^{-1}(B)) \subset B$,
(v) $f^{-1}(f(A)) \supset A$. ▮

The proofs of these easily verified statements are left for the reader.

There are special types of functions that will be encountered frequently. Corresponding to each set A is the identity function $1_A\colon A \to A$ given by

$$1_A(a) = a.$$

Corresponding to every subset A of some universal set U is the characteristic function $\chi_A\colon U \to \{0, 1\}$ given by

$$\chi_A(x) = \begin{cases} 1 & \text{if } x \in A \\ 0 & \text{if } x \notin A \end{cases}$$

If $D = \{(x, y) \in X \times X\colon x = y\}$ for some set X, then *Kronecker's δ function* corresponding to X is the function $\delta\colon X \times X \to \{0, 1\}$ given by

$$\delta_{xy} \equiv \delta(x, y) = \chi_D(x, y).$$

A function $f\colon N \to X$, where X is any set, is called a *sequence* in X. Because a sequence is uniquely and completely determined by the values x_i equal to $f(i)$ for $i \in N$, a sequence is usually denoted by $(x_i)_{i\in N}$ without explicit reference to f. The value x_i is called the ith term of the sequence $(x_i)_N$. A *finite sequence* in X or an *n-tuple in X* is a function from $\{1, 2, \ldots, n\}$ into X, denoted by $(x_i)_{i=1}^n$, where $x_i = f(i)$.

Definition 1.6. A *finite set* is any set that is empty or the range of a one-to-one correspondence on the set $\{1, 2, \ldots, n\}$ for some $n \in N$. (See Problem 1.1.6.) Any set not finite is called *infinite*. A *countably infinite* set is any set that is the range of a one-to-one correspondence on N. A *countable set* is any set that is finite or countably infinite. An *uncountable* set is any set not countable.

Two sets A and B have the same *cardinality* (written card $A =$ card B) if there is a bijection from one set onto the other. It is easy to show that for any positive integers n and m, card $\{1, 2, \ldots, m\} =$ card $\{1, 2, \ldots, n\}$ if and only if $n = m$. By virtue of this fact we can define the *cardinality* of a finite set A to be n if there is a bijection from $\{1, 2, \ldots, n\}$ onto A. The cardinality of $\varnothing$ is zero. The cardinality of a countably infinite set is $\aleph_0$, and the cardinality of R is c. Also card $A \leq$ card B if there is an injection from A to B. ∎

We have already spoken of an "indexed" family of sets in Definition 1.1. Generally we say that a nonempty set of elements A is an *indexed set* if there is a nonempty set I, called the *index set*, and an onto function f: $I \to A$. A can then be denoted by $(a_i)_{i \in I}$, where $a_i = f(i)$ for $i \in I$. Clearly an indexed family of sets is nothing more than an indexed subset of the *power set* 2^U, the set of all subsets of some universal set U.

Clearly any countably infinite set can be indexed by N. Conversely any infinite set indexed by N is countably infinite. Indeed suppose $A = (a_i)_{i \in N}$ is an infinite set. Define f: $N \to A$ recursively as follows: Let $f(1) = a_1$; assuming $f(1), f(2), \ldots, f(k)$ are chosen, denote $f(1)$ by a_{n_1}, $f(2)$ by a_{n_2}, $\ldots$, and $f(k)$ by a_{n_k}; define $f(k+1)$ to be $a_{n_{k+1}}$ where n_{k+1} is the smallest positive integer n greater than $n_1, \ldots, n_k$ such that $a_n \notin \{a_{n_1}, a_{n_2}, \ldots, a_{n_k}\}$. That n_{k+1} exists follows from the fact that every nonempty subset of N has a least element and the fact that A is an infinite set. The reader may check that f: $N \to A$ is a bijection.

The facts recorded in the following propositions are important.

Proposition 1.3. Any subset B of a countable set A is countable. ∎

Proof. Assume B is not empty. Since A is countable, A can be written as $(a_n)_N$. Let n_1 be the smallest positive integer n such that $a_n \in B$. Inductively, assuming $B - \{a_{n_1}, \ldots, a_{n_k}\}$ is not empty, let n_{k+1} be the smallest positive integer n such that $a_n \in B - \{a_{n_1}, \ldots, a_{n_k}\}$. If for some k, $B - \{a_{n_1}, a_{n_2}, \ldots, a_{n_k}\} = \varnothing$, then B is finite. If not, we have defined recursively a function $f: N \to B$ given by $f(k) = a_{n_k}$. f is easily seen to be bijective. ∎

Proposition 1.4. The Cartesian product $N \times N$ is countable. ∎

Proof. Define $f: N \to N \times N$ by $f(n) = (p, q)$, where $n = 2^{p-1}(2q - 1)$ and p is the smallest positive integer such that 2^p does not divide n. Since for any q in N, $2q - 1$ is odd, f is clearly onto and one-to-one. ∎

Proposition 1.5. The union of a countable family of countable sets is a countable set. ∎

Proof. The countable family can be written as $(A_n)_{n \in N}$, where each A_n is countable. Let $A = \bigcup_{n \in N} A_n$. Since an empty set adds nothing to the union, we may assume $A_n \neq \varnothing$ for each n. For each n there is a surjection $g_n: N \to A_n$. Define $g: N \times N \to A$ by $g(n, m) = g_n(m)$, an element of A_n. It is obvious that g is a surjection. Letting h be the composition $g \circ f$, where f is a bijection from N onto $N \times N$ as in Proposition 1.4, h is a surjection from N onto A. According to our remarks preceding Proposition 1.3, A is countable. ∎

An immediate consequence of Proposition 1.5 is the fact that the sets Z and Q are countable sets. Indeed,

$$Z = \{-n: n \in N\} \cup \{0\} \cup N$$

and

$$Q = \bigcup_{n=1}^{\infty} \{m/n: m \in Z\}.$$

In contrast the set R is uncountable. The proof of this is realized by showing that the subset $(0, 1)$ of positive real numbers less than 1 is uncountable. Since each real number in $(0, 1)$ can be written as an infinite decimal, if $(0, 1)$ were a countable set $(r_n)_{n \in N}$, then each element in $(0, 1)$ could be expressed as an infinite decimal

$$r_n = 0.s_{n_1}s_{n_2}\cdots$$

for some n and for some sequence $(s_{n_i})_{i \in N}$, where $s_{n_i} = 0, 1, \ldots,$ or 9 for each i. To see that this is impossible let

$$x = 0.x_1x_2\cdots,$$

where

$$x_n = \begin{cases} 1 & \text{if } s_{n_n} \neq 1 \\ 2 & \text{if } s_{n_n} = 1. \end{cases}$$

Then x is an infinite decimal representing a real number in $(0, 1)$ but not in $(r_n)_{n\in N}$ since $x_1 \neq s_{1_1}$, $x_2 \neq s_{2_2}$, ... Hence $(0, 1) \neq (r_n)_{n\in N}$.

As a final consideration in this section, let $(A_n)_N$ be any countable collection of subsets of a set X. The *limit superior* of $(A_n)_N$, also denoted by $\overline{\lim}\, A_n$, is given by

$$\lim_n \sup A_n \equiv \bigcap_{n=1}^{\infty} \left(\bigcup_{m=n}^{\infty} A_m \right).$$

The *limit inferior*, also denoted by $\underline{\lim}\, A_n$, is given by

$$\lim_n \inf A_n \equiv \bigcup_{n=1}^{\infty} \left(\bigcap_{m=n}^{\infty} A_m \right).$$

The reader should convince her- or himself that $\lim_n \sup A_n$ is the set of all points from X that belong to A_n for infinitely many n, whereas $\lim_n \inf A_n$ consists of all points from X belonging to A_n for all but finitely many n. The *limit of* A_n is defined as the common value of $\lim_n \sup A_n$ and $\lim_n \inf A_n$, if it exists.

Problems

1.1.1. Prove the following:

(i) $A \triangle B = A \cup B - A \cap B$.

(ii) $A = A \triangle B$ if and only if $B = \varnothing$.

(iii) $A \cap (B \triangle C) = (A \cap B) \triangle (A \cap C)$.

(iv) $\bigcup_{i=1}^{\infty} (A_i \cup B_i) = \left(\bigcup_{i=1}^{\infty} A_i\right) \triangle \left[\left(\bigcup_{i=1}^{\infty} A_i\right)^c \cap \left(\bigcup_{i=1}^{\infty} B_i\right)\right]$.

1.1.2. Suppose f: $X \to Y$ and g: $Y \to X$ are functions.

(i) If $g \circ f = 1_X$, prove f is injective.

(ii) If $f \circ g = 1_Y$, prove f is surjective.

(iii) If $g \circ f = 1_X$ and $f \circ g = 1_Y$, prove $g = f^{-1}$.

1.1.3. Show that the inclusion in part (ii) of Proposition 1.1. may be proper.

1.1.4. Show that the inclusions in parts (iv) and (v) of Proposition 1.2 may be proper and that parts (i), (ii), and (iii) of Proposition 1.2 may fail for arbitrary relations.

1.1.5. Show that if X is a set, then 2^X, the set of all subsets of X, cannot be indexed by X. [Hint: if $f\colon X \to 2^X$ consider $\{x \in X\colon x \notin f(x)\}$.]

1.1.6. (i) Prove that nonempty set A is finite if and only if there is a bijection $f\colon A \to \{1, \ldots, n\}$ for some n.

(ii) Prove that set A is countably infinite if and only if there is a bijection $f\colon A \to N$.

1.1.7. If $(A_n)_N$ is a monotone sequence of sets (that is, $A_{n+1} \subset A_n$ for all $n \in N$ or $A_{n+1} \supset A_n$ for all $n \in N$), then $\lim_n A_n$ is $\bigcup_n A_n$ or $\bigcap_n A_n$.

1.1.8. If B is a set and $(A_n)_N$ is a sequence of sets, prove

$$B - \lim_n \sup A_n = \lim_n \inf(B - A_n),$$
$$B - \lim_n \inf A_n = \lim_n \sup(B - A_n).$$

1.2. Relations, Orderings, Zorn's Lemma, and the Axiom of Choice

Given two sets X and Y, we have defined a relation from X to Y as a subset of $X \times Y$. We now wish to consider relations *in* X, that is relations from X to X. Our first definition gives the characteristics of a special type of relation in a set X.

Definition 1.7. Let X be a set. An *equivalence relation* $\wr$ in X is a relation $\wr$ in X ($\wr \subset X \times X$) such that

(i) $x \wr x$ for all x in X (reflexivity);

(ii) if $x \wr y$ for x and y in X, then $y \wr x$ (symmetry);

(iii) if $x \wr y$ and $y \wr z$ for x, y, and z in X, then $x \wr z$ (transitivity),

where $x \wr y$ is notation meaning $(x, y) \in \wr$.

If $\wr$ is an equivalence relation in X and $x \in X$, let

$$[x] = \{y \in X\colon x \wr y\}.$$

$[x]$ is called the *equivalence class* of x with respect to $\wr$. ∎

It is easy to prove the statement $[x] = [y]$ if and only if $x \wr y$. The nonequal equivalence classes partition X in the sense that X can be written as the union of disjoint equivalence classes. In fact if a *partition* of X is defined as a collection of pairwise disjoint nonempty subsets of X whose union is X, there is a one-to-one correspondence between equivalence relations on X and partitions of X. The reader is referred to Problem 1.2.1.

Another special type of relation in X is given in the next definition.

Definition 1.8. Let X be a set. A *partial ordering* $\not{p}$ on X is a relation in X such that

(i) $x \not{p}\, x$ for each x in X (reflexivity);
(ii) if $x \not{p}\, y$ and $y \not{p}\, x$, then $x = y$ (antisymmetry);
(iii) if $x \not{p}\, y$ and $y \not{p}\, z$, then $x \not{p}\, z$ (transitivity);

here again $x \not{p}\, y$ is notation meaning $(x, y) \in \not{p}$.

A *total ordering* on X is a partial ordering $\not{p}$ on X such that

(iv) if x and y are in X, then $x \not{p}\, y$ or $y \not{p}\, x$.

A *well ordering* on X is a total ordering $\not{p}$ on X such that

(v) if A is any nonempty subset of X then there exists an element a in A such that $a \not{p}\, x$ for all x in A.

A *partially ordered* (or, respectively, *totally ordered* or *well ordered*) set is a pair $(X, \not{p})$, where X is a set and $\not{p}$ is a partial (or total or well) ordering on X. ∎

Examples

1.1. Probably the most familiar example of a totally ordered set is the set of real numbers with the usual "less than or equal to" relation denoted by $\leq$. This is not a well ordering.

1.2. Likewise a very familiar example of a well-ordered set is the set of natural numbers N with the relation $\leq$. The fact that $(N, \leq)$ is well ordered is equivalent to the *principle of mathematical induction*: If A is a subset of N such that $1 \in A$ and $n + 1 \in A$ whenever $n \in A$ for any n in N, then $A = N$. (See Problem 1.2.2.)

Usually an arbitrary partial ordering on an arbitrary set is denoted by $\leq$, notation which is obviously derived from the "less than or equal to" partial ordering on R. In this case $x < y$ means $x \leq y$ but $x \neq y$.

1.3. A simple example of a partial ordering that is not a total ordering is the inclusion ordering on the set of subsets 2^X of a set X with two or more elements given by

$$A \leq B \text{ if and only if } A \subset B \text{ for } A, B \in 2^X.$$

Definition 1.9. Suppose $(X, \leq)$ is a partially ordered set. If $A \subset X$, then $x \in X$ is an *upper bound* of A if $a \leq x$ for all $a \in A$. If x is an upper bound of A and $x \leq y$ whenever y is an upper bound of A, then x is the *least upper bound* of A, written $x = \sup A$. (By antisymmetry there can be at most one.) Likewise a *lower bound* and a *greatest lower bound* of A (written $\inf A$) are defined. $(X, \leq)$ is *order complete* if every nonempty set A in X with an upper bound has a least upper bound. (See Problem 1.2.3.) ∎

Definition 1.10. Let $(X, \leq)$ be a partially ordered set. $x \in X$ is a *maximal element* in X if whenever $x \leq y$ for y in X, then $x = y$. Similarly a *minimal element* is defined. A *chain* in $(X, \leq)$ is any subset C of X that is totally ordered by $\leq$. ∎

Examples

1.4. The real numbers R with the usual partial ordering $\leq$ form an order-complete partial ordered set whereas the rational numbers Q with the same ordering do not.

1.5. If $\mathscr{A}$ is the family of sets $\{\varnothing, \{0\}, \{1\}\}$ with the inclusion relation, then $\{0\}$ and $\{1\}$ are maximal elements, $\varnothing$ is a minimal element and a lower bound for $\mathscr{A}$. A chain $\mathscr{C}$ in $\mathscr{A}$ is the class $\{\varnothing, \{0\}\}$.

Let $(X_i)_{i \in I}$ be any indexed collection of sets indexed by set I. A function $c\colon I \to \bigcup_{i \in I} X_i$ such that $c(i) \in X_i$ for each $i \in I$ is called a *choice function*. If we write $c_i = c(i)$ for each $i \in I$ the indexed set $(c_i)_I$ completely describes the action of c on set I. The set of all choice functions has a special name.

Definition 1.11. If $(X_i)_I$ is an indexed family of sets then the *Cartesian product* $\times_{i \in I} X_i$ is the set of all choice functions $c\colon I \to \bigcup_I X_i$. ∎

If $\{A_1, A_2, \ldots, A_n\}$ is a finite collection of sets then $\times_{i \in I} A_i$, where $I = \{1, 2, \ldots, n\}$, is also denoted by $A_1 \times A_2 \times \cdots \times A_n$ and identified with the set

$$\{(a_1, a_2, \ldots, a_n)\colon a_i \in A_i \text{ for } i = 1, 2, \ldots, n\}$$

when $(a_1, a_2, \ldots, a_n)$ is identified with the choice function

$$a\colon \{1, 2, \ldots, n\} \to \bigcup_{i=1}^{n} A_i$$

given by $a(i) = a_i$.

An axiom that seems entirely reasonable is the following Axiom of Choice. It was proposed by Zermelo early in the 1900's and shown to be independent of—neither derivable from nor contradictory to—other axioms of set theory by P. J. Cohen around 1963 (see [15]). Actually the axiom is very strong and has many important consequences in mathematics. We will use the axiom both implicitly and explicitly in this book.

Axiom 1.1. *Axiom of Choice.* If $I \neq \emptyset$ and $A_i \neq \emptyset$ for each $i \in I$, then $\times_I A_i \neq \emptyset$. ∎

If $\mathscr{A}$ is any nonempty family of sets, then $\mathscr{A}$ can be considered as an indexed family of sets indexed by set $\mathscr{A}$ by virtue of the function i: $\mathscr{A} \to \mathscr{A}$ given by $i(A) = A$. The Axiom of Choice thereby says that if $\mathscr{A}$ is any nonempty family of nonempty sets an element can be "chosen" from each set in $\mathscr{A}$.

Two important consequences of the Axiom of Choice are Zorn's Lemma and The Well-Ordering Principle. In fact the following is true.

Theorem 1.2. The following statements are equivalent (each is a consequence of the other):

(i) Axiom of Choice.
(ii) Zorn's Lemma: Each nonempty partially ordered set in which each chain has an upper bound has a maximal element.
(iii) Well-Ordering Principle: If X is a set, then there exists a well-ordering for X.
(iv) Hausdorff Maximal Principle: If $\mathscr{A}$ is a family of sets and $\mathscr{C}$ is a chain in $\mathscr{C}_0$, then there is a maximal chain $\mathscr{A}$ that contains $\mathscr{C}$. ∎

We do not prove the equivalence of these statements in this book. The interested reader may consult other set theoretic books such as [25], [61], or the appendix of [34] for a discussion of the Axiom of Choice and related results. An immediate consequence of the Well-Ordering Principle is the existence of an uncountable set as described in the next proposition.

Proposition 1.6. There is an uncountable set $\bar{X}$ and a well-ordering $\leq$ on $\bar{X}$ such that

(i) there is an element Ω in $\bar{X}$ such that $\alpha \leq \Omega$ for all α in $\bar{X}$, and
(ii) for any β in $\bar{X}$, $\beta \neq \Omega$, the set $\{\alpha: \alpha \leq \beta\}$ is countable. ∎

(We will use $\bar{X}$ in constructing examples in later chapters. Ω will be called the first *uncountable ordinal*, $X = \bar{X} - \{\Omega\}$ the set of *countable ordinals*, and $\bar{X}$ the set of *ordinals* less than or equal to the first uncountable ordinal.)

Proof. Let Y be any uncountable set and well-order Y with some well-ordering $\leq$. If Y does not have an element y_0 such that $y \leq y_0$ for all y in Y, adjoin an element y_0 to Y forming $Y_0 = Y \cup \{y_0\}$ and extend the partial ordering $\leq$ to Y_0 by requiring that $y \leq y_0$ for all y in Y_0. Y_0 is then uncountable and well ordered by $\leq$. Let $Z = \{y \in Y_0 : \{x \in Y_0 : x \leq y\}$ is uncountable$\}$. Since $y_0 \in Z$, $Z \neq \varnothing$ and has a least element, say Ω. Let $\bar{X} = \{x \in Y_0 : x \leq \Omega\}$. $\bar{X}$ is the required set. ∎

Earlier (in Example 1.2) we mentioned that the Principle of Mathematical Induction follows from the fact that the positive integers N are well ordered. Analogously, any well-ordered set satisfies the following Principle of Transfinite Induction. We will make use of this fact later in particular for the set X just constructed.

Theorem 1.3. *Principle of Transfinite Induction.* Let $(W, \leq)$ be a well-ordered set. For any $a \in W$ let

$$I(a) = \{x \in W : x < a\}.$$

If A is a subset of W such that $a \in A$ whenever $I(a) \subset A$, then $A = W$. ∎

Proof. Assume $A^c \cap W \neq \varnothing$ and let a be the least element of $A^c \cap W$; that is, $a \in A^c \cap W$ and $a \leq w$ for all $w \in A^c \cap W$. Then $I(a) \subset A$, so that by hypotheses $a \in A$. This contradiction shows $A^c \cap W = \varnothing$ or $A = W$. ∎

A totally ordered set with both a maximal and a minimal element that we shall encounter frequently is the set $\bar{R}$ of *extended real numbers*. It is the set of real numbers with the usual partial ordering $\leq$ together with two additional elements $-\infty$ and ∞, which are required to satisfy $-\infty < x < \infty$ for each real number x. In addition we define the binary operations $+$ and $\cdot$ on $\bar{R}$ to be the usual $+$ and $\cdot$ on R and to additionally satisfy for all real numbers x the following equations:

$$x + \infty = \infty, \qquad x + (-\infty) = -\infty,$$
$$x \cdot \infty = \infty \quad \text{if } x > 0, \qquad 0 \cdot \infty = 0, \qquad x \cdot (-\infty) = \infty \quad \text{if } x < 0,$$
$$\infty + \infty = \infty, \qquad -\infty - \infty = -\infty,$$
$$\infty \cdot (\pm\infty) = \pm\infty, \qquad -\infty \cdot (\pm\infty) = \mp\infty.$$

It is obvious that each nonempty subset of $\bar{R}$ has an upper bound and a lower bound. The least upper bound of $\varnothing$ is defined to be $-\infty$ and the greatest lower bound of $\varnothing$ to be ∞. If S is any nonempty set of real numbers with no upper bound in R, then we define $\sup S = \infty$. Similarly if S has no lower bound in R, then we define $\inf S \equiv -\infty$. Thus for each subset S of $\bar{R}$, $\sup S$ exists and $\inf S$ exists.

If $(x_n)_N$ is a sequence of real numbers we define the *limit superior* of $(x_n)_N$, denoted as $\lim \sup x_n$ or $\overline{\lim}_n x_n$, by

$$\lim_n \sup x_n \equiv \inf\{s_n : s_n \equiv \sup_{k \geq n} x_k\}.$$

The *limit inferior* of $(x_n)_N$, also denoted by $\underline{\lim}_n x_n$, is defined by

$$\lim_n \inf x_n \equiv \sup\{i_n : i_n = \inf_{k \geq n} x_k\}.$$

The following criteria are useful.

Remarks

1.1. The real number x is the limit superior of $(x_n)_N$ if and only if

(a) given $\varepsilon > 0$, there exists n such that $x_k < x + \varepsilon$ for all $k \geq n$, and

(b) given $\varepsilon > 0$ and given $n \in N$, there exists $k \geq n$ such that $x_k > x - \varepsilon$.

1.2. The extended real number ∞ is the $\lim \sup x_n$ if and only if given any $M \in R$ and $n \in N$ there is a $k \geq n$ such that $x_k > M$.

1.3. $-\infty = \lim_n \sup x_n$ if and only if given $M \in R$, there exists $m \in N$ such that for $n \geq m$, $x_n < M$.

1.4. Similar statements can be formulated for the limit inferior of $(x_n)_N$.

The proofs of the statements in Remarks 1.1–1.4 as well as the formulation in Remark 1.4 are left as exercises for the reader (see Problem 1.2.5). Other properties are recorded in Problem 1.2.9.

Finally, a sequence $(x_n)_N$ of real numbers is said to *converge* to an extended real number x if $\lim_n \inf x_n = \lim_n \sup x_n = x$. The reader is asked to show (Problem 1.2.6) that $(x_n)_N$ converges to a real number x, written $x = \lim_n x_n$, if and only if, given $\varepsilon > 0$, there exists $n \in N$ such that $|x_m - x| < \varepsilon$ whenever $m \geq n$. The notation $x_n \to x$ is also used to mean that $x = \lim_n x_n$.

If $(x_n)_N$ is a *nondecreasing sequence* of real numbers—that is, $x_n \leq x_{n+1}$ for all $n \in N$—then $\lim x_n$ exists in the extended real numbers and equals $\sup x_n$. Similarly if $(x_n)_N$ is a nonincreasing sequence, $\lim x_n = \inf x_n$. The notations $x_n \uparrow x$ or $x_n \downarrow x$ are used sometimes to mean that $(x_n)_N$ is a nondecreasing or nonincreasing sequence, respectively, converging to x.

Problems

1.2.1. Prove that there is a one-to-one correspondence between partitions of a nonempty set X and equivalence relations defined in X.

1.2.2. Prove that the fact that $(N, \leq)$ is a well-ordered set where $\leq$ is the usual "less than or equal to" relation is equivalent to the fact that the Principle of Mathematical Induction holds.

1.2.3. Prove that a partially ordered set $(X, \leq)$ is order complete if and only if every nonempty set A in X with a lower bound has a greatest lower bound.

1.2.4. Let $X = \bar{X} - \{\Omega\}$, the set of ordinals less than the first uncountable ordinal Ω. Show that every countable subset E of X has a least upper bound in X.

1.2.5. (i) Verify Remarks 1.1–1.3.

(ii) Formulate statements similar to Remarks 1.1–1.3 for $\liminf x_n$.

1.2.6. (i) Prove that the real number $x = \lim x_n$ if and only if, given $\varepsilon > 0$, there exists $n \in N$ such that $|x_m - x| < \varepsilon$ whenever $m \geq n$.

(ii) Prove that the real sequence $(x_n)_N$ converges to ∞ (also called "diverging to ∞") if and only if, for any $M \in R$, there exists an $n \in N$ for $m \geq n$, $x_m \geq M$.

(iii) Prove that the real sequence $(x_n)_N$ converges to $-\infty$ (also termed "diverging to $-\infty$") if for any $M \in R$ there exists $n \in N$ such that $x_m < M$ whenever $m \geq n$.

1.2.7. Prove that if $(x_n)_N$ is a nondecreasing sequence of real numbers then (x_n) converges to $\sup x_n$ in the extended real numbers.

1.2.8. A sequence (x_n) of real numbers is a *Cauchy sequence* if for each $\varepsilon > 0$, there exists a natural number n_0 such that for $n, m \geq n_0$, $|x_n - x_m| < \varepsilon$. Using the order completeness of $(R, \leq)$ show that every Cauchy sequence in R converges to some real number x.

1.2.9. Let $(x_n)_N$ be a sequence of real numbers.

(i) Prove $\liminf x_n \leq \limsup x_n$.

(ii) Prove $\limsup(-x_n) = -\liminf x_n$.

(iii) If $(y_n)_N$ is another sequence of real numbers, prove

$$\begin{aligned}\liminf x_n + \liminf y_n &\leq \liminf (x_n + y_n)\\ &\leq \limsup x_n + \liminf y_n\\ &\leq \limsup (x_n + y_n)\\ &\leq \limsup x_n + \limsup y_n,\end{aligned}$$

provided that none of the sums here is of the form $\infty - \infty$.

1.3. Algebras, σ-Algebras, and Monotone Classes of Sets

Of primary importance in the definition and consideration of measures, which are essentially functions on certain classes of sets, is the domain of these functions. In Chapter 2 we will consider measures whose domains are algebras or σ-algebras of sets. In Chapter 7 it will be most instructive to consider measures on σ-rings as well as σ-algebras. The purpose of this section is to introduce these various classes of sets, compare them, and give essential properties.

Given a set X there are six types of classes of subsets of X we wish to examine. They are rings, algebras, σ-rings, σ-algebras, monotone classes, and Dynkin systems. The latter of these systems we will only encounter in Chapter 7. Let us begin with the definition of a ring and σ-ring.

Definition 1.12. A *ring of sets* $\mathscr{R}$ is a nonempty family of sets such that whenever A and B are in $\mathscr{R}$ then

$$A \cup B \text{ is in } \mathscr{R}$$

and

$$A - B \text{ is in } \mathscr{R}.$$

A *σ-ring of sets* is a ring $\mathscr{R}$ satisfying the additional condition that whenever $(A_i)_N$ is a countable class of sets in $\mathscr{R}$ then $\bigcup_{i=1}^{\infty} A_i$ is in $\mathscr{R}$. ∎

Remarks

1.5. If $\mathscr{R}$ is a ring, $\varnothing \in \mathscr{R}$ since $\mathscr{R}$ is a nonempty family containing a set A and $\varnothing = A - A$.

1.6. If $\mathscr{R}$ is a ring and A and B are in $\mathscr{R}$, then $A \triangle B$ and $A \cap B$ are in $\mathscr{R}$ since

$$A \triangle B = (A - B) \cup (B - A)$$
$$A \cap B = (A \cup B) - (A \triangle B).$$

1.7. If $\mathscr{R}$ is a ring and $(A_i)_{i=1}^{n}$ is any finite collection in $\mathscr{R}$, then

$$\bigcup_{i=1}^{n} A_i \in \mathscr{R} \qquad \text{and} \qquad \bigcap_{i=1}^{n} A_i \in \mathscr{R}.$$

1.8. If $\mathscr{R}$ is a σ-ring and $(A_i)_N$ is a countable collection of sets in $\mathscr{R}$, then the identity

$$\bigcap_{i=1}^{\infty} A_i = A - \bigcup_{i=1}^{\infty} (A - A_i),$$

where $A = \bigcup_{i=1}^{\infty} A_i$, implies that $\bigcap_{i=1}^{\infty} A_i \in \mathscr{R}$.

1.9. If $\mathscr{R}$ is a σ-ring and $(A_i)_N$ is a countable collection of sets in $\mathscr{R}$, then

$$\lim_i \inf A_i \in \mathscr{R} \qquad \text{and} \qquad \lim_i \sup A_i \in \mathscr{R}.$$

Examples

1.6. $\mathscr{R} = \{\varnothing\}$ and $\mathscr{R} = 2^X$, where X is any set, are examples of σ-rings.

1.7. If X is any set, then the class of all finite subsets of X is a ring. This class is a σ-ring if and only if X is itself a finite set. The class of all countable subsets of X is a σ-ring.

1.8. The class $\mathscr{R}$ of finite unions of "half-open" intervals $(a, b]$ for real numbers a and b is a ring.

1.9. If $\mathscr{C}$ is a collection of rings (or a collection of σ-rings) then $\bigcap_{\mathscr{R} \in \mathscr{C}} \mathscr{R}$ is a ring (or a σ-ring, respectively); that is, the intersection of a collection of rings (σ-rings) is a ring (σ-ring).

The definition of an algebra of sets has one essential difference from that of a ring of sets. More precisely, we have the following.

Definition 1.13. Let X be a set. An *algebra of sets* $\mathscr{A}$ *in* X is a nonempty family of sets such that

$$A \cup B \text{ is in } \mathscr{A} \text{ whenever } A \text{ and } B \text{ are in } \mathscr{A}$$

and

$$A^c \text{ is in } \mathscr{A} \text{ whenever } A \text{ is in } \mathscr{A}.$$

A *σ-algebra* in X is an algebra of sets $\mathscr{A}$ in X such that $\bigcup_{n=1}^{\infty} A_n$ is in $\mathscr{A}$ whenever $(A_n)_N$ is a countable collection from $\mathscr{A}$. ∎

Remarks

1.10. Each algebra (σ-algebra) $\mathscr{A}$ is a ring (σ-ring) because if A and B are in $\mathscr{A}$ then $A - B \in \mathscr{A}$ since

$$A - B = (A^c \cup B)^c.$$

That the converse is false is evident from Example 1.7 if X is an infinite set and from Example 1.8.

1.11. If X is a set, then a ring $\mathscr{R}$ of subsets of X is an algebra in X if and only if $X \in \mathscr{R}$. Indeed, if $\mathscr{R}$ is an algebra then $X = A^c \cup A$, where A is some element of $\mathscr{R}$ so that $X \in \mathscr{R}$. Conversely, if $X \in \mathscr{R}$ then whenever $A \in \mathscr{A}$, $A^c \in \mathscr{A}$, since

$$A^c = X - A$$

so that $\mathscr{R}$ is an algebra. Similarly a σ-ring $\mathscr{R}$ of subsets of X is a σ-algebra if and only if $X \in \mathscr{R}$.

1.12. The intersection of any collection of algebras (σ-algebras) of subsets of set X is again an algebra (σ-algebra).

Examples

1.10. As mentioned above, Example 1.8 is not an algebra. However, if we consider the collection of all finite unions of intervals of the type $(a, b]$ for $-\infty \leq a < b < \infty$ or (b, ∞) for $-\infty \leq b < \infty$, then the collection is an algebra.

1.11. If X is any set, the class of countable subsets of X is a σ-algebra if and only if X is countable. If X is any set, the class of all subsets A of X such that A or A^c is countable is a σ-algebra.

Of significance will be rings, σ-rings, algebras, and σ-algebras related to a given class of sets. More precisely, if X is a set and $\mathscr{E}$ is a class of subsets of X we will be interested in the ring, σ-ring, algebra, or σ-algebra *generated* by $\mathscr{E}$. By definition the *ring generated by* $\mathscr{E}$ is the "smallest" ring of subsets of X *containing* $\mathscr{E}$ and is denoted by $\mathscr{R}(\mathscr{E})$. (Ring $\mathscr{R}$ is smaller than ring $\mathscr{R}'$ if $\mathscr{R} \subset \mathscr{R}'$.) Similarly, *the σ-ring generated by* $\mathscr{E}$ is defined as the smallest σ-ring $\sigma_r(\mathscr{E})$ containing $\mathscr{E}$, the *algebra generated by* $\mathscr{E}$ as the smallest algebra $\mathscr{A}(\mathscr{E})$ containing $\mathscr{E}$, and the *σ-algebra generated by* $\mathscr{E}$ as the smallest σ-algebra $\sigma(\mathscr{E})$ containing $\mathscr{E}$.

Proposition 1.7. If $\mathscr{E}$ is any class of subsets of X, then $\mathscr{R}(\mathscr{E})$, $\sigma_r(\mathscr{E})$, $\mathscr{A}(\mathscr{E})$, and $\sigma(\mathscr{E})$ exist. ∎

Proof. Let

$$\mathscr{C} = \{\mathscr{R}: \mathscr{R} \text{ is a ring of subsets of } X \text{ and } \mathscr{E} \subset \mathscr{R}\}.$$

Since $2^X \in \mathscr{C}$, $\mathscr{C}$ is a nonempty collection, and by Example 1.9, $\bigcap_{\mathscr{R} \in \mathscr{C}} \mathscr{R}$ is a ring in $\mathscr{C}$. Clearly $\bigcap_{\mathscr{R} \in \mathscr{C}} \mathscr{R} = \mathscr{R}(\mathscr{E})$. Similarly $\sigma_r(\mathscr{E})$, $\mathscr{A}(\mathscr{E})$, and $\sigma(\mathscr{E})$ exist. ∎

If $\mathscr{E}$ is any class of subsets of X, then the following diagram is valid:

$$\begin{array}{ccc} \mathscr{R}(\mathscr{E}) & \subset & \mathscr{A}(\mathscr{E}) \\ \cap & & \cap \\ \sigma_r(\mathscr{E}) & \subset & \sigma(\mathscr{E}) \end{array}$$

If $\mathscr{E}$ is any class of subsets of set X and $A \subset X$, then $\mathscr{E} \cap A$ denotes the class of all sets of the form $E \cap A$, where $E \in \mathscr{E}$. The equalities

$$\left(\bigcup E_i\right) \cap A = \bigcup (E_i \cap A)$$

and

$$(E_1 - E_2) \cap A = (E_1 \cap A) - (E_2 \cap A)$$

for arbitrary sets show in particular that if $\mathscr{E}$ is a ring or σ-ring of subsets of X, then $\mathscr{E} \cap A$ is a ring or σ-ring, respectively, of subsets of A. Moreover, $\mathscr{E} \cap A$ is an algebra of subsets of A whenever $\mathscr{E}$ is an algebra in X or $\mathscr{E}$ is a ring and $A \subset E$ for some $E \in \mathscr{E}$. The following proposition says even more.

Proposition 1.8. If $\mathscr{E}$ is any class of subsets of a set X and $A \subset X$, then

(i) $\mathscr{R}(\mathscr{E}) \cap A = \mathscr{R}(\mathscr{E} \cap A)$,

(ii) $\sigma_r(\mathscr{E}) \cap A = \sigma_r(\mathscr{E} \cap A)$,

(iii) $\mathscr{A}(\mathscr{E}) \cap A = \mathscr{A}(\mathscr{E} \cap A)$ (as an algebra of subsets of A),

(iv) $\sigma(\mathscr{E}) \cap A = \sigma(\mathscr{E} \cap A)$ (as a σ-algebra of subsets of A). ∎

Proof. We prove (ii) and leave the other proofs as an exercise. Clearly

$$\sigma_r(\mathscr{E} \cap A) \subset \sigma_r(\mathscr{E}) \cap A$$

since $\sigma_r(\mathscr{E}) \cap A$ is a σ-ring containing $\mathscr{E} \cap A$ and $\sigma_r(\mathscr{E} \cap A)$ is the smallest such σ-ring by definition. Now let $B \in \sigma_r(\mathscr{E} \cap A)$ and let $C \in \sigma_r(\mathscr{E})$. Since B is a subset of A, we have

$$B = B \cap A = [B \cup (C - A)] \cap A.$$

However, the class $\mathscr{C}$ of all sets of the form $B \cup (C - A)$, where $B \in \sigma_r(\mathscr{E} \cap A)$ and $C \in \sigma_r(\mathscr{E})$, is a σ-ring (verify!) that contains $\mathscr{E}$, since if $E \in \mathscr{E}$ then

$$E = (E \cap A) \cup (E - A)$$

with $E \cap A \in \sigma_r(\mathscr{E} \cap A)$ and $E \in \sigma_r(\mathscr{E})$. This means $\mathscr{C} \supset \sigma_r(\mathscr{E})$ and

$$\sigma_r(\mathscr{E} \cap A) = \mathscr{C} \cap A \supset \sigma_r(\mathscr{E}) \cap A. \qquad ∎$$

The next two results will be of special interest in Chapter 7.

Proposition 1.9. If $\mathscr{E}$ is any class of sets and $A \in \sigma_r(\mathscr{E})$ [$A \in \sigma(\mathscr{E})$], then there exists a countable subclass of $\mathscr{E}$, call it $\mathscr{C}$, such that $A \in \sigma_r(\mathscr{C})$ [$A \in \sigma(\mathscr{C})$]. ∎

Proof. The union of all σ-rings $\sigma_r(\mathscr{C})$ for every countable subclass $\mathscr{C}$ of $\mathscr{E}$ is a σ-ring. Since this union contains $\mathscr{E}$ and is contained in $\sigma_r(\mathscr{E})$, it must be equal to $\sigma_r(\mathscr{E})$. ∎

Proposition 1.10. Let $\mathscr{E}$ be a class of subsets of some set Y. The following statements are true:

(i) If $\mathscr{E}$ is a countable class, then $\mathscr{A}(\mathscr{E})$ is countable.
(ii) If card $\mathscr{E} \leq c$, then card $\sigma_r(\mathscr{E}) \leq c$ and card $\sigma(\mathscr{E}) \leq c$. ∎

Remark. To prove part (ii), there are additional facts from set theory we must use but do not prove. They are as follows:

(1) If $(A_i)_I$ is an indexed family of sets with card $A_i \leq c$ for each $i \in I$ and card $I \leq \aleph_0$, then card $(\times_I A_i) \leq c$.
(2) If $(A_i)_I$ is an indexed family of sets with card $A_i \leq c$ for each $i \in I$ and card $I \leq c$, then card $(\bigcup_I A_i) \leq c$.

In light of (2) the set X constructed in Proposition 1.6 can be seen to have cardinality less than or equal to c since Y in the proof of Proposition 1.6 can be chosen to be R so that Y_0 is the set $R \cup \{y_0\}$ and card $Y_0 \leq c$.

Proof. We prove (ii) and leave the analogous proof of (i) for the reader (Problem 1.3.3). For each class of sets $\mathscr{C}$ let $\mathscr{C}^*$ denote the class of countable unions of differences of sets of $\mathscr{C}$, that is,

$$\mathscr{C}^* = \{\bigcup_{i=1}^{\infty} A_i - B_i \colon A_i, B_i \in \mathscr{C}\}.$$

It is clear that if $\mathscr{C} \subset \sigma(\mathscr{E})$, then $\mathscr{C}^* \subset \sigma(\mathscr{E})$. Moreover if $\varnothing$ and Y are in $\mathscr{C}$, then obviously $\varnothing$ and Y are in $\mathscr{C}^*$ and in particular $\mathscr{C} \subset \mathscr{C}^*$. In addition, from the preceding remark it follows that if card $\mathscr{C} \leq c$, then card $\mathscr{C}^* \leq c$.

To prove (ii) it suffices to show card $\sigma(\mathscr{E}) \leq c$, and to this end we may assume that $\varnothing$ and Y are in $\mathscr{E}$. Letting X be the well-ordered set $\{\alpha\colon \alpha < \Omega\}$ of countable ordinals constructed in Proposition 1.6, we will show that for each element α in X there exists a class $\mathscr{E}_\alpha$ of subsets of Y such that

(a) if $\alpha < \beta$, then $\mathscr{E} \subset \mathscr{E}_\alpha \subset \mathscr{E}_\beta \subset \sigma(\mathscr{E})$,

(b) card $\mathscr{E}_\alpha \leq c$,

(c) $\sigma(\mathscr{E}) = \bigcup_{\alpha \in X} \mathscr{E}_\alpha$.

For the least element 1 of X, let $\mathscr{E}_1 = \mathscr{E}$. Assuming that for each element $\alpha < \delta$, $\mathscr{E}_\alpha$ has been defined satisfying (a) and (b), let

$$\mathscr{E}_\delta = \Big(\bigcup_{\alpha<\delta} \mathscr{E}_\alpha\Big)^*.$$

Clearly (a) and (b) are satisfied for all $\alpha \leq \delta$, and using the Principle of Transfinite Induction (Theorem 1.3), a class $\mathscr{E}_\alpha$ exists for each α in X so that (a) and (b) are satisfied. Statement (c) is proved by showing that $\bigcup_{\alpha \in X} \mathscr{E}_\alpha$ is a σ-algebra. To this end let $(E_i)_N$ be a countable collection of sets from $\bigcup_{\alpha \in X} \mathscr{E}_\alpha$. For each $i \in N$, there exists an $\mathscr{E}_{\alpha_i}$ such that $E_i \in \mathscr{E}_{\alpha_i}$. Let $\alpha = \sup \alpha_i$, an element of X by Problem 1.2.4. Since $\mathscr{E}_{\alpha_i} \subset \mathscr{E}_\alpha$ by (a), $E_i \in \mathscr{E}_\alpha$ for each $i \in N$. Since $\varnothing \in \mathscr{E}_\alpha$, this means that

$$\bigcup_{i=1}^{\infty} E_i = \bigcup_{i=1}^{\infty} (E_i - \varnothing) \in \mathscr{E}_\alpha^* \subset \bigcup_{\alpha \in X} \mathscr{E}_\alpha.$$

In a similar fashion, it can be shown that whenever E_1 and E_2 are in $\bigcup_{\alpha \in X} E_\alpha$, then $E_1 - E_2 \in \bigcup_{\alpha \in X} E_\alpha$.

Since card $\mathscr{E}_\alpha \leq c$ and card $X \leq c$, card $\sigma(\mathscr{E}) =$ card $(\bigcup_{\alpha \in X} \mathscr{E}_\alpha) \leq c$ by our preceding remark. ∎

Having discussed σ-rings and σ-algebras, our next consideration is that of monotone classes and their relation to σ-rings.

Definition 1.14. A nonempty class $\mathscr{M}$ of sets is called a *monotone class* if for every monotone sequence (i.e., a nonincreasing or nondecreasing sequence) of sets (E_n),

$$\lim_n E_n \in \mathscr{M}.$$ ∎

Remarks

1.13. If X is a set, $\{\varnothing, X\}$ and 2^X are examples of monotone classes.

1.14. Any σ-ring $\mathscr{R}$ is a monotone class. A monotone class $\mathscr{M}$ is a σ-ring if and only if $\mathscr{M}$ is a ring (Problem 1.3.4).

1.15. The intersection of any collection of monotone classes is a monotone class.

1.16. If X is a set and $\mathscr{E}$ is a collection of subsets of X, then there exists a smallest monotone class $\mathscr{M}(\mathscr{E})$ containing $\mathscr{E}$, $\mathscr{M}(\mathscr{E}) \subset \sigma_r(\mathscr{E})$.

The next result compares monotone classes with σ-rings and σ-algebras.

Theorem 1.4. *Monotone Class Theorem.* If $\mathscr{R}$ is a ring, then $\sigma_r(\mathscr{R}) = \mathscr{M}(\mathscr{R})$. If $\mathscr{R}$ is an algebra, then $\sigma(\mathscr{R}) = \mathscr{M}(\mathscr{R})$. ∎

Proof. Suppose $\mathscr{R}$ is a ring. Since $\sigma_r(\mathscr{R})$ is a monotone class, clearly $\mathscr{M}(\mathscr{R}) \subset \sigma_r(\mathscr{R})$. The proof is accomplished by showing $\mathscr{M}(\mathscr{R})$ is a σ-ring, for then $\mathscr{M}(\mathscr{R}) \supset \sigma_r(\mathscr{R})$. From Remark 1.14 it is sufficient to show that $\mathscr{M}(\mathscr{R})$ is a ring. For each A in $\mathscr{M}(\mathscr{R})$ let

$$\mathscr{M}_A = \{B \in \mathscr{M}(\mathscr{R}): A \cup B,\ A - B, \text{ and } B - A \in \mathscr{M}(\mathscr{R})\}.$$

It is trivial to see (using Problem 1.1.8) that $\mathscr{M}_A$ is a monotone class contained in $\mathscr{M}(\mathscr{R})$. Note also that $B \in \mathscr{M}_A$ if and only if $A \in \mathscr{M}_B$. Inasmuch as $\mathscr{R} \subset \mathscr{M}_A$ whenever $A \in \mathscr{R}$, we have $\mathscr{M}(\mathscr{R}) \subset \mathscr{M}_A$ whenever $A \in \mathscr{R}$. Therefore if $B \in \mathscr{M}(\mathscr{R})$, then $B \in \mathscr{M}_A$ and hence $A \in \mathscr{M}_B$. Since A is an arbitrary element of $\mathscr{R}$, $\mathscr{R} \subset \mathscr{M}_B$ and hence $\mathscr{M}(\mathscr{R}) \subset \mathscr{M}_B$ for every B in $\mathscr{M}(\mathscr{R})$. The inclusion $\mathscr{M}(\mathscr{R}) \subset \mathscr{M}_B$ for every B in $\mathscr{M}(\mathscr{R})$ implies that $\mathscr{M}(\mathscr{R})$ is a ring.

The proof of the second statement follows immediately upon noting that if $\mathscr{R}$ is an algebra, $\mathscr{M}(\mathscr{R})$ is not only a σ-ring but also a σ-algebra. ∎

Our final consideration in this section is that of a Dynkin system.

• **Definition 1.15.** Let $\mathscr{D}$ be a class of subsets of set X. $\mathscr{D}$ is a *Dynkin system* if

(i) $X \in \mathscr{D}$;
(ii) whenever $A, B \in \mathscr{D}$ and $B \subset A$, then $A - B \in \mathscr{D}$;
(iii) for every sequence $(A_i)_N$ of pairwise disjoint sets in $\mathscr{D}$, $\bigcup_{i=1}^{\infty} A_i \in \mathscr{D}$. ∎

• **Remarks**

1.17. Every σ-algebra of subsets of X is a Dynkin system.

1.18. Every Dynkin system is a monotone class (Problem 1.3.7).

1.19. For every collection $\mathscr{E}$ of subsets of X, there is a smallest Dynkin system $\mathscr{D}(\mathscr{E})$ containing $\mathscr{E}$ (Problem 1.3.5).

• **Proposition 1.11.** A Dynkin system $\mathscr{D}$ is a σ-algebra if and only if $A \cap B \in \mathscr{D}$ whenever $A, B \in \mathscr{D}$. ∎

Proof. The necessity is obvious, so we prove the sufficiency. Clearly Definition 1.15(i) and (ii) imply that if $A \in \mathscr{D}$ then $A^c \in \mathscr{D}$. If A and B are in $\mathscr{D}$ then the equations

$$A \cup B = A \cup [B - (A \cap B)] \text{ and } A \cap [B - (A \cap B)] = \varnothing$$

imply that $A \cup B \in \mathscr{D}$. By induction, for any finite sequence $(A_i)_{i=1}^n$ from $\mathscr{D}$, $\bigcup_{i=1}^n A_i \in \mathscr{D}$. Now let $(A_i)_N$ be any countable collection from $\mathscr{D}$. Letting $(B_i)_N$ be the collection in $\mathscr{D}$ given by

$$\begin{aligned} B_1 &= A_1, \\ B_2 &= (A_1 \cup A_2) - A_1, \\ B_3 &= (A_1 \cup A_2 \cup A_3) - (A_1 \cup A_2), \\ &\vdots \\ B_n &= (A_1 \cup A_2 \cdots \cup A_n) - (A_1 \cup A_2 \cdots \cup A_{n-1}), \end{aligned}$$

we have a pairwise disjoint sequence in $\mathscr{D}$ and $\bigcup_{i=1}^\infty A_i = \bigcup_{i=1}^\infty B_i$. ∎

• **Proposition 1.12.** If $\mathscr{E}$ is a class of subsets of X for which $A \cap B \in \mathscr{E}$ whenever A and B are in $\mathscr{E}$, then

$$\mathscr{D}(\mathscr{E}) = \sigma(\mathscr{E}).$$ ∎

Proof. Clearly $\mathscr{D}(\mathscr{E}) \subset \sigma(\mathscr{E})$ as $\sigma(\mathscr{E})$ is a Dynkin system. To show $\sigma(\mathscr{E}) \subset \mathscr{D}(\mathscr{E})$ it suffices to show that $\mathscr{D}(\mathscr{E})$ is a σ-algebra. By Proposition 1.11 it suffices to show that $A \cap B \in \mathscr{D}(\mathscr{E})$ whenever A and B are in $\mathscr{D}(\mathscr{E})$. If $A \in \mathscr{D}(\mathscr{E})$ define

$$\mathscr{D}_A = \{B \in 2^X: A \cap B \in \mathscr{D}(\mathscr{E})\}.$$

It is easy to see that $\mathscr{D}_A$ is a Dynkin system. Inasmuch as $\mathscr{D}_A \supset \mathscr{E}$ whenever $A \in \mathscr{E}$ according to our hypothesis on $\mathscr{E}$, we have $\mathscr{D}_A \supset \mathscr{D}(\mathscr{E})$ whenever $A \in \mathscr{E}$. This means that for every B in $\mathscr{D}(\mathscr{E})$ and every $A \in \mathscr{E}$, $B \cap A \in \mathscr{D}(\mathscr{E})$ so that $A \in \mathscr{D}_B$. This means $\mathscr{E} \subset \mathscr{D}_B$ and hence $\mathscr{D}(\mathscr{E}) \subset \mathscr{D}_B$. However, $\mathscr{D}(\mathscr{E}) \subset \mathscr{D}_B$ for arbitrary B in $\mathscr{D}(\mathscr{E})$ means $A \cap B \in \mathscr{D}(\mathscr{E})$ whenever A and B are in $\mathscr{D}(\mathscr{E})$. ∎

Problems

1.3.1. If $\mathscr{A}$ is an algebra and $(A_i)_N$ is a countable collection of sets in $\mathscr{A}$, prove there is a sequence $(B_i)_N$ from $\mathscr{A}$ such that $B_i \cap B_j = \varnothing$ if $i \neq j$ and $\bigcup_{i=1}^\infty A_i = \bigcup_{i=1}^\infty B_i$.

1.3.2. (i) Prove that if $\mathscr{E}$ is a collection of subsets of set X and $A \in \mathscr{R}(\mathscr{E})$, then there is a finite collection of sets in $\mathscr{E}$ that cover A, that is, their union contains A. (Hint: Show the class of sets that can be covered by a finite collection from $\mathscr{E}$ is a ring.)

(ii) Prove that if $A \in \sigma_r(\mathscr{E})$, then there is a countable collection of sets in $\mathscr{E}$ covering A.

1.3.3. Prove Proposition 1.10 (i).

1.3.4. Prove Remark 1.14.

• **1.3.5.** If $\mathscr{E}$ is a collection of subsets of X, prove that a smallest Dynkin system $\mathscr{D}(\mathscr{E})$ containing $\mathscr{E}$ exists.

1.3.6. Prove that a nonempty collection of subsets $\mathscr{A}$ of set X is an algebra if and only if (a) $A^c \in \mathscr{A}$ whenever $A \in \mathscr{A}$ and (b) $A \cap B \in \mathscr{A}$ whenever $A, B \in \mathscr{A}$.

• **1.3.7.** Prove that a collection of subsets $\mathscr{D}$ of set X is a Dynkin system if and only if (a) $X \in \mathscr{D}$, (b) $A, B \in \mathscr{D}$ with $A \subset B$ implies $B - A \in \mathscr{D}$, (c) $A, B \in \mathscr{D}$ and $A \cap B = \varnothing$ implies $A \cup B \in \mathscr{D}$, and (d) $\bigcup_n A_n \in \mathscr{D}$ for every increasing sequence $(A_n)_N$ in $\mathscr{D}$.

1.3.8. Suppose $\mathscr{E}$ and $\mathscr{F}$ are classes of subsets of X and Y, respectively. Let $\mathscr{G} = \{E \times F: E \in \mathscr{E}$ and $F \in \mathscr{F}\}$. Let $\sigma_r(\mathscr{E}) \times \sigma_r(\mathscr{F})$ be the smallest σ-ring containing sets of the form $A \times B$, where $A \in \sigma_r(\mathscr{E})$ and $B \in \sigma_r(\mathscr{F})$. Prove $\sigma_r(\mathscr{G}) = \sigma_r(\mathscr{E}) \times \sigma_r(\mathscr{F})$.

1.3.9. (i) If $f: X \to Y$ is a function and $\mathscr{F}$ is a σ-ring of subsets of Y, then prove

$$\mathscr{E} = \{E \subset X: E = f^{-1}(F) \text{ for } F \in \mathscr{F}\}$$

is a σ-ring. What happens if $\mathscr{F}$ is a σ-algebra?

(ii) If $f: X \to Y$ and $\mathscr{E}$ is a σ-ring of subsets of X, prove

$$\mathscr{F} = \{F \subset Y: f^{-1}(F) \in \mathscr{E}\}$$

is a σ-ring in Y. What happens if $\mathscr{E}$ is a σ-algebra?

(iii) Prove that $\mathscr{F} = \{E \subset X: A \in \mathscr{E} \Rightarrow A \cap E \in \mathscr{E}\}$ is a σ-algebra in X containing the σ-ring $\mathscr{E}$ in X.

1.3.10. Let $\mathscr{F}$ be a σ-ring of subsets of X such that $X \notin \mathscr{F}$. Show that the smallest σ-algebra containing $\mathscr{F}$ is $\mathscr{A} = \{A \subset X: A \in \mathscr{F} \text{ or } X - A \in \mathscr{F}\}$.

★ **1.3.11.** *On σ-Classes and σ-Algebras*: A collection $\mathscr{F}$ of subsets of a nonempty set X containing X is called a *σ-class* if it is closed under countable *disjoint* unions and complementations. Suppose $\mathscr{A}$ is the σ-class generated by a collection $\mathscr{C}$ of subsets of X. Then prove the following result due to T. Neubrunn: $\mathscr{A}$ is a σ-algebra if and only if one of the following conditions holds:

(i) If $A, B \in \mathscr{C}$, then $A - B \in \mathscr{A}$.

(ii) If $A, B \in \mathscr{C}$, then $A \cap B \in \mathscr{A}$.

[Hint: Notice that (1) $A, B \in \mathscr{A}$ and $A \subset B$ imply $B - A = (A \cup B^c)^c \in \mathscr{A}$, and (2) if $A_1 \supset A_2 \supset \cdots$, $A_i \in \mathscr{A}$, then $\bigcap_{i=1}^{\infty} \in \mathscr{A}$. Now assume (i) above. It is sufficient to show that $A - B \in \mathscr{A}$ whenever A, $B \in \mathscr{A}$. Let $D \in \mathscr{C}$ and $\varphi(D) = \{E \in \mathscr{A}: D - E \in \mathscr{A}\}$. Show that $\varphi(D)$ is a σ-class. Then since $\mathscr{C} \subset \varphi(D)$, $\mathscr{A} = \varphi(D)$. Again for fixed $A \in \mathscr{A}$, let $\psi(A) = \{B \in \mathscr{A}: B - A \in \mathscr{A}\}$. As before, show that $\psi(A)$ is a σ-class containing $\mathscr{C}$ so that $\psi(A) = \mathscr{A}$. The proof of (i) is now clear.]

★ **1.3.12.** *Cardinalities of σ-Algebras.* Regarding cardinal numbers, the following are facts:

(i) For any cardinal number a, $2^a > a$.

(ii) $b \leq a$ and a infinite implies $a + b = a$.

(iii) $0 < b \leq a$ and a infinite implies $a \cdot b = a$.

Use these to prove that card $\sigma(\mathscr{F}) \leq a^{\aleph_0}$ whenever card $\mathscr{F} = a \geq 2$, where $\mathscr{F}$ is a family of subsets of a set X containing $\varnothing$. [Hint: An argument similar to the one used in the proof of Proposition 1.10 is helpful.]

1.3.13. *The Borel Subsets of R.* The sets in the smallest σ-algebra containing the open subsets of R are called the Borel subsets of R. (The definition for the Borel sets in a metric space is similar.) Show that the Borel subsets of R are also in the smallest σ-algebra containing the closed and bounded intervals of R, and that they have cardinality c.

1.3.14. *The Smallest σ-Algebra Containing a Given σ-Algebra $\mathscr{A}$ and a Given Set E* (all sets are here subsets of a set X) is the class of all sets of the form $(A \cap E) \cup (B \cap E^c)$, A and $B \in \mathscr{A}$. Verify this and then use this to prove that card $\sigma(\mathscr{F}) \leq 2^{2^n}$ if $\mathscr{F}$ is a class of subsets of X with cardinality at most n. Can the reader find an alternative proof?

• 1.4. Topological Spaces

One of the primary reasons for including this section in the text is to make the book self-contained with respect to topological notions that are used frequently throughout the text. The student should treat this section as a synopsis of topology that is encountered in the text. For completeness, topological concepts closely related to normality, local compactness, compactness, and separability are also considered.

One of the main aims of this text is to present a comprehensive account of the basic theory of integration. In Chapter 7, it will be seen how this theory has an ultimate connection with some topological concepts. Though a detailed study of topology is not possible here, we present many of the basic results of the theory in a form resembling a typical "Texas-style" course: The reader is required to supply many of the proofs of theorems and remarks, which in most cases is easy and straightforward. However, propositions, theorems, and examples that are not straightforward and demand a little more than routine arguments have all been proved or given in detail. Readers interested in a detailed account of this discipline should consult [34].

Topological spaces are generalizations of the spaces R^n ($n \geq 1$). One of the main reasons for considering such generalizations is to study the concept of limit of functions on spaces other than R^n. The definition of limit and the theory of convergence in R^n are based on the notion of distance between points, and use is made of only the following few properties of this distance:

(i) $d(x, y) = 0$ if and only if $x = y$,

(ii) $0 \leq d(x, y) = d(y, x) < \infty$,

(iii) $d(x, z) \leq d(x, y) + d(y, z)$,

for every $x, y, z \in R^n$. These properties alone provide the framework for a class of topological spaces called the metric spaces. Nevertheless, there are situations that demand an even more general approach—for instance, the study of pointwise convergence in a space of bounded real-valued functions. This is one motivation behind the consideration of general topological spaces.

Definition 1.16. A *topology* $\mathscr{T}$ for a set X is a collection of subsets of X, called *open sets*, such that

(i) $\varnothing \in \mathscr{T}$, $X \in \mathscr{T}$.

(ii) $A \cap B \in \mathscr{T}$ whenever $A \in \mathscr{T}$ and $B \in \mathscr{T}$.

(iii) $\cup A_\alpha \in \mathscr{T}$ whenever $A_\alpha \in \mathscr{T}$ for each α. ∎

Examples

1.12. If $\mathscr{T} = 2^X$ (the class of all subsets of X), then $\mathscr{T}$ is a topology, called the *discrete topology* for X.

1.13. If $\mathscr{T} = \{\varnothing, X\}$, then $\mathscr{T}$ is a topology called the *indiscrete topology* for X.

1.14. Let $X = R$ and $\mathscr{T}$ be the collection of all subsets G of R such that $x \in G$ implies that for some $\mu > 0$, $\{y \in R: |y - x| < \mu\} \subset G$. Then $\mathscr{T}$ is a topology for R, called its *usual topology*.

1.15. Let $A \subset X$ and $\mathscr{T}$ be a topology for X. Let $\mathscr{T}|_A = \{A \cap T: T \in \mathscr{T}\}$. Then $\mathscr{T}|_A$ is a topology for A, called the *relative topology* for A.

Definition 1.17. Let $\mathscr{T}$ be a topology for X. A point $x \in X$ is a *limit point* of $A \subset X$ if whenever $x \in T$ and $T \in \mathscr{T}$ then $T \cap (A - \{x\}) \neq \emptyset$. The union of A and the set of all its limit points is called its closure, $\bar{A}$. A set A is called closed if $A = \bar{A}$. (Note that $A = \bar{A}$ if and only if $X - A \in \mathscr{T}$.) ▮

Remarks. Let $A, B \subset X$ and $\mathscr{T}$ be a topology for X. Then

1.20. $A \subset B \Rightarrow \bar{A} \subset \bar{B}$

1.21. $\overline{A \cup B} = \bar{A} \cup \bar{B}$

1.22. $\overline{A \cap B} \subset \bar{A} \cap \bar{B}$.

Definition 1.18. $(X, \mathscr{T})$ is called a *topological space* if $\mathscr{T}$ is a topology for X. A point $x \in X$ is called an *interior point* of $A \subset X$ if there is $T \in \mathscr{T}$ such that $x \in T \subset A$. The set of all interior points of A is called its *interior* and denoted by A°. ▮

Remarks. Let $(X, \mathscr{T})$ be a topological space and $A, B \subset X$. Then

1.23. $A \subset B \Rightarrow A^\circ \subset B^\circ$.

1.24. $(A \cap B)^\circ = A^\circ \cap B^\circ$.

1.25. $A = A^\circ$ if and only if $A \in \mathscr{T}$.

Definition 1.19. In a topological space $(X, \mathscr{T})$, $x \in X$ is called a *boundary point* of $A \subset X$ if

$$x \in T \in \mathscr{T} \Rightarrow T \cap A \neq \emptyset$$

and

$$T \cap (X - A) \neq \emptyset.$$

The *boundary* A_b of A is the set of all its boundary points. ▮

Remarks

1.26. $A_b = \bar{A} - A^\circ$.

1.27. $(A \cup B)_b \subset A_b \cup B_b$.

1.28. $(\bar{A})_b = A_b$.

Definition 1.20. A mapping $f\colon X \to Y$ where $(X, \mathscr{T}_1)$ and $(Y, \mathscr{T}_2)$ are topological spaces, is called *continuous* (relative to $\mathscr{T}_1$ and $\mathscr{T}_2$) if $f^{-1}(V) \in \mathscr{T}_1$ for each $V \in \mathscr{T}_2$. The mapping f is called *open* (or *closed*, respectively) if $f(T)$ is open (closed) for each open (closed) T. ▮

Definition 1.21. Let $\mathscr{L}$, $\mathscr{B}$ be collections of sets such that $\mathscr{B}$ consists of all the finite intersections of sets in $\mathscr{L}$. Let $\mathscr{T}$ consist of all sets that are arbitrary unions of sets in $\mathscr{B}$. Then $\mathscr{L}$ is called a *subbase* and $\mathscr{B}$ a *base* for $\mathscr{T}$. ∎

The reader can easily verify that $\mathscr{T}$ is a topology for the set $X = \cup\{A: A \in \mathscr{B}\}$. Note that if $(X, \mathscr{T})$ is a topological space and $\mathscr{B}$ is a base for $\mathscr{T}$, then given $x \in V \in \mathscr{T}$, there exists $U \in \mathscr{B}$ with $x \in U \subset V$.

Example 1.16. Let $(X_\lambda, \mathscr{T}_\lambda)$ be a topological space for each $\lambda \in \Lambda$, an indexed set. Let $X = \times_{\lambda \in \Lambda} X_\lambda$ (see Definition 1.11 for definition) and $\pi_\lambda: X \to X_\lambda$ be defined by $\pi_\lambda(x) = x(\lambda)$. Consider the topology $\mathscr{T}$ generated by $\{\pi_\lambda^{-1}(V): V \in \mathscr{T}_\lambda, \lambda \in \Lambda\}$ as a subbase. Then $(X, \mathscr{T})$ is a topological space, called *the product of the spaces* $(X_\lambda, \mathscr{T}_\lambda)$, $\lambda \in \Lambda$. *The functions* π_λ *are all open and continuous.* It is also clear that if $(X_1, \mathscr{T}_1)$ and $(X_2, \mathscr{T}_2)$ are two topological spaces, then $\{U \times V: U \in \mathscr{T}_1, V \in \mathscr{T}_2\}$ is *a base for the product* topology for $X \times Y$.

Proposition 1.13. Let $(X_1, \mathscr{T}_1)$ and $(X_2, \mathscr{T}_2)$ be topological spaces and $f: X_1 \to X_2$. Then the following are equivalent:

(i) f is continuous.

(ii) If $x \in X_1$, and $f(x) \in V \in \mathscr{T}_2$, then there is $U \in \mathscr{T}_1$ such that $x \in U$ and $f(U) \subset V$.

(iii) $f(\bar{A}) \subset \overline{f(A)}$ for all $A \subset X_1$.

(iv) $f^{-1}(B)$ is closed whenever B is closed. ∎

Proposition 1.13 (ii) motivates us to define *continuity at a point of X*. We say f is continuous at $x \in X$ if for any open V containing $f(x)$, there is an open U containing x such that $f(U) \subset V$.

Definition 1.22. A class $\mathscr{A}$ of subsets of X is a cover of X if $X = \cup\{A: A \in \mathscr{A}\}$. A cover $\mathscr{A}$ of a topological space $(X, \mathscr{T})$ is *open* if $A \in \mathscr{A}$ implies $A \in \mathscr{T}$. $(X, \mathscr{T})$ is called *compact* if each open cover of X contains a finite subcover. If $A \subset X$, then A is called compact if A is compact in its relative topology. ∎

Remarks on Compactness

1.29. The continuous image of a compact set is compact.

1.30. A closed subset of a compact set is compact.

1.31. Any nonempty collection of closed subsets in a compact topological space X with *finite intersection property* (a collection of subsets has the finite intersection property if any finite number of sets in this collection has nonempty intersection) has a nonempty intersection. [To prove this, suppose $\mathscr{A}$ is the collection and $\bigcap_{A\in\mathscr{A}} A = \varnothing$. Then $X = \bigcup_{A\in\mathscr{A}}(X - A)$ and by compactness of X, there are $A_1, A_2, \ldots, A_n \in \mathscr{A}$ such that $X = \bigcup_{x=1}^{n}(X - A_i)$ and therefore, $\bigcap_{i=1}^{n} A_i = \varnothing$, a contradiction.]

1.32. A topological space $(X, \mathscr{C})$ is compact if and only if each nonempty collection of closed subsets of X having f.i.p. (finite intersection property) has nonempty intersection.

1.33. If $\mathscr{A}$ is a nonempty family of subsets of X having f.i.p., then by the Hausdorff's Maximal Principle, $\mathscr{A}$ is contained in $\mathscr{B}$ ($\subset 2^X$), which is maximal relative to having f.i.p. Because of maximality, $\mathscr{B}$ has the following properties:

(a) $A, B \in \mathscr{B} \Rightarrow A \cap B \in \mathscr{B}$,

(b) $B \cap A \neq \varnothing$ for each $A \in \mathscr{B} \Rightarrow B \in \mathscr{B}$.

1.34. X is compact if and only if every nonempty collection of *subsets* of X with f.i.p. has the property $\cap\{\bar{A}: A \in \mathscr{A}\} \neq \varnothing$. This statement remains true if the "collection" above is replaced by a "collection" maximal relative to having f.i.p.

Theorem 1.5. (*Tychonoff*) The product of compact spaces is compact. (This theorem was first proved by Tychonoff (or Tihonov) in the case where the compact spaces are all [0, 1].† Later on, E. Čech proved it in the general case.‡) ∎

Proof. Let $X = \times_{\lambda\in\Lambda} X_\lambda$ and each X_λ, $\lambda \in \Lambda$, be compact. Let $\mathscr{A}$ be a collection of subsets of X maximal relative to having f.i.p. Since each X_λ is compact, $\cap\{\overline{\pi_\lambda(A)}: A \in \mathscr{A}\}$ is nonempty and must contain some element $x(\lambda) \in X_\lambda$. Let $x \in X$ such that $\pi_\lambda(x) = x(\lambda)$. We claim that $x \in \bar{A}$, for each $A \in \mathscr{A}$. Let $\bigcap_{i=1}^{n}\{\pi_{\lambda_i}^{-1}(U_{\lambda_i}): U_{\lambda_i}$ open in $X_{\lambda_i}\}$ be an open set containing x. Then since $x_{\lambda_i} = \pi_{\lambda_i}(x) \in U_{\lambda_i}$, $U_{\lambda_i} \cap \pi_{\lambda_i}(A) \neq \varnothing$ for each $A \in \mathscr{A}$. This means that $\pi_{\lambda_i}^{-1}(U_{\lambda_i}) \cap A \neq \varnothing$ for each $A \in \mathscr{A}$. By Remark 1.33(b), $\pi_{\lambda_i}^{-1}(U_{\lambda_i}) \in \mathscr{A}$ and by Remark 1.33(a), $\bigcap_{i=1}^{n}\pi_{\lambda_i}^{-1}(U_{\lambda_i}) \in \mathscr{A}$, and therefore by f.i.p. of $\mathscr{A}$, $\bigcap_{i=1}^{n}\pi_{\lambda_i}^{-1}(U_{\lambda_i})$ has nonempty intersection with each $A \in \mathscr{A}$. The claim now follows and the theorem follows by Remark 1.34. ∎

† A. Tychonoff, *Math. Ann.* **102**, 544 (1930).
‡ E. Čech, *Ann. Math.* **38**(2), 823 (1937).

The reader may very well recall now the following theorem, from his or her first course in "Real Analysis."

Theorem 1.6. (*Heine–Borel–Bolzano–Weierstrass*) A subset A of R^n is compact if and only if it is closed and bounded. ∎

Definition 1.23. Let $(X, \mathscr{T})$ be a topological space. Then we have the following:

(i) X is called T_1 if each singleton $\{x\}$ in X is closed.

(ii) X is T_2 (or Hausdorff) if $x \neq y$, $x, y \in X$ imply that there are open sets V, W such that $x \in V$, $y \in W$, and $V \cap W = \varnothing$.

(iii) X is called *regular* if for any closed set A and $x \notin A$, there are open sets V containing x and $W \supset A$ with $V \cap W = \varnothing$.

(iv) X is called *normal* if for any two disjoint closed sets A and B, there are open sets $V \supset A$ and $W \supset B$ with $V \cap W = \varnothing$.

(v) A regular T_1 space is called T_3 and a normal T_1 space is called T_4. ∎

Remarks

1.35. $T_4 \Rightarrow T_3 \Rightarrow T_2 \Rightarrow T_1$. The set of real numbers with the usual topology (see Example 1.14) is a T_4 space.

1.36. $(X, \mathscr{T})$ is Hausdorff if and only if $\{(x, x): x \in X\}$ is closed in the product topology for $X \times X$.

1.37. A compact subset of a Hausdorff space is closed. A compact Hausdorff space is normal.

1.38. Let A be a compact subset of a topological space $(X, \mathscr{T})$ and B be a compact subset of a topological space $(Y, \mathscr{T}')$ such that $A \times B \subset G$, an open set in $X \times Y$. Then there are open sets $V \in \mathscr{T}$, $W \in \mathscr{T}'$ such that $A \times B \subset V \times W \subset G$.

1.39. Let $(X_\lambda, \mathscr{T}_\lambda)$, $\lambda \in \Lambda$ be a family of topological spaces and let $X = \times_{\lambda \in \Lambda} X_\lambda$, their product with the product topology. Then (a) if each X_λ is T_1, then X is T_1; (b) if each X_λ is T_2, then X is T_2; (c) if each X_λ is regular, then X is regular.

1.40. If $(X, \mathscr{T}_1)$ is a Hausdorff space and $(X, \mathscr{T}_2)$ is compact, then $\mathscr{T}_1 \subset \mathscr{T}_2$ implies $\mathscr{T}_1 = \mathscr{T}_2$. For, if i: $(X, \mathscr{T}_2) \to (X, \mathscr{T}_1)$ is the identity mapping, then i is one-to-one, onto, continuous, and closed; this means that i is open and therefore $\mathscr{T}_2 \subset \mathscr{T}_1$.

1.41. If $(X, \mathscr{T})$ is a compact Hausdorff space, then X is not compact with any topology $\mathscr{T}_1$ satisfying $\mathscr{T}_1 \supsetneqq \mathscr{T}$ and X is not Hausdorff with any topology $\mathscr{T}_2$ satisfying $\mathscr{T}_2 \subsetneqq \mathscr{T}$.

Examples

1.17. *A T_1 Space That Is Not T_2.* Let X be an uncountable set and $\mathscr{T} = \{\emptyset\} \cup \{U \subset X: X - U$ is at most countable$\}$. Then $\mathscr{T}$ is clearly a T_1 topology that is not T_2. [At most countable = finite or countable.]

1.18. *A T_2 Space That Is Not Regular.* Let X be the set of real numbers. Let all sets of the form $\{x\} \cup I_x$, where I_x contains only the rationals in an open interval around x, form a base for a topology $\mathscr{T}$. $\mathscr{T}$ is clearly Hausdorff. But $\mathscr{T}$ is *not* regular, since the irrationals in X form a closed set, and cannot be separated from a rational number by open sets.

1.19. *Regularity or Normality Does Not Imply T_2.* The indiscrete topology provides an example.

1.20. *Normality Need Not Imply Regularity.* A trivial example is the topology $\{\emptyset, \{x\}, \{y\}, \{x, y\}, \{x, y, z\}\}$ for the set $X = \{x, y, z\}$.

1.21. *T_3 Need Not Imply T_4.* The following example is due to Niemytzki. Let $X = \{(x, y) \in R^2: y \geq 0\}$. We consider that topology $\mathscr{T}$ on X which has as a base the set of all open disks in X along with all sets of the form $\{(x, 0)\} \cup D$, where D is an open disk in X touching the x axis at $(x, 0)$. Then $\mathscr{T}$ contains the relative topology of R^2 on X. Let $A \subset X$ be a closed set and $a \in X - A$. If a is not on the x axis, then there is an open disk U such that $a \in U \subset X - A$. Since the relative topology of R^2 is regular by Remarks 1.35 and 1.39, a and $X - U$ (and therefore, a and A) can be separated by sets open in this relative topology and therefore, in $\mathscr{T}$. If $a = (a_1, 0)$ and D be an open disk in X with radius r and tangent at a to the x axis such that $\{a\} \cup D \subset X - A$, then we can define the function f by $f(a) = 0$, $f(x) = 1$ if $x \notin \{a\} \cup D$ and $f(z, w) = [(z - a_1)^2 + w^2]/2rw$ if $(z, w) \in D$. It can be easily verified that f is a continuous mapping from X into R. Since $a \in f^{-1}((-\infty, \frac{1}{2}))$ and $A \subset f^{-1}((\frac{1}{2}, \infty))$, it follows that X is regular (in fact, *completely regular*[†]). But X is not normal. The reason is that the rationals and the irrationals on the x axis are disjoint and closed, but do not have any disjoint open sets containing them.

1.22. A compact Hausdorff topology on $[0, 1]$ different from the usual topology is the topology $\mathscr{T} = \{A: A \subset [0, 1)\} \cup \{B: B \subset [0, 1], 1 \in B$ and $[0, 1] - B$ is finite$\}$.

By definition, a sequence (x_n) in a topological space $(X, \mathscr{T})$ converges to x in X if for every open U with $x \in U$ there is an integer n_0 such that $x_n \in U$ whenever $n > n_0$.

[†] A topological space $(X, \mathscr{T})$ is called completely regular if for $x \notin A$ and A closed, there exists a continuous function $f: X \to [0, 1]$ with $f(x) = 0$ and $f(A) = 1$.

Proposition 1.14. In a Hausdorff space, a convergent sequence has a unique limit. ∎

Proposition 1.15. Let X be the product of the topological spaces X_λ, $\lambda \in \Lambda$. Let f_λ, for each λ be a continuous mapping from a topological space Y into X_λ. If $f: Y \to X$ is defined by $[f(y)]\ (\lambda) = f_\lambda(y)$, then f is continuous. ∎

Proposition 1.16. Let X be a nonempty compact (or only countably compact, to be defined in Definition 1.28) space and f be a real-valued *lower-semicontinuous* function on X, i.e., the set $\{x \in X: f(x) \leq r\}$ is closed for every real r. Then there is $y \in X$ such that $f(y) \leq f(x)$ for each $x \in X$. ∎

Proof. Since $X = \bigcup_{n=1}^{\infty} \{x: f(x) > -n\}$ and X is compact, there is a k such that $X = \bigcup_{n=1}^{k} \{x: f(x) > -k\}$ and therefore f is bounded from below. Let $c =$ g.l.b. $\{f(x): x \in X\}$. If $c \notin f(X)$, then $X = \bigcup_{n=1}^{\infty} \{x: f(x) > c + 1/n\}$ and by compactness, there is an integer m such that $X = \bigcup_{n=1}^{m} \{x: f(x) > c + 1/n\}$. This means that $c + 1/m \leq$ g.l.b. $f(X) = c$, which is a contradiction. ∎

Definition 1.24. A topological space X is said to be *first countable* if for each $x \in X$ there exists a countable family $[V_n(x)]$ of open sets containing x such that whenever G is an open set containing x, $V_n(x) \subset G$ for some n. ∎

(Example 1.22 gives an example of a compact Hausdorff space, that is not first countable; the requirement for first countability fails at 1. Note that [0, 1], with the usual topology, is first countable.)

Remarks

1.42. A first countable topological space is Hausdorff if and only if every convergent sequence has a unique limit.

1.43. If X is first countable and T_1, then x is a limit point of E if and only if there exists a sequence of *distinct* points in E converging to x.

1.44. Let f be a mapping from a first countable topological space X into a topological space Y. Then f is continuous at $x \in X$ if and only if $f(x_n)$ converges to $f(x)$ whenever x_n converges to x.

Definition 1.25. A topological space is called *second countable* if there is a countable base for its topology. ∎

Remarks

1.45. Every second countable space is first countable, but not conversely. The discrete topology on an uncountable set illustrates this. The real numbers, with the usual topology, is second countable since all the open intervals with rational endpoints form a countable base.

1.46. In a second countable space, every open cover of a subset has a countable subcover.

To see this, let $\mathscr{A}$ be an open cover for $E \subset X$, which has (B_n) as a countable base. Let $N = \{n: B_n \subset A \text{ for some } A \in \mathscr{A}\}$. Then N is countable and $E \subset \bigcup_{n \in N} B_n$. For each $n \in N$, let $A_n \in \mathscr{A}$ such that $B_n \subset A_n$. Then it can be easily verified that $E \subset \bigcup_{n \in N} A_n$.

1.47. If $(X, \mathscr{C})$ is an infinite, second countable, and Hausdorff space, then card $\mathscr{C} = c$ and card $X \leq c$.

To see this, let (B_n) be a countable base for X. Then for $x \in X$, we have $\{x\} = \bigcap\{B_n: x \in B_n\}$. This means that the subset of all those n for which $x \in B_n$ determines the point x. Therefore, card $X \leq c$. By a similar argument (since each open set is a union of the B_n's), card $\mathscr{C} \leq c$. Now we observe that an infinite Hausdorff space contains an infinite sequence of pairwise disjoint nonempty open sets. [If X has no limit points, X has the discrete topology and it is easy to prove. If x is a limit point of X, then one can use an inductive argument and the Hausdorff property to prove this.] Now the union of each distinct subsequence (finite or infinite) of these open sets is a different open set. It follows that card $\mathscr{C} \geq c$. Hence, card $\mathscr{C} = c$.

Definition 1.26. A topological space is called *Lindelöf* if every open cover of the space has a countable subcover. ∎

Remarks

1.48. Every compact space is a Lindelöf space. A subspace of a compact Hausdorff space need not be Lindelöf; for instance, see Example 1.22,—which also shows that "Lindelöf" need not imply "second countable."

1.49. Closed subspaces of Lindelöf spaces are Lindelöf.

1.50. Every regular Lindelöf space X is normal.

To see this, let A and B be disjoint closed sets. For $x \in A$, by regularity there is an open set $V(x)$ such that $x \in V(x) \subset \overline{V(x)} \subset X - B$. By the Lindelöf property, there exist a sequence of open sets (V_n) such that $A \subset \bigcup_{n=1}^{\infty} V_n$ and for each n, $\bar{V}_n \subset X - B$. Similarly, there exists a sequence of open sets (W_n) such that $B \subset \bigcup_{n=1}^{\infty} W_n$ and for each n, $\overline{W_n} \subset X - A$. Let

$$G = \bigcup_{n=1}^{\infty} \left(V_n - \bigcup_{i=1}^{n} \bar{W}_i \right)$$

and

$$H = \bigcup_{n=1}^{\infty} \left(W_n - \bigcup_{i=1}^{n} \bar{V}_i \right).$$

Then $G \cap H = \varnothing$, $A \subset G$, $B \subset H$ and G, H are open.

Definition 1.27. A subset A of a topological space X is *dense* if $\bar{A} = X$. If X has a countable dense subset, then X is called *separable*. ▮

Remarks

1.51. $R^n (n \geq 1)$ with its usual topology is separable. Any uncountable set with discrete topology, though first countable, is not separable or Lindelöf.

1.52. Every second countable space is separable. The converse need not be true. For instance, let X be an uncountable set and $\mathscr{C} = \varnothing \cup \{A \subset X: X - A \text{ is finite}\}$. Then $(X, \mathscr{C})$ is a *separable space*, since every infinite set is dense in X. But this space is *not even* first countable, since in that case each $x \in X$ is the intersection of a countable number of open sets, a contradiction.

1.53. Separability need not imply the Lindelöf property. Also, a closed subspace of a separable space need not be separable. (See Problem 1.4.10.)

We next present an important characterization of normality often useful in analysis. For a proof of this result, we refer the reader to [34].

Lemma 1.1. *Urysohn's Lemma.* A topological space X is normal if and only if given any two disjoint closed subsets A and B of X, there exists a continuous real-valued mapping $f: X \to [0, 1]$ such that $f(A) = \{0\}$ and $f(B) = \{1\}$. ▮

Theorem 1.7. *Tietze Extension Theorem.* A topological space X is normal if and only if every real-valued continuous function f from a closed subset A of X into $[-1, 1]$ can be extended to a continuous function $g: X \to [-1, 1]$. ▮

Proof. The "if" part follows easily using Urysohn's Lemma. For the "only if" part, let X be normal and f be a nonconstant continuous function from a closed subset A of X into $[-1, 1]$. Suppose $[a, b]$ is the smallest closed subinterval of $[-1, 1]$ containing $f(A)$. Then if

$$h(x) = \frac{2}{b-a} x - \frac{b+a}{b-a},$$

the smallest closed interval containing $h \circ f(A)$ is $[-1, 1]$. Since h: $[a, b]$

$\to [-1, 1]$ is a homeomorphism, the theorem will be proven for f if it is proven for $h \circ f$. Therefore, with no loss of generality, we assume that the smallest closed interval containing $f(A)$ is $[-1, 1]$.

Now let $C = \{x \in A: f(x) \leq -1/3\}$ and $D = \{x \in A: f(x) \geq 1/3\}$. Then C and D are nonempty disjoint closed subsets of X. Noting that Urysohn's Lemma holds easily for any two real numbers in place of the numbers 0 and 1, there exists a continuous f_1: $X \to [-1/3, 1/3]$ such that $f_1(C) = \{-1/3\}$ and $f_1(D) = \{1/3\}$. Then for each x in A, $|f(x) - f_1(x)| \leq 2/3$. By a similar argument applied to $f - f_1$, there is a continuous f_2: $X \to [-2/3^2, 2/3^2]$ such that $|f(x) - f_1(x) - f_2(x)| \leq (2/3)^2$ for each x in A. Continuing, we can find a sequence of continuous functions f_n: $X \to [-2/3^n, 2/3^n]$ such that $|f(x) - \sum_{i=1}^n f_i(x)| \leq (2/3)^n$ for each x in A. Since $\sum_{i=1}^\infty |f_i(x)| \leq 1$ for all x in X, it follows that $\sum_{i=1}^\infty f_i(x)$ is uniformly convergent to some continuous function $g(x)$. Then g is the desired extension of f. ∎

Next, we introduce some notions related to compactness.

Definition 1.28. A topological space X is called *countably compact* if every countable open cover of X has a finite subcover. ∎

Proposition 1.17. A topological space X is countably compact if and only if every sequence (x_n) in X has at least one *cluster point* (i.e., a point x such that every open set containing x contains x_n for infinitely many n.) ∎

Proof. It can be verified easily that X is countably compact if and only if every countable family of closed sets with f.i.p. has a nonempty intersection.

For the "if" part, let (A_i) be a sequence of closed sets with f.i.p. For each n, let $x_n \in \bigcap_{i=1}^n A_i$. Let x be a cluster point of (x_n). Then clearly $x \in \bigcap_{i=1}^\infty A_i$. For the "only if" part, let X be countably compact and (x_n) be a sequence in X. Let $A_n = \{x_n, x_{n+1}, \ldots\}$. Then (A_n) has f.i.p. and therefore, $\bigcap_{n=1}^\infty \bar{A}_n$ is nonempty and contains some point x. Then x is a cluster point of (x_n). ∎

Definition 1.29. A topological space X is called *sequentially compact* if every infinite sequence in X has a convergent subsequence. ∎

The following result is obvious.

Proposition 1.18. A sequentially compact space is countably compact. Also a first countable, countably compact space is sequentially compact. ∎

Example 1.23. *A First Countable Sequentially Compact Space That Is Not Compact.* Let $(X, \leq)$ be the well-ordered set of all ordinals that are less than Ω, the first uncountable ordinal. (See Proposition 1.6.) Let all sets of the form $\{x: x < a\}$ or $\{x: a < x\}$ for some $a \in X$ be a subbase for a topology, called the *order topology* for X. Then this topology is *first countable*. For, if $y \in X$, then there is $z \in X$ such that $y < z < \Omega$ and $\{(x_1, x_2): x_1 < y < x_2 \leq z\}$ is countable by Proposition 1.6 (ii). A similar argument shows that this topology is *not second countable*.

Also, X has the following property. If $A \subset X$, A countable, then sup A (which exists by the well-ordering of $X \cup \{\Omega\}$) is less than Ω. This is because sup A has only a countable number of predecessors relative to the ordering and, therefore, sup $A < \Omega$. It is clear, therefore, that X is countably compact.

To prove that X is not compact, consider the open cover (for X) $\bigcup_{x \in X}\{y: y < x\}$. Since every countable union of these open sets is countable, it does not have a finite subcover and therefore X is *not compact*.

Example 1.24. *A Compact Space That Is Not Sequentially Compact.* Consider $X = I^I$, the uncountable cartesian product of $[0, 1] = I$, with product topology. Since I is compact and Hausdorff (with usual topology), X is also compact and Hausdorff. Points of X are really functions $f: I \to I$. A sequence (f_n) converges to f if and only if for each $x \in I$, $f_n(x)$ converges to $f(x)$. This follows from the definition of product topology. To show that X is not sequentially compact, let $f_n(x)$ be the nth digit in the binary expansion of x. Suppose (f_{n_k}) is a convergent subsequence. There is $x \in I$ such that the sequence $f_{n_k}(x)$ is a sequence of infinitely many 0's and infinitely many 1's, so that the f_{n_k}'s cannot converge.

Definition 1.30. A topological space X is called *locally compact* if for each $x \in X$, there is an open set V such that $x \in V$ and $\bar{V}$ is compact. ▌

Remarks on Local Compactness

1.54. Every compact space is locally compact; $R^n (n \geq 1)$ (with usual topology) are locally compact, but not compact.

1.55. Let X be a locally compact Hausdorff space. We can compactify X by considering $X^* = X \cup \{\infty\}$ and taking a set in X^* to be open if and only if it is either an open subset of X or the complement of a compact subset of X. Clearly, the topological space X^* (thus formed) is a compact Hausdorff space. X^* is called the *one-point compactification* of X. Since every compact Hausdorff space is T_4 (Remark 1.37) and $T_4 \Rightarrow T_3$, X^* is T_3. Since any subspace of a T_3 space is also T_3, X is T_3.

1.56. By Remark 1.55, a locally compact Hausdorff space is regular. Therefore, in such a space given any open set V containing x, there exists open W with $x \in W$, $\overline{W}$ compact and $\overline{W} \subset V$.

1.57. Every closed or open subspace of a locally compact space is locally compact.

1.58. The continuous open image of a locally compact space is locally compact.

1.59. $\times\{X_\lambda: \lambda \in \Lambda\}$ is locally compact if and only if all the X_λ are locally compact and all but finitely many X_λ are compact.

The "if" part is easy to prove. For the "only if" part, let the product space be locally compact. Then each projection π_λ is a continuous, open, onto mapping and therefore each X_λ is locally compact. Also, if V is open in $\times_{\lambda \in \Lambda} X_\lambda$ and $\bar{V}$ is compact, then $\pi_\lambda(\bar{V})$ is compact and $\pi_\lambda(\bar{V}) = X_\lambda$ for all but finitely many λ.

1.60. If X is Hausdorff and Y is a dense locally compact subspace, then Y is open.

To prove this, let $x \in Y$. Then there is open V such that $x \in V$ and $V \cap Y$ has its closure (in Y) compact. Suppose $V \cap (X - Y)$ is nonempty and contains z. Then if $\{W_\lambda(z)\}$ is the class of all open sets ($\subset V$) containing z, clearly $\{z\} = \cap\, W_\lambda(z)$. Now $\{\overline{W_\lambda(z) \cap Y}\}$ is a family of closed sets in Y with f.i.p. (since Y is dense). But it is not difficult to see that $\cap \{\overline{W_\lambda(z) \cap Y}\}$ is empty, contradicting the compactness of $\bar{V} \cap Y$ (= the closure of $V \cap Y$ in Y). Hence $V \subset Y$.

Proposition 1.19. (*Baire*) The intersection of any countable family of open dense sets in a locally compact Hausdorff space X is dense. ▮

Proof. Let (A_i) be a sequence of open dense sets. Let V be an open set. Then $V \cap A_1 \neq \varnothing$. By Remark 1.56 there is open B_1, such that $\bar{B}_1$ is compact and $\bar{B}_1 \subset V \cap A_1$. Similarly, working with B_1 and A_2, we see that there is open B_2 with $\bar{B}_2$ compact and $\bar{B}_2 \subset B_1 \cap A_2$. By induction, we can find a sequence of open sets (B_n), $\bar{B}_n \subset B_{n-1} \cap A_n$ for each n. Since $\bar{B}_1$ is compact and $(\bar{B}_n)$ have f.i.p., the proposition follows. ▮

Definition 1.31. A topological space X is called a *Baire* space if the intersection of any countable family of open dense sets is dense. ▮

Remarks on a Baire Space

1.61. Every locally compact Hausdorff space is a Baire space.

1.62. If X is a Baire space and $X = \bigcup_{n=1}^{\infty} A_n$, A_n closed, then for some

n the interior of A_n is nonempty. The reason is that since $\bigcap_{n=1}^{\infty} X - A_n = \varnothing$, $X - A_n$ is not dense for some n and therefore $A_n^\circ \neq \varnothing$.

1.63. Because of Remark 1.62 the set of rationals with relative topology (in R) is *not* a Baire space.

1.64. A set A in a topological space is called *a set of the first category* if and only if it is a countable union of *nowhere dense sets* (i.e., sets whose closure have an empty interior). The interior of a set of the first category in a Baire space is empty. A Baire space is of the *second category*, i.e., a space that is not of the first category.

Problems

1.4.1. Let f be a mapping from a topological space X into a topological space Y. Then show that the following are equivalent:

(i) f is continuous.

(ii) If $B \subset Y$, then $\overline{f^{-1}(B)} \subset f^{-1}(\bar{B})$.

(iii) If $B \subset Y$, then $f^{-1}(B^\circ) \subset [f^{-1}(B)]^\circ$.

1.4.2. If Y is Hausdorff, X is any topological space and $f, g\colon X \to Y$ are continuous, then show that $\{x \in X\colon f(x) = g(x)\}$ is a closed set.

1.4.3. Show that a one-to-one, onto, and continuous mapping f from a compact space into a Hausdorff space is also open. [A continuous, open, one-to-one, and onto map from one topological space into another is called a *homeomorphism*.]

1.4.4. Given A, a closed G_δ in a normal topological space X, show that there exists a continuous real-valued function f such that $A = f^{-1}\{0\}$. Note that a set $B \subset X$ is a G_δ if and only if $B = \bigcap_{n=1}^{\infty} G_n$, G_n open in X for each n. [Hint: Write $A = \bigcap_{n=1}^{\infty} G_n$, $G_n \supset G_{n+1}$ and G_n open for each n. By Urysohn's Lemma, there is a continuous function f_n with $f_n(A) = \{0\}$ and $f_n(X - G_n) = 1/n^2$. Then consider $f = \sum_{n=1}^{\infty} f_n$.]

1.4.5. Show that closures are preserved by homeomorphisms, but not necessarily by continuous mappings.

1.4.6. Show that the graph $G = \{(x, f(x))\colon x \in X\}$ (with relative product topology) of a continuous mapping $f\colon X \to Y$ is homeomorphic with X.

1.4.7. A point x in a topological space X is called *isolated* if and only if $\{x\}$ is open. Show that there are an infinite number of isolated points in a compact Hausdorff space X if and only if there exists $x_0 \in X$ such that X and $X - \{x_0\}$ are homeomorphic.

1.4.8. Show that normality is preserved by closed continuous mappings.

1.4.9. *Half-Open Interval Topology.* Show that the family of all sets of the form $[a, b)$ (a and b real numbers) form a basis for a topology in R (the reals). Show also that this topology is T_4, first countable, separable, *not* second countable, but Lindelöf. [Hint for T_4: For A and B closed, $A \cap B = \emptyset$, let $[a, x_a) \subset X - B$ and $[b, x_b) \subset X - A$; then $(\bigcup_{a \in A}[a, x_a)) \cap (\bigcup_{b \in B}[b, x_b)) = \emptyset$.]

Furthermore, show that any compact set in this topology is a nowhere dense set in the usual topology of R. [Hint: If A is not nowhere dense, then $A \cap [a, b]$ is dense in some interval $[a, b]$; this means that if $a < b_i < b_{i+1} < b$ and $\lim b_i = b$, then $\{(-\infty, a), [a, b_1), [b_i, b_{i+1})$ for $1 \leq i < \infty$; $[b, \infty)\}$ is an open cover of A with no finite subcover.]

1.4.10. Let X be the reals with the half-open interval topology. Show the following:

(i) $S = \{(x, y): x = -y\}$ is a closed subset in $X \times X$ (with the product topology) and *discrete* in its relative topology.

(ii) S is neither separable nor Lindelöf; hence $X \times X$ is not Lindelöf.

(iii) $X \times X$ is not normal.

[Hint: Since $X \times X$ is separable, there are $2^{\aleph_0}$ continuous real-valued functions on $X \times X$. But there are 2^c continuous real-valued functions on S. Since S is closed, this contradicts normality for $X \times X$.]

1.4.11. A topological space X is called *pseudocompact* if and only if every continuous real-valued function on X is bounded. Show that every countably compact space is pseudocompact whereas a pseudocompact T_4 space is countably compact. [Hint: If X is not countably compact, then there is an infinite closed set $A \subset X$, discrete in the relative topology. If $(x_n) \subset A$ and $f(x_n) = n$ for each n, then the function f continuous on A cannot be extended to a bounded function on X.]

1.4.12. Consider an infinite set X and $\mathscr{E} = \emptyset \cup \{A \subset X: a \in A\}$, where a is some fixed point of X. Show that $(X, \mathscr{E})$ is a pseudocompact space that is *not* countably compact.

★ **1.4.13.** *Paracompact Spaces.* A cover (A_α) is called a *refinement* of a cover (B_β) if each $A_\alpha \subset B_\beta$ for some β. A cover is called *locally finite* if each point has an open set (containing it) intersecting only finitely many members of the cover. A topological space is called *paracompact* if every open cover has an open locally finite refinement. Show the following:

(i) Every paracompact Hausdorff (or regular) space is normal.

[Hint: Let X be paracompact regular, A and B closed disjoint subsets of X. For $x \in A$, let $V(x)$ be open, $x \in V(x)$, and $\overline{V(x)} \subset X - B$. Then if $\mathscr{F}$ is a locally finite refinement of $\{V(x): x \in A\} \cup \{X - A\}$, and if $G = \cup \{H:$

$H \cap A \neq \emptyset$ and $H \in \mathscr{F}\}$, then G is open and $\supset A$. If $J = X - \cup \{\bar{H}$: $H \cap A \neq \emptyset$ and $H \in \mathscr{F}\}$, then J will be open and $\supset B$. Note that if $\mathscr{E}$ is a locally finite family of subsets of a topological space, then $\cup\{\bar{E}: E \in \mathscr{E}\} = \overline{\cup\{E: E \in \mathscr{E}\}}$.]

(ii) Every compact (or metric) topological space is paracompact. (The metric case will be considered in Section 1.5.)

1.4.14. Let X be a locally compact Hausdorff space and K a compact subset of X. Show that there is a continuous real-valued function f on X such that $f(x) = 1$, $x \in K$, and $\{x: f(x) \neq 0\}$ has compact closure.

1.4.15. Show that in a regular space, the closure of a compact set is compact.

★ **1.4.16.** Uncountable products of Z^+ (the positive integers with discrete topology) need *not* be *separable*. Let $X_\lambda = \times_{k \in A} Z_k^+$, where card $A = \lambda$ and $Z_k^+ = Z^+$. Show the following:

(i) If $\lambda \leq 2^{\aleph_0}$, then X_λ is separable.

(ii) If $\lambda > 2^{\aleph_0}$, then X_λ is *not* separable.

[Hint: Let $\lambda \leq 2^{\aleph_0}$. Let f be a bijection from A onto a subset of [0, 1]. For every finite collection of disjoint closed subintervals $I_1, I_2, \ldots, I_n$ of [0, 1] with rational endpoints and for every finite subset $p_1, p_2, \ldots, p_n$ of positive integers, we choose $(x_k) \in X_\lambda$ such that $x_k = p_i$ if $f(k) \in I_i$, $= 1$ otherwise. Then the set of all these points so chosen is a countable dense set in X_λ. Suppose now that $\lambda > 2^{\aleph_0}$. For each, $\alpha, \beta \in A$, $S \cap \pi_\alpha^{-1}(1)$ and $S \cap \pi_\beta^{-1}(1)$ are distinct nonempty subsets of S, if S is a dense set in X_λ. Then $g: A \to 2^S$ defined by $g(\alpha) = S \cap \pi_\alpha^{-1}(1)$ is one-to-one; hence, card $A \leq 2^{\aleph_0}$ if S is countable.]

1.4.17. *Product of Second (First) Countable Spaces.* Show that the product topology is second (first) countable if and only if the topology of each coordinate space is second (first) countable and all but a countable number of the coordinate spaces have the indiscrete topology.

1.4.18. *Functions with Compact Graphs.* Let f be a map from a topological space X into a topological space Y. Then show the following:

(i) Compactness of the graph of f implies that of X and $f(X)$.

(ii) The graph of f is compact whenever X is compact, Y is T_2, and f is continuous.

(iii) If the graph of f is compact and X is T_2, then f is continuous.

1.4.19. *Nets*: A directed system is a set A with a relation $<$ satisfying

(i) $a < b$ and $b < c$ imply $a < c$, and

(ii) for $a, b \in A$, there is $c \in A$ with $a < c$ and $b < c$.

[The family of all open sets containing a point x is a directed system, with

$G_1 < G_2$, if $G_2 \subset G_1$.] A net (x_α) is a mapping of a directed system into a topological space X, where x_α is the value of the net at α. A net (x_α) is said to converge to $x \in X$ if given an open set G, $x \in G$, there is an $\alpha_0 \in A$ such that for all $\alpha > \alpha_0$, $x_\alpha \in G$. Prove the following.

(i) For $E \subset X$, $x \in \bar{E}$ if and only if it is the limit of a net (x_α) in E.

(ii) X is T_2 if and only if every net in X has at most one limit.

(iii) $f: X \to Y$ is continuous at x if and only if for each net (x_α) converging to x, the net $[f(x_\alpha)]$ converges to $f(x)$.

★ **1.4.20.** *The Stone–Čech Compactification.* Let X be a completely regular Hausdorff space and $\mathscr{F}$ the family of continuous real-valued functions f on X with $|f| \leq 1$. Let $I = [-1, 1]$ and for each $f \in \mathscr{F}$, let $I_f = I$. Consider $Y = \times_{f \in \mathscr{F}} I_f$ with product topology. Let φ: $X \to Y$ be defined such that the f coordinate of $\varphi(x)$ is $f(x)$. Show that X and $\varphi(X)$ are homeomorphic. $\overline{\varphi(X)}$, a compact Hausdorff space containing X [identifying X with $\varphi(X)$] as a dense subspace, is called the Stone–Čech compactification of X and is denoted by $\beta(X)$. Prove the following: (i) Each bounded continuous real-valued function on X extends to a unique continuous function on $\beta(X)$. (ii) If X is a dense subspace of a compact Hausdorff space W with property (i) above, then there is a homeomorphism ψ from W onto $\beta(X)$ such that $\psi(x) = x$ for all $x \in X$. (It may be noted that [0, 1] is not the Stone–Čech compactification of (0, 1], since sin $(1/x)$ on (0, 1] does not have a continuous extension to [0, 1].)

★ **1.4.21.** Prove that R is not homeomorphic to $X \times X$ for any topological space X.

• 1.5. Connected Spaces, Metric Spaces, and Fixed Point Theorems

In this text, as will be found later, metric spaces are referred to often and are of utmost importance. They are studied in this section. Although connectedness does not appear explicitly in the theme of this text, for completeness we also study connectedness, an important topological concept. We also study in this section fixed point theorems specifying conditions when a mapping of a topological space into itself leaves one or more points invariant or fixed. Fixed point theorems are of immense importance in analysis; they are used often in the existence theory of differential and integral equations. In Chapter 7, we present the well-known Kakutani Fixed Point Theorem and then use it to show the existence of a Haar measure in a compact topological group, a result of fundamental importance in analysis.

Definition 1.32. A topological space X is called *connected* if X and $\varnothing$ are the *only* subsets of X that are both open and closed. X is called *disconnected* if X is *not* connected. A subset A of X is called connected if A with its relative topology is connected. ∎

Remarks

1.65. A subset A of the reals with the usual topology, which contains at least two distinct points, is connected if and only if it is an interval.

To see this, suppose A is not an interval; then there are a, b in A and c not in A with $a < c < b$. This means $A \cap (-\infty, c)$ is open and closed in A, and A is not connected. Conversely, suppose A is an interval, and $B \subset A$ with B both open and closed in A, $B \neq \varnothing$ and $A - B \neq \varnothing$. Then there is $x \in A - B$ such that $x > y$ for some $y \in B$; otherwise for some $z \in A - B$, $B = (z, \infty) \cap A$ and is therefore not closed. Let $c = \text{l.u.b. } \{y \in B: y < x\}$. Then $c \in A$ (since A is an interval) and c is a limit point of B and so $c \in B$. Clearly, $c < x$ and $[c, x] \subset A$. But since B is open in A, there is d in B with $c < d < x$. This contradicts the l.u.b. property of c.

1.66. A topological space X is connected if and only if the only continuous mappings $f: X \to \{0, 1\}$ (with discrete topology) are the constant mappings. The reason is that if f is nonconstant, then $f^{-1}\{0\}$ is an open and closed subset different from both X and $\varnothing$; also if A is a nontrivial open and closed subset of X, then f defined by $f(A) = \{0\}$ and $f(X - A) = \{1\}$ is a continuous map.

1.67. Connectedness is preserved under continuous mappings. The intermediate value property of a continuous function follows easily from this fact.

1.68. If A is a connected set in a topological space and $A \subset B \subset \bar{A}$, then B is also connected. The reason is that if $D \subset B$ and D is open and closed in B, then $D \cap A$ is also open and closed in A.

1.69. The union of any family (A_λ) of connected sets, having a nonempty intersection, is connected. To see this, let $x \in \bigcap_{\lambda \in \Lambda} A_\lambda$ and H be a closed and open subset of $\bigcup_{\lambda \in \Lambda} A_\lambda = A$. If $x \in H$, then for each λ, $A_\lambda \cap H \neq \varnothing$ and is both open and closed in A_λ. Since A_λ is connected, $A_\lambda \cap H = A_\lambda$ or $A_\lambda \subset H$ for each λ. On the other hand, if $x \in A - H$, then $A_\lambda \cap H^c \neq \varnothing$ and is both open and closed in A_λ. Since A_λ is connected, $A_\lambda \cap H^c = A_\lambda$ or $A_\lambda \subset H^c$ for all λ, meaning $H = \varnothing$. In either case $H = A$ or $H = \varnothing$.

1.70. $X_1 \times X_2$ (with the product topology) is connected if X_1 and X_2 are connected. Notice that for $x_1 \in X_1$ and $x_2 \in X_2$, $\{x_1\} \times X_2$ and $X_1 \times \{x_2\}$ are both connected, $X_1 \times X_2 = \bigcup_{x_2 \in X_2} [\{x_1\} \times X_2 \cup X_1 \times \{x_2\}]$, which is connected by Remark 1.69 above.

1.71. If $n > 1$ and $A \subset R^n$ is countable, then $R^n - A$ is connected. The reason is simple: For simplicity, let $n = 2$ and $x, y \in R^2 - A$. For any line segment I_y meeting the segment $\overline{xy}$ in only one point (see Figure 1), if $z, z^1 \in I_y$, then $(\overline{xz} \cup \overline{zy}) \cap (\overline{xz^1} \cup \overline{z^1y}) = \{x, y\}$. Clearly, since I_y is uncountable and A is countable, $A \cap (\overline{xz_y} \cup \overline{z_yy}) = \varnothing$ for some $z_y \in I_y$. Hence, $R^2 - A = \cup\{(\overline{xz_y} \cup \overline{z_yy})\colon y \in R^2 - A\}$ is connected by Remark 1.69.

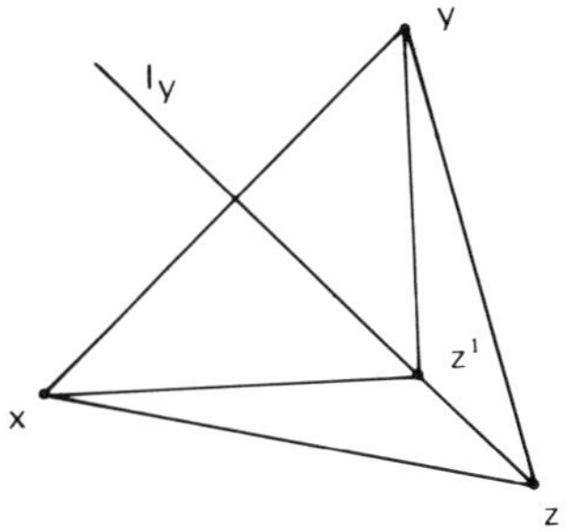

Fig. 1

1.72. R^1 and R^n $(n > 1)$ are not homeomorphic.† For if $f\colon R^1 \to R^n$ is a homeomorphism, then for $x \in R$, $R^1 - \{x\}$ is not connected, despite the fact that $f(R^1 - \{x\}) = R^n - \{f(x)\}$, which is connected, by Remark 1.71. (It is relevant to point out here that for $n \neq m$, R^n and R^m are also *not* homeomorphic; the proof is more involved and omitted.)

Theorem 1.8. *A Fixed Point Theorem.* Let f be a continuous function from $[0, 1]$ into itself. Then there is $x \in [0, 1]$ with $f(x) = x$. ∎

Proof. We may and do assume that $f(0) > 0$ and $f(1) < 1$. Let $g(x) = x - f(x)$. Then g is continuous, $g(0) < 0 < g(1)$. By the intermediate-value property of g, there is $x \in (0, 1)$ with $g(x) = 0$. ∎

This theorem is a special case of the famous Brouwer's Fixed Point Theorem (the earliest fixed point theorem, proven in 1912 by L. E. J. Brouwer).

Theorem 1.9. *The Brouwer Fixed Point Theorem.* Let $f\colon I^n \to I^n$ be a continuous map. Then there is $x \in I^n$ with $f(x) = x$. [Here $I^n = \{(x_1, x_2, \ldots, x_n)\colon 0 \leq x_i \leq 1, 1 \leq i \leq n\}$.] ∎

† Two topological spaces are called homeomorphic if there is a homeomorphism (i.e., a one-to-one, onto, continuous, and open map) from one into the other.

An elementary proof of this theorem is recently given by Kiyoshi Kuga.† The proof is too lengthy to be included here.

Proposition 1.20. Let $\{X_\lambda\}$, $\lambda \in \Lambda$ be an indexed family of topological spaces. Then $X = \times_{\lambda \in \Lambda} X_\lambda$ is connected if and only if each X_λ is connected. ∎

Proof. The "only if" part follows by Remark 1.67, since the projection π_λ is a continuous map from X onto X_λ.

In the "if" part, let f be a continuous map from X into the discrete space $\{0, 1\}$. Suppose $x, y \in X$ and $f(x) = 0$. We claim that $f(y) = 0$. Let V be open in X, $x \in V$ and $f(V) = 0$. Let $\lambda_1, \lambda_2, \ldots, \lambda_n \in \Lambda$ be such that $\bigcap_{i=1}^n \pi_{\lambda_i}^{-1}(V_{\lambda_i}) \subset V$, V_{λ_i} open in X_{λ_i} and $x(\lambda_i) \in V_{\lambda_i}$. Let $z \in X$ be defined by $z(\lambda_i) = x(\lambda_i)$, $1 \leq i \leq n$ and $z(\lambda) = y(\lambda)$ for all other $\lambda \in \Lambda$. Then $z \in V$ and $f(z) = 0$. Now $W \equiv \{w \in X\colon w(\lambda) = y(\lambda)$ for all $\lambda \neq \lambda_i$, $1 \leq i \leq n\}$ is clearly homeomorphic to $X_{\lambda_1} \times X_{\lambda_2} \times \cdots \times X_{\lambda_n}$, which is connected by Remark 1.70. Since $z \in W$ and $y \in W$, $f(z) = f(y) = 0$. ∎

A disconnected space can be partitioned uniquely by maximal connected sets or components. The number of components gives a rough idea of how disconnected the space is.

Definition 1.33. The *component* C_x of $x \in X$, a topological space, is the union of all connected subsets of X containing x. ∎

Clearly, C_x is a *maximal connected* set, which is closed, by Remark 1.68.

Examples

1.25. Let $X = \{1/n\colon n$ is a positive integer$\} \cup \{0\}$ with the relative topology from R. Then $\{0\}$ is a component, which is *not* open.

1.26. In $R - \{x\}$, with the relative topology, $(-\infty, x)$ and (x, ∞) are the components.

1.27. In R^2, the subspace $\{(x, y)\colon$ either $xy = 1$ or $y = 0$ or $x = 0\}$ has three components. Note that the map $f\colon (0, \infty) \to R^2$ defined by $f(x) = (x, 1/x)$ is continuous and therefore $\{(x, 1/x)\colon x > 0\}$ is a connected subset of R^2.

A topological concept called "*local connectedness*" provides examples of nonconnected spaces, where the *components are open*.

† Kiyoshi Kuga, *SIAM J. Math. Anal.* **5**(6) (1974).

Definition 1.34. A topological space X is called *locally connected* at a point $x \in X$ if given V open with $x \in V$, there exists an open U, a connected set W such that $x \in U \subset W \subset V$. X is called locally connected if it is locally connected at each point in X. Equivalently, X is locally connected if components of open subsets of X are open in X. ∎

Clearly R and every interval of R is locally connected, whereas the set of rationals (with relative topology) is not locally connected at any of its points. The rationals are, in fact, *totally disconnected* (i.e., a space, where every singleton is a component). Note that a discrete topological space, though not connected, is locally connected.

Example 1.28. *A Connected Space That Is Not Locally Connected.* Consider f: $(0, 1] \to R$ defined by $f(x) = \sin(1/x)$. Let G be the graph of f. Then G, being the image of a continuous map of $(0, 1]$ into R^2, is a connected subset of R^2. Clearly, its closure in R^2, that is, $G \cup \{(0, y)$: $-1 \leq y \leq 1\}$ is also connected. Since $\bar{G} \cap \{R \times (-\frac{1}{2}, \frac{1}{2})\}$ consists of separated line segments (verify by drawing a neat picture), the component of $(0, 0)$ in $\bar{G} \cap \{R \times (-\frac{1}{2}, \frac{1}{2})\}$ is the segment $0 \times (-\frac{1}{2}, \frac{1}{2})$, which is not open. Hence $\bar{G}$ is *not* locally connected.

One can easily verify that for any two points in R^2 there is a continuous function f: $[0, 1] \to R^2$ mapping 0 and 1 to these two points. In fact, in any open connected set in R^2, any two points can be joined[†] by such a function or *a path*. This motivates the following stronger concept of connectedness.

Definition 1.35. A topological space X is called *path connected* if for any $x_1, x_2 \in X$, there is a continuous map f: $[0, 1] \to X$ such that $f(0) = x_1$ and $f(1) = x_2$. ∎

Remarks

1.73. Path connectedness implies connectedness. The reason is that if A is a proper open and closed subset of X (path connected) then there is a continuous f: $[0, 1] \to X$ such that $f(0) \in A$ and $f(1) \in X - A$. This means that $f^{-1}(A)$ is a proper open and closed subset of $[0, 1]$, contradicting the connectedness of $[0, 1]$.

1.74. The set $\bar{G}$ in Example 1.28, though connected, is *not* path connected.

[†] This will be clarified in Problem 1.5.4.

1.75. Every continuous image of a path-connected space is path connected.

1.76. For every family of path-connected sets (taken with the relative topology) having nonempty intersection, their union is also path connected.

Now we turn our attention to the most widely used topological space—metric spaces.

Definition 1.36. A metric space (X, d) is a nonempty set X with a nonnegative real-valued function d on $X \times X$ such that, for all x, y, and $z \in X$, the following properties hold:

(i) $d(x, y) = 0 \Leftrightarrow x = y$,

(ii) $d(x, y) = d(y, x)$,

(iii) $d(x, z) \leq d(x, y) + d(y, z)$.

The function d is called a *metric* (or *distance*). ∎

Examples

1.29. For $x = (x_1, x_2, \ldots, x_n)$ and $y = (y_1, y_2, \ldots, x_n) \in R^n$, let

$$d(x, y) = \left[\sum_{i=1}^{n} (x_i - y_i)^2 \right]^{1/2}.$$

Then d defines a metric in R^n. To see this, it is sufficient to observe that

(a) $$\left(\sum_{i=1}^{n} a_i b_i \right)^2 \leq \left(\sum_{i=1}^{n} a_i^2 \right) \left(\sum_{i=1}^{n} b_i^2 \right)$$

and

(b) $$\left[\sum_{i=1}^{n} (a_i + b_i)^2 \right]^{1/2} \leq \left(\sum_{i=1}^{n} a_i^2 \right)^{1/2} + \left(\sum_{i=1}^{n} b_i^2 \right)^{1/2}.$$

1.30. Let (X_n, d_n), $1 \leq n < \infty$ be an infinite sequence of metric spaces. For $x = (x_1, x_2, \ldots,)$ and $y = (y_1, y_2, \ldots)$ in $X = \times_{n=1}^{\infty} X_n$, let

$$d(x, y) = \sum_{n=1}^{\infty} \frac{1}{2^n} \cdot \frac{d_n(x_n, y_n)}{1 + d_n(x_n, y_n)}.$$

Then (X, d) is a metric space.

Proposition 1.21. In any metric space (X, d), the family of all sets $S_x(\varepsilon) = \{y\colon d(x, y) < \varepsilon\}$, where $x \in X$ and ε is a positive real number is a base for a topology (called *the topology induced by d*). ∎

Proof. If $z \in S_x(\varepsilon_1) \cap S_y(\varepsilon_2)$ and $\varepsilon = \min\{\varepsilon_1 - d(z, x), \varepsilon_2 - d(z, y)\}$, then $S_z(\varepsilon) \subset S_x(\varepsilon_1) \cap S_y(\varepsilon_2)$. This means that the family of all unions of sets of the forms $S_x(\varepsilon)$ is a topology. ∎

In what follows, a metric space will always be considered a topological space with topology induced by d.

Remarks

1.77. Two different metrics can, sometimes, induce the same topology on a set. For instance, in R^2, the metric d_1 defined by

$$d_1((x_1, y_1), (x_2, y_2)) = |x_1 - x_2| + |y_1 - y_2|$$

and the metric d_2 defined by

$$d_2((x_1, y_1), (x_2, y_2)) = \max\{|x_1 - x_2|, |y_1 - y_2|\}$$

induce the same topology as induced by the metric in Example 1.29.

1.78. The metric d defined by

$$d(x, y) = \begin{cases} 0, & x = y \\ 1, & x \neq y \end{cases}$$

induces a discrete topology on any nonempty set X.

1.79. Every metric space is a first countable, T_4 space. To see this, it is sufficient to observe the following:

(a) For $x \neq y$ and $0 < 2\varepsilon < d(x, y)$, $S_x(\varepsilon) \cap S_y(\varepsilon) = \varnothing$.

(b) Given any open V containing x, there is a positive integer n such that $S_x(1/n) \subset V$.

(c) Given any two closed disjoint sets A and B, for $x \in A$ and $y \in B$ we can choose $\varepsilon_x > 0$ and $\varepsilon_y > 0$ such that $S_x(2\varepsilon_x) \cap B = \varnothing$ and $S_y(2\varepsilon_y) \cap A = \varnothing$.

If $G = \bigcup_{x \in A} S_x(\varepsilon_x)$ and $H = \bigcup_{y \in B} S_y(\varepsilon_y)$, then G and H are open disjoint sets containing A and B, respectively.

1.80. A metric space is separable if and only if it is second countable if and only if it is Lindelöf.

To see this, we need to observe only the following, since "second countable" implies easily "separable" and "Lindelöf."

(a) If (x_n) is a countable dense set, then the family of all sets of the form $S_{x_n}(1/k)$, where k is a positive integer, is a base for the topology.

(b) If the space is Lindelöf, then, for each positive integer k, the open cover $\{S_x(1/k): x \in X\}$ has a countable subcover, $\{S_{x_{n,k}}(1/k): x_{n,k} \in X\}$. Then the family $\{S_{x_{n,k}}(1/k): n \text{ and } k \text{ are positive integers}\}$ is a base for the topology.

Definition 1.37. A metric space (X, d) is called *bounded* if there exists a real number k such that $X = S_x(k)$ for each $x \in X$. It is called *totally bounded* if, given $\varepsilon > 0$, there is a finite set $A_\varepsilon \subset X$ such that $X = \bigcup_{x \in A_\varepsilon} S_x(\varepsilon)$. ▮

Remarks

1.81. A totally bounded metric space is separable. The reason is that the set $A = \bigcup_{n=1}^{\infty} A_{1/n}$ ($A_{1/n}$ is as in Definition 1.37) is a countable dense set.

1.82. Every countably compact metric space is totally bounded. For, if X is countably compact and not totally bounded, then there is $\varepsilon > 0$ and a sequence (x_n) in X such that for each n, $d(x_n, x_i) \geq \varepsilon$, $1 \leq i \leq n$; but this contradicts that (x_n) has a limit point in X.

Proposition 1.22. The following are equivalent for a metric space X:

(a) X is compact,
(b) X is countably compact,
(c) X is sequentially compact. ▮

Proof. By Remark 1.79, a metric space is first countable. Because of Proposition 1.18, we only need to show that (b) ⇒ (a). By Remarks 1.80–1.82 a countably compact metric space is Lindelöf and therefore, compact. ▮

Definition 1.38. A sequence (x_n) in a metric space (X, d) is called a *Cauchy sequence*, if given $\varepsilon > 0$ there is a positive integer $N(\varepsilon)$ such that for $m, n > N(\varepsilon)$, $d(x_m, x_n) < \varepsilon$. (X, d) is called *complete* if every Cauchy sequence in X converges to some point in X. ▮

Remarks

1.83. Completeness need not be preserved by homeomorphisms. For example, R and $(0, 1)$ are homeomorphic, whereas R is complete and $(0, 1)$ is not, with their usual distance metric.

1.84. The class $C_b(X)$ of bounded, real-valued continuous functions on an arbitrary topological space X is a complete metric space with the metric d_0 defined by $d_0(f, g) = \sup_{x \in X} | f(x) - g(x) |$. Completeness follows from the fact that for any Cauchy sequence (f_n) in $C_b(X)$ and for each $x \in X$, $(f_n(x))$ is a Cauchy sequence of real numbers; then if $f(x) = \lim_{n \to \infty} f_n(x)$, it follows easily that $f \in C_b(X)$ and $d_0(f_n, f) \to 0$ as $n \to \infty$.

Definition 1.39. Two metric spaces (X, d_1) and (Y, d_2) are called *isometric* if there is a mapping f from X onto Y such that $d_1(x, y) = d_2(f(x), f(y))$. Here f is called an *isometry*. (Clearly an isometry is a homeomorphism.) ▮

Proposition 1.23. Every metric space (X, d) is isometric to a dense subset of a complete metric space. ∎

Proof. Let $z \in X$. For each $x \in X$, let us define the function f_x by $f_x(t) = d(t, x) - d(t, z)$. Clearly $|f_x(t)| \leq d(x, z)$ and $f_x \in C(X)$. We will now consider the "sup" metric d_0 in $C(X)$ [as in Remark 1.84]. Then $d_0(f_x, f_y) = \sup_{t \in X} |d(t, x) - d(t, y)|$. By taking $t = x$ here, it follows that $d_0(f_x, f_y) \geq d(x, y)$. If $d_0(f_x, f_y) > d(x, y)$, then for some $t \in X$ we have either

$$d(t, x) - d(t, y) > d(x, y)$$

or

$$d(t, x) - d(t, y) < - d(x, y),$$

both of which contradict the triangular inequality property of d. Therefore $d_0(f_x, f_y) = d(x, y)$. Since the closure of $\{f_x : x \in X\}$, being a closed subset of $C(X)$, is a complete metric space, $f: x \to f_x$ is the desired isometry. ∎

Theorem 1.10. A metric space is compact if and only if it is both complete and totally bounded. ∎

Proof. The "only if" part follows by Remark 1.82 and the fact that a Cauchy sequence always converges to its limit point.

For the "if" part, suppose (X, d) is complete and totally bounded. Let (x_n) be an infinite sequence. By the total boundedness property, there exists y_1 in X such that $S_{y_1}(1)$ contains infinitely many terms of (x_n). For the same reason, there exists y_2 in X such that $S_{y_1}(1) \cap S_{y_2}(\frac{1}{2})$ contains infinitely many terms of the sequence (x_n). In this way, we can get a sequence of open sets $S_{y_n}(1/n)$ such that for every positive integer k, the set $B_k = \bigcap_{i=1}^{k} S_{y_i}(1/i)$ contains infinitely many terms of (x_n). Then we choose a subsequence (x_{n_k}) such that $x_{n_k} \in B_k$ for each k. The subsequence (x_{n_k}) is clearly Cauchy and therefore convergent. The rest follows by Proposition 1.22. ∎

Theorem 1.11. *The Baire-Category Theorem.* In a complete metric space X, $\bigcap_{n=1}^{\infty} O_n$, where each O_n is open and dense, is dense in X. Hence a complete metric space is of the second category. ∎

Proof. Suffice it to show that $O \cap (\bigcap_{n=1}^{\infty} O_n)$ is nonempty for any nonempty open O in X. Since O_1 is dense and open, there exist $x_1 \in O \cap O_1$ and $\varepsilon_1 > O$ such that $\overline{S_{x_1}(\varepsilon_1)} \subset O \cap O_1$. Similarly, we can find $x_2 \in S_{x_1}(\varepsilon_1) \cap O_2$ and $0 < \varepsilon_2 < \frac{1}{2}$ such that $\overline{S_{x_2}(\varepsilon_2)} \subset S_{x_1}(\varepsilon_1) \cap O_2$. By induction, we can find sequences $(x_n) \in X$ and real numbers (ε_n) such that $0 < \varepsilon_n < 1/2^{n-1}$ and $\overline{S_{x_{n+1}}(\varepsilon_{n+1})} \subset S_{x_n}(\varepsilon_n) \cap O_n$. Since $\varepsilon_n \to 0$, the sequence (x_n) is Cauchy and has a limit $x \in X$. Then $x \in \bigcap_{n=1}^{\infty} \overline{S_{x_n}(\varepsilon_n)}$, for otherwise there exists some positive integer N such that $x \notin \overline{S_{x_n}(\varepsilon_n)}$ for all $n > N$, which means that $d(x, x_n) > \varepsilon_N$ for all $n > N$, which is a contradiction. Clearly $x \in O \cap (\bigcap_{n=1}^{\infty} O_n)$. ∎

Remarks

1.85. If a complete metric space is the union of a countable number of closed sets, then at least one of them is *not* nowhere dense and therefore must contain a nonempty open set.

1.86. *A Principle of Uniform Boundedness.* Let $\mathscr{F}$ be a family of real-valued continuous functions on a complete metric space X. Suppose that for each $x \in X$ there is a real number $k(x)$ such that $|f(x)| \leq k(x)$ for all $f \in \mathscr{F}$. Then there is a nonempty open set O and a constant K such that $|f(x)| \leq K$ for all $f \in \mathscr{F}$ and $x \in O$.

To see the validity of the principle, we observe that $X = \bigcup_{m=1}^{\infty} E_m$, where $E_m = \{x: |f(x)| \leq m \text{ for all } f \in \mathscr{F}\}$. Since E_m is closed for each m, the principle follows by Remark 1.85 above.

In the context of metric spaces, we now introduce one last important topological concept, that of metrizability. A topological space $(X, \mathscr{C})$ is called *metrizable* if there exists a metric d for X that induces $\mathscr{C}$. A natural question is: When is a topological space metrizable? The metrization problem was solved by P. Urysohn in 1924 for second countable topological spaces. He proved the following theorem.

Theorem 1.12. *The Urysohn Metrization Theorem.* Every second countable T_3 space is metrizable. ∎

For the proofs of this and the next two theorems, we refer the reader to [34].

R. H. Bing (1951), J. Nagata (1950), and Y. M. Smirnov (1951) all independently solved the general metrization problem. To state their results, we introduce the following concept: A family $\mathscr{A}$ of subsets of a topological space X is called *discrete* (or *locally finite*, respectively) if every point of X has an open set containing it that intersects *at most one* member (or only finitely many members, respectively) of $\mathscr{A}$. A family $\mathscr{A}$ is called σ *discrete* (or σ *locally finite*, respectively) if it is a countable union of discrete (locally finite) subfamilies.

Theorem 1.13. *The Nagata–Smirnov Metrization Theorem.* A topological space is metrizable if and only if it is a T_3 space with a σ locally finite base. ∎

Theorem 1.14. *The Bing Metrization Theorem.* A topological space is metrizable if and only if it is a T_3 space with a σ discrete base. ∎

Now that we are familiar with the basic theorems in metric spaces, we can study some of the fundamental and interesting fixed point theorems in such spaces. First, we need a definition. In what follows, d will always stand for the metric in any metric space.

Definition 1.40. Let X and Y be metric spaces and T be a mapping from X into Y. Then T is said to be *Lipschitz* if there exists a real number M such that for all x, y in X, we have

$$d(Tx, Ty) \leq Md(x, y).$$

T is said to be *nonexpansive* if $M = 1$ and a *contraction* if $M < 1$. We call T *contractive* if for all x, y, in X and $x \neq y$, we have

$$d(Tx, Ty) < d(x, y).$$ ∎

Observe that contraction $\Rightarrow$ contractive $\Rightarrow$ nonexpansive $\Rightarrow$ Lipschitz, and a mapping satisfying any of these conditions is continuous.

Examples

1.31. *A Contractive T That Is Not a Contraction.* Let $X = [1, \infty)$ and $Tx = x + 1/x$. Then for x, y in X and $x \neq y$, we have

$$d(Tx, Ty) = |x - y| (xy - 1)/xy < d(x, y).$$

Since $\lim_{y\to\infty}(xy - 1)/xy = 1$, T is not a contraction. It may be noted here that T does not have a fixed point.

1.32. *A Contraction.* A simple example is $Tx = x/2$ for x in $(-\infty,\infty)$. A more interesting example is the following: Let $X = C[0, 1]$ with the usual "sup" metric. Then the mapping T from X into X defined by

$$T[f](t) = k \int_0^t f(x)\, dx, \qquad 0 \leqq t \leqq 1$$

is a contraction for $0 < k < 1$.

1.33. *A Mapping That Is a Contraction Only after Iteration.* Consider the mapping T from $C[a, b]$ (with the "sup" metric and $-\infty < a < b < \infty$) into itself defined by

$$T[f](t) = \int_a^t f(x)\,dx.$$

Then we have

$$T^m[f](t) = \frac{1}{(m-1)!}\int_a^t (t-x)^{m-1} f(x)\,dx.$$

Now it is clear that for sufficiently large values of m the mapping T^m is a contraction, whereas T itself need not be a contraction if $b - a > 1$.

A very frequently used theorem in proving the existence and uniqueness of solutions of an equation is the following result of S. Banach proven in 1922.

Theorem 1.15. *The Banach Fixed Point Theorem.* Let T be a contraction from a complete metric space X into itself. Then there exists a unique x_0 in X satisfying (i) $Tx_0 = x_0$ and (ii) for any x in X, $\lim_{n\to\infty} T^n x = x_0$. ∎

The reader is encouraged to prove Theorem 1.15. This theorem can also be deduced as an easy corollary to the next theorem.

Theorem 1.16. Suppose T is a continuous mapping from a complete metric space X into itself such that there exists x_0 in X and a real number $k < 1$ satisfying the inequality

$$d(T^{n+1}x_0, T^{n+2}x_0) \leqq k \cdot d(T^n x_0, T^{n+1}x_0)$$

for all non-negative integers n. Then the sequence $T^n x_0$ converges to a fixed point u of T as n tends to infinity. ∎

Proof. First we show that the sequence $T^n x_0$ is a Cauchy sequence. For any positive integer m,

$$\begin{aligned} d(T^n x_0, T^{n+m}x_0) &\leqq \sum_{i=0}^{m-1} d(T^{n+i}x_0, T^{n+i+1}x_0) \\ &\leqq d(x_0, Tx_0) \sum_{i=0}^{m-1} k^{n+i} \\ &\leqq d(x_0, Tx_0)\frac{k^n}{(1-k)}, \end{aligned}$$

which goes to zero as $n \to \infty$. Thus the sequence $T^n x_0$ is a Cauchy sequence and has a limit u in X. Since T is continuous,

$$u = \lim_{n\to\infty} T^{n+1}x_0 = T(\lim_{n\to\infty} T^n x_0) = Tu. \qquad \blacksquare$$

The next theorem slightly improves on Theorem 1.15.

Theorem 1.17. If one of the iterates of a mapping T from a complete metric space into itself is a contraction, then T has a unique fixed point. ▮

Proof. If T^n is a contraction, then by Theorem 1.15 there is a unique x_0 such that $T^n x_0 = x_0$. Since $T^n(Tx_0) = T(T^n x_0) = Tx_0$, the uniqueness of the fixed point of T^n implies that $Tx_0 = x_0$. ▮

Now we present an interesting fixed point theorem of Sehgal [57] for a class of mappings that includes the contractive mappings. Note that Example 1.31 shows that additional conditions are necessary on such mappings to obtain fixed point theorems. Sehgal's theorem was proven for contractive mappings by M. Edelstein in 1962.

Theorem 1.18. Let T a continuous mapping from a metric space X into itself such that for all x, y in X with $x \neq y$,

$$d(Tx, Ty) < \max\{d(x, Tx), d(y, Ty), d(x, y)\}. \tag{1.1}$$

Suppose that, for some z in X, the sequence $T^n z$ has a cluster point u. Then the sequence $T^n z$ converges to u and u is the unique fixed point of T. ▮

Proof. If $T^n z = T^{n+1} z$ for some nonnegative integer n, then it is clear that $\lim_{n\to\infty} T^n z = u$, and by condition (1.1) u is the unique fixed point of T. Therefore, we may assume that for all nonnegative integers n, $d(T^n z, T^{n+1} z) > 0$. Now, let $V(y) = d(y, Ty)$. Then V is a continuous function on X and by condition (1.1), for all positive integers n, $V(T^n z) < V(T^{n-1} z) < V(z)$. Let $a = \lim_{n\to\infty} V(T^n z)$. Let (n_i) be a sequence of positive integers such that $T^{n_i} z$ converges to u. Then by the continuity of V and T, $V(u) = V(Tu) = a$. It follows by condition (1.1) that $u = Tu$ and $a = 0$. We now show that the sequence $T^n z$ converges to u. Given $\varepsilon > 0$, there exists a positive integer k such that

$$\max\{V(T^k z), d(T^k z, u)\} < \varepsilon.$$

Using condition (1.1), we have for all positive integers $n \geq k$,

$$\begin{aligned} d(T^n z, u) &= d(T^n z, T^n u) \\ &\leq \max\{V(T^{n-1}z), d(T^{n-1}z, u)\} \\ &\leq \max\{V(T^{n-2}z), d(T^{n-2}z, u)\} \\ &\leq \max\{V(T^k z), d(T^k z, u)\} \\ &< \varepsilon. \end{aligned}$$

Hence $\lim_{n\to\infty} T^n z = u$ and the theorem follows. ∎

Clearly, a contractive T satisfies condition (1.1). But the converse is not true. For example, let T: $[0, 5] \to [0, 5]$ be defined by $Tx = x/2$ id $x \in [0, 4]$ and $= -2x + 10$ if $x \in [4, 5]$. Since $T4 - T5 = 2$, T is not contractive. However, T satisfies condition (1.1).

We now remark that the fixed point u of a contractive mapping, even if it exists, may not be approximated by the successive iterates of the mapping at an arbitrary point in the space. For example, consider $X = \{0, \frac{3}{2}, \frac{4}{3}, \frac{5}{4}, \ldots\}$ with the absolute-value metric, and the mapping T with $T0 = 0$, $T(n + 1)/n = (n + 2)/(n + 1)$ for $n = 2, 3, \ldots$ Our next theorem, due to Sam B. Nadler, Jr., shows that such a situation cannot arise in a locally compact connected metric space.

Theorem 1.19. Let T be a contractive mapping on a locally compact connected metric space X. Suppose that $Tu = u$ for some u in X. Then for each x in X, the sequence $T^n x$ converges to u as n tends to infinity. ∎

Proof. Let $C = \{x: \lim_{n\to\infty} T^n x = u\}$. Then $u \in C$. It suffices to show that the set C is both closed and open.

First, let $x_i \in C$ and $\lim_{i\to\infty} x_i = x_0$ Let $\varepsilon > 0$. Choose integers j and N such that $d(x_j, x_0) < \varepsilon/2$ and $d(T^n x_j, u) < \varepsilon/2$ for all $n \geq N$. Then for $n \geq N$,

$$d(T^n x_0, u) \leq d(T^n x_0, T^n x_j) + \varepsilon/2 \leq d(x_0, x_j) + \varepsilon/2 < \varepsilon.$$

Hence x_0 is in C and C is closed. Now to prove that C is open, let $x \in C$. Choose $r > 0$ such that $K = \{y: d(y, u) \leq r\}$ is compact. Choose a positive integer N and any point y such that $d(x, y) < r/2$ and $d(T^n x, u) < r/2$ for all $n \geq N$. Then since T is contractive,

$$d(T^n y, u) \leq d(T^n y, T^n x) + d(T^n x, u) < r/2 + r/2 = r.$$

Hence $T^n y$ belongs to K for all $n \geq N$. Since K is compact, the sequence $T^n y$

has a cluster point z. By Theorem 1.18, $z = u$ and $\lim_{n\to\infty} T^n y = u$. Hence C is open. ∎

Recently there have been numerous attempts to weaken the hypothesis of the Banach Fixed Point Theorem, but at the same time retaining the convergence property of the successive iterates to the unique fixed point of the mapping. We present two of these generalizations. The first one is due to Sehgal and later improved upon by Guseman and the second one is due to Boyd and Wong. The motivation of the Sehgal–Guseman result comes from the fact that there are mappings T that have at each point an iterate that is a contraction, and yet none of the iterates of T is a contraction. The following example will illustrate this.

Example 1.34. Let $X = [0, 1]$ and for every positive integer n, $I_n = [1/2^n, 1/2^{n-1}]$. Then $X = \{0\} \cup (\bigcup_{n=1}^{\infty} I_n)$. We define $T: X \to X$ as follows. Let

$$Tx = \begin{cases} \dfrac{n+2}{n+3}\left(x - \dfrac{1}{2^{n-1}}\right) + \dfrac{1}{2^n}, & \text{if } x \in \left[\dfrac{3n+5}{2^{n+1}(n+2)}, \dfrac{1}{2^{n-1}}\right] \\ \dfrac{1}{2^{n+1}}, & \text{if } x \in \left[\dfrac{1}{2^n}, \dfrac{3n+5}{2^{n+1}(n+2)}\right] \end{cases}$$

and $T0 = 0$. Then the following properties of T follow easily:

(i) T is a nondecreasing continuous function on $[0, 1]$.

(ii) For x in I_n and any y in X,

$$| Tx - Ty | \leqq \frac{n+3}{n+4} | x - y |.$$

By taking the $(n + 3)$th iterate of T, we have for x in I_n and any y in X,

$$| T^{n+3}x - T^{n+3}y | \leqq \tfrac{1}{2} | x - y |$$

Also $| T^2 0 - T^2 y | \leqq \frac{1}{2} | 0 - y |$. This means that at every point of X, T has an iterate that is a contraction. Now we show that none of the iterates of T is a contraction. Let $0 \leqq k \leqq 1$ and N be a given positive integer. Let $n > (Nk/(1 - k)) - 2$. By the uniform continuity of the iterates of T, there is a $\delta > 0$ such that for $| x - y | < \delta$ and $1 \leqq i \leqq N$, we have

$$| T^i x - T^i y | < \frac{n+N+3}{2^{n+N+1}(n+N+2)}.$$

Setting $x = 1/2^{n-1}$ and y any member of I_n such that $0 < | x - y | < \delta$, it can be verified that $T^i x$ and $T^i y$ are both members of

$$\left[\frac{3(n+i)+5}{2^{n+i+1}(n+i+2)}, \frac{1}{2^{n+i-1}}\right]$$

for $i = 1, 2, \ldots, N$. Thus we have

$$| Tx - Ty | = \frac{n+2}{n+3} | x - y |,$$

$$| T^2 x - T^2 y | = \frac{n+2}{n+4} | x - y |,$$

$$\cdots$$

$$| T^N x - T^N y | = \frac{n+2}{n+2+N} | x - y | > k | x - y |.$$

This shows that none of the iterates of T can be a contraction.

We now present the theorem due to Sehgal and Guseman for mappings having a contractive iterate at each point.

Theorem 1.20. (*Sehgal–Guseman*) Let T be a mapping from a complete metric space X into itself. Suppose there exists $B \subset X$ such that the following hold.

(i) $TB \subset B$.

(ii) There is k, $0 < k < 1$ such that for each $x \in B$, there is a positive integer $n(x)$ with

$$d(T^{n(x)}x, T^{n(x)}y) \leq kd(x, y) \qquad \text{for all } y \in B. \tag{1.2}$$

(iii) For some $z \in B$, $\overline{\{T^n z: n \geq 1\}} \subset B$.

Then there is a unique $u \in B$ such that $Tu = u$ and

$$T^n x \to u \text{ as } n \to \infty \text{ for each } x \in B. \qquad \blacksquare$$

Proof. First, we claim that for $x \in B$

$$r(x) = \sup_n \{d(T^n x, x)\} < \infty.$$

To see this, let

$$t(x) = \sup \{d(T^n x, x): 1 \leqq n \leqq n(x)\}.$$

If n is any positive integer, there is an integer $s \geq 0$ such that $s \cdot n(x) < n$

$\leq (s+1) \cdot n(x)$. Then

$$\begin{aligned} d(T^n x, x) &\leq d\big(T^{n(x)}(T^{n-n(x)}x), T^{n(x)}x\big) + d(T^{n(x)}x, x) \\ &\leq kd(T^{n-n(x)}x, x) + t(x) \\ &\leq t(x) + kt(x) + \cdots + k^s t(x) \\ &< t(x)/(1-k). \end{aligned}$$

This proves our claim.

Next, let $z_1 = T^{n(z)}z$ and $z_{i+1} = T^{n(z_i)}z_i$. Then it follows by routine calculation that

$$d(z_{i+1}, z_i) \leq k^i d(T^{n(z_i)}z, z) \leq k^i r(z)$$

and

$$d(z_j, z_i) \leq \sum_{l=i}^{j-1} d(z_{l+1}, z_l) \leq \frac{k^i}{1-k} r(z) \qquad (\text{for } j > i).$$

Then (z_i) is Cauchy. By (iii) and completeness of X, $z_i \to u \in B$ as $i \to \infty$. Clearly, $T^{n(u)}z_i \to T^{n(u)}u$ as $i \to \infty$.

But

$$\begin{aligned} d(T^{n(u)}z_i, z_i) &= d\big(T^{n(z_{i-1})}(T^{n(u)}z_{i-1}), T^{n(z_{i-1})}z_{i-1}\big) \\ &\leq kd(T^{n(u)}z_{i-1}, z_{i-1}) \\ &\quad \cdots \\ &\leqq k^i\, d(T^{n(u)}z, z) \to 0 \qquad \text{as } i \to \infty. \end{aligned}$$

Hence

$$d(T^{n(u)}u, u) = \lim_{i\to\infty} d(T^{n(u)}z_i, z_i) = 0.$$

It is clear that u is the unique fixed point for $T^{n(u)}$ in B. Therefore since $Tu = T(T^{n(u)}u) = T^{n(u)}(Tu)$, $Tu = u$.

The theorem will be proved if we show $T^n x \to u$ as $n \to \infty$ for each $x \subset B$. To show this, let

$$A(x) = \sup\,\{d(T^m x, u)\colon 1 \leq m \leq n(u) - 1\}.$$

Now, if $n = an(u) + s$ with integers a and s such that $a > 0, 0 \leq s < n(u)$, then

$$\begin{aligned} d(T^n x, u) &= d(T^{an(u)+s}x, T^{n(u)}u) \\ &\leq kd(T^{(a-1)n(u)+s}x, u) \\ &\quad \cdots \\ &\leqq k^a A(x). \end{aligned}$$

It follows that $T^n x \to u$ as $n \to \infty$. ∎

The next theorem is a slightly simplified version of a theorem of Boyd and Wong.

Theorem 1.21. Let T be a mapping from a complete metrix space X into itself. Suppose there exists a right continuous mapping f from R^+, the nonnegative reals into itself such that for all x, y in X

$$d(Tx, Ty) \leqq f(d(x, y)).$$

If $f(t) < t$ for each $t > 0$, then T has a unique fixed point u in X and for each x in X, $\lim_{n\to\infty} T^n x = u$. ▮

Proof. Let $x \in X$, $x_n = T^n x$ and $d_n = d(x_n, x_{n+1})$. We may and do assume that $d_n > 0$ for $n \geq 0$. Then for $n > 1$,

$$d_n = d(Tx_{n-1}, Tx_n) \leqq f(d_{n-1}) < d_{n-1},$$

so that the sequence d_n is decreasing. Let $d = \lim_{n\to\infty} d_n$. Then $d = 0$, since the preceding inequality implies that $d \leqq f(d)$, which is less than d unless $d = 0$. Now we show that the sequence x_n is Cauchy. If not, then there exists $\varepsilon > 0$ and for each positive integer k there exist positive integers n_k and m_k with $k \leqq m_k < n_k$ such that $d(x_{m_k}, x_{n_k}) \geqq \varepsilon$. With no loss of generality, we can also assume that n_k is the smallest integer greater than m_k satisfying the above inequality. Let $t_k = d(x_{m_k}, x_{n_k})$. Then

$$\varepsilon \leqq t_k \leqq d(x_{m_k}, x_{n_k-1}) + d(x_{n_k-1}, x_{n_k}) \leqq \varepsilon + d_{n_k-1},$$

and therefore $\varepsilon \leqq t_k$ and $\lim_{k\to\infty} t_k = \varepsilon$. We have also

$$\begin{aligned} \varepsilon \leqq t_k &\leqq d(x_{m_k}, x_{m_k+1}) + d(x_{m_k+1}, x_{n_k+1}) + d(x_{n_k+1}, x_{n_k}) \\ &\leqq d_{m_k} + f(t_k) + d_{n_k} \to f(\varepsilon) \qquad \text{as } k \to \infty. \end{aligned}$$

This is a contradiction since $f(\varepsilon) < \varepsilon$. Therefore, the sequence x_n is Cauchy with some limit u in X. Clearly by the continuity of T, $Tu = u$. The uniqueness of the fixed point follows immediately from the definition of f. ▮

In Chapter 7, we shall present a common fixed point theorem (due to Kakutani) for commuting linear mappings. But for two commuting mappings on [0, 1], a simple theorem of this type can be very easily stated and proved.

Theorem 1.22. Let f and g both map [0, 1] into itself such that

(i) $f(g(x)) = g(f(x))$ for each x in [0, 1];
(ii) f is nonexpansive;
(iii) g is continuous.

Then there is a common fixed point for both f and g. ▮

Proof. By Theorem 1.8 the set A of fixed points for f is nonempty. By condition (i), $g(A) \subset A$. Since f is continuous, A is closed. It is sufficient to show that A is an interval since then g has a fixed point in A by Theorem 1.8.

Let $m = \inf A$ and $M = \sup A$. Let $x \in [m, M]$. Then

$$f(x) - m \leqq |f(x) - f(m)| \leqq x - m,$$

which means that $f(x) \leqq x$. Similarly, $f(x) \geq x$. This means that x is a fixed point of f and therefore, $x \in A$. Hence $A = [m, M]$. The theorem now follows. ∎

Remark 1.87. Theorem 1.22 was proved by R. DeMarr† who also proved the following theorem.

Theorem. (*DeMarr*). Let f and g be two commuting mappings from $[0, 1]$ into itself such that for all $x, y \in [0, 1]$

(i) $|f(x) - f(y)| \leq \alpha |x - y|$,

(ii) $|g(x) - g(y)| \leq \beta |x - y|$, where $\beta < (\alpha + 1)/(\alpha - 1)$.

Then there exists a common fixed point for both f and g. ∎

Our last fixed point theorem is another interesting extension of the Banach Fixed Point Theorem. It involves mappings that are contractions in a different sense. To clarify this, let us consider the following variants of the contraction condition. (In what follows, T is a mapping from a complete metric space X into itself.)

Condition (K). There is a constant $c < \frac{1}{2}$ such that for all x, y in X

$$d(Tx, Ty) \leq c \cdot [d(x, Tx) + d(y, Ty)].$$

Condition (K_1). There is a constant $c < \frac{1}{2}$ such that for all x, y in X

$$d(Tx, Ty) \leqq c \cdot [d(x, Ty) + d(y, Tx)].$$

Condition (K_2). There exist nonnegative constants a, b, and c such that $2a + 2b + c < 1$ and for all x, y in X

$$d(Tx, Ty) \leqq a \cdot [d(x, Tx) + d(y, Ty)] + b \cdot [d(x, Ty) + d(y, Tx)] + cd(x, y).$$

† R. DeMarr, *Amer. Math. Monthly*, May (1963).

Condition (K) was first studied by R. Kannan in 1968. Since then, the other two conditions have been studied independently by various mathematicians. It is clear that the first two conditions imply the third. But none of these conditions, though seemingly contractions of some kind, are contractions. They need not even imply continuity. The following examples clarify this and their relationships to one another.

Examples.† Mappings T satisfying the conditions (K), (K_1), *or* (K_2).

1.35. An example of a discontinuous T satisfying condition (K), but not condition (K_1), is the following: Define $Tx = -x/2$ for x in $(-1, 1)$ and $T1 = T(-1) = 0$.

1.36. An example of a discontinuous T satisfying condition (K_1), but not condition (K) is the following: Define $Tx = x/2$ if $x \in [0, 1)$ and $T1 = 0$. Then T: $[0, 1] \to [0, 1]$.

1.37. The mapping $Tx = -x/2$, $x \in [-1, 1]$ is the example of a contraction that does not satisfy condition (K_1).

1.38. The mapping $Tx = x/2$, $x \in [0, 1]$ is a contraction that does not satisfy condition (K).

1.39. An example of a continuous T satisfying condition (K) but neither (K_1) nor the contraction condition is the following: $Tx = -x/2$ if $x \in [-4, 4]$, $= 2x - 10$ if $x \in [4, 5]$, $= 2x + 10$ if $x \in [-5, -4]$. One can verify that T satisfies (K) with c in $(\frac{2}{5}, \frac{1}{2})$.

1.40. An example of a discontinuous T satisfying condition (K_2) but neither condition (K) nor condition (K_1) is the following: $Tx = x/2$ if $x \in [0, 1)$, $= -\frac{1}{8}$ if $x = 1$, $= -x/2$ if $x \in [-2, 0)$.

Now we present our last fixed point theorem in this section.

Theorem 1.23. Let T be a mapping from a complete metric space X into itself satisfying condition (K_2). Then for any x in X, $\lim T^n x$ exists as n tends to infinity and this limit is the unique fixed point of T. ∎

Proof. For all x, y in X, T satisfies the inequality

$$d(Tx, Ty) \leq a[d(x, Tx) + d(y, Ty)] + b[d(x, Ty) + d(y, Tx)] + cd(x, y), \quad (1.3)$$

where $2a + 2b + c < 1$.

† Some easy computations are needed to verify these examples. These are left to the reader.

For each positive integer n and x in X, we replace x by $T^{n-1}x$ and y by $T^n x$ in (1.3). Then using the triangle inequality, we have

$$\begin{aligned} d(T^n x, T^{n+1}x) &\leqq a[d(T^{n-1}x, T^n x) + d(T^n x, T^{n+1}x)] + b[d(T^{n-1}x, T^{n+1}x) \\ &\quad + d(T^n x, T^n x)] + cd(T^{n-1}x, T^n x) \\ &\leqq a[d(T^{n-1}x, T^n x) + d(T^n x, T^{n+1}x)] + b[d(T^{n-1}x, T^n x) \\ &\quad + d(T^n x, T^{n+1}x)] + cd(T^{n-1}x, T^n x). \end{aligned}$$

Simplifying, we have

$$d(T^n x, T^{n+1}x) \leqq kd(T^{n-1}x, T^n x)$$

where $k = (a + b + c)/(1 - a - b)$, which is less than 1. Repeating the above process n times, we have

$$d(T^n x, T^{n+1}x) \leqq k^n d(x, Tx)$$

and therefore, for each positive integer m we can write

$$\begin{aligned} d(T^n x, T^{n+m}x) &\leqq \sum_{i=0}^{m-1} d(T^{n+i}x, T^{n+i+1}x) \\ &\leqq \sum_{i=0}^{m-1} k^{n+i} d(x, Tx). \end{aligned}$$

Thus the sequence $(T^n x)$ is a Cauchy sequence, and since X is complete, the sequence has a limit ξ.

To prove that $T\xi = \xi$, we use condition (1.3) and find that

$$\begin{aligned} d(\xi, T\xi) &\leqq d(\xi, T^n x) + d(T^n x, T\xi) \\ &\leqq d(\xi, T^n x) + a[d(T^{n-1}x, T^n x) + d(\xi, T\xi)] \\ &\quad + b[d(T^{n-1}x, T\xi) + d(\xi, T^n x)] + cd(T^{n-1}x, \xi). \end{aligned} \tag{1.4}$$

Noting that $d(T^{n-1}x, T\xi) \leqq d(T^{n-1}x, \xi) + d(\xi, T\xi)$, we see from condition (1.4) that $(1 - a - b)d(\xi, T\xi)$ can be made as small as we please. Hence $\xi = T\xi$. To prove the uniqueness of the fixed point, let $T\xi = \xi$ and $T\eta = \eta$. Then from condition (1.3), writing $x = \xi$ and $y = \eta$, we have

$$d(\xi, \eta) = d(T\xi, T\eta) \leqq (2b + c)d(\xi, \eta).$$

Since $2b + c$ is less than 1, $\xi = \eta$. ∎

Theorem 1.23 can be further extended, as the reader can verify in Problem 1.5.45. Though condition (K_2) is quite different from the contraction condition, as Examples 1.35–1.40 demonstrated, it can be shown that in a compact metric space, a continuous mapping T satisfying this condition is a contraction with respect to an equivalent metric, i.e., a metric generating the same topology as the original metric. The reader can verify this in Problem 1.5.43.

Before we close this section, we present two applications of the Banach Fixed Point Theorem. First, we prove one form of the classical implicit function theorem.

Theorem 1.24. Let f be a continuous real-valued function defined on $|x| \leq a$, $|y| \leq b$ such that

(i) $f(0, 0) = 0$,

(ii) $|f(x_1, y_1) - f(x_2, y_2)| \leq k(|x_1 - x_2| + |y_1 - y_2|)$,

where k is a fixed number in (0, 1). Then the equation $y = cx + f(x, y)$ has a unique solution $y = h(x)$ with $h(0) = 0$ and h defined in

$$|x| \leq s < \min\left\{a, \frac{1-k}{|c|+k} \cdot b\right\}.$$

Furthermore,

$$|h(x_2) - h(x_1)| \leq \frac{|c|+k}{1-k}|x_2 - x_1|. \qquad \blacksquare$$

Proof. Let X be the family of all continuous functions g defined on $[-s, s]$ (s defined as in the theorem) such that $g(0) = 0$ and $|g(x)| \leqq b$. With the usual "sup" metric d, X is a complete metric space.

We define $T: X \to X$ by

$$T(g)(x) = cx + f(x, g(x)).$$

Then T is well defined since

(i) $T(g)(0) = f(0, 0) = 0$,

(ii) $T(g)$ is continuous,

(iii) $$\begin{aligned} |T(g)(x)| &\leqq |cx| + |f(x, g(x))| \\ &\leq |cs| + k(s + b) \\ &< b, \qquad \text{by the definition of } s. \end{aligned}$$

Moreover,

$$\begin{aligned} |T(g_1)(x) - T(g_2)(x)| &= |f(x, g_1(x)) - f(x, g_2(x))| \\ &\leq k\,|g_1(x) - g_2(x)|. \end{aligned}$$

This means that $d(T(g_1), T(g_2)) \leq kd(g_1, g_2)$. By the Banach Fixed Point Theorem, T has a unique fixed point $h(x) \in X$. The rest of the proof is now clear. ▮

Our next application of the Banach Fixed Point Theorem deals with the question of existence and uniqueness of solutions to ordinary differential equations. Let us consider the initial-value problem

$$x'(t) = f(t, x(t)), \qquad x(0) = x_0 \tag{1.5}$$

where $x'(t)$ denotes the derivative of $x(t)$. We are interested in the existence of a function $x(t)$ satisfying Eq. (1.5) on some closed interval $[-\sigma, \sigma]$, $\sigma > 0$.

Theorem 1.25. *The Picard Existence Theorem.* Let $f(t, s)$ be a continuous function from $D = [-a, a] \times [-b, b]$ into R. Suppose $|f(t, s)| \leq M$ in D and

$$|f(t, s_1) - f(t, s_2)| \leq K |s_1 - s_2|$$

for some positive number K, $t \in [-a, a]$ and $s_1, s_2 \in [-b, b]$. Then Equation (1.5) with $x_0 = 0$ has a unique solution defined for $|t| \leq r < \min(a, b/M, 1/K)$. ▮

Proof. Let $X = \{g \in C[-r, r]: g(0) = 0$ and $|g(t)| \leq b$ if $-r \leq t \leq r\}$. Then X is a complete metric space with the usual "sup" metric d. Clearly, the integral

$$T(g)(s) = \int_0^s f(t, g(t))\, dt$$

is well defined for $g \in X$ and is a continuous function of s for $s \in [-r, r]$. Also, by the choice of r, $|T(g)(s)| \leq M|s| \leq Mr < b$, and therefore $T(g) \in X$. Now

$$\begin{aligned} |T(g_1)(s) - T(g_2)(s)| &\leq \int_0^s |f(t, g_1(t)) - f(t, g_2(t))|\, dt \\ &\leq K \int_0^s |g_1(t) - g_2(t)|\, dt \end{aligned}$$

so that

$$d(T(g_1), T(g_2)) \leq Krd(g_1, g_2).$$

Since $Kr < 1$, T is a contraction, and therefore it has a unique fixed point

$x(t) \in X$ so that

$$x(s) = \int_0^s f(t, x(t))\, dt.$$

Clearly $x(0) = 0$, and by the continuity of the integrand, $x'(s)$ exists and

$$x'(s) = f(s, x(s)), \qquad s \in (-r, r). \qquad \blacksquare$$

Problems

1.5.1. Show that the sum of two real-valued functions on the reals need not have the intermediate-value property, though each one has this property. [Hint: Consider $[f'(x)]^2 + [g'(x)]^2$, where $f(x) = x^2 \sin(1/x)$ for $x \neq 0$ and $f(x) = 0$ for $x = 0$, and $g(x) = x^2 \cos(1/x)$ for $x \neq 0$, $g(x) = 0$ for $x = 0$.] Is this true if one of the functions is also continuous?

1.5.2. (Rottmann). Suppose $d(x, y) = 0$ for $x = y$ and $= \max\{|x|, |y|\}$ for $x \neq y$. Show that d defines a metric for the reals R and that (R, d) is a disconnected but complete metric space. Show also that the mapping $x \to -x$ is continuous, whereas the mapping $(x, y) \to x + y$ from $R \times R$ into R is not continuous.

1.5.3. Let X be a topological space. Suppose every point of X has a path-connected open set containing it. Show that every path-connected component (i.e., maximal path-connected set) is then open and therefore closed. Use this to show that any open set in R^n is an at most countable disjoint union of open connected sets.

1.5.4. Use (1.5.3) to show that X is path connected if and only if X is connected and every point of X has a path-connected open set containing it. Use this to show that an open set in R^2 is connected if and only if it is path connected.

1.5.5. Let X be a completely regular, Hausdorff, and connected topological space consisting of more than one point. Then show that every nonempty open subset of X is uncountable.

1.5.6. Show that the product of two locally connected topological spaces is locally connected.

1.5.7. Prove the following result (due to S. S. Mitra): If there is a continuous open map from $I = [0, 1]$ (with usual topology) *onto* a Hausdorff space X containing at least two elements, then X and I are homeomorphic. [Hint: Let a be the smallest number $0 < a \leq 1$ such that $f([0, a]) = X$; show that $f|_{[0,a]}$ is one-to-one.]

1.5.8. Let f be a real-valued, strictly increasing function defined on a set D of real numbers. Show that f^{-1} is a continuous, strictly increasing function if D is an open, closed, or connected set.

1.5.9. Show that any real-valued continuous, one-to-one function defined on $[a, b]$ is strictly monotonic.

★ **1.5.10.** Prove that a metric space is connected if and only if every nonempty proper subset has a nonempty boundary.

★ **1.5.11.** Prove that a metric space X is complete if there is a positive number ε such that for each $x \in X$, the set $\{y: d(x, y) \leq \varepsilon\}$ is compact.

1.5.12. Let D be the set of all nondecreasing, left-continuous functions f defined on R with $0 \leq f \leq 1$. Suppose for $f, g \in D$, $d(f, g) = \inf\{\delta: f(x - \delta) - \delta \leq g(x) \leq f(x + \delta) + \delta$ for every $x\}$. Show that

(i) d is a metric on D;

(ii) $d(f_n, f) \to 0$ as $n \to \infty$ if and only if $f_n(x) \to f(x)$ as $n \to \infty$ whenever x is a point of continuity of f.

[This metric (useful in probability theory) is called the Lèvy metric, named after the great French probabilist Paul Lèvy.]

1.5.13. Let (X, d) be a metric space. If $d_1(x, y) = d(x, y)/[1 + d(x, y)]$, then show that (X, d_1) is also a metric space having the same topology as that of (X, d).

1.5.14. Suppose X and Y are any two metric spaces and the metric in Y is bounded. Let $\mathscr{F}$ be the set of all maps from X into Y. Define for $f, g \in \mathscr{F}$, $d(f, g) = \sup_{x \in X} d(f(x), g(x))$. Show that this d defines a metric in $\mathscr{F}$ that is complete if Y is complete.

1.5.15. Show that every continuous map f of a compact metric space into another metric space is uniformly continuous [i.e., given $\varepsilon > 0$, there exists $\delta > 0$ such that $d(x, y) < \delta \Rightarrow d(f(x), f(y)) < \varepsilon$].

1.5.16. Let X and Y be metric spaces and $f: X \to Y$ be a homeomorphism. Show that if f is uniformly continuous (see Problem 1.5.15) and Y is complete, then X is complete; but completeness of Y need not be implied by that of X and the uniform continuity of f.

1.5.17. Let E be a dense subset of a metric space X and f be a uniformly continuous mapping of X into a complete metric space Y. Show that there exists a unique uniformly continuous mapping $g: X \to Y$ such that for $x \in E$, $g(x) = f(x)$.

1.5.18. Prove *Dini's Theorem*: Let X be a compact metric space and $f_n \in C(X)$ such that $f_n(x) \leq f_{n+1}(x)$ and $f(x) = \lim_{n\to\infty} f_n(x)$, for each $x \in X$. Then $f_n(x) \to f(x)$ uniformly as $n \to \infty$, if $f \in C(X)$.

★ **1.5.19.** *Limit Points of Sequences and Permutations.* Let (X, d) be a compact metric space and (x_n), (y_n) two sequences. If $d(x_n, y_n) \to 0$, then these sequences have the same set of limit points. Prove the converse (due to Von Neumann): If the sequences have the same set of limit points, then there exists a permutation π of the natural numbers such that $d(x_n, y_{\pi(n)}) \to 0$ as $n \to \infty$. [The following hint is due to P. R. Halmos: Let $\varepsilon_n = 1/n + d(x_n, C)$, where C is the common set of limit points. Then $S_{x_n}(\varepsilon_n)$ contains infinitely many y_m. Let $\sigma(1) > 1$ be the smallest positive integer with $y_{\sigma(1)} \in S_{x_1}(\varepsilon_1)$. Inductively, $\sigma(n+1)$ is the smallest positive integer $> \sigma(n)$ and $y_{\sigma(n+1)} \in S_{x_{n+1}}(\varepsilon_{n+1})$. Then $d(x_n, y_{\sigma(n)}) \to 0$ as $n \to \infty$. Similarly, there is a one-to-one map τ of the naturals into itself such that $d(x_{\tau(n)}, y_n) \to 0$ as $n \to \infty$. Now by the Schröder–Bernstein theorem (see [5], pp. 88, 89), there is a permutation π of the naturals such that for each n, $\pi(n) = \sigma(n)$ or $\tau^{-1}(n)$. Then $d(x_n, y_{\pi(n)}) \to 0$ as $n \to \infty$.]

1.5.20. *The Cantor Set and the Cantor Function.* (a) The *Cantor set* C is a subset of [0, 1] obtained by first removing the middle third (1/3, 2/3) from [0, 1], then removing the middle thirds (1/9, 2/9) and (7/9, 8/9) of the remaining intervals, and continuing indefinitely. Prove the following assertions:

(i) C is compact, nowhere dense, and has no isolated points.

(ii) $C = \{\sum_{n=1}^{\infty} x_n/3^n\colon x_n = 0 \text{ or } 2 \text{ for each } n\}$ and card $C = c$.

(b) Define a function f on [0, 1] as follows: $f(1/3, 2/3) = 1/2$, $f(1/9, 2/9) = 1/4$, $f(7/9, 8/9) = 3/4$, $f(1/27, 2/27) = 1/8$, $f(7/27, 8/27) = 3/8$, $f(19/27, 20/27) = 5/8$, $f(25/27, 26/27) = 7/8$, and so on; and for $x \in C$, let $f(x) = \sup\{f(y)\colon y \in [0, 1] - C,\ y < x\}$. Show that

(i) f is a continuous nondecreasing function from [0, 1] *onto* [0, 1];

(ii) for $x \in C$ and $x = \sum_{n=1}^{\infty} 2x_n/3^n$, where each $x_n = 0$ or 1, $f(x) = \sum_{n=1}^{\infty} x_n/2^n$.

This f is called the *Cantor function.*

★ **1.5.21.** *Convex sets.* A subset $A \subset R^2$ is called *convex* if for *all* t in (0, 1), A has the condition (c): $x, y \in A \Rightarrow tx + (1 - t)y \in A$. Prove that if A is closed and if the condition (c) holds for only a single fixed t in (0, 1), then A is convex. What can you say if the set A is open? (These results also hold in much more general topological spaces with a vector space structure.)

★ **1.5.22.** *Convex Functions.* A real-valued function f defined on (a, b) is said to be *convex* if for all $t \in (0, 1)$, f satisfies the condition (c'): $f(tx + (1 - t)y) \leq tf(x) + (1 - t)f(y)$ for all $x, y \in (a, b)$. Prove the following assertions.

(i) If f is convex then f is continuous.

(ii) Suppose f is continuous and the condition (c′) holds for only a single fixed t in (0, 1), then f is convex.

(iii) Suppose f is convex on (a, b). Then f has a left-hand derivative

$D^-f(x)$ as well as a right-hand derivative $D^+f(x)$ at each point x of (a, b) and $D^-f(x) \leq D^+f(x)$. Furthermore, on any $[c, d] \subset (a, b)$, f satisfies the Lipschitz condition. [Hint: For $x < y < z$, note that

$$\frac{f(y) - f(x)}{y - x} \leq \frac{f(z) - f(x)}{z - x} \leq \frac{f(z) - f(y)}{z - y};$$

to see this, write

$$y = \frac{z - y}{z - x} x + \frac{y - x}{z - x} z.$$

Then for $c < x < y < d$, we easily have

$$\frac{f(x) - f(c)}{x - c} \leq \frac{f(y) - f(x)}{y - x} \leq \frac{f(d) - f(y)}{d - y}.$$

Now show that $|\,[f(y) - f(x)]/(y - x)\,| \leq \max\{|\,D^-f(c)\,|, |\,D^+f(d)\,|\}$.]

(iv) If f has a nonnegative second derivative in (a, b), then f is convex. The converse is also true if f is twice differentiable.

1.5.23. Find a complete metric for $X = \{0\} \cup [1, \infty)$ such that $[1, \infty)$ does not contain a closest point to 0. [Hint: Define for $x, y \in [1, \infty)$, $d(x, y) = |\,x - y\,|/[1 + |\,x - y\,|]$ and $d(0, x) = [2 + |\,x\,|]/[1 + |\,x\,|]$. Then d is the desired metric.]

1.5.24. Let X be a compact metric space and $T: X \to X$ such that $d(x, y) = d(Tx, Ty)$ for all $x, y \in X$. Show that T is *onto*.

1.5.25. *A Fixed-Set Theorem.* Let X be a compact Hausdorff space and $T: X \to X$ be a continuous map. Show that there exists a nonempty closed set $A \subset X$ such that $T(A) = A$.

1.5.26. *The Topology of Pointwise Convergence.* Let $\mathscr{F}(X, Y)$ be the family of all maps from a topological space X into a topological space Y. Then $\mathscr{F}(X, Y) = \times_{x \in X} Y_x$, where $Y_x = Y$ for each x. The product topology from this product is called the topology of pointwise convergence. Show that $f_n \to f$ as $n \to \infty$ in $\mathscr{F}(X, Y)$ with this topology if and only if for each x in X, $f_n(x) \to f(x)$ as $n \to \infty$ in Y. Also show that $\mathscr{F}(X, Y)$ with the topology of pointwise convergence is compact, T_2, T_3, or connected if and only if Y has the same property.

1.5.27. *The Quotient Topology.* The largest topology for a set Y that would make a given mapping f from a topological space X onto Y continuous is called the *quotient topology* for Y (relative to f). Show the following.

(i) If f is a continuous open map from a topological space X onto a topological space Y, then the topology for Y is its quotient topology (relative to f).

(ii) If X is compact, separable, connected, or locally connected, then so is Y with the quotient topology.

1.5.28. Suppose there is an equivalence relation r on a topological space X, X/r is the set of all equivalence classes, and $\psi(x) = [x]$, the equivalence class containing x. X/r with the quotient topology (relative to ψ) is called the *quotient space*. Show the following.

(i) X/r is T_1 if and only if the equivalence classes in X are closed.

(ii) A set A in X/r is open if and only if the union of all the equivalence classes in A is open in X.

1.5.29. *A Pseudometric Space.* A mapping $d\colon X \times X \to R^+$ (the nonnegative reals) is called a *pseudometric* for X if for all x, y and $z \in X$, d satisfies (i) $d(x, x) = 0$, (ii) $d(x, y) = d(y, x)$, and (iii) $d(x, y) + d(y, z) \geq d(x, z)$. A pseudometric d induces a natural topology for X, by considering as a base the family of all sets of the form $\{y\colon d(y, x) < k\}$; and X, with this topology, is called a *pseudometric space*. Show that in a pseudometric space (X, d)

(i) the relation xry if and only if $d(x, y) = 0$ is an equivalence relation; and

(ii) the quotient space X/r is metrizable with metric d_0 well defined by $d_0([x], [y]) = d(x, y)$. (See 1.5.28.) [It is relevant to mention here that *pseudometric spaces are paracompact*. For a proof of this nontrivial fact, the reader should consult [34], p. 160.]

1.5.30. Prove that a contractive mapping T of a complete metric space X into a *compact* subset of itself has a fixed point x_0 such that for every $x \in X$, $\lim_{n\to\infty} T^n x = x_0$.

1.5.31. Prove the following result due to V. M. Sehgal and J. W. Thomas:

A Common Fixed Point Theorem. Let M be a closed subset of a complete metric space X and let $\mathscr{F}$ be a commutative semigroup of mappings of M into M such that for each $x \in M$ there is $f_x \in \mathscr{F}$ with

$$d(f_x(y), f_x(x)) \leq \psi(d(y, x))$$

for all $y \in M$, where ψ is a nondecreasing, right-continuous function such that $\psi(r) < r$ for all $r > 0$. Suppose for some $x_0 \in M$, $\sup\{d(f(x_0), x_0)\colon f \in \mathscr{F}\} = d_0 < \infty$. Then there exists a unique $\xi \in M$ such that $f(\xi) = \xi$ for each $f \in \mathscr{F}$. Moreover, there is a sequence $g_n \in \mathscr{F}$ such that for each $x \in M$, $\lim_{n\to\infty} g_n(x) = \xi$. [Hint: First show that $\lim_{n\to\infty} \psi^n(d_0) = 0$. Let $f_0 = f_{x_0}$, $f_n = f_{x_n}$ where $x_{n+1} = f_n(x_n)$; show that (x_n) is Cauchy and $\lim_{n\to\infty} x_n = \xi$ with $f_\xi(\xi) = \xi$. Then ξ is the unique fixed point for f_ξ and hence for each $f \in \mathscr{F}$.]

1.5.32. Let X be a topological space with relative topology and $Y \subset X$.

Then Y is called a *retract* of X if there exists a continuous map $f\colon X \to Y$ such that $f(y) = y$, $y \in Y$. Prove the following.

(i) If Y is a retract of X and X is T_2, then Y is closed in X.

(ii) A necessary and sufficient condition that Y be a retract of X is that every continuous map from Y to any topological space Z can be extended to X.

(iii) If Y is a retract of X and X has the fixed point property (i.e., any continuous map from X into X has a fixed point), then Y also has this property.

1.5.33. Prove the following result due to M. Edelstein. Let $T\colon X \to X$, a complete metric space. Suppose there exists $\varepsilon > 0$ such that $0 < d(x, y) < \varepsilon$ implies $d(Tx, Ty) < d(x, y)$. If ξ is a limit point of the sequence $T^n x$, then ξ is a fixed point for some T^k.

★ **1.5.34.** [*Definition.* Let $T\colon X \to X$, a complete metric space. Let $O(x)$ denote the orbit of $x \in X$, that is the set $\{T^n x\colon n \geq 0,\ T^0 x = x\}$ and $\varrho(x) = \sup\{d(z, w)\colon z, w \in O(x)\}$. Then the mapping T is said to be of *diminishing orbital diameter* if $\varrho(x) < \infty$ for each $x \in X$ and $\varrho(x) > 0 \Rightarrow \lim_{n\to\infty} \varrho(T^n x) < \varrho(x)$.] Prove the following assertions.

(i) Let $X = \{(x, y)\colon x \geq 0,\ y \geq 0\}$ with the usual "distance" metric and let $T\colon X \to X$ be defined by $T(x, y) = (x, x^2)$. Show that T has diminishing orbital diameter and $\varrho(x)$ is continuous on X.

(ii) *An Extension of a Result of W. A. Kirk by V. M. Sehgal.* Suppose T is continuous and has diminishing orbital diameter. If $\varrho(x)$ is continuous on X, then each limit point $\xi \in X$ of the sequence $(T^n x)$, $x \in X$, is a fixed point of T and $\xi = \lim_{n\to\infty} T^n x$.

1.5.35. *A Limited Contraction Fixed Point Theorem* (S. Weingram). Prove that if $f\colon R^n \to R^n$ is a continuous map such that for some compact set K there is a constant $k (0 < k < 1)$ satisfying

$$| f(x) - f(y) | \leq k \,| x - y |$$

for any two points x, y *not* in K, then f has a fixed point.

1.5.36. Let T be a mapping from a complete metric space (X, d) into itself such that

(i) the map $x \to d(x, Tx)$ is lower semicontinuous;
(ii) there exists $(x_n) \in X$ such that $d(x_n, Tx_n) \to 0$ as $n \to \infty$; and
(iii) there exist $a > 0$, $b > 0$ and $0 < c < 1$ such that

$$d(Tx, Ty) \leq ad(x, Tx) + bd(y, Ty) + cd(x, y).$$

Show that T has a unique fixed point and none of the conditions (i), (ii), or (iii) can be omitted.

1.5.37. *A Common Fixed Point Theorem* (J. Cano). Suppose f and g are continuous commuting mappings from a metric space (X, d) into itself. Suppose that $\{x: g(x) = x\}$ is a nonempty compact set and whenever $f(y) \neq y$, $d(f^2(y), f(y)) < d(f(y), y)$. Then f and g have a common fixed point.

1.5.38. *Necessary Condition for a Set of Reals to Have the Fixed Point Property* (Dotson). If every continuous function from $S(\subset R)$ into S has a fixed point, then S is either a singleton or a closed bounded interval.

1.5.39. *A Fixed Point Theorem in* R^n (Reich). For $x = (x_1, x_2, \ldots, x_n) \in R^n$, let $\| x \| = \sup_{1 \leq i \leq n} | x_i |$. Suppose f is a continuous map from K^n into R^n, where $K^n = \{x \in R^n: \| x \| \leq 1\}$ such that for each y in $S^{n-1} = \{x \in R^n: \| x \| = 1\}$, there is *no* $m > 1$ with $f(y) = my$. Then f has a fixed point. [Hint: Suppose f has no fixed point; then the mapping g defined by $g(x) = [f(x) - x]/\| f(x) - x \|$ is continuous. Use Brouwer's Fixed Point Theorem to get a contradiction.]

1.5.40. *Cesaro Means and Fixed Points* (Dotson). Suppose T is a function from R into R such that for some x, the sequence $(1/n)\sum_{k=0}^{n-1} T^k x$ is bounded. Then T has a fixed point. Here the fixed point need not be the limit of the sequence $(1/n)\sum_{k=0}^{n-1} T^k y$ for any y different from the fixed point.

1.5.41. Prove the following result due to J. Cano. Let f be a continuous function from a compact set $K(\subset R)$ into intself. Suppose that there is some $x \in K$ such that every cluster point of $[f^n(x)]$ is a fixed point of f. Then the sequence $[f^n(x)]$ is convergent. This result fails in R^2.

★ **1.5.42.** *A Converse of the Banach Fixed Point Theorem.* It follows from Theorem 1.15 that in a bounded complete metric space (X, d), $\bigcap_{n=1}^{\infty} T^n X = \{a\}$ for every contraction T with unique fixed point a. Show that the following converse holds: Suppose (X, d) is a compact metric space. T is a continuous map from X into itself such that $\bigcap_{n=1}^{\infty} T^n X = \{a\}$, a singleton. Then there is a metric d_1, generating the given topology of X, such that T is a contraction with respect to d_1. [Hint: Define $\bar{d}(x, y) = \sup\{d(T^n x, T^n y): 0 \leqq n < \infty\}$. Then $(X, \bar{d})$ is a metric space. T is nonexpansive with respect to $\bar{d}$, which induces the same topology as d. Now let $n(y) = \sup\{n: y \in T^n X\}$ and $n(x, y) = \inf\{n(x), n(y)\}$. Given k in $(0, 1)$, define $p(x, y) = k^{n(x,y)} \cdot \bar{d}(x, y)$ and let $d_1(x, y) = \inf \sum_{n=1}^{m} p(x_i, x_{i+1})$, where the infimum is taken over all finite subsets $x_1, x_2, \ldots, x_{m+1}$ in X such that $x = x_1$ and $y = x_{m+1}$. Then d_1 is the desired metric.] This result is due to L. Janos. For a more general result, the reader can consult the 1967 paper of P. R. Meyers.[†] He proved the following: Let T be a continuous mapping from

[†] P. R. Meyers, *J. Res. Nat. Bur. Standards Sect. B*, **71** (1967).

a complete metric space into itself with a unique fixed point u such that $\lim_{n\to\infty} T^n x = u$ for any x. Then if u has an open neighborhood with compact closure, T is a contraction with respect to an equivalent metric. *This result is false if u does not have such a neighborhood.* Can the reader find an example?

1.5.43. Let (X, d) be a compact metric space and T be a continuous map from X into itself. Suppose that T satisfies condition (K_2) as in Theorem 1.23. Show that there is a metric d_1 on X, generating the same topology as d, such that T is a contraction with respect to d_1. (Hint: Use Problem 1.5.42.)

1.5.44. *Another Extension of the Banach Fixed Point Theorem.* Let T be a map from a complete metric space X into itself satisfying the following condition. Given $\varepsilon > 0$, there exists $\delta > 0$ such that $\varepsilon \leqq d(x, y) < \varepsilon + \delta$ implies $d(Tx, Ty) < \varepsilon$. Then for any x in X, $\lim T^n x$ exists and this limit is the unique fixed point of T. [Hint: Notice that T is contractive and $c_n = d(T^n x, T^{n+1} x)$ is a decreasing sequence with limit 0. If the sequence $T^n x$ is not Cauchy, then for infinitely many n and m, $d(T^n x, T^m x)$ is greater than 2ε (> 0). For this ε, let δ be as in the hypothesis above. Let N be such that for $n > N$, $c_n < \delta'/3$, where $\delta' = \min\{\delta, \varepsilon\}$. Let n ($> N$) and k be such that $d(T^n x, T^{n+k} x) > 2\varepsilon$. Since for all j with $1 \leqq j \leqq k$, $| d(T^n x, T^{n+j} x) - d(T^n x, T^{n+j+1} x) | \leqq c_{n+j} < \delta'/3$, it follows that for some j, $1 \leqq j \leqq k$, we must have $\varepsilon + 2\delta'/3 < d(T^n x, T^{n+j} x) < \varepsilon + \delta'$. But for this j, $d(T^n x, T^{n+j} x) \leqq d(T^n x, T^{n+1} x) + d(T^{n+1} x, T^{n+j+1} x) + d(T^{n+j} x, T^{n+j+1} x) < \delta'/3 + \varepsilon + \delta'/3$, which is a contradiction.] This result is due to A. Meir and E. Keeler.

1.5.45. *An Extension of Theorem 1.23.* Suppose T maps a complete metric space X into itself such that for some constant $c < 1$ and for all x, y in X, $d(Tx, Ty) \leqq c \max\{d(x, Tx), d(y, Ty), d(x, Ty), d(y, Tx), d(x, y)\}$. Then for any x in X, $\lim T^n x$ exists and this limit is the unique fixed point of T. This result is due to L. B. Ćirić.

1.5.46. *A Common Fixed Point Theorem.* Let (T_n) be a sequence of mappings of a complete metric space X into itself. Suppose that for each pair T_i, T_j, there are nonnegative constants a, b, and c (depending upon i and j) such that for all x, y in X,

$$d(T_i x, T_j y) \leqq a[d(x, T_i x) + d(y, T_j y)] + b[d(x, T_j y) + d(y, T_i x)] + cd(x, y),$$

where $2a + 2b + c < 1$. Then the sequence T_n has a unique common fixed point. [Hint: Let $x_0 \in X$ and $x_n = T_n x_{n-1}$, $n = 1, 2, \ldots$. Then the sequence (x_n) is Cauchy and its limit is the unique common fixed point of the T_n's.] This is a result of B. N. Roy extended by K. Iseki. This result can be extended to an arbitrary family of mappings.

1.6. The Stone–Weierstrass Theorem and the Ascoli Theorem

The two theorems in the title are extremely important tools for the analyst for the proof of many general results on continuous functions. The Ascoli theorem provides the basis for most proofs of compactness in function spaces, whereas the Stone–Weierstrass theorem helps us to extend certain results holding for functions of a special type to all continuous functions by an approximation argument.

Around 1883, G. Ascoli introduced the notion of equicontinuity for a set of continuous functions and then used this concept to find a sufficient condition for the compactness of a set of continuous functions with "sup" metric. Later on in 1889, C. Arzela proved the necessity of this condition.

In regard to the other main theorem in this section, it was K. Weierstrass who, around 1900, first showed that the continuous functions on [0, 1] could be approximated uniformly by the polynomials. This beautiful result was then extended in many different directions by various mathematicians. The most important of them all was achieved by M. H. Stone, first in 1937 and then, more completely, in 1948.

We will begin with two definitions and then proceed to prove Stone's extension of Weierstrass' theorem and several important consequences.

Definitions 1.41. (1) A linear subspace S of $C_b(X)$, the space of real-valued bounded continuous functions on a topological space X, is called an *algebra* if the product of any two elements in S is again in S.

(2) A family $\mathscr{F}$ of real-valued functions on X is said *to separate points* of X if, given x, y in X with $x \neq y$, there is $f \in \mathscr{F}$ with $f(x) \neq f(y)$. ∎

In what follows, $C_b(X)$ is always considered a complete metric space with the metric $d(f, g) = \sup\{|f(x) - g(x)|: x \in X\}$.

Theorem 1.26. *The Stone–Weierstrass Theorem.* Suppose X is compact and $\mathscr{A}$ is an algebra contained in $C(X)$ that separates points of X and contains the constant functions; then $\bar{\mathscr{A}} = C(X)$. [$C(X)$ denotes the continuous functions on X, which coincides with $C_b(X)$ when X is compact.] ∎

Proof. First, we notice that given $\varepsilon > 0$, there is a polynomial $P(x)$ such that $|P(x) - |x|| < \varepsilon$ for all $x \in [-1, 1]$. The reason is that considering the binomial series† for $(1 - x)^{1/2}$, we can find a positive integer N

† Notice that for any $\delta > 0$, the binomial series for $(1 + \delta - x)^{1/2}$ has radius of convergence $1 + \delta$ and therefore this series converges uniformly on $[-1, 1]$.

and constants C_n such that for all $x \in [-1, 1]$,

$$\left| (1 - x)^{1/2} - \sum_{n=0}^{N} C_n x^n \right| < \varepsilon;$$

and then replacing $(1 - x)$ with x^2, we have

$$\left| |x| - \sum_{n=0}^{N} C_n (1 - x^2)^n \right| < \varepsilon.$$

Because of the above, given $g \in \mathscr{A}$ and $\varepsilon > 0$, we can find a polynomial $P(g)$ in g such that $\| g(x) | - P(g)(x) | < \varepsilon$ for all $x \in X$. In other words, $g \in \mathscr{A} \Rightarrow | g | \in \bar{\mathscr{A}}$. Hence for $f, g \in \mathscr{A}$, the functions

$$f \vee g = \tfrac{1}{2}[f + g + | f - g |]$$

and

$$f \wedge g = \tfrac{1}{2}[f + g - | f - g |]$$

are also in $\bar{\mathscr{A}}$.

Next we note that given $f \in C(X)$ and any $x, y \in X$, we can find $g_{x,y} \in \mathscr{A}$ such that $g_{x,y}(x) = f(x)$ and $g_{x,y}(y) = f(y)$. Here one simply needs to take

$$g_{x,y} = \frac{f(x) - f(y)}{h(x) - h(y)} \cdot h + \frac{f(x)h(y) - f(y)h(x)}{h(y) - h(x)}$$

where $h \in \mathscr{A}$ such that $h(x) \neq h(y)$.

Now let $f \in C(X)$ and $\varepsilon > 0$. Let $y \in X$. For each $x \in X$, there is an open set V_x containing x such that $g_{x,y}(t) > f(t) - \varepsilon$ for $t \in V_x$. Let $X = \bigcup_{i=1}^{p} V_{x_i}$ and $\bigvee_{i=1}^{p} g_{x_i,y} = g_y$. Then $g_y(t) > f(t) - \varepsilon$ for all $t \in X$, $g_y \in \bar{\mathscr{A}}$, and $g_y(y) = f(y)$. Again, for each $y \in X$, there is an open set U_y containing y such that $g_y(t) < f(t) + \varepsilon$, $t \in U_y$. Let $X = \bigcup_{i=1}^{q} U_{y_i}$ and $\bigwedge_{i=1}^{q} g_{y_i} = g$. Then $g \in \bar{\mathscr{A}}$ and $| g(t) - f(t) | < \varepsilon$ for all t $\in X$. The proof is complete. ∎

Corollary 1.1. Any real-valued continuous function on a compact subset E of R^n is the uniform limit of a sequence of polynomials (in the n coordinates) on E. ∎

Proof. Corollary 1.1 follows easily from Theorem 1.26, since the polynomials clearly form an algebra, and also, they separate points, since for two distinct points on E at least one of the coordinates will be different for both. ∎

We note that Theorem 1.26 does *not* hold for complex-valued functions. The reason is that any sequence of polynomials in the complex variable z converging uniformly on $\{z: |z| \leq 1\}$ must have its limit differentiable in the open unit disk. However, we present the following weaker result.

Theorem 1.27. Let $\mathscr{A}$ be an algebra $\subset C_1(X)$, the complex-valued continuous functions on a compact space X, such that

(i) $\mathscr{A}$ contains the constant functions,
(ii) $\mathscr{A}$ separates points of X,
(iii) $f \in \mathscr{A} \Rightarrow$ the conjugate $\bar{f}$ is in $\mathscr{A}$.

Then $\bar{\mathscr{A}} = C_1(X)$. ∎

Proof. For $f \in \mathscr{A}$, $R(f) = \frac{1}{2}(f + \bar{f})$ and $I(f) = (1/2i)(f - \bar{f})$ are both real valued and in $\mathscr{A}$. Let $\mathscr{A}_0$ consist of the real-valued functions in $\mathscr{A}$. Then $\mathscr{A}_0$ is a sub algebra of $C(X)$. Since $f(x) \neq f(y)$ implies either $R(f)(x) \neq R(f)(y)$ or $I(f)(x) \neq I(f)(y)$, $\mathscr{A}_0$ separates points of X. By Theorem 1.26, $\bar{\mathscr{A}}_0 = C(X)$. Clearly $\bar{\mathscr{A}} = C_1(X)$. ∎

Corollary 1.2. Let f be a continuous periodic real-valued function on R with period 2π, i.e., $f(x + 2\pi) = f(x)$ for all $x \in R$. Then given $\varepsilon > 0$, there exist constants a_n, b_n such that

$$\left| f(x) - \sum_{n=0}^{N} (a_n \cos nx + b_n \sin nx) \right| < \varepsilon$$

for all $x \in R$. ∎

Proof. For $x \in R$, let $e^{ix} = z$ and let $\Gamma = \{z: |z| = 1\}$ be the unit circle on the complex plane. Then the mapping Φ: $x \to e^{ix}$ is a homeomorphism of $(0, 2\pi)$ onto $\Gamma - e^{2\pi i}$. Define the continuous function $g: \Gamma \to R$ by $g(\Phi(x)) = f(x)$ and $g(e^{2\pi i}) = f(2\pi) = f(0)$. Now the algebra generated by $\{e^{ix}, e^{-ix}, 1\}$ satisfies the conditions of Theorem 1.27. Hence there exist complex numbers (α_n) such that

$$\left| g(e^{ix}) - \sum_{k=-N}^{N} \alpha_k e^{ikx} \right| < \varepsilon$$

for all $x \in [0, 2\pi]$ and therefore for all $x \in R$ (by periodicity). The corollary follows by considering the real parts inside the "absolute value" expression. ∎

Remarks

1.88. By Weierstrass' theorem, all real linear combinations of the functions $1, x, x^2, \ldots, x^n, \ldots$ are dense in $C[0, 1]$. This theorem remains true if instead of considering all positive powers of x, we only consider the infinite set $1, x^{n_1}, x^{n_2}, \ldots, x^{n_k}, \ldots$, where (n_k) is a strictly increasing sequence of positive integers and $\sum_{k=1}^{\infty}(1/n_k)$ diverges. This remarkable result is due to Ch. H. Muntz, obtained in 1914. A form of this theorem can be found in a paper by Clarkson and Erdös.[†]

1.89. $C(X)$ is separable if X is a compact metric space. To see this, let (V_n) be a countable basis for the topology of X and let $f_n(x) = d(x, X - V_n)$. Clearly, if $x \neq y$ and $x \in V_n$, $y \notin V_n$, then $f_n(x) \neq 0$ and $f_n(y) = 0$. Now to apply Theorem 1.26, consider the algebra of all finite linear combinations of $f_1^{k_1} f_2^{k_2} \ldots f_n^{k_n}$ (where k_i's are nonnegative integers) which separates points of X. This algebra, by Theorem 1.26, is dense in $C(X)$. Clearly, the above finite linear combinations with rational coefficients form a countable dense set for the algebra and hence for $C(X)$.

1.90. Let $\mathscr{A}$ be an algebra $\subset C(X)$, X a compact Hausdorff space. Suppose $\mathscr{A}$ separates points of X and given $x \in X$, there exists $f \in \mathscr{A}$ such that $f(x) \neq 0$. Then $\bar{\mathscr{A}} = C(X)$. [This will follow exactly as in the proof of Theorem 1.26, noting that $|x|$ on $[-1, 1]$ can be uniformly approximated by polynomials in x having no constant term.]

1.91. Let $\mathscr{A}$ be as in Remark 1.90, separating points of X but all $f \in \mathscr{A}$ vanishing at some $x_0 \in X$. Then $\bar{\mathscr{A}} = \{f \in C(X): f(x_0) = 0\}$. The reason is that $\mathscr{A}_0 = \{f + c: f \in \mathscr{A}$ and c is a constant$\}$ is an algebra satisfying the conditions in Remark 1.90 and therefore $\bar{\mathscr{A}}_0 = C(X)$. So given $\varepsilon > 0$ and $g \in C(X)$ with $g(x_0) = 0$, there is $f \in \mathscr{A}$ and some constant c such that $|f(x) + c - g(x)| < \varepsilon/2$ for all $x \in X$. Since $f(x_0) = g(x_0) = 0$, $|c| < \varepsilon/2$; and therefore $|f(x) - g(x)| < \varepsilon$ for all $x \in X$.

1.92. Let $C_0(X)$ be the real-valued continuous functions f on a *locally compact Hausdorff* space X that vanish at infinity (i.e., for each $\varepsilon > 0$, there is compact $K \subset X$ such that $|f(x)| < \varepsilon$ if $x \in X - K$). If $X = R$, $1 \notin C_0(R)$ but $1/(1 + x^2) \in C_0(R)$. The reader can easily verify the following facts:

(a) $C_0(X)$ is a closed subalgebra of $C(X)$.

(b) If $X_\infty = X \cup \{\infty\}$ is the one-point compactification of X, then $C_0(X) = \{f|_X: f \in C(X_\infty)$ and $f(\infty) = 0\}$.

(c) If $\mathscr{A}$ is an algebra $[\subset C_0(X)]$ separating points of X and contains for each x a function that does not vanish at x, then $\bar{\mathscr{A}} = C_0(X)$.

[†] J. A. Clarkson and P. Erdös, *Duke Math. J.* **10**, 5 (1943).

Next we present a version of the Ascoli theorem. For a more general version of the theorem, the reader should consult [34].

First we need a definition.

Definition 1.42. Let X be a topological space and Y a metric space. A family $\mathscr{F}$ of continuous maps from X into Y is called *equicontinuous* at a point $x_0 \in X$ if given $\varepsilon > 0$, there exists an open set $V(x_0)$ such that $x_0 \in V(x_0)$ and $d(f(x), f(x_0)) < \varepsilon$ whenever $x \in V(x_0)$ and $f \in \mathscr{F}$. If $\mathscr{F}$ is equicontinuous at every point on X, then $\mathscr{F}$ is called equicontinuous on X. ∎

In what follows, X is a compact Hausdorff space and Y is a complete metric space. Then $C(X, Y)$, the set of all continuous maps from X into Y, is a complete metric space with the metric $d(f, g) = \sup_{x \in X} d(f(x), g(x))$.

Theorem 1.28. *The Arzela–Ascoli Theorem.* A subset $\mathscr{F}$ of $C(X, Y)$ has compact closure if and only if $\mathscr{F}$ is equicontinuous on X and for each $x \in X$ the set $\{f(x): f \in \mathscr{F}\}$ has compact closure in Y. ∎

Proof. The "only if" part is left to the reader. For the "if" part, it is sufficient to prove that $\mathscr{F}$ is totally bounded.

Let $\varepsilon > 0$. By compactness of X and equicontinuity of $\mathscr{F}$, there exist open sets $V(x_i)$, $1 \leq i \leq n$, whose union is X such that $x_i \in V(x_i)$ and $d(f(y), f(x_i)) < \varepsilon$ for any $y \in V(x_i)$ and all $f \in \mathscr{F}$. Now the set $B = \{f(x_i): 1 \leq i \leq n$ and $f \in \mathscr{F}\}$ is totally bounded and therefore there exist $y_j \in Y$, $1 \leq j \leq m$, such that $B \subset \bigcup_{j=1}^{m} S_{y_j}(\varepsilon)$. For any mapping $\pi: \{1, 2, \ldots, n\} \to \{1, 2, \ldots, m\}$, let $\mathscr{F}_\pi = \{f \in \mathscr{F}: f(x_i) \in S_{y_{\pi(i)}}(\varepsilon), 1 \leq i \leq n\}$. Clearly for $f_1, f_2 \in \mathscr{F}_\pi$, if $x \in V(x_i)$, then

$$d(f_1(x), f_2(x)) \leq d(f_1(x), f_1(x_i)) + d(f_1(x_i), y_{\pi(i)}) + d(y_{\pi(i)}, f_2(x_i)) + d(f_2(x), f_2(x_i)) < 4\varepsilon;$$

this means that diameter $(\mathscr{F}_\pi) \leq 4\varepsilon$. Since $\mathscr{F}$ is the union of finitely many such $\mathscr{F}_\pi$'s, the proof is complete. ∎

Problems

1.6.1. Prove the following result due to J. L. Walsh: Let $\Phi \in C[0, 1]$. Then a necessary and sufficient condition that an arbitrary $f \in C[0, 1]$ can be uniformly approximated by a polynomial in Φ is that Φ be *strictly monotonic* on [0, 1].

1.6.2. Let X be a locally compact Hausdorff space and $C_c(X)$ be the set of all real-valued continuous functions on X that vanish outside compact sets. Suppose $\mathscr{A}$ is an algebra $\subset C_c(X)$, separating points of X, and is such that for every compact set K there is $f \in \mathscr{A}$ such that $f(x) = 1$ for $x \in K$. Show that $\mathscr{A}$ is dense in $C_c(X)$ with the "sup" metric.

1.6.3. Let X and Y be locally compact Hausdorff spaces and let $f \in C_c(X \times Y)$. Show that given $\varepsilon > 0$ there exist $g_i \in C_c(X)$ and $h_i \in C_c(Y)$, $1 \le i \le n$, such that

$$\left| f(x, y) - \sum_{i=1}^{n} g_i(x)h_i(y) \right| < \varepsilon$$

for all $(x, y) \in X \times Y$. (See Problem 1.6.2.)

1.6.4. Prove (with details) Remark 1.92.

★ **1.6.5.** Let $f \in C[0, 1]$ and $f(0) = 0$. Show that f can be uniformly approximated by a sequence of polynomials $P_n(x) = \sum_{m=0}^{k_n} a_{nm} x^m$ such that for each m, $\lim_{n\to\infty} a_{nm} = 0$.

★ **1.6.6.** Prove the following extension of Problem 1.6.1—also due to Walsh. Let Φ be a real-valued function on [0, 1]. Then an arbitrary $f \in C[0, 1]$ can be uniformly approximated by a polynomial in Φ if and only if Φ is one-to-one and bounded and Φ^{-1} can be extended to a continuous function on the closure of the range of Φ.

1.6.7. Let f be a real-valued continuous function on R, and let h be any real-valued function that has positive infimum on each compact subset of R. Show that there exists an infinitely differentiable function g on R such that $|f(x) - g(x)| < h(x)$ for all $x \in R$. [Hint: Let p_n be a polynomial such that $|p_n(x) - f(x)| \le \frac{1}{3}h(x)$ for all x with $|x| \le n + \frac{1}{2}$. Find infinitely differentiable g_n such that $0 \le g_n \le 1$, $g_n(x) = 1$ if $n - 1 \le |x| \le n$ and $= 0$ if $|x| \le n - \frac{3}{2}$ or $|x| \ge n + \frac{1}{2}$. Take $g = \sum_{n=1}^{\infty} (1 - g_{n-1}) g_n p_n$.] This result is due to T. Carleman. There is an extension of this result to R^n due to A. Boghossian.

1.6.8. Prove the "only if" part of Theorem 1.28.

1.6.9. Let $\mathscr{F}_k = \{f\colon [0, 1] \to [0, 1]$ and $|f(x) - f(y)| \le k|x - y|$ for $0 \le x, y \le 1\}$, where $k > 0$. Show that $\mathscr{F}_k$ is compact as a subset of $C([0, 1], [0, 1])$.

1.6.10. Show that the family of functions $1/[1 + (x - n)^2]$ of R into [0, 1] is equicontinuous but not compact.

1.6.11. Let X, Y be compact metric spaces and $f\colon X \times Y \to Z$ (a metric space). For $x \in X$, let $f_x(y) = f(x, y)$. Show that if f is continuous, then the mapping $x \to f_x$ of X into $C(Y, Z)$ is continuous and the family $\{f_x: x \in X\}$ is equicontinuous.

★ **1.6.12.** *Stone–Weierstrass Theorem for Lattices* (Kakutani–Krein): Let X be a compact Hausdorff space. Then a subset $S \subset C(X)$ is called a *lattice* if $f \vee g$ and $f \wedge g$ are in S for all $f, g \in S$. Prove that (i) any closed subalgebra of $C(X)$ containing 1 is a lattice, and (ii) any lattice $S \subset C(X)$ that is a closed subspace, contains 1, and separates points of X, is all of $C(X)$.

★ **1.6.13.** Let f be a continuous real-valued function on [0, 1]. Show that there is a monotonic increasing sequence (P_n) of polynomials converging uniformly to f on [0, 1].

2

Measure

The concept of measure is an extension of the concept of length: The measure of an interval in R is usually its length; the measure of a polygon in R^2 is usually its area; and so on. The problem of extending the classes of sets for which the notions of length, area, etc. are defined to larger classes of sets for which these notions are still defined gave rise to the general theory of measure.

The first definitions of the measure of an arbitrary set in R^n seem to have been given by G. Cantor (in 1883), O. Stolz (in 1884), and A Harnack (in 1885). These definitions were substantially improved later by G. Peano (in 1887) and C. Jordan (in 1892). By means of his concept of measure, Peano determined a necessary and sufficient condition for a bounded nonnegative function on a closed and bounded interval in R to be Riemann integrable. In 1898, in his book *Leçons sur la Théorie des Fonctions*, E. Borel formulated the following postulates (which outline essential properties of length) for defining measures of sets:

(1) A measure is always nonnegative.

(2) The measure of the union of a finite number of nonoverlapping sets equals the sum of their measures.

(3) The measure of the difference of a set and a subset is equal to the difference of their measures.

(4) Every set whose measure is not zero is uncountable.

H. Lebesgue[†] presented a mathematically rigorous description of the class of sets for which he defined a measure satisfying Borel's postulates. This

[†] H. Lebesgue, Intégrale, longueur, aire, *Ann. Mat. Pura Appl.* **7**(3), 231–359 (1902).

measure is now well known as Lebesgue measure on R^n and is perhaps the most important and useful concept of a measure that can be found in R^n to date. A theory of measure similar to that of Lebesgue was also independently developed by W. H. Young.†

It is now well known that measure theory and its techniques are indispensible tools for many parts of modern analysis. Measure theory provides the proper framework for the study of discontinuous and nondifferentiable functions. For instance, the concept of measure provides us with a precise idea about the set of nondifferentiable points of a monotonic function or the set of points of discontinuity of a Rieman-integrable function. Different concepts of integrals are either based on or inseparably connected with the concept of a measure; and these, in turn, play an important part in the application of mathematical analysis to present day science, including theoretical physics and probability theory.

This chapter presents the theory roughly as follows.

First, a measure is defined on an algebra of subsets of an arbitrary abstract nonempty set and different properties of the measure based on its definition are derived. Then, as concrete examples, we construct Lebesgue measure on intervals in R. Next, with a view to obtaining a measure on a σ-algebra, we introduce outer measures and measurable sets. We show how to construct outer measures (and therefore measures in general) and extend a measure on an algebra to a σ-algebra. In order to obtain a large class of measurable sets, we introduce the concept of a metric outer measure on a metric space. These ideas of constructing a measure from an outer measure, considering a metric outer measure to obtain a large class of measurable sets as the domain space of a measure, etc., were noted first by C. Carathéodory.‡ Finally, to take a concrete example, we discuss Lebesgue measure in R and its properties at length.

Throughout this chapter, X is a nonempty abstract set.

2.1. Measure on an Algebra

In this section, we will define a measure on an algebra and establish several properties of a measure based upon its definition.

† W. H. Young, Open sets and the theory of content, *Proc. London Math. Soc.* **2**(2), 16–51 (1904).

‡ C. Carathéodory, *Nachr. Ges. Wiss. Göttingen*, 404–426 (1914).

Definition 2.1. Let $\mathscr{A}$ be an algebra† of subsets of X and μ be an extended real-valued function on $\mathscr{A}$. Then μ is called a measure on $\mathscr{A}$ if

(a) $\mu(\emptyset) = 0$,

(b) $\mu(A) \geq 0 \quad$ if $A \in \mathscr{A}$,

(c) $\mu(\bigcup_{n=1}^{\infty} A_n) = \sum_{n=1}^{\infty} \mu(A_n)$, for every sequence A_n of pairwise disjoint sets of $\mathscr{A}$ with $\bigcup_{n=1}^{\infty} A_n \in \mathscr{A}$.

The property (c) is known as the *countable additivity* property of μ. μ is called a *finitely additive measure* if (c) is replaced by

(c′) $\mu(\bigcup_{n=1}^{k} A_n) = \sum_{n=1}^{k} \mu(A_n)$, for every finite sequence (A_n) of pairwise disjoint sets in $\mathscr{A}$. ∎

Example 2.1. Let $\mathscr{A} = 2^X$. Let μ be defined on $\mathscr{A}$ by

$$\mu(A) = \begin{cases} \text{number of points in } A, & \text{if } A \text{ is finite} \\ \infty, & \text{if } A \text{ is infinite.} \end{cases}$$

Then μ is a measure. (Prove this assertion.) This measure is known as the *counting measure* in X.

Example 2.2. Let X be uncountable and $\mathscr{A}$ be the class defined by

$$\mathscr{A} = \{A \subset X\colon \text{either } A \text{ or } X - A \text{ is countable}\}.$$

Then $\mathscr{A}$ is an algebra (in fact, a σ-algebra). Let μ be defined on $\mathscr{A}$ by

$$\mu(A) = \begin{cases} 0, & \text{if } A \text{ is countable} \\ 1, & \text{if } X - A \text{ is countable.} \end{cases}$$

Then μ is a measure on $\mathscr{A}$. (Prove this assertion.)

Example 2.3. Let $X = Z^+$, $\mathscr{A} = 2^X$, and $\sum_{n=1}^{\infty} a_n$ be a convergent series of positive real numbers. Let μ be defined on $\mathscr{A}$ by

$$\mu(A) = \begin{cases} \sum_{n \in A} a_n, & \text{if } A \text{ is nonempty and finite} \\ \infty, & \text{if } A \text{ is infinite} \\ 0, & \text{if } A \text{ is empty.} \end{cases}$$

† If $\mathscr{A}$ is a ring, then this definition defines a measure on a ring.

Then μ is not a measure, since

$$X = \bigcup_{n=1}^{\infty} \{n\}, \qquad \mu(X) = \infty,$$

but

$$\sum_{n=1}^{\infty} \mu\{n\} = \sum_{n=1}^{\infty} a_n < \infty.$$

However, μ is clearly a finitely additive measure.

Several important properties of a measure that follow as consequences of its definition are given in the next two propositions.

Proposition 2.1. Let μ be a measure on an algebra $\mathscr{A}$. Then

(a) $A \subset B$, $A \in \mathscr{A}$, $B \in \mathscr{A}$ imply $\mu(A) \leq \mu(B)$.

(b) $\mu(\bigcup_{n=1}^{\infty} A_n) \leq \sum_{n=1}^{\infty} \mu(A_n)$ (the *countable subadditive property*), if $A_n \in \mathscr{A}$, $1 \leq n < \infty$ and $\bigcup_{n=1}^{\infty} A_n \in \mathscr{A}$. ∎

Proof. (a) $B = A \cup (B - A)$; so, if $A_1 = A$, $A_2 = B - A$, and $A_n = \varnothing$ for $n > 2$, then with μ being countably additive, we have $\mu(B) = \mu(A) + \mu(B - A)$. Hence $\mu(B) \geq \mu(A)$.

(b) Let $B_1 = A_1$ and $B_n = A_n - \bigcup_{i=1}^{n-1} A_i$, for $n > 1$. Then $B_n \in \mathscr{A}$ for each n, $\bigcup_{n=1}^{\infty} B_n = \bigcup_{n=1}^{\infty} A_n \in \mathscr{A}$, $B_n \subset A_n$, and the B_n's are pairwise disjoint. Hence

$$\mu\left(\bigcup_{n=1}^{\infty} A_n\right) = \mu\left(\bigcup_{n=1}^{\infty} B_n\right) = \sum_{n=1}^{\infty} \mu(B_n) \leqq \sum_{n=1}^{\infty} \mu(A_n).$$ ∎

Proposition 2.2. Let μ be a measure on an algebra $\mathscr{A}$. Then

(a) If $A_n \subset A_{n+1}$, $A_n \in \mathscr{A}$ for $1 \leqq n < \infty$ and $\bigcup_{n=1}^{\infty} A_n \in \mathscr{A}$, then

$$\mu\left(\bigcup_{n=1}^{\infty} A_n\right) = \lim_{n \to \infty} \mu(A_n).$$

(b) If $A_n \supset A_{n+1}$, $A_n \in \mathscr{A}$ for $1 \leqq n < \infty$, $\mu(A_1) < \infty$ and $\bigcap_{n=1}^{\infty} A_n \in \mathscr{A}$, then

$$\mu\left(\bigcap_{n=1}^{\infty} A_n\right) = \lim_{n \to \infty} \mu(A_n).$$

Furthermore, for a finitely additive μ, countable additivity is implied by (a). ∎

Proof. First, we prove the last assertion. So let μ be a finitely additive measure with property (a). Let $(A_n)_{n=1}^{\infty}$ be a sequence of pairwise disjoint sets in $\mathscr{A}$ with $\bigcup_{n=1}^{\infty} A_n \in \mathscr{A}$. Let $B_k = \bigcup_{n=1}^{k} A_n$. Then $B_k \subset B_{k+1}$, $B_k \in \mathscr{A}$, and $\bigcup_{k=1}^{\infty} B_k = \bigcup_{n=1}^{\infty} A_n \in \mathscr{A}$. Hence by the assumption of (a),

$$\mu\left(\bigcup_{n=1}^{\infty} A_n\right) = \mu\left(\bigcup_{k=1}^{\infty} B_k\right) = \lim_{k\to\infty} \mu(B_k) = \lim_{k\to\infty} \sum_{n=1}^{k} \mu(A_n) = \sum_{n=1}^{\infty} \mu(A_n).$$

Next, we establish properties (a) and (b) of a measure. To prove (a), let $A_n \subset A_{n+1}$ and $\bigcup_{n=1}^{\infty} A_n \in \mathscr{A}$. Let $B_1 = A_1$, $B_n = A_n - A_{n-1}$ for $n > 1$. Then $B_n \cap B_m = \varnothing$ $(n \neq m)$ and $\bigcup_{n=1}^{\infty} B_n = \bigcup_{n=1}^{\infty} A_n \in \mathscr{A}$. Then

$$\begin{aligned}\mu\left(\bigcup_{n=1}^{\infty} A_n\right) &= \mu\left(\bigcup_{n=1}^{\infty} B_n\right) = \sum_{n=1}^{\infty} \mu(B_n) \\ &= \lim_{k\to\infty} \sum_{n=1}^{k} \mu(B_n) = \lim_{k\to\infty} \mu(A_k), \text{ since } A_k = \bigcup_{n=1}^{k} B_n.\end{aligned}$$

To prove (b), let $A_n \supset A_{n+1}$, $\mu(A_1) < \infty$, $\bigcap_{n=1}^{\infty} A_n \in \mathscr{A}$. Then

$$A_1 - \bigcap_{n=1}^{\infty} A_n = \bigcup_{n=1}^{\infty} (A_1 - A_n), \qquad A_1 - A_n \subset A_1 - A_{n+1}.$$

Therefore by (a),

$$\mu\left(A_1 - \bigcap_{n=1}^{\infty} A_n\right) = \lim_{n\to\infty} \mu(A_1 - A_n).$$

Since $\mu(A_1) < \infty$,

$$\mu\left(A_1 - \bigcap_{n=1}^{\infty} A_n\right) = \mu(A_1) - \mu\left(\bigcap_{n=1}^{\infty} A_n\right)$$

and

$$\mu(A_1 - A_n) = \mu(A_1) - \mu(A_n).$$

Now part (b) follows. ∎

A measure μ on an algebra $\mathscr{A}$ of subsets of X is called *finite* if $\mu(X) < \infty$. It is called *σ-finite* if there is a sequence (X_n) of sets in $\mathscr{A}$ with $\mu(X_n) < \infty$ and $X = \bigcup_{n=1}^{\infty} X_n$. As was the case in the proof of Proposition 2.1(b), it is always possible to take $(X_n)_{n=1}^{\infty}$ as pairwise disjoint. This finiteness condition on a measure is often, though not always, needed for many parts of the general theory. The most important example of a σ-finite measure on an algebra of subsets of R is that of Lebesgue measure on intervals (which is discussed in the next section). Of course not every measure is σ-finite. For instance, the counting measure on 2^X (X uncountable) is not σ-finite.

Another useful concept is that of *semifiniteness*—a notion weaker than that of σ-finiteness. A measure μ on an algebra $\mathscr{A}$ of subsets of X is called *semifinite* if for each $A \in \mathscr{A}$, with $\mu(A) = \infty$, there is $B \subset A$, with $B \in \mathscr{A}$ and $0 < \mu(B) < \infty$. Thus every σ-finite measure is semifinite (Problem 2.1.7). The counting measure is always semifinite. But the 0–∞ measure (which is zero on countable subsets of X and infinity on uncountable subsets of X) is not semifinite. Measures that are not semifinite are very wild when restricted to certain sets. Every measure is, in a sense, semifinite, once its 0–∞ part (the wild part) is taken away. The next proposition demonstrates this.

Proposition 2.3. Let μ be a measure on a σ-algebra $\mathscr{A}$ of subsets of X.

(a) If μ is semifinite, then every set in $\mathscr{A}$ with infinite measure contains sets in $\mathscr{A}$ with arbitrarily large finite measure.

(b) Each measure μ on $\mathscr{A}$ is the sum $\mu_1 + \mu_2$ of a semifinite measure μ_1 and a measure μ_2 that assumes only the values 0 and ∞. The measure μ_1 need not be unique, but there is a smallest μ_2. [Here $\mu_2 \leq \mu_4$ means $\mu_2(E) \leq \mu_4(E)\ \forall\ E \in \mathscr{A}$.] ∎

Proof. (a) Let $A \in \mathscr{A}$ with $\mu(A) = \infty$. If the assertion is false, then

$$0 < \sup\{\mu(B)\colon B \subset A,\ B \in \mathscr{A},\ \mu(B) < \infty\} = a < \infty. \tag{2.1}$$

Now since $a < \infty$, we can find $B_n \subset A$, $B_n \in \mathscr{A}$, and $B_n \subset B_{n+1}$ such that $a \geqq \mu(B_n) > a - 1/n$. Let $B = \bigcup_{n=1}^{\infty} B_n$. Then $B \subset A$ and $B \in \mathscr{A}$. By Proposition 2.2, $\mu(B) = \lim_{n\to\infty}\mu(B_n)$. Therefore $\mu(B) = a$. Since $\mu(A) = \infty$, $\mu(A - B) = \infty$. Since μ is semifinite, there is $C \subset A - B$ such that $0 < \mu(C) < \infty$. Then $B \cup C \subset A$ and $\mu(B \cup C) = \mu(B) + \mu(C) > a$, in contradiction to equation (2.1). The assertion now follows.

(b) We define

$$\mu_1(E) = \sup\{\mu(B)\colon\ B \subset E,\ B \in \mathscr{A},\ \mu(B) < \infty\}$$

and $\mu_2(E) = \sup\{\mu(B) - \mu_1(B)\colon B \subset E,\ B \in \mathscr{A}, \mu_1(B) < \infty\}$. These μ_1 and μ_2 will meet the requirements of the theorem. We will only show that μ_1 is a measure and leave the rest of the proof to the reader (Problem 2.1.10).

Notice that we can always find a sequence (B_n) of elements of $\mathscr{A}$ such that $B_n \subset B_{n+1} \subset E$, $\mu(B_n) < \infty$ and $\lim_{n\to\infty}\mu(B_n) = \mu_1(E)$; hence if $B = \bigcup_{n=1}^{\infty} B_n$, then $B \in \mathscr{A}$, $B \subset E$, and $\mu(B) = \mu_1(E)$. Also if $\mu(E) < \infty$, then $\mu(E) = \mu_1(E)$. Let $(E_i)_{i=1}^{\infty}$ be a sequence of pairwise disjoint sets in $\mathscr{A}$. Then if $\mu_1(\bigcup_{i=1}^{\infty} E_i) < \infty$, we can find $B_i \subset E_i$, $B_i \in \mathscr{A}$, such that $\mu(B_i)$

$= \mu_1(E_i) < \infty$; therefore,

$$\mu_1\Big(\bigcup_{i=1}^{\infty} E_i\Big) \geqq \mu_1\Big(\bigcup_{i=1}^{n} E_i\Big) \geqq \mu\Big(\bigcup_{i=1}^{n} B_i\Big) = \sum_{i=1}^{n} \mu(B_i) = \sum_{i=1}^{n} \mu_1(E_i),$$

for each n. This means that $\mu_1(\bigcup_{i=1}^{\infty} E_i) \geqq \sum_{i=1}^{\infty} \mu_1(E_i)$. To prove the converse inequality, we notice that there exist $B_n \subset \bigcup_{i=1}^{\infty} E_i$, $B_n \in \mathscr{A}$, $\mu(B_n) < \infty$, such that $\mu_1(\bigcup_{i=1}^{\infty} E_i) = \lim_{n\to\infty} \mu(B_n)$. Since $B_n = \bigcup_{i=1}^{\infty}(B_n \cap E_i)$ and μ is a measure, $\mu(B_n) = \sum_{i=1}^{\infty} \mu_1(B_n \cap E_i) \leqq \sum_{i=1}^{\infty} \mu_1(E_i)$. Hence $\mu_1(\bigcup_{i=1}^{\infty} E_i) \leqq \sum_{i=1}^{\infty} \mu_1(E_i)$ and μ_1 is a measure. ∎

Problems

2.1.1. Let μ be a finitely additive measure on an algebra $\mathscr{A}$. For A, $B \in \mathscr{A}$, show that

(a) $\mu(A) \leqq \mu(B)$, if $A \subset B$,

(b) $\mu(A \cup B) + \mu(A \cap B) = \mu(A) + \mu(B)$,

(c) $\mu(B - A) = \mu(B) - \mu(A)$, if $A \subset B$ and $\mu(A) < \infty$.

2.1.2. Show that in Proposition 2.2, (b) is false unless $\mu(A_n) < \infty$ for some n. [Hint: Take $X = R$, $A_n = (n, \infty)$, $\mathscr{A} = 2^X$, $\mu =$ counting measure.]

2.1.3. Let $X = Z^+$, $\mathscr{A} = \{A \subset X$: either A or $X - A$ is finite$\}$. Let $\mu(A) = 0$, if A is finite, $= \infty$, if A is infinite. Then show that (a) $\mathscr{A}$ is an algebra and (b) μ is a finitely additive measure, which is not a measure.

2.1.4. Show that in Proposition 2.2, for a finitely additive μ, (b) does not imply countable additivity. (Hint: Use problem 2.1.3.)

2.1.5. Let $(\mu_n)_{n=1}^{\infty}$ be a sequence of measures on an algebra $\mathscr{A}$ such that for $A \in \mathscr{A}$, $\mu_n(A) \leq \mu_{n+1}(A)$. Show that μ defined on $\mathscr{A}$ by $\mu(A) = \lim_{n\to\infty} \mu_n(A)$ is a measure on $\mathscr{A}$.

2.1.6. Let μ be a measure on a σ-algebra $\mathscr{A}$. Let $E_i \in \mathscr{A}$, $1 \leq i < \infty$, and $\sum_{i=1}^{\infty} \mu(E_i) < \infty$. Show that $\mu(\overline{\lim}_{i\to\infty} E_i) = 0$. (This is known as the Borel–Cantelli Lemma and is widely used in probability theory.)

2.1.7. Show that every σ-finite measure is semifinite.

2.1.8. Let μ be a measure on an algebra $\mathscr{A}$. Suppose $A \in \mathscr{A}$ and for every $E \in \mathscr{A}$, $\mu_A(E) = \mu(A \cap E)$. Show that μ_A is a measure on the algebra $\{E \in \mathscr{A}: E \subset A\}$. ($\mu_A$ is called the restriction of μ to A.)

2.1.9. Suppose μ_1 and μ_2 are two measures on a σ-algebra $\mathscr{A}$ such that for each $E \in \mathscr{A}$, $\mu_1(E) \geq \mu_2(E)$. Show that there is a measure μ_3 on $\mathscr{A}$ such that $\mu_1(E) = \mu_2(E) + \mu_3(E)$ for $E \in \mathscr{A}$. This μ_3 is unique if μ_2 is σ-finite. [Hint: Take $\mu_3(E) = \sup\{\mu_1(B) - \mu_2(B): B \subset E, \mu_2(B) < \infty\}$.]

2.1.10. Complete the proof of Proposition 2.3(b).

2.2. Lebesgue Measure on Intervals

The most important example of a measure is, perhaps, the Lebesgue measure on R^n ($n \geq 1$), named after the celebrated French mathematician Henri Lebesgue (1875–1941), one of the original founders of the theory of measure and integration. Lebesgue measure is the subject of this section. Though usually the domain of Lebesgue measure is a σ-algebra containing all open sets and numerous other complicated sets, in this section we restrict ourselves to the simpler problem of defining Lebesgue measure on an algebra of half-open intervals.

Let $X = R$ and $\mathscr{A}$ be the class of all finite disjoint unions of right-closed, left-open intervals including intervals of the form $(-\infty, a]$, (a, ∞) and $(-\infty, \infty)$, $-\infty < a < \infty$, and the empty set. These types of intervals are considered because, unlike open intervals, they form an algebra.

Lemma 2.1. $\mathscr{A}$ is an algebra. (See Problem 2.2.1.) ∎

To define a measure m_0 on $\mathscr{A}$, we need the following observation. If $(I_n)_{n=1}^k$ and $(J_m)_{m=1}^s$ are two finite sequences of pairwise disjoint intervals in $\mathscr{A}$ such that $\bigcup_{n=1}^k I_n = \bigcup_{m=1}^s J_m = A$, then $\sum_{n=1}^k l(I_n) = \sum_{m=1}^s l(J_m)$, where l denotes the length, $l((a, b]) = b - a$ and the length of an infinite interval is infinite. This follows from the equalities

$$l(I_n) = \sum_{m=1}^{s} l(I_n \cap J_m), \qquad l(J_m) = \sum_{n=1}^{k} l(I_n \cap J_m).$$

Thus for $A \in \mathscr{A}$, we can define

$$m_0(A) = \sum_{n=1}^{k} l(I_n) \qquad \text{if } A = \bigcup_{n=1}^{k} I_n.$$

We also note that if $A = \bigcup_{n=1}^k I_n$, $B = \bigcup_{j=1}^s J_m$, and $A \subset B$ (the intervals I_n and J_m being in $\mathscr{A}$), then we have for A and B in $\mathscr{A}$,

$$\text{(i)}\quad I_n = \bigcup_{m=1}^{s} (I_n \cap J_m) \qquad \text{and } J_m \supset \bigcup_{n=1}^{k} (I_n \cap J_m),$$

$$\text{(ii)}\quad l(I_n) = \sum_{m=1}^{s} l(I_n \cap J_m) \qquad \text{and } l(J_m) \geq \sum_{n=1}^{k} l(I_n \cap J_m),$$

and therefore it follows easily that $m_0(A) \leq m_0(B)$. Thus we have the following lemma.

Lemma 2.2. For $A \in \mathscr{A}$, let $m_0(A) = \sum_{n=1}^k l(I_n)$ if $A = \bigcup_{n=1}^k I_n$. Then m_0 is well defined, and for $B, C \in \mathscr{A}$ and $B \subset C$, $m_0(B) \leq m_0(C)$. ∎

Now we can prove the following theorem.

Theorem 2.1. m_0 is a measure on $\mathscr{A}$. (m_0 is called the Lebesgue measure on intervals.) ▮

Proof. Let $(A_i)_{i=1}^\infty$ be a sequence of pairwise disjoint sets in $\mathscr{A}$ such that $\bigcup_{i=1}^\infty A_i \in \mathscr{A}$. Then for each i, $1 \leq i < \infty$, there is a finite sequence of pairwise disjoint intervals $I_{ij} \in \mathscr{A}$ such that $A_i = \bigcup_{j=1}^{m_i} I_{ij}$. Also since $\bigcup_{i=1}^\infty A_i \in \mathscr{A}$, there are pairwise disjoint intervals $I_k \in \mathscr{A}$ such that $\bigcup_{i=1}^\infty A_i = \bigcup_{k=1}^n I_k$.

Now for each k,

$$\bigcup_{i=1}^{\infty} A_i \supset \bigcup_{i=1}^{k} A_i = \bigcup_{i=1}^{k} \bigcup_{j=1}^{m_i} I_{ij}.$$

So by Lemma 2.2.

$$m_0\left(\bigcup_{i=1}^{\infty} A_i\right) \geq m_0\left(\bigcup_{i=1}^{k} A_i\right) = \sum_{i=1}^{k} \sum_{j=1}^{m_i} l(I_{ij}) = \sum_{i=1}^{k} m_0(A_i).$$

Hence $m_0(\bigcup_{i=1}^\infty A_i) \geq \sum_{i=1}^\infty m_0(A_i)$. To prove the converse inequality, we observe that $I_k = \bigcup_{i=1}^\infty \bigcup_{j=1}^{m_i} (I_k \cap I_{ij})$. If $l(I_k \cap I_{ij}) = \infty$ for some i, j, then $m_0(A_i) = \infty$ and consequently $m_0(\bigcup_{i=1}^\infty A_i) = \sum_{i=1}^\infty m_0(A_i)$. So then we assume that $l(I_k \cap I_{ij}) < \infty$ for all i, j, and k.

Let $\varepsilon > 0$. We write $I_k \cap I_{ij} = (a_{ij}^k, b_{ij}^k]$. Let $[a_k, b_k] \subset I_k$. Then

$$[a_k, b_k] \subset \bigcup_{i=1}^{\infty} \bigcup_{j=1}^{m_i} \left(a_{ij}^k, b_{ij}^k + \frac{\varepsilon}{2^i m_i}\right).$$

By compactness of $[a_k, b_k]$, there is a positive integer p such that

$$[a_k, b_k] \subset \bigcup_{i=1}^{p} \bigcup_{j=1}^{m_i} \left(a_{ij}^k, b_{ij}^k + \frac{\varepsilon}{2^i m_i}\right).$$

Therefore,

$$b_k - a_k \leq \sum_{i=1}^{p} \sum_{j=1}^{m_i} \left(b_{ij}^k - a_{ij}^k + \frac{\varepsilon}{2^i m_i}\right) \leq \sum_{i=1}^{\infty} \sum_{j=1}^{m_i} l(I_k \cap I_{ij}) + \varepsilon.$$

Since $\varepsilon(> 0)$ is arbitrary and $[a_k, b_k]$ is an arbitrary closed subinterval of I_k,

$$l(I_k) \leq \sum_{i=1}^{\infty} \sum_{j=1}^{m_i} l(I_k \cap I_{ij}).$$

Hence

$$
\begin{aligned}
m_0\left(\bigcup_{i=1}^{\infty} A_i\right) &= \sum_{k=1}^{n} l(I_k) \\
&\le \sum_{k=1}^{n} \sum_{i=1}^{\infty} \sum_{j=1}^{m_i} l(I_k \cap I_{ij}) \\
&= \sum_{i=1}^{\infty} \sum_{j=1}^{m_i} \sum_{k=1}^{n} l(I_k \cap I_{ij}) \\
&= \sum_{i=1}^{\infty} \sum_{j=1}^{m_i} l(I_{ij}), \qquad I_{ij} = \bigcup_{k=1}^{n} I_k \cap I_{ij} \\
&= \sum_{i=1}^{\infty} m_0(A_i).
\end{aligned}
$$

∎

Remarks

2.1. In R^n $(n > 1)$, the class of all finite disjoint unions of sets of the form

$$I = \{(x_1, x_2, \ldots, x_n): x_i \in I_i, \text{ an interval in } \mathscr{A} \text{ above}, 1 \le i \le n\}$$

form an algebra, as before. I is generally called a generalized interval and the interval I_j in its definiton is called its jth side. By defining $m_0(I) = l(I_1) \cdots l(I_n)$, we can again obtain (as before) a measure (known as n-dimensional Lebesgue measure on the generalized intervals). (Serious readers are encouraged to construct proofs for these facts. The proofs should be similar to those in the case of $n = 1$.)

2.2. For any given monotonic increasing function F on R that is continuous on the right, it is now possible to construct a measure m_F on the algebra $\mathscr{A}$ such that $m_F(a, b] = F(b) - F(a)$. This measure m_F is usually called the Lebesgue–Stieltjes measure on $\mathscr{A}$ induced by F. (See Problem 2.2.5.)

Problems

(In the following problems, $\mathscr{A}$ is always the algebra as defined in the beginning of this section.)

2.2.1. Prove Lemma 2.1.

2.2.2. Prove that the Lebesgue measure m_0 is translation invariant on the intervals in $\mathscr{A}$ [that is, $m_0(I) = m_0(I + x)$ for each interval I in $\mathscr{A}$ and any point $x \in R$].

2.2.3. Let λ be a measure on $\mathscr{A}$ such that λ is translation invariant (Problem 2.2.2). Let λ be finite on each finite interval. Then show that there is a constant k such that $\lambda(A) = k \cdot m_0(A)$ for all $A \in \mathscr{A}$, m_0 being the Lebesgue measure. [Hint: If $\lambda(0, 1] = k$, then $\lambda(0, p] = k \cdot p$ for every positive integer p. Thus if $l(I)$ is a rational number, $\lambda(I) = k \cdot l(I)$.]

2.2.4. Let λ be a measure on $\mathscr{A}$ such that $\lambda(I) < \infty$ for every finite interval I. Let $y \in R$ and F_y be the function on R given by

$$F_y(x) = \begin{cases} -\lambda(x, y], & \text{if } x < y \\ 0, & \text{if } x = y \\ \lambda(y, x], & \text{if } x > y. \end{cases}$$

Show that F_y is a monotonic increasing function, which is continuous on the right. (F_y is usually called a distribution function induced by λ.)

2.2.5. Let F be a monotonic increasing function on R that is continuous on the right. Then show that

(a) If $(a, b] \subset \bigcup_{i=1}^{\infty} (a_i, b_i)$, then

$$F(b) - F(a) \leq \sum_{i=1}^{\infty} F(b_i) - F(a_i)$$

(b) Let

$$m_F(a, b] = F(b) - F(a),$$

$$m_F(a, \infty) = \lim_{x\to\infty} F(x) - F(a),$$

$$m_F(-\infty, a] = F(a) - \lim_{x\to-\infty} F(x).$$

Show that m_F defines a measure on $\mathscr{A}$ in an obvious manner. [This measure is usually called the *Lebesgue–Stieltjes measure* on $\mathscr{A}$ induced by F. Note that $F(x) = x$ yields the Lebesgue measure on $\mathscr{A}$.]

2.2.6. Let $Z_1 = \{1/n \colon n \in Z^+\}$. For $B \in \mathscr{A}$, let $\mu_1(B) = m(B \cap Z^+)$ and $\mu_2(B) = m(B \cap Z_1)$, where m is the counting measure. Show that μ_1 is a Lebesgue–Stieltjes measure, whereas μ_2 is not.

2.2.7. *The Setwise Limit of a Sequence of Finite Measures on an Algebra Need Not Be a Measure.* Consider the following example to show this. For each positive integer n, define $\mu_n(E) = n \cdot m_0\big(E \cap (0, 1/n]\big)$. Then each μ_n is a measure and for each $E \in \mathscr{A}$, $\lim_{n\to\infty}\mu_n(E) = \mu(E)$ exists, where μ is a set function having only the values 0 or 1. This μ is not countably additive. (Compare Problem 2.1.5. If the algebra is replaced by a σ-algebra, then the limit must be a measure; see Problem 4.3.17.)

2.3. Construction of Measures: Outer Measures and Measurable Sets

In the previous two sections, we defined the concept of a measure on an algebra and then constructed Lebesgue measure on the algebra generated by the left-open, right-closed intervals in R. To develop a suitable theory of integration, whether on R or on any abstract set, we need a measure to be defined on a σ-algebra. This is usually done through the introduction of outer measures, a concept somewhat different from that of measure and due to C. Carathéodory, a distinguished Greek mathematician.

Our main concerns in this section are (1) how to construct an outer measure on 2^X and then to derive from it a measure on a σ-algebra, and (2) to extend a given measure on an algebra to one on a σ-algebra containing it and, in particular, to extend the domain of Lebesgue measure from the algebra generated by the half-open intervals to a much larger class of sets, a σ-algebra of Lebesgue-measurable sets.

Definition 2.2. By an outer measure μ^* we mean an extended real-valued set function defined on 2^X, having the following properties:

(a) $\mu^*(\varnothing) = 0$,

(b) $\mu^*(A) \leq \mu^*(B) \qquad \text{for } A \subset B$,

(c) $\mu^*(\bigcup_{n=1}^{\infty} E_n) \leq \sum_{n=1}^{\infty} \mu^*(E_n)$. ∎

Note that an outer measure is always nonnegative.

Examples

2.4. Suppose

$$\mu^*(E) = \begin{cases} 0, & E = \varnothing \\ 1, & E \neq \varnothing. \end{cases}$$

Then μ^* is an outer measure, which is not a measure on 2^X, if X has at least two points.

2.5. Let X be uncountable. Suppose

$$\mu^*(E) = \begin{cases} 0, & E \text{ countable} \\ 1, & E \text{ uncountable}. \end{cases}$$

Then μ^* is an outer measure.

More examples of outer measures will follow the next proposition, which outlines a method of constructing an outer measure on 2^X.

Proposition 2.4. Let $\mathscr{F}$ be a class of subsets of X containing the empty set such that for every $A \subset X$ there exists a sequence $(B_n)_{n=1}^{\infty}$ from $\mathscr{F}$ such that $A \subset \bigcup_{n=1}^{\infty} B_n$. Let τ be an extended real-valued function on $\mathscr{F}$ such that $\tau(\varnothing) = 0$ and $\tau(A) \geq 0$ for $A \in \mathscr{F}$. Then μ^* defined on 2^X by

$$\mu^*(A) = \inf\left\{ \sum_{n=1}^{\infty} \tau(B_n) : B_n \in \mathscr{F},\ A \subset \bigcup_{n=1}^{\infty} B_n \right\}$$

is an outer measure. ∎

Proof. First, clearly $\mu^*(\varnothing) = 0$. Next, if $A_1 \subset A_2$ and $A_2 \subset \bigcup_{n=1}^{\infty} B_n$, then $A_1 \subset \bigcup_{n=1}^{\infty} B_n$. This means that $\mu^*(A_1) \leq \mu^*(A_2)$.

Finally, let $E_n \subset X$ for each natural number n. Then if $\mu^*(E_n) = \infty$ for some n, $\mu^*(\bigcup_{n=1}^{\infty} E_n) \leq \sum_{n=1}^{\infty} \mu^*(E_n)$. Suppose that $\mu^*(E_n) < \infty$ for each n. Then given $\varepsilon > 0$, there exists $(B_{nm})_{m=1}^{\infty}$ from $\mathscr{F}$ such that

$$E_n \subset \bigcup_{m=1}^{\infty} B_{nm}$$

and

$$\sum_{m=1}^{\infty} \tau(B_{nm}) < \mu^*(E_n) + \varepsilon/2^n.$$

Now

$$\bigcup_{n=1}^{\infty} E_n \subset \bigcup_{n=1}^{\infty} \bigcup_{m=1}^{\infty} B_{nm}$$

and, therefore,

$$\mu^*\left(\bigcup_{n=1}^{\infty} E_n \right) \leq \sum_{n=1}^{\infty} \sum_{m=1}^{\infty} \tau(B_{nm}) < \sum_{n=1}^{\infty} \mu^*(E_n) + \varepsilon.$$

Since $\varepsilon > 0$ is arbitrary, the proposition follows. ∎

In the next four examples, $\mathscr{F}$, τ, and μ^* are defined as in Proposition 2.4. These examples will illustrate this proposition.

Examples

2.6. Let $X = Z_+$, $\mathscr{F} = \{\{x\} : x \in X\} \cup \varnothing$.

Suppose

$$\tau(E) = \begin{cases} 0, & E = \varnothing \\ 1, & E \neq \varnothing. \end{cases}$$

Then

$$\mu^*(A) = \begin{cases} \infty, & \text{if } A \text{ is infinite} \\ \text{the number of points in } A, & \text{if } A \text{ is finite.} \end{cases}$$

2.7. In the previous example, let $\mathscr{F} = \{X, \varnothing\}$ and $\tau(X) = 1$, $\tau(\varnothing) = 0$. Then

$$\mu^*(A) = \begin{cases} 1, & \text{if } A \neq \varnothing \\ 0, & \text{if } A = \varnothing. \end{cases}$$

2.8. Let $X = R$ and $\mathscr{F} = \{A: A \neq X, A \subset X\}$.
Suppose

$$\tau(A) = \begin{cases} 0, & \text{if } A = \varnothing \\ 1, & \text{if } A \neq \varnothing. \end{cases}$$

Then

$$\mu^*(B) = \begin{cases} 2, & \text{if } B = X \\ 1, & \text{if } B \neq \varnothing, \quad B \neq X \\ 0, & \text{if } B = \varnothing \end{cases}$$

2.9. *Lebesgue Outer Measure.* Let $X = R$ and $\mathscr{F}$ be the class of all left-open, right-closed intervals {finite or infinite of the form $(-\infty, a]$, $(a, b]$, or (a, ∞)}, including the empty set.

Let $\tau(\varnothing) = 0$ and $\tau(I) = l(I)$ for every interval $I \in \mathscr{F}$. Then the outer measure μ^*, induced by τ, on 2^R is called the Lebesgue outer measure on R and

$$\mu^*(A) = \inf\left\{ \sum_{n=1}^{\infty} l(I_n): A \subset \bigcup_{n=1}^{\infty} I_n, I_n \in \mathscr{F} \right\}.$$

Later in this section, it will follow that $\mu^*(I) = l(I)$ for every interval $I \in \mathscr{F}$ and the restriction of μ^* to a special class of subsets of R will yield a measure known as the Lebesgue measure on R, which will be an extension of the Lebesgue measure on intervals discussed in Section 2.2.

The purpose of introducing outer measures is to construct a measure on a σ-algebra. Outer measures are, in general, not measures on 2^X, as can be seen in Examples 2.4 and 2.5. But it so happens that an outer measure when restricted to a suitable class of subsets (usually called measurable sets) becomes a measure on a σ-algebra. The next theorem will demonstrate this fact. First, we need the following definition. (In what follows μ^* is always an outer measure.)

Definition 2.3. $E \subset X$ is called μ^*-measurable if for every $A \subset X$

$$\mu^*(A) = \mu^*(A \cap E) + \mu^*(A \cap E^c).$$

[This is equivalent to requiring only $\mu^*(A) \geq \mu^*(A \cap E) + \mu^*(A \cap E^c)$, since the converse inequality is obvious from the subadditive property of μ^*.] ∎

Notice that a μ^*-measurable set E splits no set A in such a way that μ^* fails to be additive on $\{A \cap E, A \cap E^c\}$. Since we are looking for a class of sets on which μ^* can act at least additively, the preceding definition seems to be meaningful to achieve that end. The next theorem justifies its meaningfulness.

Theorem 2.2. The class $\mathscr{B}$ of μ^*-measurable sets is a σ-algebra. Also $\bar{\mu}$, the restriction of μ^* to $\mathscr{B}$ is a measure. ▮

Proof. First, $\varnothing \in \mathscr{B}$; also $E \in \mathscr{B}$ imlies $E^c \in \mathscr{B}$, by the symmetry of Definition 2.3.

Next, we wish to show that $\mathscr{B}$ is closed under finite unions. Let $E \in \mathscr{B}$, $F \in \mathscr{B}$, and $A \subset X$. Since $F \in \mathscr{B}$,

$$\mu^*(A \cap E^c) = \mu^*(A \cap E^c \cap F) + \mu^*(A \cap E^c \cap F^c).$$

Also

$$A \cap (E \cup F) = (A \cap E) \cup (A \cap E^c \cap F)$$

and therefore

$$\mu^*(A \cap (E \cup F)) \leq \mu^*(A \cap E) + \mu^*(A \cap E^c \cap F),$$

so that

$$\begin{aligned}
\mu^*(A \cap (E \cup F)) &+ \mu^*(A \cap (E \cup F)^c) \\
&\leq \mu^*(A \cap E) + \mu^*(A \cap E^c \cap F) + \mu^*(A \cap E^c \cap F^c) \\
&= \mu^*(A \cap E) + \mu^*(A \cap E^c) \qquad \text{(using the first equality)} \\
&= \mu^*(A), \qquad \text{since } E \in \mathscr{B}.
\end{aligned}$$

This proves that $E \cup F \in \mathscr{B}$ and $\mathscr{B}$ is now an algebra.

Now we show that μ^* is a measure on the algebra $\mathscr{B}$. Let $(E_n)_{n=1}^\infty$ be a sequence of pairwise disjoint sets from $\mathscr{B}$. Then if $A \subset X$,

$$\mu^*(A \cap (E_1 \cup E_2)) = \mu^*(A \cap E_1) + \mu^*(A \cap E_2)$$

(using the fact that $E_2 \in \mathscr{B}$). Hence by induction (verify this), for each n

$$\mu^*\left[A \cap \left(\bigcup_{i=1}^n E_i\right)\right] = \sum_{i=1}^n \mu^*(A \cap E_i) \tag{2.2}$$

Now if $A = X$, for each n

$$\mu^*\left(\bigcup_{i=1}^\infty E_i\right) \geq \mu^*\left(\bigcup_{i=1}^n E_i\right) = \sum_{i=1}^n \mu^*(E_i).$$

This means that $\mu^*(\bigcup_{i=1}^\infty E_i) \geq \sum_{i=1}^\infty \mu^*(E_i)$ or μ^* is countably additive on $\mathscr{B}$ and hence a measure on $\mathscr{B}$.

Finally, we show that $\mathscr{B}$ is a σ-algebra. Let $(F_n)_{n=1}^\infty$ be a sequence of sets in $\mathscr{B}$. Write $G_1 = F_1$, $G_n = F_n - \bigcup_{i=1}^{n-1} F_i$ for $n > 1$. Then G_n's are pairwise disjoint sets in $\mathscr{B}$ and $\bigcup_{n=1}^\infty G_n = \bigcup_{n=1}^\infty F_n$. (Recall that $\mathscr{B}$ is already an algebra.) Let $A \subset X$. Then since $\bigcup_{n=1}^m G_n \in \mathscr{B}$, we have [using equation (2.2)]

$$\begin{aligned}\mu^*(A) &= \mu^*\left[A \cap \left(\bigcup_{n=1}^m G_n\right)\right] + \mu^*\left[A \cap \left(\bigcup_{n=1}^m G_n\right)^c\right] \\ &\geq \sum_{n=1}^m \mu^*(A \cap G_n) + \mu^*\left[A \cap \left(\bigcup_{n=1}^\infty G_n\right)^c\right]\end{aligned}$$

for every positive integer m. This means that

$$\begin{aligned}\mu^*(A) &\geq \sum_{n=1}^\infty \mu^*(A \cap G_n) + \mu^*\left[A \cap \left(\bigcup_{n=1}^\infty G_n\right)^c\right] \\ &\geq \mu^*\left[A \cap \left(\bigcup_{n=1}^\infty G_n\right)\right] + \mu^*\left[A \cap \left(\bigcup_{n=1}^\infty G_n\right)^c\right]\end{aligned}$$

and so

$$\bigcup_{n=1}^\infty G_n = \bigcup_{n=1}^\infty F_n \in \mathscr{B}. \qquad ▮$$

Remark 2.3. Note that $\mu^*(A) = 0$ implies A is μ^* measurable; and therefore the measure $\bar{\mu}$ (the restriction of μ^* on $\mathscr{B}$) has the following property: $E \in \mathscr{B}$, $\bar{\mu}(E) = 0$, and $F \subset E$ imply $F \in \mathscr{B}$. A measure having this property is called a *complete measure*. Completeness is a useful property for many technical considerations. (This will be seen when we discuss measurable functions in the next chapter.) Not every measure on a σ-algebra is complete (see Problems 2.4.3 and 2.4.4.) However, every measure on a σ-algebra can be completed in the following sense.

Proposition 2.5. Let μ be a measure on a σ-algebra $\mathscr{A}$. Suppose

$$\mathscr{A}_0 = \{A \cup B\colon A \in \mathscr{A}, B \subset C, C \in \mathscr{A}, \text{ and } \mu(C) = 0\}$$

and $\mu_0(A \cup B) = \mu(A)$ for $A \cup B \in \mathscr{A}_0$. Then $\mathscr{A}_0$ is a σ-algebra and μ_0 is a complete measure on $\mathscr{A}_0$, which is an extension of μ. ▮

The proof of this proposition is left to the reader.

The following examples will now illustrate Theorem 2.2.

Examples

2.10. Consider the outer measure μ^* in Example 2.7. Then if E is a nonempty proper subset of X and $A = X$,

$$\mu^*(A) = 1 \neq 2 = \mu^*(A \cap E) + \mu^*(A \cap E^c).$$

Hence $\{\emptyset, X\}$ is the class of all μ^*measurable subsets of X.

2.11. If μ^* is the counting measure on 2^X, then the class of all μ^*-measurable sets is 2^X.

If the class of μ^*-measurable sets turns out to be trivial or small, the measure $\bar{\mu}$ (induced by μ^*) does not seem to be very useful for many practical purposes. So it is important to know when sufficiently many sets will be μ^*-measurable. The next theorem provides an interesting result in this direction. We need first the following definition illustrated by an example.

Definition 2.4. Let (X, d) be a metric space and μ^* be an outer measure on 2^X such that $\mu^*(A \cup B) = \mu^*(A) + \mu^*(B)$ whenever $d(A, B) > 0$. Then μ^* is called a *metric outer measure.* ∎

Example 2.12. The *Lebesgue outer measure* is a metric outer measure. To show this, let $X = R$, $A \subset R$, $B \subset R$, $d(A, B) > 0$ (d being the usual real-line distance) and μ^*, Lebesgue outer measure defined in Example 2.9.

It suffices to show that $\mu^*(A \cup B) \geq \mu^*(A) + \mu^*(B)$. If $\mu^*(A \cup B) = \infty$, we are done. Suppose $\mu^*(A \cup B) < \infty$. Then given $\varepsilon > 0$, there exists a sequence of intervals $((a_n, b_n])_{n=1}^{\infty}$ such that $A \cup B \subset \bigcup_{n=1}^{\infty}(a_n, b_n]$ and $\sum_{n=1}^{\infty}(b_n - a_n) \leq \mu^*(A \cup B) + \varepsilon$. Now since $d(A, B) > 0$, there is a positive integer n_0 such that $d(A, B) > 1/n_0$. Now for each n, we can write $(a_n, b_n] = \bigcup_{i=1}^{k_n} I_{ni}$; where $(I_{ni})_{i=1}^{k_n}$ are pairwise disjoint left-open, right-closed intervals and $l(I_{ni}) \leq 1/n_0$. Then $A \cup B \subset \bigcup_{n=1}^{\infty}\bigcup_{i=1}^{k_n} I_{ni}$. Since $d(A, B) > 1/n_0$, each I_{ni} can only intersect one of the two sets A and B. Hence, some of the I_{ni}'s will cover A while others will cover B. Therefore,

$$\begin{aligned}\mu^*(A) + \mu^*(B) &\leq \sum_{n=1}^{\infty} \sum_{i=1}^{k_n} l(I_{ni}) \\ &= \sum_{n=1}^{\infty} (b_n - a_n) \leq \mu^*(A \cup B) + \varepsilon.\end{aligned}$$

Since $\varepsilon > 0$ is arbitrary, the claim is proven.

Theorem 2.3. Every Borel set in a metric space (X, d) is μ^*measurable with respect to an outer measure μ^* on 2^X if and only if μ^* is a metric outer measure. (Note that the class of Borel sets is the smallest σ-algebra containing the open sets of X. See Problem 1.3.13 in Chap. 1.) ∎

Proof. For the "only if" part, suppose every open set in X is μ^*-measurable. Let $d(A, B) = \delta > 0$. Then $A \subset G = \bigcup_{x \in A} \{y: d(x, y) < \delta\}$, which is an open set and therefore μ^*-measurable. Clearly $G \cap B = \varnothing$. Therefore,

$$\mu^*(A \cup B) = \mu^*\big((A \cup B) \cap G\big) + \mu^*\big((A \cup B) \cap G^c\big) = \mu^*(A) + \mu^*(B).$$

Hence μ^* is a metric outer measure.

For the "if" part, suppose μ^* is a metric outer measure. Since the class of μ^*-measurable sets is a σ-algebra by Theorem 2.2, it is sufficient to show that every closed set is μ^*-measurable. Let B be a closed set and A any set. We must show that

$$\mu^*(A) \geq \mu^*(A \cap B) + \mu^*(A \cap B^c).$$

To do this, we consider $B_n = \{x \in A \cap B^c: d(x, B) \geq 1/n\}$. Clearly, $B_n \subset B_{n+1} \subset A \cap B^c$ and $d(B_n, B) \geq 1/n$. Since

$$\mu^*(A) \geq \mu^*\big((A \cap B) \cup B_n\big) = \mu^*(A \cap B) + \mu^*(B_n),$$

we only need to show that

$$\lim_{n \to \infty} \mu^*(B_n) \geq \mu^*(A \cap B^c).$$

To show this, we observe that

$$A \cap B^c = \bigcup_{n=1}^{\infty} B_n = B_n \cup (B_{n+1} - B_n) \cup (B_{n+2} - B_{n+1}) \cup \cdots$$

Hence

$$\mu^*(A \cap B^c) \leq \mu^*(B_n) + \sum_{k=n}^{\infty} \mu^*(B_{k+1} - B_k).$$

If

$$\sum_{k=1}^{\infty} \mu^*(B_{k+1} - B_k) < \infty,$$

then clearly

$$\mu^*(A \cap B^c) \leq \lim_{n \to \infty} \mu^*(B_n)$$

and we are done.

Therefore, we suppose that

$$\sum_{k=1}^{\infty} \mu^*(B_{k+1} - B_k) = \infty.$$

We notice that, for $n \geq 2$,

$$B_1 \cup (B_3 - B_2) \cup \cdots \cup (B_{2n-1} - B_{2n-2}) \subset B_{2n}$$

and

$$(B_2 - B_1) \cup \cdots \cup (B_{2n} - B_{2n-1}) \subset B_{2n};$$

Also

$$d(B_{2n-1} - B_{2n-2}, B_{2n-3} - B_{2n-4}) > 0$$

as well as

$$d(B_{2n} - B_{2n-1}, B_{2n-2} - B_{2n-3}) > 0.$$

Since μ^* is a metric outer measure, this means that we have, for every positive integer n,

$$\begin{aligned} 2\mu^*(B_{2n}) &\geq \sum_{k=1}^{n-1} \mu^*(B_{2k+1} - B_{2k}) + \sum_{k=1}^{n} \mu^*(B_{2k} - B_{2k-1}) \\ &= \sum_{k=1}^{2n-1} \mu^*(B_{k+1} - B_k) \end{aligned}$$

and hence $\lim_{n\to\infty}\mu^*(B_{2n}) = \infty$. Therefore, $\lim_{n\to\infty}\mu^*(B_n) = \infty$ and

$$\mu^*(A \cap B^c) \leq \lim_{n\to\infty} \mu^*(B_n),$$

which was to be shown. ∎

Now we recall the definition of *Lebesgue outer measure* from Example 2.9. Since it is a very special (being perhaps the most important outer measure on R) outer measure on R, we will denote it, from now on, by m^*. Hence for $A \subset R$,

$$m^*(A) = \inf\left\{ \sum_{n=1}^{\infty} l(I_n) \colon A \subset \bigcup_{n=1}^{\infty} I_n, \text{ where the } I_n \text{ are intervals of the form } (-\infty, a], (a, b] \text{ or } (a, \infty) \right\}.$$

Let $\mathscr{A}$ be the algebra of finite disjoint union of these intervals and m_0 be the Lebesgue measure on $\mathscr{A}$, as discussed in Section 2.2. Then it is clear that

$$m^*(A) = \inf\left\{ \sum_{n=1}^{\infty} m_0(B_n) \colon A \subset \bigcup_{n=1}^{\infty} B_n, B_n \in \mathscr{A} \right\}.$$

In other words, the outer measure m^* is induced on 2^R by the measure m_0 on $\mathscr{A}$. By Theorem 2.2, m^* (restricted to the m^*-measurable sets, called the

Lebesgue-measurable sets and denoted by $\mathscr{M}$) is a measure. This is the well-known *Lebesgue measure* on R. We denote it by m. Now by Example 2.12, m^* is a metric outer measure, and by Theorem 2.3, $\mathscr{M}$ contains all Borel sets on R. (Caution: Not every set in $\mathscr{M}$ is a Borel set and not every set in 2^R is in $\mathscr{M}$. See Problem 2.3.3, Theorem 2.5, and Problem 2.4.3.) Since m^* is induced by m_0 and $\mathscr{M} \supset \mathscr{A}$, a natural question is whether m is an extension of m_0. That indeed this is so even in the general situation is demonstrated by the following important theorem.

Theorem 2.4. *Carathéodory Extension Theorem.* Let μ be a measure on an algebra[†] $\mathscr{A} \subset 2^X$. Suppose for $E \subset X$

$$\mu^*(E) = \inf\left\{ \sum_{i=1}^{\infty} \mu(E_i) \colon E \subset \bigcup_{i=1}^{\infty} E_i,\ E_i \in \mathscr{A} \right\}.$$

Then the following properties hold:

(a) μ^* is an outer measure.
(b) $E \in \mathscr{A}$ implies $\mu(E) = \mu^*(E)$.
(c) $E \in \mathscr{A}$ implies E is μ^*-measurable.
(d) The restriction $\bar{\mu}$ of μ^* to the μ^*-measurable sets is an extension of μ to a measure on a σ-algebra containing $\mathscr{A}$.
(e) If μ is σ-finite,[‡] then $\bar{\mu}$ is the only measure (on the smallest σ-algebra containing $\mathscr{A}$) that is an extension of μ. ∎

Proof. (a) This follows from Proposition 2.4.

(b) Let $E \in \mathscr{A}$. Clearly $\mu^*(E) \leq \mu(E)$. Conversely, given $\varepsilon > 0$, there exists $E_i \in \mathscr{A}$, $1 \leq i < \infty$ such that

$$E \subset \bigcup_{i=1}^{\infty} E_i \qquad \text{and} \qquad \sum_{i=1}^{\infty} \mu(E_i) \leq \mu^*(E) + \varepsilon.$$

But

$$E = \bigcup_{i=1}^{\infty} (E \cap E_i).$$

Therefore,

$$\mu(E) \leq \sum_{i=1}^{\infty} \mu(E \cap E_i).$$

This means that $\mu(E) \leq \mu^*(E) + \varepsilon$. Since $\varepsilon > 0$ is arbitrary, (b) follows.

† It is sufficient to let μ be a nonnegative countably additive extended real-valued set function and $\mathscr{A}$ be a ring of sets satisfying Proposition 2.4.

‡ Problem 2.3.20 describes what happens when μ is semifinite.

(c) Let $E \in \mathscr{A}$. To prove that E is μ^*measurable, it suffices to show the following:

$$\mu^*(A) \geq \mu^*(A \cap E) + \mu^*(A \cap E^c) \qquad \text{for } A \subset X. \tag{2.3}$$

Given $\varepsilon > 0$, there exists $A_i \in \mathscr{A}$, $1 \leq i < \infty$ such that

$$\sum_{i=1}^{\infty} \mu(A_i) \leq \mu^*(A) + \varepsilon \qquad \left(A \subset \bigcup_{i=1}^{\infty} A_i\right). \tag{2.4}$$

Now

$$A \cap E \subset \bigcup_{i=1}^{\infty} (A_i \cap E) \qquad \text{and} \qquad A \cap E^c \subset \bigcup_{i=1}^{\infty} (A_i \cap E^c).$$

Therefore,

$$\mu^*(A \cap E) \leq \sum_{i=1}^{\infty} \mu(A_i \cap E) \tag{2.5}$$

and

$$\mu^*(A \cap E^c) \leq \sum_{i=1}^{\infty} \mu(A_i \cap E^c) \tag{2.6}$$

From inequalities (2.4)–(2.6), the inequality (2.3) follows.

(d) This assertion follows from above and Theorem 2.2.

(e) Let $\mathscr{B}$ be the smallest σ-algebra containing $\mathscr{A}$ and μ_1 be another measure on $\mathscr{B}$ such that $\mu_1(E) = \mu(E)$ for $E \in \mathscr{A}$. We need to show the following:

$$\mu_1(A) = \bar{\mu}(A) \qquad \text{for } A \in \mathscr{B}. \tag{2.7}$$

Since μ is σ-finite, we can write $X = \bigcup_{i=1}^{\infty} E_i$, $E_i \in \mathscr{A}$, $E_i \cap E_j = \varnothing$ $(i \neq j)$ and $\mu(E_i) < \infty$, $1 \leq i < \infty$. For $A \in \mathscr{B}$,

$$\bar{\mu}(A) = \sum_{i=1}^{\infty} \bar{\mu}(A \cap E_i) \qquad \text{and} \qquad \mu_1(A) = \sum_{i=1}^{\infty} \mu_1(A \cap E_i).$$

So to prove equation (2.7), it is sufficient to show the following:

$$\mu_1(A) = \bar{\mu}(A) \qquad \text{for } A \in \mathscr{B}, \text{ whenever } \bar{\mu}(A) < \infty. \tag{2.8}$$

Let $A \in \mathscr{B}$ with $\bar{\mu}(A) < \infty$. Given $\varepsilon > 0$, there are $E_i \in \mathscr{A}$, $1 \leq i < \infty$, $A \subset \bigcup_{i=1}^{\infty} E_i$, and

$$\bar{\mu}\left(\bigcup_{i=1}^{\infty} E_i\right) \leq \sum_{i=1}^{\infty} \mu(E_i) < \bar{\mu}(A) + \varepsilon. \tag{2.9}$$

Since

$$\mu_1(A) \leq \mu_1\left(\bigcup_{i=1}^{\infty} E_i \right) \leq \sum_{i=1}^{\infty} \mu_1(E_i) = \sum_{i=1}^{\infty} \mu(E_i),$$

it follows from inequality (2.9) that

$$\mu_1(A) \leq \bar{\mu}(A). \tag{2.10}$$

Now considering the sets E_i from inequality (2.9), $F = \bigcup_{i=1}^{\infty} E_i \in \mathscr{B}$ and so F is μ^*-measurable. Since $A \subset F$, $\bar{\mu}(F) = \bar{\mu}(A) + \bar{\mu}(F - A)$ or $\bar{\mu}(F - A) = \bar{\mu}(F) - \bar{\mu}(A) < \varepsilon$ [from inequality (2.9)]. Since $\mu_1(E) = \bar{\mu}(E)$ for each $E \in \mathscr{A}$, $\mu_1(F) = \bar{\mu}(F)$. (Why?) Then

$$\bar{\mu}(A) \leq \bar{\mu}(F) = \mu_1(F) = \mu_1(A) + \mu_1(F - A) \leq \mu_1(A) + \bar{\mu}(F - A) \tag{2.11}$$

[by inequality (2.10), since inequality (2.10) is true if A is replaced by any set in $\mathscr{B}$ with finite $\bar{\mu}$-measure]. Then from inequality (2.11) it follows that $\bar{\mu}(A) \leq \mu_1(A)$. This inequality along with inequality (2.10) completes the proof. ∎

We are now in a position to summarize the main properties of m, the Lebesgue measure on R:

(1) m is a measure on the σ-algebra of Lebesgue-measurable sets, which includes all Borel sets properly. (An example of a Lebesgue-measurable set that is not a Borel set is given in the next section. See also Problem 2.3.13.) Also not every set on R is Lebesgue measurable, which is the subject matter of the next section.

(2) $m(I) = l(I)$ for every interval I (open, half-open, or closed), since m is an extension of m_0 (Theorem 2.4) and $m(\{x\}) = 0$ for every singleton $\{x\}$ (Problem 2.3.4).

(3) m is translation invariant (Problems 2.3.5 and 2.3.6), i.e., for each Lebesgue-measurable set $A \subset R$ and $x \in R$, $A + x$ is Lebesgue measurable and $m(A + x) = m(A)$.

(4) The Lebesgue-measurable sets can be approximated in Lebesgue measure by open sets containing them and by closed sets contained in them. (See Problem 2.3.9.)

If we now recall Problem 2.2.5, we see that the Lebesgue measure m_0 on intervals is only a special case of the more general (and very useful in probability theory) measure called the Lebesgue–Stieltjes measure m_F on intervals induced by a monotonic increasing function F that is continuous on the right. Moreover

$$m_F(a, b] = F(b) - F(a),$$

$$m_F(a, \infty) = \lim_{x\to\infty} F(x) - F(a),$$

and

$$m_F(-\infty, a] = F(a) - \lim_{x\to-\infty} F(x).$$

Now this m_F is a measure on the algebra $\mathscr{A}$ of finite disjoint union of the above half-open intervals. Following the Extension Theorem, Theorem 2.4, we can form m_F^*, the outer measure induced by m_F. Consequently, this outer measure yields a unique measure $\bar{m}_F$ (since m_F is σ-finite on $\mathscr{A}$) on the Borel sets of R, which is an extension of m_F. By Proposition 2.5, the Borel measure $\bar{m}_F$ (that means $\bar{m}_F$ restricted to the Borel sets) can be completed. This completion is called the Lebesgue–Stieltjes measure on R induced by F.

Problems

2.3.1. Let $X = Z^+$. For $A \subset Z^+$, let $a = \sup A$. Suppose

$$\mu^*(A) = \begin{cases} \dfrac{a}{(a+1)}, & \text{if } A \text{ is finite} \\ 0, & \text{if } A = \varnothing \\ 1, & \text{if } A \text{ is infinite.} \end{cases}$$

Show that μ^* is an outer measure. Find the μ^*-measurable sets.

2.3.2. Suppose μ^* is an outer measure on 2^X, and for every $A \subset X$ there is a μ^*-measurable set $E \supset A$ such that $\mu^*(A) = \mu^*(E)$. (E is called a *μ^*-measurable cover* of A.) Show that

(a) if $A_n \subset A_{n+1} \subset X$, then $\mu^*(\bigcup_{n=1}^{\infty} A_n) = \lim_{n\to\infty} \mu^*(A_n)$;

(b) if $B_n \subset X$, then $\mu^*(\underline{\lim}_n B_n) \leq \underline{\lim}_n \mu^*(B_n)$.

2.3.3. Suppose μ^* is an outer measure on 2^X, where X is a topological space. Show that every Borel set is μ^*-measurable if and only if $\mu^*(A \cup B) = \mu^*(A) + \mu^*(B)$ whenever $A \cap \bar{B}$ is empty.

2.3.4. Show that $m^*(A) = 0$ if A is countable. (Recall that m^* is the Lebesgue outer measure.)

2.3.5. Show that m^* is translation invariant, i.e., for $A \subset R$ and $x \in R$, $m^*(A) = m^*(A + x)$.

2.3.6. For each Lebesgue-measurable set $E \subset R$, show that $E + x$ is also Lebesgue measurable for $x \in R$. (Hint: Use Problem 2.3.5.)

2.3.7. (a) Show that the Cantor set has Lebesgue measure zero. (Hint: Compute the measure of its complement in [0, 1].)

(b) Show that for every $1 > \varepsilon > 0$, there exists a nowhere dense perfect† set in [0, 1] that has Lebesgue measure greater than $1 - \varepsilon$. (Hint: Construct the set in the same manner as the usual Cantor set except that each of the intervals removed at the nth step has length $\varepsilon \cdot 3^{-n}$.)

(c) Show that there is a set of Lebesgue measure zero that is of second category in [0, 1].

2.3.8. Show that for $A \subset R$

$$m^*(A) = \inf\left\{ \sum_{n=1}^{\infty} l(I_n)\colon\ A \subset \bigcup_{n=1}^{\infty} I_n,\ I_n\text{'s are open intervals}\right\}.$$

(Recall that in the definition of m^* in Example 2.9, I_n's are left-open right-closed intervals.)

2.3.9. (i) Suppose $E \subset R$ and $m^*(E) < \infty$. Then show that the following conditions are equivalent:

(a) E is Lebesgue measurable

(b) Given $\varepsilon > 0$, there is an open set $0 \supset E$ with $m^*(0 - E) < \varepsilon$.

(c) Given $\varepsilon > 0$, there is a finite union U of open intervals such that $m^*(U \triangle E) < \varepsilon$.

(ii) For arbitrary $E \subset R$, each of the first two statements in (i) is equivalent to each of the following statements.

(d) Given $\varepsilon > 0$, there is a closed set $F \subset E$ with $m^*(E - F) < \varepsilon$.

(e) There is a G_δ set‡ $G \supset E$ such that $m^*(G - E) = 0$.

(f) There is a F_σ set§ $F \subset E$ such that $m^*(E - F) = 0$.

2.3.10. Show that every set on R has a m^*-measurable cover (Problem 2.3.2) and therefore $m^*(\bigcup_{n=1}^{\infty} A_n) = \lim_{k\to\infty} m^*(\bigcup_{n=1}^{k} A_n)$, $A_n \subset R$.

2.3.11. Show that the σ-finiteness assumption is essential in Theorem 2.4 for the uniqueness of the extension of μ on the smallest σ-algebra containing $\mathscr{A}$. {Hint: Let $X = (0, 1]$, and let $\mathscr{A}$ be the algebra of all finite unions of intervals of the form $(a, b] \subset (0, 1]$ and $\mu(A) = \infty$ if $A \neq \varnothing$, $= 0$ if $A = \varnothing$.}

2.3.12. (a) Suppose μ is a finite measure on the Borel sets of R. Then show that for each Borel set B

$$\begin{aligned}\mu(B) &= \sup\{\mu(K)\colon\ K \subset B,\ K \text{ compact}\}\\ &= \inf\{\mu(V)\colon\ B \subset V,\ V \text{ open}\}.\end{aligned}$$

† For the definition and properties of perfect sets, see Problem 2.3.17.

‡ A G_δ set is a countable intersection of open sets.

§ A F_σ set is a countable union of closed sets.

(Hint: The class $[B: \mu(B) = \sup\{\mu(K): K \subset B, K \text{ compact}\}]$ is a monotone class containing all half-open intervals. Use Theorem 1.4.)

(b) If $A_n = \{x \in (0, 1)$: the nth digit in the binary expansion of x is $1\}$, then find $m(\bigcap_{i=1}^{k} A_{n_i})$ for $n_1 < n_2 < \cdots < n_k$.

2.3.13. Show that there exists a Lebesgue-measurable set on R that is not a Borel set. (Hint: Every subset with Lebesgue outer measure zero is Lebesgue measurable. Since c is the cardinality of the Cantor set, there are 2^c Lebesgue-measurable sets, whereas there are c Borel sets. See Problem 1.3.13.)

2.3.14. Let φ be an isometry of R into R (i.e., $|\varphi(x) - \varphi(y)| = |x - y|$, $\forall x, y \in R$). Then show that

(a) $\varphi(x) = x + d(\forall x \in R)$ or $\varphi(x) = -x + d(\forall x \in R)$, for some $d \in R$.

(b) If $A \subset R$ is Lebesgue measurable, then $\varphi(A)$ is Lebesgue measurable and $m(A) = m(\varphi(A))$.

2.3.15. Construct a real-valued function on [0, 1] whose set of discontinuities has Lebesgue measure zero but has an uncountable intersection with every open subinterval. {Hint: Let K_1 be the Cantor set $\subset [0, 1]$ of Lebesgue measure zero. Let K_2 be the union of similar Cantor sets constructed in each of the intervals of $[0, 1] - K_1$. The sequence (K_n) is constructed inductively, and let $K = \bigcup_{n=1}^{\infty} K_n$. Define $f(x) = 2^{-n}$ for $x \in K_n$, $= 0$ for $x \notin K$.}

2.3.16. Suppose E is a Lebesgue measurable subset of R, and for each x in a dense set of R, $m(E \triangle (E + x)) = 0$. Prove that $m(E) = 0$ or $m(R - E) = 0$.

2.3.17. *Perfect Sets and Measures on R.* A set $A \subset R$ is called *perfect* if it is closed and every $x \in A$ is a limit point of A. Verify the following assertions:

(i) Every (uncountable) closed set of real numbers is the union of a perfect set and an at most countable set.

(ii) A set of real numbers is perfect if and only if it is the complement of an at most countable number of disjoint open intervals, no two of which have a common endpoint.

(iii) Every nonempty perfect set of real numbers is uncountable.

(iv) The Cantor sets (in Problem 2.3.7) are perfect.

★ (v) Every perfect set of real numbers contains a perfect subset of Lebesgue measure zero.

★ (vi) Every closed set of positive Lebesgue measure contains a perfect subset of Lebesgue measure zero.

(vii) Let μ be a nonzero measure defined on the Borel sets of R such that $\mu([-n, n]) < \infty$ for every positive integer n. Let $S_\mu = \{x \in R: \mu(V_x) > 0$

for every open set V_x containing $x\}$. Then (a) S_μ is closed, (b) $\mu(R - S_\mu) = 0$, and (c) S_μ is perfect, when points have zero measure. [Note that $\mu(R - S_\mu) = \sup\{\mu(K)\colon K \subset R - S_\mu,\ K \text{ compact}\}$ by Problem 2.3.12.]

2.3.18. *Representation of a Finite Borel Measure in Terms of Its Distribution Function and the Lebesgue Measure* (J. J. Higgins). Let μ be a finite measure on the Borel sets of R. Let f be a real-valued, bounded, nondecreasing function defined on R such that it is continuous from the right and $f(x) \to 0$ as $x \to -\infty$. Such a function is sometimes called a *distribution function.* (Compare Problem 2.2.4.) Show the following:

(i) If f is a distribution function as above, then for any Borel set A, $f(A)$ is a Borel set; also $m(f(A)) = \lambda(A)$ (m being the Lebesgue measure) defines a finite Borel measure. [Hint: $A \cap B = \varnothing$ means that $f(A) \cap f(B)$ is at most countable.]

(ii) Let μ be a finite Borel measure. Then $f(x) = \mu((-\infty, x])$ is a distribution function. If

$$j(x_i) = f(x_i) - f(x_i-),$$

where the (x_i) are the discontinuities of f and if

$$f_c(x) = f(x) - \sum_{x_i \le x} j(x_i),$$

then

$$\mu(A) = m(f_c(A)) + \sum_{x_i \in A} j(x_i)$$

for all Borel sets A.

2.3.19. *Completion of μ and the Outer Measure μ^*.* Let μ be a measure on a σ-algebra $\mathscr{A}$ of subsets of X, and let $\bar{\mu}$ be its completion on the σ-algebra $\bar{\mathscr{A}}$. Let $\mathscr{B}$ denote the μ^*-measurable subsets of X. Then $\mathscr{A} \subset \bar{\mathscr{A}} \subset \mathscr{B}$. Show that $\bar{\mathscr{A}} = \mathscr{B}$ when μ is σ-finite, and that this equality need not be valid otherwise.

2.3.20. *Semifiniteness and the Extension Theorem.* Let μ be a semifinite measure on an algebra $\mathscr{A}$ of subsets of X. Show that there is always an extension of μ to a semifinite measure on $\sigma(\mathscr{A})$, and the extension of μ to a measure on $\sigma(\mathscr{A})$ is unique if and only if μ^* is semifinite on $\sigma(\mathscr{A})$. Give an example of a semifinite measure on an algebra $\mathscr{A}$ with infinitely many semifinite extensions on $\sigma(\mathscr{A})$. [Hint: Take X = the reals, Q = the rationals, and $\mathscr{A} = \{A \subset X \mid \text{either } A \text{ or } X - A \text{ is finite}\}$. Let $\mu(A) = \text{card}(A \cap Q)$. For any nonnegative real number s, define ν_s on $\sigma(\mathscr{A})$ by

$$\nu_s(A) = \begin{cases} \text{card}(A \cap Q), & \text{if } A \cap (X - Q) \text{ is at most countable}, \\ s + \text{card}(A \cap Q), & \text{if } (X - A) \cap (X - Q) \text{ is at most countable}. \end{cases}$$

Then ν_s is a semifinite extension of μ on $\sigma(\mathscr{A})$.]

2.4. Non-Lebesgue-Measurable Sets and Inner Measure

In this section, we show the existence of a *non-Lebesgue-measurable set* on R by invoking the Axiom of Choice. Such sets, indeed, rarely arise naturally in practical situations. R. Solovay has recently shown in a 1970 paper† that the existence of such a set on R implies the Axiom of Choice. Often nonmeasurable sets are used to construct different counterexamples to understand different aspects of the theory well. For instance, using the existence of a nonmeasurable set one can construct a Lebesgue-measurable set on R that is not a Borel set (Problem 2.4.3).

We will also introduce in this section the concept of inner measure. This concept has some historical significance since the concept of measurability was originally characterized in terms of both inner and outer measure. Aside from this it is also useful for the purpose of extending a measure on an algebra to an algebra containing the given algebra and *any* given set. Using this concept, we will also present a proper simple extension of the Lebesgue measure that is complete and translation invariant.

First we give an example of a non-Lebesgue-measurable set.

Theorem 2.5. Let A be a Lebesgue-measurable set with $m(A) > 0$. Then there exists $E \subset A$ such that E is not Lebesgue measurable. ∎

Proof. Since $A = \bigcup_{n=1}^{\infty}(A \cap [-n, n])$, there is a positive integer n_0 such that

$$m(A \cap [-n_0, n_0]) > 0.$$

We write $B = A \cap [-n_0, n_0]$. Let $x \in B$ and $B_x = \{y \in B\colon y - x \text{ is rational}\}$. Then $B = \bigcup_{x\in B} B_x$. For $x_1, x_2 \in B$, $B_{x_1} = B_{x_2}$ if $x_1 - x_2$ is rational; otherwise, $B_{x_1} \cap B_{x_2} = \emptyset$. By the Axiom of Choice, there exists a set $E \subset B$ such that E contains exactly one point from each of the distinct sets $\{B_x\}$. We claim that E is not Lebesgue-measurable.

To prove this claim, let $(r_n)_{n=1}^{\infty}$ be the rationals in $[-2n_0, 2n_0]$. Then $E + r_n$ and $E + r_l$ are disjoint (if $n \neq l$); for if $r_n \neq r_l$, $e_n, e_l \in E$ and $e_n + r_n = e_l + r_l$, then $e_n \neq e_l$ and so $e_n - e_l$ is irrational, which is a contradiction. Also $\bigcup_{n=1}^{\infty}(E + r_n) \subset [-3n_0, 3n_0]$ and $B \subset \bigcup_{n=1}^{\infty}(E + r_n)$; for if $x \in B$, then there is some $e \in E$ such that $x \in B_e$ or $x - e = r_n$ (for some n), a rational in $[-2n_0, 2n_0]$. If E is Lebesgue measurable, $E + r_n$ is so for each n, and therefore $0 < m(B) \leq \sum_{n=1}^{\infty} m(E + r_n) = \sum_{n=1}^{\infty} m(E) \leq 6n_0$ (m being

† R. Solovay, *Ann. of Math.* (2) 92, 1–56 (1970).

translation invariant). If $m(E) = 0$, then $m(B) = 0$, which is not possible. If $m(E) > 0$, then $\sum_{n=1}^{\infty} m(E) = \infty \leq 6n_0$, which is an absurdity. Hence E is not Lebesgue measurable. ∎

Remark 2.4. It is clear from the proof of Theorem 2.5 that if, instead of m, we consider an arbitrary nonzero measure μ that is translation invariant, defined on a σ-algebra containing the Borel sets of R, and finite on finite intervals, then its domain cannot contain the set E. In other words, it is impossible to define a translation-invariant, countably additive nonzero finite measure on the class of all subsets on [0, 1]. However, there exists a finitely additive, translation-invariant measure on the class of all subsets of [0, 1] such that the measure of every subinterval in [0, 1] is its length. Such a measure (which is finitely additive and congruence invariant, i.e., two sets that are congruent or isometric have the same measure) also exists on the class of all subsets of $[0, 1] \times [0, 1]$ in R^2. This was first shown by S. Banach.[†] We will sketch the construction on [0, 1] (as an application of the famous Hahn–Banach Theorem) in the chapter on Banach spaces. Since in R^n, in general, the group of isometries becomes increasingly larger with increase in the number of dimensions, it is natural to be less hopeful of finding finitely additive, congruence-invariant measures on the class of all subsets of a general n-dimensional unit cube. Indeed, it has been shown by F. Hausdorff that there does not exist any such measure for $n > 2$. In the context of different extensions of Lebesgue measure to larger classes of sets, we would also mention the works of S. Kakutani and J. C. Oxtoby.[‡] They obtained countably additive, translation-invariant extensions of Lebesgue measure to very large σ-algebras containing properly the class of all Lebesgue-measurable sets.

Actually it can be shown (via the continuum hypothesis)[§] that it is *impossible* to have a finite nonzero measure that is zero for points and is defined on all subsets of a set of cardinality c. This follows immediately from a well-known theorem of S. M. Ulam.[||]

† S. Banach, *Fund. Math.* **4**, 7–33 (1923).

‡ S. Kakutani and J. C. Oxtoby, *Ann. of Math.* **52**(2), 580–590 (1950).

§ The continuum hypothesis is the assertion that each infinite subset of R is either countable or of cardinal number c. P. J. Cohen has shown recently that this hypothesis is independent of the Zermelo–Fraenkel axioms of set theory. See P. J. Cohen, *Proc. Natl. Acad. Sci. USA* **50**, 1143–1148 (1963) and **51**, 105–110 (1964).

|| S. M. Ulam, *Fund. Math.* **16**, 141–150 (1930).

• **Theorem 2.6.** *The Ulam Theorem.* Let Ω be the first uncountable ordinal and $X = [0, \Omega)$. Then a finite measure μ defined for all subsets of X and zero for points must be a zero measure. ▮

Proof.[†] Suppose that μ is a finite measure defined for all subsets of X and zero for points. Let $y \in X$ and $A_y = \{x: x < y\}$. Then A_y is countable and there is a one-to-one correspondence $f(x, y)$ from A_y onto the natural numbers. Let us define

$$B_{x,n} = \{y: x < y, f(x, y) = n\}$$

for each $x \in X$ and each natural number n. Then these sets satisfy

(i) $B_{x,n} \cap B_{z,n} = \varnothing$ if $x \neq z$,
(ii) $X - \bigcup_{n=1}^{\infty} B_{x,n}$ is countable for each $x \in X$.

We establish only (ii). For $x \in X$, $y > x$ implies $f(x, y) = n$ for some n, and hence $\{y: x < y\} \subset \bigcup_{n=1}^{\infty} B_{x,n}$. Since $\{y: y \leq x\}$ is countable, (ii) follows.

By (i) and since $\mu(X) < \infty$, for each natural number n $\mu(B_{x,n}) > 0$ for at most countably many x. Therefore since X is uncountable, there is $x \in X$ such that $\mu(B_{x,n}) = 0$ for all natural numbers n. By (ii), $\mu(X) = 0$ if μ is zero for points. ▮

Next in this section we will introduce and discuss *inner measures* with a view to extending a measure on an algebra $\mathscr{A}$ to a measure on an algebra containing $\mathscr{A}$ and *any* given set E. Through inner measures, we will also obtain a translation-invariant proper extension of Lebesgue measure.

• **Definition 2.5.** Let μ be a measure on an algebra $\mathscr{A}$ and μ^* the induced outer measure as in Theorem 2.4. Then the inner measure μ_* is defined by

$$\mu_*(E) = \sup\{\mu(A) - \mu^*(A - E)\},$$

where the supremum is taken over all sets $A \in \mathscr{A}$ with $\mu^*(A - E) < \infty$. ▮

It follows easily from the definition that

$$\mu_*(E) \leq \mu^*(E), \tag{2.12}$$

$$E \subset F \Rightarrow \mu_*(E) \leq \mu_*(F), \tag{2.13}$$

$$E \in \mathscr{A} \Rightarrow \mu_*(E) = \mu^*(E) = \mu(E). \tag{2.14}$$

[†] Another proof of this theorem is indicated in Problem 3.4.13 in Chap. 3.

But the most interesting and the less obvious properties of the inner measure are perhaps the following.

• **Remarks** *Properties of the Inner Measure.*

2.5. For certain sets the inner measure has a more convenient expression, as in equation (2.15) below.

$$\mu_*(E) = \mu(A) - \mu^*(A - E),$$
$$\text{whenever } E \subset A,\ A \in \mathscr{A}, \text{ and } \mu^*(A - E) < \infty. \tag{2.15}$$

If $B \in \mathscr{A}$, then for each $C \subset X$, the set $B \cap C \subset B$ and it follows from equation (2.15) that

$$\mu(B) = \mu_*(B \cap C) + \mu^*(B - C). \tag{2.16}$$

Proof of Equation (2.15). Let $E \subset A$, $A \in \mathscr{A}$, and $\mu^*(A - E) < \infty$. For $B \in \mathscr{A}$ and $\mu^*(B - E) < \infty$, using the μ^*-measurability of $A \cup B - A$ and the set equality $A \cup B - E = (A \cup B - A) \cup (A - E)$, we have

$$\mu^*(A \cup B - E) = \mu(A \cup B - A) + \mu^*(A - E).$$

This implies that

$$\begin{aligned}\mu(A) - \mu^*(A - E) &= \mu(A \cup B - A) + \mu(A) - \mu^*(A \cup B - E)\\ &= [\mu(A \cup B - A) + \mu(A \cap B)]\\ &\quad + [\mu(A - B) + \mu^*(B - E)\\ &\quad - \mu^*(A \cup B - E)] - \mu^*(B - E)\\ &\geq \mu(B) - \mu^*(B - E).\end{aligned}$$

[Note that $A \cup B - E \subset (A - B) \cup (B - E)$.] The equality (2.15) now follows easily.

2.6. Every set E has a measurable kernel C. We clarify this below.

If $\mathscr{A}$ is a σ-algebra, then for each $E \subset X$ there exists $C \subset E$ and $C \in \mathscr{A}$ such that $\mu_*(E) = \mu(C)$. (2.17)

Proof of Equation (2.17). First, let $\mu_*(E) < \infty$. There exist $A_n \in \mathscr{A}$ with $\mu^*(A_n - E) < \infty$ and $\mu(A_n) - \mu^*(A_n - E) > \mu_*(E) - 1/n$. By the definition of μ^*, there exist $B_n \in \mathscr{A}_{\sigma\delta}$ $(= \mathscr{A})$ such that $A_n - E \subset B_n$ and $\mu^*(A_n - E) = \mu(B_n)$. Since $A_n - B_n \subset E$ and $A_n \subset (A_n - B_n) \cup B_n$, we have

$$\mu(A_n - B_n) \geq \mu(A_n) - \mu(B_n) > \mu_*(E) - 1/n.$$

Let $C = \bigcup_{n=1}^{\infty}(A_n - B_n)$. Then $C \subset E$ and $\mu(C) = \mu_*(E)$, by equations (2.12)–(2.14). Clearly C is the measurable kernel of E. In case $\mu_*(E) = \infty$, equation (2.17) follows easily if we replace above the expression $\mu_*(E) - 1/n$ by n.

2.7. For Lebesgue measure m, m_* has the following approximation property: $m_*(A) = \sup\{m(F)\colon F \text{ closed} \subset A\}$.

This follows from Remark 2.6 and Problem 2.3.9(d).

2.8. The inner measure is countably additive in the following sense. For any set E and a disjoint sequence of sets A_n in $\mathscr{A}$,

$$\mu_*\left(E \cap \bigcup_{n=1}^{\infty} A_n\right) = \sum_{n=1}^{\infty} \mu_*(E \cap A_n).$$

Proof of Remark 2.8. We may and do assume that $E \subset \bigcup_{n=1}^{\infty} A_n$. Then $E = \bigcup_{n=1}^{\infty}(E \cap A_n)$. First we show that $\mu_*(E) \leq \sum_{n=1}^{\infty}\mu_*(E \cap A_n)$. Recalling the definition of μ_*, we consider $B \in \mathscr{A}$ with $\mu^*(B - E) < \infty$. Then since $\bigcup_{n=1}^{\infty} A_n$ is μ^* measurable, we have

$$\begin{aligned} \mu(B) &= \mu^*\left[B \cap \left(\bigcup_{n=1}^{\infty} A_n\right)\right] + \mu^*\left(B - \bigcup_{n=1}^{\infty} A_n\right), \\ \mu^*(B - E) &= \mu^*\left[(B - E) \cap \left(\bigcup_{n=1}^{\infty} A_n\right)\right] + \mu^*\left[(B - E) - \left(\bigcup_{n=1}^{\infty} A_n\right)\right]. \end{aligned} \tag{2.18}$$

From equation (2.18), we have

$$\begin{aligned} \mu(B) - \mu^*(B - E) &= \mu^*\left[B \cap \left(\bigcup_{n=1}^{\infty} A_n\right)\right] - \mu^*\left[(B - E) \cap \left(\bigcup_{n=1}^{\infty} A_n\right)\right] \\ &= \sum_{n=1}^{\infty} \mu(B \cap A_n) - \mu^*(B \cap A_n \cap E^c), \quad \text{by Thm. 2.2.} \\ &\leq \sum_{n=1}^{\infty} \mu_*(E \cap A_n), \qquad \text{using the definition of } \mu_*. \end{aligned}$$

It follows that $\mu_*(E) \leq \sum_{n=1}^{\infty}\mu_*(E \cap A_n)$. To prove the converse inequality, we now consider the definition of $\mu_*(E \cap A_n)$ for each n. Let $B_n \in \mathscr{A}$ and $\mu^*(B_n - (E \cap A_n)) < \infty$. Then since $B_n - A_n$ is μ^*-measurable and

$$B_n - (E \cap A_n) = (B_n - A_n) \cup (B_n \cap A_n \cap E^c),$$

we have

$$\mu^*(B_n - (E \cap A_n)) = \mu(B_n - A_n) + \mu^*(B_n \cap A_n \cap E^c)$$

and therefore

$$\begin{aligned}\sum_{n=1}^{k} \mu(B_n) - \mu^*(B_n - (E \cap A_n)) &= \sum_{n=1}^{k} \mu(B_n \cap A_n) - \mu^*(B_n \cap A_n \cap E^c) \\ &= \mu\left[\bigcup_{n=1}^{k} (B_n \cap A_n)\right] \\ &\quad - \mu^*\left[\left(\bigcup_{n=1}^{k} (B_n \cap A_n)\right) \cap E^c\right] \\ &\leq \mu_*(E). \qquad (2.19)\end{aligned}$$

By taking "sup" over all such choices of B_n, we have

$$\sum_{n=1}^{k} \mu_*(E \cap A_n) \leq \mu_*(E) \qquad \text{for all } k. \qquad (2.20)$$

The rest of the proof is clear. ∎

2.9. If $\mu^*(E) < \infty$, then E is μ^*-measurable if and only if $\mu_*(E) = \mu^*(E)$.

Proof of Remark 2.9. First we prove the "if" part. By Remark 2.6, there is a measurable kernel $C \in \mathscr{A}$, $C \subset E$ such that $\mu_*(E) = \mu(C)$. If $\mu^*(E) = \mu_*(E)$, then it follows that $\mu(C) = \mu^*(E) = \mu^*(E \cap C) + \mu^*(E - C)$ implying $\mu^*(E - C) = 0$. Thus $E - C$ is μ^*-measurable and so is $E = C \cup (E - C)$. To prove the "only if" part, let E be μ^*-measurable and $\mu^*(E) < \infty$. Then there is a disjoint sequence of sets A_n in $\mathscr{A}$ such that $E \subset \bigcup_{n=1}^{\infty} A_n$ and $\mu^*(\bigcup_{n=1}^{\infty} A_n) < \infty$. By Remark 2.8, we have

$$\begin{aligned}\mu_*(E) = \sum_{n=1}^{\infty} \mu_*(E \cap A_n) &= \sum_{n=1}^{\infty} [\mu(A_n) - \mu^*(A_n - E)], \text{ by equation (2.15)} \\ &= \sum_{n=1}^{\infty} \mu^*(A_n \cap E) = \mu^*(E).\end{aligned}$$

∎

We note that in Remark 2.6 it is necessary that $\mathscr{A}$ be a σ-algebra. The reason is that a Lebesgue-measurable set of positive Lebesgue measure, which is nowhere dense in the reals, cannot have a measurable kernel in the algebra generated by the half-open intervals $(a, b]$. We also remark that if μ is σ-finite and E is μ^*-measurable, then $\mu_*(E) = \mu^*(E)$. The converse is of course not true. For, if A is not Lebesgue measurable and $A \subset (0, 1)$, then $E = A \cup (1, \infty)$ is not Lebesgue measurable, despite having the same Lebesgue inner and outer measure.

Now we state and prove the promised extension theorem.

• **Theorem 2.7.** Let μ be a measure on an algebra $\mathscr{A}$ of subsets of X and let E be any subset of X. Let $\mathscr{B}$ be the algebra generated by $\mathscr{A}$ and E. Then if for $B \in \mathscr{B}$

$$\bar{\mu}(B) = \mu^*(B \cap E) + \mu_*(B - E) \tag{2.21}$$

and

$$\underline{\mu}(B) = \mu_*(B \cap E) + \mu^*(B - E), \tag{2.22}$$

then $\bar{\mu}$ and $\underline{\mu}$ are measures on $\mathscr{B}$ such that $\bar{\mu}(A) = \underline{\mu}(A) = \mu(A)$ for $A \in \mathscr{A}$. ▮

Proof. First we observe that

$$\mathscr{B} = \{(A \cap E) \cup (B \cap E^c)\colon\ A,\ B \in \mathscr{A}\}.$$

By equation (2.16), $\bar{\mu}(A) = \underline{\mu}(A) = \mu(A)$ for $A \in \mathscr{A}$. We only need to establish that $\bar{\mu}$ and $\underline{\mu}$ are measures on $\mathscr{A}$.

Suppose $D_i = (A_i \cap E) \cup (B_i \cap E^c)$, where $A_i,\ B_i \in \mathscr{A},\ D_i \cap D_j = \varnothing\ (i \neq j)$, and $\bigcup_{i=1}^\infty D_i \in \mathscr{B}$. Now

$$\bigcup_{i=1}^\infty D_i = \left[\left(\bigcup_{i=1}^\infty A_i\right) \cap E\right] \cup \left[\left(\bigcup_{i=1}^\infty B_i\right) \cap E^c\right]. \tag{2.23}$$

Clearly for $i \neq j$, $A_i \cap A_j \cap E = \varnothing$ and $B_i \cap B_j \cap E^c = \varnothing$. Let us write $P_1 = A_1$, $Q_1 = B_1$ and for $n > 1$,

$$P_n = A_n - \bigcup_{i=1}^{n-1} A_i, \qquad Q_n = B_n - \bigcup_{i=1}^{n-1} B_i.$$

Then $P_n, Q_n \in \mathscr{A}$ and $P_n \cap E = A_n \cap E$, $Q_n \cap E^c = B_n \cap E^c$; also $P_i \cap P_j = Q_i \cap Q_j = \varnothing$ for $i \neq j$. Therefore, by equation (2.23)

$$\mu^*\left(\bigcup_{n=1}^\infty D_n \cap E\right) = \mu^*\left(\bigcup_{n=1}^\infty (P_n \cap E)\right) = \sum_{n=1}^\infty \mu^*(P_n \cap E) = \sum_{n=1}^\infty \mu^*(A_n \cap E). \tag{2.24}$$

Also by Remark 2.8,

$$\mu_*\left(\bigcup_{n=1}^\infty D_n \cap E\right) = \sum_{n=1}^\infty \mu_*(P_n \cap E) = \sum_{n=1}^\infty \mu_*(A_n \cap E). \tag{2.25}$$

Similarly, we have

$$\mu^*\left(\bigcup_{n=1}^\infty D_n \cap E^c\right) = \sum_{n=1}^\infty \mu^*(B_n \cap E^c) \tag{2.26}$$

and

$$\mu_*\left(\bigcup_{n=1}^{\infty} D_n \cap E^c\right) = \sum_{n=1}^{\infty} \mu_*(B_n \cap E^c). \tag{2.27}$$

From (2.24) and (2.27),

$$\bar{\mu}\left(\bigcup_{n=1}^{\infty} D_n\right) = \sum_{n=1}^{\infty} \bar{\mu}(D_n),$$

and therefore $\bar{\mu}$ is countably additive. Similarly, by equations (2.25) and (2.26) $\underline{\mu}$ is countably additive. ▮

Now we present a simple (but proper) translation-invariant extension of Lebesgue measure that was considered in [53].

Let A be a set $\subset R$ such that both A and $R - A$ have nonempty intersection with every uncountable closed set in R. (Such a set is constructed in Problem 2.4.5.) Such a set is necessarily non-Lebesgue-measurable (see Problem 2.4.5) and has the following properties:

$$E \subset A \text{ is Lebesgue measurable if and only if } m^*(E) = 0. \tag{2.28a}$$

$$E \subset R - A \text{ is Lebesgue measurable if and only if } m^*(E) = 0. \tag{2.28b}$$

The reason for property (2.28a) is that if $E \subset A$ and E is Lebesgue measurable, then

$$m^*(E) = m(E) = \sup\{m(B)\colon\ B \text{ closed and } B \subset E\};$$

but $B \subset E \subset A \Rightarrow B \cap (R - A) = \varnothing$, and therefore B must be countable. Hence $m(B) = 0$, implying that $m^*(E) = 0$. The same reasoning applies for property (2.28b).

Let $\mathscr{M}$ be the Lebesgue-measurable sets on R and $\mathscr{M}^*$ be the σ-algebra generated by $\mathscr{M}$ and the set A above. Then since the class of sets $\{(E \cap A) \cup (F \cap A^c)\colon E, F \in \mathscr{M}\}$ is a σ-algebra containing A, we have

$$\mathscr{M}^* = \{(E \cap A) \cup (F \cap A^c)\colon\ E, F \in \mathscr{M}\}. \tag{2.29}$$

Preliminary to proving our extension theorem, we need a lemma.

• **Lemma 2.3.** Let $B \in \mathscr{M}^*$ and let $B = (E \cap A) \cup (F \cap A^c)$, where $E, F \in \mathscr{M}$. Then $m^*(B) + m_*(B) = m(E) + m(F)$. ▮

Proof. Note that

$$B = (E \cap F^c \cap A) \cup (E^c \cap F \cap A^c) \cup (E \cap F).$$

By Remark 2.8, we have

$$m_*(B) = m_*(A \cap E \cap F^c) + m_*(A^c \cap E^c \cap F) + m_*(E \cap F). \quad (2.30)$$

By Remark 2.7 and equation (2.28), it follows that

$$m_*(A \cap E \cap F^c) = m_*(A^c \cap E^c \cap F) = 0.$$

Hence $m_*(B) = m(E \cap F)$. Also, noting that Remark 2.8 remains true if the inner measure is replaced by the outer measure, we have

$$m^*(B) = m^*(A \cap E \cap F^c) + m^*(A^c \cap E^c \cap F) + m^*(E \cap F).$$

It follows from equation (2.16) that

$$m(E \cap F^c) = m_*(E \cap F^c \cap A^c) + m^*(E \cap F^c \cap A) = m^*(E \cap F^c \cap A),$$

by equation (2.28). Similarly, $m^*(A^c \cap E^c \cap F) = m(E^c \cap F)$. Hence $m^*(B) = m(E \cup F)$. The rest is clear. ∎

Theorem 2.8. Let λ be defined on $\mathscr{M}^*$ by

$$\lambda(B) = \tfrac{1}{2}[m^*(B) + m_*(B)]. \quad (2.31)$$

Then λ is a complete measure on $\mathscr{M}^*$ and a proper translation-invariant extension of m. ∎

Proof. Since m^* is translation invariant by Problem 2.3.5, m_* is so also by its definition, and therefore λ is translation invariant. Also since $m^*(B) = m_*(B)$ for Lebesgue-measurable B by Remark 2.9, $\lambda(B) = m(B)$.

We only need to establish that λ is a measure. First note that λ is subadditive since m^* and m_* are both subadditive. Let (B_n) be a disjoint sequence in $\mathscr{M}^*$ such that

$$B_n = (E_n \cap A) \cup (F_n \cap A^c), \qquad E_n, F_n \in \mathscr{M}.$$

Notice that for $n \neq k$, $E_n \cap E_k \subset A^c$ and $F_n \cap F_k \subset A$. Therefore by equation (2.28) $n \neq k$ implies that

$$m(E_n \cap E_k) = 0 = m(F_n \cap F_k).$$

Hence we have

$$m\left(\bigcup_{n=1}^{\infty} E_n\right) = \sum_{n=1}^{\infty} m(E_n)$$

and

$$m\left(\bigcup_{n=1}^{\infty} F_n\right) = \sum_{n=1}^{\infty} m(F_n).$$

Since

$$\bigcup_{n=1}^{\infty} B_n = \left[\left(\bigcup_{n=1}^{\infty} E_n\right) \cap A\right] \cup \left[\left(\bigcup_{n=1}^{\infty} F_n\right) \cap A^c\right],$$

it follows by Lemma 2.3 that

$$\lambda\left(\bigcup_{n=1}^{\infty} B_n\right) = \tfrac{1}{2}\left[m\left(\bigcup_{n=1}^{\infty} E_n\right) + m\left(\bigcup_{n=1}^{\infty} F_n\right)\right] = \sum_{n=1}^{\infty} \tfrac{1}{2}[m(E_n) + m(F_n)]$$
$$= \sum_{n=1}^{\infty} \lambda(B_n).$$

The proof is complete. ∎

Problems

2.4.1. Give an example of a sequence (E_n) of pairwise disjoint sets on R such that $m^*(\bigcup_{n=1}^{\infty} E_n) < \sum_{n=1}^{\infty} m^*(E_n)$.

2.4.2. Give an example of a sequence of sets (E_n) such that $E_n \supset E_{n+1}$, $m^*(E_n) < \infty$, and $m^*(\bigcap_{n=1}^{\infty} E_n) < \lim_{n\to\infty} m^*(E_n)$.

2.4.3. Give an example of a Lebesgue-measurable set that is not a Borel set. {Hint: Take the Cantor set $K \subset [0, 1]$ and the Cantor function L. Let $g(x) = L(x) + x$. Then g is a homeomorphism from $[0, 1]$ onto $[0, 2]$ and $g(K)$ has Lebesgue measure 1. Now there is $E \subset g(K)$, E non-Lebesgue-measurable. Show that $g^{-1}(E)$ is the desired set.}

2.4.4. Show that the Borel measure (the restriction of the Lebesgue measure on the Borel sets of R) is not complete.

2.4.5. *Another Example of a Non-Lebesgue-Measurable Set* [due to F. Bernstein (1908.)] The set $\mathscr{F}$ of all closed (but *uncountable*) subsets of R has cardinality c. (Note that every open set is a finite or countable union of open intervals with rational endpoints.) Assuming the continuum hypothesis, there is a one-to-one correspondence between $\mathscr{F}$ and $[0, \Omega)$, Ω the first uncountable ordinal. Let $(A_\alpha)_{\alpha<\Omega}$ denote a well-ordering of $\mathscr{F}$. For each $\alpha < \Omega$, there exist $a_\alpha, b_\alpha \in R$ such that $a_\alpha \in A_\alpha - \bigcup_{\beta<\alpha}\{a_\beta, b_\beta\}$, $b_\alpha \in A_\alpha - \bigcup_{\beta<\alpha}\{a_\beta, b_\beta\}$, and $a_\alpha \neq b_\alpha$. This is possible since A_α's are all uncountable. Let $A = \{a_\alpha: \alpha < \Omega\}$ and $B = \{b_\alpha: \alpha < \Omega\}$. Then A and B are disjoint and $m^*(A \cap [0, 1]) = m^*(B \cap [0, 1]) = 1$. Therefore A and B are both non-Lebesgue-measurable.

2.4.6. Let E be a Lebesgue-measurable set of positive Lebesgue measure. Show that there are disjoint non-Lebesgue-measurable sets E_1 and E_2 such that $E = E_1 \cup E_2$ and $m(E) = m^*(E_1) = m^*(E_2)$. [Hint: Let A be the set constructed in Problem 2.4.5. Let $E_1 = E \cap A$ and $E_2 = E \cap A^c$. Use equation (2.28).]

2.4.7. *Arbitrary Union of Open Sets of Measure Zero with Measure Zero* (E. Marczewski and R. Sikorski). Let (X, d) be a metric space with a base $\mathscr{B}$ with the property that every finite measure defined for all subsets of a set of cardinality equal to card $\mathscr{B}$ and zero for points is a zero measure. Let μ be a finite (weakly) Borel measure on X. Then the union of any family of open sets of measure zero has measure zero. [Hint: Let $\mathscr{F}$ be a family of open sets of measure zero and let $\{G_\alpha: \alpha \in A\}$ be a well-ordering of $\mathscr{F}$. Let $H_\alpha = G_\alpha - \bigcup_{\beta<\alpha} G_\beta$. Then H_α is an F_σ-set. Notice that $\bigcup_{\alpha \in E} H_\alpha$ is also an F_σ-set for any $E \subset A$; if C_α is a closed subset of H_α and $C_{\alpha,n} = \{x \in C_\alpha: d(x, X - G_\alpha) \geq 1/n\}$, then $C_\alpha = \bigcup_{n=1}^{\infty} C_{\alpha,n}$ and for each n the set $\bigcup_{\alpha \in E} C_{\alpha,n}$ is closed. Define $\lambda(E) = \mu(\bigcup_{\alpha \in E} H_\alpha)$. Then λ is a finite measure defined for all subsets of E and zero for points. Since card $A \leq$ card $\mathscr{B}$, $\lambda(A) = 0$ (by hypothesis).]

2.4.8. *Thick Subsets of a Measure Space.* Let μ be a measure of a σ-algebra $\mathscr{A}$ of subsets of X. A subset $E \subset X$ is called *thick* if $\mu_*(X - E) = 0$. Suppose E is thick, $\mathscr{A}_E = \{B \cap E: B \in \mathscr{A}\}$, and $\lambda(B \cap E) = \mu(B)$. Prove that λ is a measure on the σ-algebra $\mathscr{A}_E$ of subsets of E.

2.4.9. *Egoroff's Theorem*[†] *for Families of Functions.* Suppose for each y in $[2, \infty)$, $f(x, y)$ is a real-valued Lebesgue-measurable function on $[0, 1]$ such that $\lim_{y \to \infty} f(x, y) = h(x)$ exists. If for each x in $[0, 1]$, the function $f(x, y)$ is continuous in y, then the above limit is almost uniform. This result is not true in general. Consider the following example due to W. Walter.

Let $E \subset [0, \frac{1}{2})$ be a non-Lebesgue-measurable set as constructed in the proof of Theorem 2.5. Let $Q = [0, 1] \times [2, \infty)$, $Q_n = [0, 1] \times [n, n + 1]$, and the diagonal D_n of $Q_n = \{(x, n + x): 0 \leq x \leq 1\}$. Define $f: Q \to \{0, 1\}$ as follows:

$$f(x, y) = \begin{cases} 1, & \text{if for some } n \geq 2,\ x \in E + \dfrac{1}{n} \text{ and } (x, y) \in D_n; \\ 0, & \text{otherwise.} \end{cases}$$

Then $\lim_{y \to \infty} f(x, y) = 0$, but this limit is not almost uniform.

2.4.10. Let m be the Lebesgue measure on R. Show that there exists a decreasing sequence of thick sets with empty intersection.

[†] Measurable functions are discussed in Section 3.1. At this point, the reader is expected to verify only the example in this problem.

2.4.11. Let X be a separable metric space, $\mathscr{A}$ the σ-algebra of Borel sets of X and μ be a finite measure on $\mathscr{A}$. A Borel set A is called an atom if for each Borel subset B of A, $\mu(B) = 0$ or $\mu(A - B) = 0$. Prove the following assertions:

(i) $\mathscr{A}$ has no atoms if and only if every singleton has measure zero. [Hint (for the "if" part): Let $E \in \mathscr{A}$ be an atom and $\mu(\{x\}) = 0$ for all $x \in X$. Then $\mu(E) > 0$. Let (B_n) be a base for the relative topology of E. Define D_n to be B_n or $E - B_n$ according as $\mu(B_n) = \mu(E)$ or $\mu(B_n) = 0$. Consider $D = \bigcap_{n=1}^{\infty} D_n$.]

(ii) If there is a thick set whose complement is also a thick set, then $\mathscr{A}$ has no atoms.

(iii) If there is a decreasing sequence of thick sets with empty intersection, then $\mathscr{A}$ has no atoms. [Note: the converses of (ii) and (iii) are also true (and due to S. B. Rao) when X is also complete and has no isolated points.]

2.4.12. Let μ be a semifinite measure on an algebra $\mathscr{A}$. Show that μ_* is a semifinite measure on the smallest σ-algebra $\mathscr{B}$ containing $\mathscr{A}$ and the smallest extension of μ to $\mathscr{B}$.

2.4.13. Let $E_1, E_2 \subset R$ and m^*, m_* be the Lebesgue outer and inner measures, respectively. Show that if $m^*(E_1 \cup E_2) = m^*(E_1) + m^*(E_2)$, then $m_*(E_1 \cup E_2) = m_*(E_1) + m_*(E_2)$. Show also that the converse need not be true.

2.4.14. Let $E \subset R$. Suppose $m^*(E \cap I) \geq \delta \cdot m(I)$ for some $\delta > 0$ and all intervals I of R. Show that $m^*(R - E) = 0$ if E is Lebesgue–measurable and $m^*(R - E)$ may be nonzero if E is not so. [Hint: Let A be Lebesgue measurable and $m(A) > 0$. Then there is a non-Lebesgue-measurable subset B of A such that $m^*(B) \geq \frac{1}{2}m(A)$; using this fact observe that there is a non-Lebesgue-measurable subset D of A such that $m^*(D) = m(A)$. Take $E = (R - A) \cup D$.]

2.4.15. *Nonmeasurable Sets Whose Measurable Subsets Are All at Most Countable.* A set $A \subset R$ is said to have property (P) if $A \cap B$ is at most countable for every set B with Lebesgue measure zero. Prove that every uncountable subset of a set with property (P) is non-Lebesgue-measurable. Prove also the following result due to R. E. Dressler and R. B. Kirk: Assuming that the continuum hypothesis holds, there is a partition of the real numbers by an uncountable family of sets (each of which is an uncountable set of Lebesgue measure zero) such that a subset of R has property (P) if and only if it intersects each member of the partition in a set that is at most countable. [Hint: Let $\mathscr{F} = \{A \subset R$: A is an uncountable G_δ set of Lebesgue measure zero. Then the cardinal number of $\mathscr{F}$ is $2^{\aleph_0}$. Assuming the continuum hypothesis, enumerate R as $\{x_\alpha: \alpha < \Omega\}$ and $\mathscr{F}$ as $\{E_\alpha: \alpha < \Omega\}$, where Ω is the first uncountable ordinal. Using transfinite induction, show that there is a family of subsets of R denoted by $\{A_\alpha: \alpha < \Omega\}$ such that

(i) $\alpha < \Omega \Rightarrow E_\alpha - \cup\{A_\beta: \beta \leq \alpha\}$ is at most countable,

(ii) each A_α is an uncountable set of Lebesgue measure zero.

(iii) $\alpha \neq \beta \Rightarrow A_\alpha \cap A_\beta$ is empty, and

(iv) $x_\alpha \in \cup\{A_\beta: \beta \leq \alpha\}$.

Then these A_α's provide the desired partition of R.]

3

Integration

O. Cauchy (1789–1857) was perhaps the first mathematician to give a rigorous definition of an integral as the limit of a sum. But he considered only functions having at most a finite number of discontinuities.

Motivated by the needs of the theory of trigonometric series, already studied by Dirichlet in 1829 in expanding certain monotonic functions f into a series $\sum(a_n \cos nx + b_n \sin nx)$, where

$$a_n = \frac{1}{\pi} \int_0^{2\pi} f(x) \cos nx \, dx \quad \text{and} \quad b_n = \frac{1}{\pi} \int_0^{2\pi} f(x) \sin nx \, dx,$$

B. Riemann (1826–1866) continued the work of Cauchy. He defined the integral in a way similar to Cauchy's, but examined the class of all functions for which the integral could be defined. He discovered that he could even integrate functions with an everywhere dense set of points of discontinuity.

During the time of Cauchy and Riemann, mathematicians were considering mostly integrals for bounded functions. But soon, when unbounded functions started to appear in the theory of trigonometric series, mathematicians turned their attention to the possibility of defining a useful integral for such functions. Harnack (1883) and de la Vallée-Poussin (1894) are among the first mathematicians to take steps in this direction.

Motivated by diverse problems in analysis, investigations in integration theory continued and soon G. Peano (1858–1932) and C. Jordan (1838–1922) connected the concept of integration with the then recent concept of measure, already introduced by G. Cantor (1845–1918) and others. But the decisive step in the theory of integration was taken by H. Lebesgue

(1875–1941) by the discovery of a new theory of the integral contained in his thesis "Intégrale, Longueur, Aire," published in 1902. Mainly motivated by questions of applying the integral, as an effective tool in mathematical analysis, to classical problems such as the determination of curve lengths and areas of planar sets, Lebesgue took up the problem of defining an integral having all the properties that could determine this applicability. He reduced the problem to the problem of defining a countably additive, congruence-invariant measure on the class of all bounded sets on the reals such that the measure of an interval is its length. He defined a measure, now well-known as Lebesgue measure, and showed that this was the unique solution of the problem of measure for a large class of sets, now called Lebesgue-measurable sets.

An important contribution to the theory of integration was also made by W. H. Young (1863–1942). He was mainly influenced by the idea of generalizing the Riemann integral, and he wrote two important papers on this subject.[†] However, the final results of Young's and Lebesgue's contributions are the same.

After Lebesgue had laid the foundations of the modern theory of integration, more work was done later by mathematicians including F. Riesz (1880–1956). A. Denjoy (1884–   ), J. Radon (1887–1956), and others, to modify and further extend the Lebesgue definition of the integral.

This chapter presents the theory of integration in an abstract setting, containing the Lebesgue theory as a special case. In Section 3.1 we introduce and study the measurable functions, which will serve as the domain of the general integral studied in Section 3.2. In Section 3.3, we study Riemann–Stieltjes integrals from the point of view of measure theory and compare the Riemann and the Lebesgue integrals. Finally, in Section 3.4, we present Fubini's Theorem, a cornerstone in modern analysis, unavoidable in the evaluation of most multiple integrals in R^n.

Throughout this chapter, $(X, \mathscr{A}, \mu)$ will denote a measure space, that is, a triple where X is a nonempty set, $\mathscr{A}$ is a σ-algebra of subsets of X, and μ is a measure on $\mathscr{A}$. When X is R, $\mathscr{A}$ is the class of Lebesgue-measurable sets, and μ is the Lebesgue measure, then the measure space is referred to as the *Lebesgue measure space*; if here $\mathscr{A}$ is the class of the Borel sets of R, the measure space is called the *Borel measure space*. (Note that in what follows μ will always denote the Lebesgue measure whenever X is R or a subset of R.)

[†] W. H. Young, *Phil. Trans. Roy. Soc. London*, 204A, 221–252 (1905); *Proc. London Math. Soc.* **9**(2), 15–20 (1910).

Also in this chapter, we will occasionally use the following *notation*: For any extended real-valued function f on a measure space, we will write

$$f^+(x) = \sup\{f(x), 0\}, \qquad f^-(x) = \sup\{-f(x), 0\}$$

and

$$f \vee g = \sup\{f, g\}, \qquad f \wedge g = \inf\{f, g\}.$$

3.1. Measurable Functions

The Riemann integral of classical analysis is defined for any continuous real-valued function on a closed and bounded interval. More precisely, as will be shown in Section 3.3, a bounded function on $[a, b]$ is Riemann integrable if and only if the set of its points of discontinuity has Lebesgue measure zero. The integral, which is the subject of this chapter, will be shown (in Section 3.3) to be an extension of the Riemann integral and will be defined for a much larger class of functions, called *measurable functions*, which are not necessarily bounded.

Measurable functions form the subject of this section. Unlike continuous functions, the class of measurable functions will be found to be closed with respect to additional types of limiting operations; and this makes the integral applicable to a wider class of problems and naturally more useful. Despite being quite general in nature, these functions on the Lebesgue measure space are quite nice in a certain sense. It will be shown in this section that a measurable function is "almost" continuous and in a finite measure space [when $\mu(X) < \infty$], every convergent sequence of measurable functions is "almost" uniformly convergent. The meaning of the word "almost" will be clear later in this section.

Let us begin with the definition of a measurable function.

Definition 3.1. Let $E \in \mathscr{A}$ and let $\bar{R}$ be the extended reals. A function $f\colon E \to \bar{R}$ is called *measurable* if for each real number α, the set $\{x \in E\colon f(x) > \alpha\} \in \mathscr{A}$. If $\mathscr{A}$ is the class of Lebesgue-measurable subsets (or Borel sets) on R $(= X)$, a measurable function f is usually called a *Lebesgue-* (or *Borel-*) *measurable* function. ∎

This definition immediately leads to several equivalent definitions in the next proposition.

Proposition 3.1. Let f be as in Definition 3.1. Then the following are equivalent:

(i) f is measurable.

(ii) $\{x \in E: f(x) \geq \alpha\} \in \mathscr{A}$, if $\alpha \in R$.

(iii) $\{x \in E: f(x) < \alpha\} \in \mathscr{A}$, if $\alpha \in R$.

(iv) $\{x \in E: f(x) \leq \alpha\} \in \mathscr{A}$, if $\alpha \in R$.

Moreover, these statements imply

(v) $\{x \in E: f(x) = \alpha\} \in \mathscr{A}$,

for every extended real number α. ∎

Proof. The proof is obvious from the following observations:

$$\{x \in E: f(x) \geq \alpha\} = \bigcap_{n=1}^{\infty} \{x \in E: f(x) > \alpha - 1/n\}; \tag{3.1}$$

$$\{x \in E: f(x) < \alpha\} = E - \{x \in E: f(x) \geq \alpha\}; \tag{3.2}$$

$$\{x \in E: f(x) \leq \alpha\} = \bigcap_{n=1}^{\infty} \{x \in E: f(x) < \alpha + 1/n\}; \tag{3.3}$$

$$\{x \in E: f(x) > \alpha\} = E - \{x \in E: f(x) \leq \alpha\}; \tag{3.4}$$

$$\{x \in E: f(x) = \infty\} = \bigcap_{n=1}^{\infty} (x \in E: f(x) > n\}; \tag{3.5}$$

$$\{x \in E: f(x) = -\infty\} = \bigcap_{n=1}^{\infty} \{x \in E: f(x) < -n\}. \quad ∎ \tag{3.6}$$

Remarks

3.1. If $X = R$ and $\mathscr{A}$ is the class of Lebesgue-measurable sets or the Borel sets on R, then *every* continuous function f on $E \in \mathscr{A}$ is measurable, since $\{x \in E: f(x) > \alpha\}$ is the intersection of E and an open set.

3.2. A function $f: E \to \bar{R}$ is measurable if and only if $f^{-1}(B) \in \mathscr{A}$, for every Borel set $B \subset R$. To see this, it is sufficient to observe that for the measurable function f, the class

$$\{B \subset R: f^{-1}(B) \in \mathscr{A}\}$$

is a σ-algebra containing all intervals of the form (α, ∞).

3.3. The characteristic function $\chi_A(x)$ is a measurable function on X if and only if $A \in \mathscr{A}$.

Proposition 3.2.

(i) If f and g are measurable real-valued functions having the same domain, then $f \pm g$, $|f|$, $f \vee g$, $f \wedge g$ and $f \cdot g$ are measurable functions.

(ii) If (f_n) is a sequence of measurable functions (having the same domain), then $\sup_n f_n$, $\inf_n f_n$, $\overline{\lim}_n f_n$, $\underline{\lim}_n f_n$ are all measurable. ▮

Proof. (i) Since for any real number α, there is a rational number r such that

$$f(x) < r < \alpha - g(x)$$

whenever $f(x) + g(x) < \alpha$, we have

$$\{x: f(x) + g(x) < \alpha\} = \bigcup_r [\{x: f(x) < r\} \cap \{x: g(x) < \alpha - r\}].$$

The union is taken over a set of rational numbers. It follows that $f + g$ is measurable. Similarly, $f - g$ is also measurable.

Now

$$f \vee g = \tfrac{1}{2}[f + g + |f - g|] \quad \text{and} \quad f \wedge g = \tfrac{1}{2}[f + g - |f - g|].$$

Also, $|f|$ is measurable whenever f is, since

$$\{x: |f(x)| > \alpha\} = \{x: f(x) > \alpha\} \cup \{x: f(x) < -\alpha\}.$$

This means that $f \vee g$ and $f \wedge g$ are both measurable. Since for $\alpha \geq 0$,

$$\{x: f^2(x) > \alpha\} = \{x: f(x) > \sqrt{\alpha}\} \cup \{x: f(x) < -\sqrt{\alpha}\},$$

and $4fg = (f + g)^2 - (f - g)^2$, it follows that $f \cdot g$ is measurable.

(ii) The proof of (ii) is clear from the following observations:

$$\{x: \sup_n f_n(x) > \alpha\} = \bigcup_n \{x: f_n(x) > \alpha\}, \tag{3.7}$$

$$\inf_n f_n = -\sup_n(-f_n), \tag{3.8}$$

$$\overline{\lim_n} f_n = \inf_k \sup_{n \geq k} f_n, \tag{3.9}$$

$$\underline{\lim_n} f_n = \sup_k \inf_{n \geq k} f_n. \quad ▮ \tag{3.10}$$

Definition 3.2. A property is said to hold *almost everywhere* (or a.e.) if the set of points for which it fails to hold is measurable and has measure zero. ▮

Proposition 3.3. Let $(X, \mathscr{A}, \mu)$ be a complete measure space and $f = g$ a.e. If f is measurable, then g is also measurable. ∎

The proof is left to the reader.

In Chapter 2 (Problems 2.4.3 and 2.4.4) we found that the Borel measure is not complete and there is a set $A \subset K$ (where K is the Cantor set $\subset [0, 1]$) such that A is a Lebesgue-measurable set with Lebesgue measure zero, but not a Borel set. If $(X, \mathscr{A}, \mu)$ is the Borel measure space on $[0, 1]$ then the function

$$f(x) = \begin{cases} 1, & x \in A \\ 2, & x \in K - A \\ 0, & x \notin K \end{cases}$$

is equal to the zero function almost everywhere, but not measurable since

$$\{x: f(x) = 1\} = A \notin \mathscr{A}.$$

This shows that completeness is essential in Proposition 3.3.

We will now show how a bounded measurable function can be uniformly approximated by *simple functions*, that is, *measurable real-valued functions that assume only a finite number of real values.*

Proposition 3.4. A simple function f has the form $\sum_{i=1}^{n} \alpha_i \chi_{A_i}$, where $\alpha_i \in R$ and $A_i \in \mathscr{A}$. The sum, product, and difference of two simple functions are simple. ∎

The proof is left to the reader.

Proposition 3.5. Let f be a measurable function. Then f is the pointwise limit of a sequence of simple functions. If f is bounded, then the convergence is uniform. If $f \geq 0$, then the above sequence can be taken as monotonic increasing. ∎

Proof. Let $f^+ = f \vee 0$ and $f^- = (-f) \vee 0$. Then f^+ and f^- are both nonnegative and measurable by Proposition 3.2, and $f = f^+ - f^-$. Therefore, it is no loss of generality to prove the proposition for nonnegative f. Let f be nonnegative. For each integer $n \geq 1$ and $x \in X$, let

$$f_n(x) = \begin{cases} \dfrac{i-1}{2^n} & \text{if } \dfrac{i-1}{2^n} \leq f(x) < \dfrac{i}{2^n}, \text{ for } i = 1, 2, \ldots, n2^n \\ n & \text{if } f(x) \geq n. \end{cases}$$

Then the f_n's are simple functions and $f_{n+1}(x) \geq f_n(x)$. Also if $f(x) \leq n$, then $0 \leq f(x) - f_n(x) \leq 1/2^n$. Hence $f(x) = \lim_{n\to\infty} f_n(x)$. Clearly, the convergence is uniform when f is bounded. ∎

The next proposition shows that a Lebesgue-measurable function defined on $[a, b]$ is almost continuous in a certain sense. (See also Problem 3.1.11.)

Proposition 3.6. Let f be an a.e. real-valued measurable function defined on $[a, b]$. Then given $\varepsilon > 0$, there exists a continuous function g such that $\mu\{x \in [a, b]: |f(x) - g(x)| \geq \varepsilon\} < \varepsilon$ and $\sup|g(x)| \leq \sup|f(x)|$. ∎

Proof. Since $\mu(\bigcup_{n=1}^{\infty} \{x: |f(x)| < n\}) = b - a$, there is an N such that $\mu(\{x: |f(x)| < N\})$ is greater than $b - a - \varepsilon/3$. By Proposition 3.5, there is a simple function h such that $\mu\{x: |f(x) - h(x)| < \varepsilon/2\} > b - a - \varepsilon/3$. [Note that we are applying Proposition 3.5 to the bounded function $f \cdot \chi_B$, where $B = \{x: |f(x)| < N\}$.] Now let $h = \sum_{k=1}^{n} c_k \chi_{E_k}$, where c_k's are all the distinct values assumed by h on $[a, b]$ and E_k's are pairwise disjoint. Let $F_k \subset E_k$, F_k closed and $\mu(E_k - F_k) < \varepsilon/3n$. If $F = \bigcup_{k=1}^{n} F_k$, then F is closed and the function g defined on F by $g(x) = c_k$, $x \in F_k$ is continuous on F. Now we can extend g continuously on $[a, b]$ as in Problem 3.1.11(a). Then $\mu\{x \in [a, b]: |f(x) - g(x)| < \varepsilon/2\}$ is greater than $b - a - 2\varepsilon/3$. ∎

Now we show another important property of measurable functions: the equivalence of a.e. convergence (that is, pointwise convergence everywhere except on a set of measure zero) and almost uniform convergence for a sequence of a.e. real-valued measurable functions.

Definition 3.3. A sequence (f_n) of a.e. real-valued measurable functions is said to *converge almost uniformly* to a measurable function f if for any $\varepsilon > 0$ there exists $E \in \mathscr{A}$ with $\mu(E) < \varepsilon$ such that (f_n) converges to f uniformly on $X - E$. ∎

Remark 3.4. Suppose $f_n \to f$ almost uniformly. Then for each positive integer n, there exists $E_n \in \mathscr{A}$ such that $\mu(E_n) < 1/n$ and $f_n \to f$ uniformly on $X - E_n$. Let $A = \bigcup_{n=1}^{\infty}(X - E_n)$. Then $\mu(X - A) = 0$ and for $x \in A$, $f_n(x) \to f(x)$. Hence almost uniform convergence implies convergence a.e. The converse is true if $\mu(X) < \infty$, as shown by the next theorem.

Theorem 3.1. (*Egoroff*) Let $\mu(X) < \infty$ and (f_n) be a sequence of a.e. real-valued measurable functions converging a.e. to an a.e. real-valued measurable function f. Then $f_n \to f$ almost uniformly. ∎

Proof. With no loss of generality, we assume that f and the f_n are all real-valued everywhere. For each positive integer k, let

$$A_{n,k} = \bigcap_{m=n}^{\infty} \{x: |f_m(x) - f(x)| < 1/k\}.$$

Since $\lim_{n\to\infty} f_n(x) = f(x)$ a.e., we have

$$\mu\left(\bigcup_{n=1}^{\infty} A_{n,k}\right) = \mu(X)$$

Since $A_{n,k} \subset A_{n+1,k}$, $\lim_{n\to\infty}\mu(A_{n,k}) = \mu(X)$. Now given $\varepsilon > 0$, for each positive integer k there exists n_k such that for $n \geq n_k$,

$$\mu(X - A_{n,k}) < \varepsilon/2^k$$

Let $A = \bigcap_{k=1}^{\infty} A_{n_k,k}$. Then

$$\mu(X - A) \leq \sum_{k=1}^{\infty} \mu(X - A_{n_k,k}) < \varepsilon.$$

The reader can easily check that f_n converge uniformly to f on A. ▮

Remark 3.5. Theorem 3.1 need not be true if $\mu(X) = \infty$. For example, the sequence $\chi_{(n,\infty)}(x)$ converges to 0 as $n \to \infty$, but not almost uniformly with respect to the Lebesgue measure.

Finally, we introduce another concept of convergence—that of convergence in measure, a weaker concept of convergence. This concept is of basic importance in probability theory.

Definition 3.4. A sequence (f_n) of a.e. real-valued measurable functions is said to converge in measure to a measurable function f if for every $\varepsilon > 0$,

$$\lim_{n\to\infty} \mu\{x: |f_n(x) - f(x)| \geq \varepsilon\} = 0.$$ ▮

Remarks

3.6. If $f_n \to f$ in measure, then f is a.e. real-valued; for, we can find E with $\mu(E) = 0$ such that the f_n are all real-valued on $X - E$ and $X - E = \bigcup_{n=1}^{\infty} \{x: |f_n(x) - f(x)| < \varepsilon\}$.

3.7. If $f_n \to f$ in measure and $f_n \to g$ in measure, then $f = g$ a.e. For, given $\varepsilon > 0$,

$$\{x: |f(x) - g(x)| > 2\varepsilon\} \subset \{x: |f(x) - f_n(x)| \geq \varepsilon\} \cup \{x: |g(x) - f_n(x)| \geq \varepsilon\},$$

and hence $\mu\{x: |f(x) - g(x)| > 2\varepsilon\} = 0$. This means that $f = g$ a.e.

3.8. Almost uniform convergence implies convergence in measure; but the converse is not true. (Problem 3.1.7.)

3.9. Almost everywhere convergence implies convergence in measure, when $\mu(X) < \infty$. This follows from Egoroff's Theorem and the previous remark.

Definition 3.5. A sequence (f_n) of a.e. real-valued measurable functions is called a Cauchy sequence in measure, if, for every $\varepsilon > 0$,

$$\mu\{x: |f_n(x) - f_m(x)| \geq \varepsilon\} \to 0$$

as $n, m \to \infty$. ∎

Clearly a sequence (f_n), which converges in measure, is a Cauchy sequence in measure. The converse follows from the next result.

Proposition 3.7. Let (f_n) be a Cauchy sequence in measure. Then there is a measurable function f and a subsequence (f_{n_k}) such that $f_n \to f$ in measure and $f_{n_k} \to f$ almost uniformly and hence almost everywhere. ∎

Proof. We choose (n_k) such that $n_{k+1} > n_k$ and

$$\mu\left\{x: |f_{n_k}(x) - f_{n_{k+1}}(x)| > \frac{1}{2^k}\right\} < \frac{1}{2^k}.$$

Let

$$A_m = \bigcup_{k=m+1}^{\infty} \left\{x: |f_{n_k}(x) - f_{n_{k+1}}(x)| > \frac{1}{2^k}\right\}.$$

Then $\mu(A_m) < 1/2^m$ and on $X - A_m$, the subsequence (f_{n_k}) is uniformly Cauchy. Let $A = \bigcap_{m=1}^{\infty} A_m$. Then $\mu(A) = 0$. Clearly for each m, there exist g_m such that $f_{n_k} \to g_m$ uniformly on $X - A_m$. Since $X - A_m \subset X - A_{m+1}$, $g_m = g_{m+1}$ on $X - A_m$. We define $f(x) = 0$ on A, $= g_m(x)$ on $X - A_m$, $1 \leq m < \infty$. Then $f_{n_k} \to f$ almost uniformly. Hence $f_{n_k} \to f$ a.e. and in measure by Remarks 3.4 and 3.8. Finally, since

$$\{x: |f_n(x) - f(x)| \geq 2\varepsilon\} \subset \{x: |f_n(x) - f_{n_k}(x)| \geq \varepsilon\} \cup \{x: |f_{n_k}(x) - f(x)| \geq \varepsilon\},$$

it follows that $f_n \to f$ in measure. ∎

Problems

3.1.1. Let f be an extended real-valued function such that $\{x: f(x) > \alpha\} \in \mathscr{A}$ for every $\alpha \in D$, a dense set of real numbers. Show that f is measurable.

3.1.2. Give an example of a nonmeasurable function f such that $|f|$ is measurable.

3.1.3. Show that a function of bounded variation on $[a, b]$ is Lebesgue measurable.

3.1.4. Show that $g \circ f$ is measurable whenever f is measurable and g is real-valued and continuous.

3.1.5. Let $(X, \mathscr{A}, \mu)$ be a finite measure space and $d\colon \mathscr{A} \times \mathscr{A} \to R$ is defined by $d(A, B) = \mu(A \triangle B)$. Show that $(\mathscr{A}, d)$ is a complete pseudometric space. [Hint: For a Cauchy sequence (A_n), there is a subsequence (A_{n_k}) such that $\mu(A_{n_k} \triangle B_{n_k}) \to 0$ as $k \to \infty$ where $B_{n_k} = \bigcup_{m=n_k}^{\infty} A_m$; then $\lim_{n\to\infty} A_n = \bigcap_{m=1}^{\infty} B_m$.]

3.1.6. Suppose in a finite measure space, $f_n \to f$ in measure and $g_n \to g$ in measure. Then show that (a) $f_n g_n \to fg$ in measure and (b) if for all x and each n, $f(x) \neq 0$ and $f_n(x) \neq 0$, then $1/f_n \to 1/f$ in measure.

3.1.7. Form the sequence (f_n) on $[0, 1]$ as follows:

$$f_1 = \chi_{[0,1]};\, f_2 = \chi_{[0,1/2]};\, f_3 = \chi_{[1/2,1]};\, f_4 = \chi_{[0,1/3]};\, f_5 = \chi_{[1/3,2/3]};\, f_6 = \chi_{[2/3,1]};$$

etc. Then show that $f_n \to 0$ in measure, but $f_n \not\to 0$ a.e.

3.1.8. Give an example of a continuous function g on $[0, 1]$ and a Lebesgue-measurable function h such that $h \circ g$ is not Lebesgue–measurable. {Hint: Let $f(x) = f_1(x) + x$, where $f_1(x)$ is the Cantor function on $[0, 1]$. Then f is a homeomorphism from $[0, 1]$ onto $[0, 2]$ and $f(K)$, K being the Cantor set, has positive Lebesgue measure. Let E be a nonmeasurable set $\subset f(K)$. Take $g = f^{-1}$ and $h = \chi_{f^{-1}(E)}$.}

3.1.9. Show that $f_n \to f$ almost uniformly if and only if, for each $\varepsilon > 0$,

$$\lim_{n\to\infty} \mu\left(\bigcup_{m=n}^{\infty} \{x\colon |f_m(x) - f(x)| \geq \varepsilon\}\right) = 0.$$

Find a similar condition for almost everywhere convergence.

3.1.10. Show that in a finite measure space, for any sequence (f_n) of a.e. real-valued measurable functions, there exist positive numbers a_n such that $(a_n f_n) \to 0$ a.e. [Hint: Choose positive constants b_n such that $\mu(A_n{}^c) < 1/2^{n+1}$, where $A_n = \{x\colon |f_n(x)| \leq b_n\}$. Take $a_n = 1/nb_n$ and consider the set $\bigcup_{m=1}^{\infty} \bigcap_{n=m}^{\infty} A_n$.]

3.1.11. (a) If f is a continuous real-valued function on a closed set $A \subset R$, then show that there is a continuous extension g of f to R such that

$$\sup_{x \in R} |g(x)| \leq \sup_{x \in A} |f(x)|.$$

[Hint. Writing A^c as the union of disjoint open intervals, define g as f on points of A and linearly on these intervals.]

(b) For any $\varepsilon > 0$ and a simple function f on R, show that there exists a continuous function g and a closed set $A \subset R$ such that $\mu(R - A) < \varepsilon$ and $g(x) = f(x)$, $x \in A$. Here μ is the Lebesgue measure.

(c) *Lusin's Theorem*. Let f be an a.e. real-valued measurable function on R. Then given $\varepsilon > 0$, there exists a continuous function g on R such that $\mu\{x \in R: f(x) \neq g(x)\} < \varepsilon$. {Hint: First restrict f to $E = [a, b]$. Let $f = \lim_{n\to\infty} f_n$, f_n simple. Then by (b), there exists closed $F_n \subset E$ such that $\mu(E - F_n) < \varepsilon/2^{n+1}$ and f_n (for each n) is continuous on F_n. Let $F = \bigcap_{n=1}^{\infty} F_n$. Then $\mu(E - F) \leq \varepsilon/2$ and f_n (for each n) is continuous on F. By Egoroff's Theorem, there exists a closed set $G \subset F$, $\mu(F - G) < \varepsilon/2$ and $f_n \to f$ uniformly on G. Hence f is continuous on G and $\mu(E - G) < \varepsilon$. By (a), the theorem follows. To extend this result from E to R, write $R = \bigcup_{-\infty}^{\infty} [n, n + 1) = \bigcup_{-\infty}^{\infty} E_n$ (say); then there is a closed set $G_n \subset E_n$ such that f is continuous on G_n and $\mu(E_n - G_n) < \varepsilon/2^n$. Let $G = \bigcup_{-\infty}^{\infty} G_n$. Then G is closed (why?) and f is continuous on G. The rest is clear by (a).}

3.1.12. Let f be a real-valued function on R and $f(a + b) = f(a) + f(b)$ for all $a, b \in R$. Show that (a) f is continuous if and only if f is continuous at a single point, and (b) f is continuous if f is Lebesgue measurable. [Hint: Use Lusin's Theorem (Problem 3.1.11).]

3.1.13. Prove the following version of Lusin's theorem: Let $(X, \mathscr{A}, \mu)$ be a finite measure space, where X is a locally compact Hausdorff space, $\mathscr{A}$ the σ-algebra of all Borel sets (the smallest σ-algebra containing the open sets), and for $A \in \mathscr{A}$, $\mu(A) = \sup\{\mu(K): K \subset A, K \text{ compact}\}$. Let f be an a.e. real-valued measurable function. Show that, given $\varepsilon > 0$, there exists a continuous function g vanishing outside a compact set such that $\mu\{x: f(x) \neq g(x)\} < \varepsilon$. [Hint: Assume with no loss of generality, $0 \leq f \leq 1$ and X compact. There exist simple functions f_n, $0 \leq f_n \leq f_{n+1}$, with $\lim_{n\to\infty} f_n = f$. Let $h_n = f_n - f_{n-1}$, $n \geq 2$ and $h_1 = f_1$. Then $f = \sum_{n=1}^{\infty} h_n$ and h_n is of the form $2^{-n} \cdot \chi_{A_n}$, $A_n \in \mathscr{A}$. There exist compact K_n and open V_n such that $K_n \subset A_n \subset V_n$ and $\mu(V_n - K_n) < \varepsilon/2^n$. By Urysohn's Lemma, there is a continuous function g_n, $0 \leq g_n \leq 1$, $g_n(x) = 1$ if $x \in K_n$ and $g_n(x) = 0$ if $x \in X - V_n$. Then $g(x) = \sum_{n=1}^{\infty} 2^{-n} g_n(x)$ is continuous and $g(x) = f(x)$ except on $\cup(V_n - K_n)$.]

3.1.14. Let $\mathscr{A}$ be a *σ-ring* of subsets of X. Then a real-valued function f on X is called *$\mathscr{A}$-measurable* if and only if $\{x: f(x) \neq 0\} \cap f^{-1}(B) \in \mathscr{A}$ for every Borel set $B \subset R$. Prove the following result of Bagdanowicz and McCloskey: f is $\mathscr{A}$-measurable if and only if $\{x: f(x) < -r\} \in \mathscr{A}$ and $\{x: f(x) > r\} \in \mathscr{A}$ for every positive number r.

★ **3.1.15.** Prove the following result due to G. Letac: The constants are the only measurable functions f on $(0, \infty)$ such that for $x, y > 0$, $f(x + y)$ lies in the closed interval joining $f(x)$ and $f(y)$.

★ **3.1.16.** Continuation of Problem 3.1.12: Show that the functions $f: R \to R$ satisfying

(i) $f(x) + f(y) = f(x + y)$

and

(ii) $f(p(x)) = p(f(x))$ for *some* polynomial $p(x)$ of degree ≥ 2

are of the form cx, where $c = 0, 1$, or -1.

★ **3.1.17.** A real-valued function f defined on an open convex set in R^n $(n \geq 1)$ is called *d-convex* $(0 < d < 1)$ if

$$f(dx + (1 - d)y) \leq df(x) + (1 - d)f(y)$$

for all x and y in its domain. Show that a d-convex measurable (with respect to n-dimensional Lebesgue measure) function is continuous and therefore is c-convex *for all* c in $[0, 1]$.

3.2. Definition and Properties of the Integral

The integral, which will be introduced and studied in this section, will extend the classical Riemann integral to a much wider class of functions, the class of measurable functions. The integral will be first defined for non-negative measurable functions and then the definition will be extended to an arbitrary measurable function.

In what follows, we will consider only measurable functions and see that the simple functions play a basic role in the definition of the integral. We start this section with a lemma concerning them.

Lemma 3.1. Let $f = \sum_{i=1}^n \alpha_i \chi_{A_i} = \sum_{j=1}^m \beta_j \chi_{B_j}$, where the α_i and the β_j are real numbers and the A_i and the B_j are in $\mathscr{A}$. Suppose that $\mu(A_i) < \infty$ whenever $\alpha_i \neq 0$ and $\mu(B_j) < \infty$ whenever $\beta_j \neq 0$. Then

$$\sum_{i=1}^n \alpha_i \mu(A_i) = \sum_{j=1}^m \beta_j \mu(B_j). \qquad \blacksquare$$

Proof. The lemma is trivial if the A_i's as well as the B_j's are disjoint. In the general case, the lemma will be proven if we can write

$$\sum_{i=1}^n \alpha_i \chi_{A_i}(x) = \sum_{j=1}^p \gamma_j \chi_{C_j}(x),$$

where the C_j's are disjoint and

$$\sum_{i=1}^{n} \alpha_i \mu(A_i) = \sum_{j=1}^{p} \gamma_j \mu(C_j).$$

This is clear if $n = 1$. To use induction on n, we assume for some positive integer q

$$\sum_{i=1}^{n-1} \alpha_i \chi_{A_i}(x) = \sum_{j=1}^{q} \gamma_j \chi_{C_j}(x)$$

and

$$\sum_{i=1}^{n-1} \alpha_i \mu(A_i) = \sum_{j=1}^{q} \gamma_j \mu(C_j),$$

where the C_j's are disjoint. We write

$$\begin{aligned}\sum_{i=1}^{n} \alpha_i \chi_{A_i}(x) &= \sum_{j=1}^{q} \gamma_j \chi_{C_j}(x) + \alpha_n \chi_{A_n}(x)\\ &= \sum_{j=1}^{q} \gamma_j \chi_{C_j - A_n}(x) + \sum_{j=1}^{q} (\gamma_j + \alpha_n) \chi_{C_j \cap A_n}(x) + \alpha_n \chi_{A_n - \cup_{j=1}^{q} C_j}(x)\end{aligned}$$

Then

$$\begin{aligned}\sum_{i=1}^{n} \alpha_i \mu(A_i) &= \sum_{j=1}^{q} \gamma_j \mu(C_j) + \alpha_n \mu(A_n)\\ &= \sum_{j=1}^{q} \gamma_j \mu(C_j - A_n) + \sum_{j=1}^{q} (\gamma_j + \alpha_n) \mu(C_j \cap A_n) + \alpha_n \mu\left(A_n - \bigcup_{j=1}^{q} C_j\right).\end{aligned}$$

The lemma now follows by induction. ∎

This lemma makes the following definition possible.

Definition 3.6. A simple function $f = \sum_{i=1}^{n} \alpha_i \chi_{A_i}$ is said to be *integrable* if $\mu(A_i) < \infty$ whenever $\alpha_i \neq 0$. If f is *integrable or nonnegative*, we write $\int f\, d\mu = \sum_{i=1}^{n} \alpha_i \mu(A_i)$. If $E \in \mathscr{A}$, we write $\int_E f\, d\mu = \int f \cdot \chi_E\, d\mu$. ∎

Remarks. Suppose f and g are *both* integrable (or *both* nonnegative) simple functions. Then for $\alpha, \beta \in R$ (α and β are assumed both nonnegative when f and g are assumed so), we have the following:

3.10. $\int (\alpha f + \beta g)\, d\mu = \alpha \int f\, d\mu + \beta \int g\, d\mu$.

3.11. $f \leq g$ a.e. implies $\int f\, d\mu \leq \int g\, d\mu$.

3.12. $\int |f + g|\, d\mu \leq \int |f|\, d\mu + \int |g|\, d\mu$.

3.13. If $A_i \in \mathscr{A}$, $1 \leq i < \infty$ and $A_i \cap A_j$ is empty for $i \neq j$, then

$$\int_{\cup_{i=1}^{\infty} A_i} f\, d\mu = \sum_{i=1}^{\infty} \int_{A_i} f\, d\mu.$$

The reader should become convinced of the validity of these remarks by proving each one of them.

Definition 3.7. Let f be a *nonnegative* measurable function. Let $E \in \mathscr{A}$ and $E \subset$ the domain of f. Then we define

$$\int_E f\,d\mu = \sup\left\{\int_E g\,d\mu \colon 0 \leq g \leq f \text{ and } g \text{ is a simple function}\right\}.$$

If $\int_E f\,d\mu < \infty$, then f is called *integrable* on E. ▮

Remarks

3.14. Definition 3.7 is compatible with Definition 3.6 when f is simple.

3.15. If $\int_E f\,d\mu < \infty$, then $\mu(A) = 0$, where $A = \{x \in E \colon f(x) = \infty\}$. The reason is that for each positive integer n, $0 \leq n \cdot \chi_A \leq f$ and therefore, $n\mu(A) \leq \int_E f\,d\mu$.

3.16. If $\mu(E) < \infty$ and $m \leq f \leq M$, m and M being two nonnegative real numbers, then $m\mu(E) \leq \int_E f\,d\mu \leq M\mu(E)$.

The following proposition contains some basic results on integrals. The proofs are left to the reader.

Proposition 3.8. Let f and g be nonnegative measurable functions whose domains contain $E \in \mathscr{A}$. Then

(i) If $f \leq g$ a.e., then $\int_E f\,d\mu \leq \int_E g\,d\mu$.
(ii) If $f = g$ a.e., then $\int_E f\,d\mu = \int_E g\,d\mu$.
(iii) If $\int_E f\,d\mu = 0$, then $f = 0$ a.e. on E. ▮

To prove (iii), the reader may observe that

$$\{x \in E\colon f(x) > 0\} = \bigcup_{n=1}^{\infty} \{x \in E\colon f(x) > 1/n\};$$

if $A_n = \{x \in E\colon f(x) > 1/n\}$ and $\mu(A_n) > 0$ for some n, then $\int_E f\,d\mu \geq (1/n)\mu(A_n) > 0$, which is a contradiction.

An important feature of Definition 3.7 is the fact that limits and integrals can be interchanged for every increasing sequence of functions. This is demonstrated by the next theorem, which also establishes the additivity of the integral in Definition 3.7, which is not evident otherwise.

Theorem 3.2. *The Monotone Convergence Theorem.* If (f_n) is an increasing sequence of nonnegative measurable functions converging to a measurable function f almost everywhere, then for any measurable set E, $\int_E f\,d\mu = \lim_{n\to\infty} \int_E f_n\,d\mu$. ▮

Proof. With no loss of generality, we assume that for all x, $\lim_{n\to\infty} f_n(x) = f(x)$. Since $f_n \leq f_{n+1} \leq f$, we have by Proposition 3.8,

$$\int_E f_n \, d\mu \leq \int_E f_{n+1} \, d\mu \leq \int_E f \, d\mu. \tag{3.11}$$

Let $0 < k < 1$ and g be a simple function $\sum_{i=1}^{m} c_i \chi_{B_i}$ such that $0 \leq g \leq f$. Then if

$$A_n = \{x \in E\colon \; k \cdot g(x) \leq f_n(x)\},$$

then

$$\bigcup_{n=1}^{\infty} A_n = E, \qquad A_n \subset A_{n+1}$$

and therefore $\mu(E) = \lim_{n\to\infty} \mu(A_n)$. Now we have for each positive integer n

$$\int_{A_n} k \cdot g(x) \, d\mu \leq \int_{A_n} f_n(x) \, d\mu \leq \int_E f_n(x) \, d\mu. \tag{3.12}$$

But

$$\int_{A_n} k \cdot g(x) \, d\mu = k \sum_{i=1}^{m} c_i \mu(B_i \cap A_n);$$

therefore

$$\lim_{n\to\infty} \int_{A_n} k \cdot g(x) \, d\mu = k \sum_{i=1}^{m} c_i \mu(B_i \cap E) = k \int_E g(x) \, d\mu, \tag{3.13}$$

which means

$$\int_E k \cdot g(x) \, d\mu \leq \lim_{n\to\infty} \int_E f_n(x) \, d\mu.$$

Since this inequality is true for all k such that $0 < k < 1$, it follows that

$$\int_E g(x) \, d\mu \leq \lim_{n\to\infty} \int_E f_n(x) \, d\mu \tag{3.14}$$

for every simple function g with $0 \leq g \leq f$. The theorem follows now from equations (3.11) and (3.14). ∎

Corollary 3.1. If $\alpha \geq 0$, $\beta \geq 0$ and f, g are nonnegative measurable functions whose domains contain $E \in \mathscr{A}$,

$$\int_E (\alpha f + \beta g) \, d\mu = \alpha \int_E f \, d\mu + \beta \int_E g \, d\mu.$$ ∎

Proof. By Proposition 3.5, there exist sequences (f_n) and (g_n) of simple functions such that

$$0 \leq f_n \leq f_{n+1}, \qquad 0 \leq g_n \leq g_{n+1};$$

and

$$\lim_{n\to\infty} f_n = f, \qquad \lim_{n\to\infty} g_n = g.$$

By Remark 3.10,

$$\int (\alpha f_n + \beta g_n)\, d\mu = \alpha \int f_n\, d\mu + \beta \int g_n\, d\mu.$$

The corollary now follows by an application of the Monotone Convergence Theorem. ∎

The Monotone Convergence Theorem easily leads to the following very useful result.

Lemma 3.2. *Fatou's Lemma.* If (f_n) is a sequence of nonnegative measurable functions whose domains contain $E \in \mathscr{A}$, then

$$\int_E (\varliminf_{n\to\infty} f_n)\, d\mu \leq \varliminf_{n\to\infty} \int_E f_n\, d\mu. \qquad ∎$$

Proof. Let $g_k = \inf\{f_n\colon n \geq k\}$. Then $0 \leq g_k \leq g_{k+1}$ and $\lim_{k\to\infty} g_k = \varliminf_{n\to\infty} f_n$. By the Monotone Convergence Theorem,

$$\int_E \varliminf_{n\to\infty} f_n\, d\mu = \lim_{k\to\infty} \int_E g_k\, d\mu = \varliminf_{k\to\infty} \int_E g_k\, d\mu \leq \varliminf_{k\to\infty} \int_E f_k\, d\mu. \qquad ∎$$

Now Lemma 3.2 yields easily the following extension of the Monotone Convergence Theorem.

Corollary 3.2. Let (f_n) be a sequence of nonnegative measurable functions that converge a.e. to a measurable function f, such that $f_n \leq f$ for all n. Then

$$\int_E f\, d\mu = \lim_{n\to\infty} \int_E f_n\, d\mu,$$

where $E \in \mathscr{A}$ is contained in the domain of each f_n and f. ∎

Now we define the integral for an arbitrary measurable function.

Definition 3.8. Let f be any measurable function whose domain contains $E \in \mathscr{A}$. Then if at least one of the integrals $\int_E f^+ \, d\mu$ and $\int_E f^- \, d\mu$ is finite, we define

$$\int_E f \, d\mu = \int_E f^+ \, d\mu - \int_E f^- \, d\mu.$$

f is called integrable on E if $| \int_E f \, d\mu | < \infty$. (Note that f is integrable on E if and only if f^+ and f^- are both integrable on E, which is true if and only if $|f|$ is integrable on E.) ∎

If $(X, \mathscr{A}, \mu)$ is the Lebesgue measure space on R, then the integral defined above is called the *Lebesgue integral.* In the next section, we will show that the Lebesgue integral is a proper extension of the classical Riemann integral. The reader can easily see that $\chi_A(x)$ is integrable whenever $\mu(A) < \infty$; this means that $\chi_{\{\text{rationals}\}}(x)$ is Lebesgue integrable, though not Riemann integrable. One should, of course, be careful enough not to overlook the fact that there are many "nice" functions that are not Lebesgue integrable. For instance, if μ is the Lebesgue measure on $X = (0, 1)$, then $1/x$, despite being a continuous function on X, is not Lebesgue integrable; the reason is that for each positive integer k, the simple function

$$\sum_{n=1}^{k} n \chi_{(1/(n+1), 1/n]}(x) \leq \frac{1}{x}$$

and therefore

$$\int_0^1 \frac{1}{x} \, d\mu \geq \int_0^1 \left[\sum_{n=1}^{k} n \chi_{(1/(n+1), 1/n]}(x) \right] d\mu$$
$$= \sum_{n=1}^{k} \frac{1}{n+1}$$

so that

$$\int_0^1 \frac{1}{x} \, d\mu = \infty.$$

Despite the existence of many such nonintegrable functions, the class of integrable functions is indeed very large. Clearly any bounded measurable function f vanishing outside a set A of finite measure is integrable, since $|f| \leq k\chi_A$ (k being the bound of f) and therefore $\int |f| \, d\mu \leq k\mu(A) < \infty$. Some of the basic properties of the integral are contained in the next few remarks.

Remarks

3.17. If f is integrable on $E \in \mathscr{A}$ and if $E = \bigcup_{n=1}^{\infty} E_n$, $E_n \cap E_m = \varnothing$ if $n \neq m$, and $E_n \in \mathscr{A}$, then

$$\int_E f\,d\mu = \sum_{n=1}^{\infty} \int_{E_n} f\,d\mu.$$

This follows easily by writing $f = f^+ - f^-$ and then applying the Monotone Convergence Theorem.

3.18. If f and g are integrable on $E \in \mathscr{A}$ and $f = g$ a.e., then

$$\int_E f\,d\mu = \int_E g\,d\mu.$$

3.19. If f is integrable on $E \in \mathscr{A}$, then $A = \{x \in E\colon f(x) \neq 0\}$ has σ-finite measure; for, if $A_n = \{x \in E\colon |f(x)| > 1/n\}$ then $A = \bigcup_{n=1}^{\infty} A_n$ and $\int (1/n)\chi_{A_n}\,d\mu \leq \int_E |f|\,d\mu$, which means that $\mu(A_n) < \infty$ for each positive integer n.

3.20. The integral is linear, i.e., if f and g are integrable on $E \in \mathscr{A}$ and α, β are real numbers, then

$$\int_E (\alpha f + \beta g)\,d\mu = \alpha \int_E f\,d\mu + \beta \int_E g\,d\mu.$$

To show this, one needs to observe the following:
For $\alpha \geq 0$,

$$\begin{aligned}\int_E (\alpha f)\,d\mu &= \int_E (\alpha f)^+\,d\mu - \int_E (\alpha f)^-\,d\mu \\ &= \alpha \int_E f^+\,d\mu - \alpha \int_E f^-\,d\mu = \alpha \int_E f\,d\mu. \qquad (3.15)\end{aligned}$$

For $\alpha < 0$,

$$\begin{aligned}\int_E (\alpha f)\,d\mu &= \int_E (\alpha f)^+\,d\mu - \int_E (\alpha f)^-\,d\mu \\ &= \int_E (-\alpha) f^-\,d\mu - \int_E (-\alpha) f^+\,d\mu = \alpha \int_E f\,d\mu. \qquad (3.16)\end{aligned}$$

If $f \geq 0$, $g \geq 0$, then an application of the Monotone Convergence Theorem yields easily (Corollary 3.1)

$$\int_E (f + g)\,d\mu = \int_E f\,d\mu + \int_E g\,d\mu. \qquad (3.17)$$

In the general case, let

$$E_1 = \{x \in E: f(x) \geq 0,\ g(x) \geq 0\},$$
$$E_2 = \{x \in E: f(x) < 0,\ g(x) < 0\},$$
$$E_3 = \{x \in E: f(x) \geq 0,\ g(x) < 0,\ f(x) + g(x) < 0\},$$
$$E_4 = \{x \in E: f(x) < 0,\ g(x) \geq 0,\ f(x) + g(x) < 0\},$$
$$E_5 = \{x \in E: f(x) \geq 0,\ g(x) < 0,\ f(x) + g(x) \geq 0\},$$
$$E_6 = \{x \in E: f(x) < 0,\ g(x) \geq 0,\ f(x) + g(x) \geq 0\}.$$

The reader can easily show by using equation (3.17), that

$$\int_{E_i} (f + g)\, d\mu = \int_{E_i} f\, d\mu + \int_{E_i} g\, d\mu, \qquad 1 \leq i \leq 6. \tag{3.18}$$

The linearity of the integral now follows easily from equations (3.15), (3.16), and (3.18).

3.21. If f and g are integrable on $E \in \mathscr{A}$, and $f \leq g$ a.e., then

$$\int_E f\, d\mu \leq \int_E g\, d\mu.$$

Indeed since $|f|$ and $|g|$, being integrable, are finite a.e.,

$$f \leq g \text{ a.e.} \Rightarrow f^+ + g^- \leq g^+ + f^- \text{ a.e.}$$

Then by Proposition 3.8,

$$\int (f^+ + g^-)\, d\mu \leq \int (g^+ + f^-)\, d\mu.$$

which means that

$$\int f^+\, d\mu - \int f^-\, d\mu \leq \int g^+\, d\mu - \int g^-\, d\mu.$$

3.22. If f is integrable on $E \in \mathscr{A}$ and g is a measurable function such that on E, $|g| \leq f$, then g is integrable.

To see this, we need to observe that $g^+ \leq f$ and $g^- \leq f$ so that

$$0 \leq \int_E g^+\, d\mu \leq \int_E f\, d\mu \quad \text{and} \quad 0 \leq \int_E g^-\, d\mu \leq \int_E f\, d\mu$$

and therefore g^+ and g^- are both integrable.

Now we will present the most important theorem of this section, the Dominated Convergence Theorem, providing a very useful criterion for the interchange of limit and integrals.

Theorem 3.3. *The Dominated Convergence Theorem.*† If (f_n) is a sequence of measurable functions converging a.e. to a measurable function f such that $|f_n| \leq g$ a.e. on $E \in \mathscr{A}$, where g is an integrable function on E, then

$$\int_E f \, d\mu = \lim_{n\to\infty} \int_E f_n \, d\mu. \qquad \blacksquare$$

Proof. We will apply Fatou's Lemma to the sequence $g + f_n$ and the sequence $g - f_n$. Clearly, $g \pm f_n \geq 0$ for each n. Therefore, by Fatou's Lemma,

$$\int_E (g + f) \, d\mu \leq \varliminf_{n\to\infty} \int_E (g + f_n) \, d\mu, \tag{3.19}$$

$$\int_E (g - f) \, d\mu \leq \varliminf_{n\to\infty} \int_E (g - f_n) \, d\mu. \tag{3.20}$$

From equation (3.20), and the integrability of f, we have

$$\int_E g \, d\mu - \int_E f \, d\mu \leq \int_E g \, d\mu - \varlimsup_{n\to\infty} \int_E f_n \, d\mu. \tag{3.21}$$

From equations (3.19) and (3.21), we have

$$\varlimsup_{n\to\infty} \int_E f_n \, d\mu \leq \int_E f \, d\mu \leq \varliminf_{n\to\infty} \int_E f_n \, d\mu.$$

The theorem now follows easily. $\blacksquare$

We now present a few examples to show that convergence a.e. or even uniform convergence need not imply that the integral and the limit can be interchanged.

Examples

3.1. If $(X, \mathscr{A}, \mu)$ is the Lebesgue measure space on [0, 1] and $f_n = n\chi_{[0,1/n]}$, then $f_n \to 0$ a.e. and in measure; but $\int_0^1 f_n \, d\mu = 1 \neq 0$.

3.2. If $(X, \mathscr{A}, \mu)$ is the Lebesgue measure space on R and $f_n = (1/n)\chi_{[0,n]}$, then $f_n \to 0$ uniformly on R; but $\int_R f_n \, d\mu = 1 \neq 0$.

† This theorem is also known as the *Lebesgue Convergence Theorem*.

3.3. If μ is the counting measure on X, the positive integers, and $\mathscr{A} = 2^X$, then the sequence

$$f_n(k) = \begin{cases} 1/k, & 1 \le k \le n \\ 0, & k > n \end{cases}$$

converges uniformly to $f(k) = 1/k$, $k \ge 1$. Note that the f_n are all integrable, whereas f is not integrable since $\int f\,d\mu = \sum_{k=1}^{\infty} 1/k = \infty$. However, the reader may easily become convinced of the following:

If $(X, \mathscr{A}, \mu)$ is any measure space with $\mu(X) < \infty$, then uniform convergence of a sequence of integrable functions f_n to a function f implies that f is integrable and $\lim_{n\to\infty} \int_E f_n\,d\mu = \int_E f\,d\mu$ for all $E \in \mathscr{A}$.

Before we close this section, let us mention an often useful, but very simple, generalization of the Dominated Convergence Theorem.

Theorem 3.4. Let (f_n) and (g_n) be two sequences of measurable functions which converge a.e. to the measurable functions f and g, respectively. Suppose $|f_n| \le g_n$ and

$$\lim_{n\to\infty} \int_E g_n\,d\mu = \int_E g\,d\mu < \infty.$$

Then

$$\lim_{n\to\infty} \int_E f_n\,d\mu = \int_E f\,d\mu. \qquad \blacksquare$$

The proof follows immediately by applying Fatou's lemma to the sequences $(g_n - f_n)$ and $(g_n + f_n)$.

Corollary 3.3. Let (f_n) be a sequence of integrable functions such that $f_n \to f$ a.e. with f integrable. Then $\int_E |f - f_n|\,d\mu \to 0$ if and only if

$$\int_E |f_n|\,d\mu \to \int_E |f|\,d\mu,$$

with $E \in \mathscr{A}$. ∎

Proof. The "only if" part is easy. We prove the "if" part. Suppose $\int_E |f_n|\,d\mu \to \int_E |f|\,d\mu$. Then since $|f_n - f| \le |f_n| + |f|$ and

$$\lim_{n\to\infty} \int_E (|f_n| + |f|)\,d\mu = 2\int_E |f|\,d\mu < \infty,$$

the corollary follows from Theorem 3.4. ∎

Finally we point out that the Lebesgue integral $\int f\,d\mu$ is often written as $\int f(x)\,dx$. This notation is consistent with the fact that a Riemann integrable function on a closed and bounded interval $[a, b] \subset R$ is Lebesgue integrable and the integrals are equal. This will be discussed at length in the next section.

Problems

3.2.1. Show that for an integrable function f, for every positive number ε there is a positive number δ such that $\int_A |f|\,d\mu < \varepsilon$ whenever $\mu(A) < \delta$.

3.2.2. Show that for a Lebesgue integrable function f, the function $g(x) = \int_{-\infty}^{x} f\,d\mu$ is continuous. Is f^2 integrable when f is? Does the converse hold?

3.2.3. Let f be integrable on $E \in \mathscr{A}$. Then given $\varepsilon > 0$, there is a simple function g vanishing outside a set of finite measure such that $\int_E |f - g|\,d\mu < \varepsilon$.

3.2.4. Suppose that $\lambda(E) = \int_E f\,d\mu$, where E is a measurable set and f is a nonnegative measurable function. Show that λ is a measure and $\int g\,d\lambda = \int fg\,d\mu$ for each nonnegative measurable function g. (Hint: Prove the result first for simple functions and then use the Monotone Convergence Theorem.)

3.2.5. Prove the Monotone Convergence Theorem, Corollary 3.2, and the Dominated Convergence Theorem by replacing almost everywhere convergence by convergence in measure.

3.2.6. *Translation Invariance of the Lebesgue Integral.* Show that for any Lebesgue-integrable function f on R and any real number t, (i) $\int f\,d\mu = \int f_t\,d\mu$, $f_t(x) = f(x + t)$ and (ii) $\int f\,d\mu = \int f_-\,d\mu$, $f_-(x) = f(-x)$. (Hint: Prove the results first for simple functions.)

3.2.7. Let (f_n) be a sequence of integrable functions such that for some integrable function f, $\lim_{n\to\infty} \int |f_n - f|\,d\mu = 0$. Show that (i)

$$\lim_{n\to\infty} \int_{A_n} f_n\,d\mu = \int_A f\,d\mu \text{ if } \lim_{n\to\infty} \mu(A_n \triangle A) = 0;$$

(ii) f_n converges to f in measure and if $A_{nk} = \{x\colon |f_n(x)| > k\}$, then

$$\lim_{k\to\infty} \sup_n \int_{A_{nk}} |f_n|\,d\mu = 0.$$

This "sup" condition is known as the condition of *uniform integrability*. Also show that (ii) implies that $\lim_{n\to\infty} \int |f_n - f|\,d\mu = 0$.

3.2.8. Use Proposition 3.6 to show that, given $\varepsilon > 0$ and any Lebesgue integrable function f on E (Lebesgue measurable), there is a continuous function g on R vanishing outside a finite interval such that $\int_E |f - g| \, d\mu < \varepsilon$.

3.2.9. Show that $f(x) = (\sin x)/x$ on $[0, \infty)$, $f(0)$ being 1, is not Lebesgue integrable. (Notice that $\lim_{n\to\infty} \int_0^n [(\sin x)/x] \, dx$ exists.)

3.2.10. Let (f_n) be a sequence of integrable functions. Prove that if

$$\sum_{n=1}^{\infty} \int |f_n| \, d\mu < \infty,$$

then

$$\sum_{n=1}^{\infty} f_n(x)$$

is convergent a.e. to an integrable function f and

$$\int f \, d\mu = \sum_{n=1}^{\infty} \int f_n \, d\mu.$$

3.2.11. Let $\mu(X) < \infty$. Show that a sequence (f_n) of a.e. real-valued measurable functions is convergent in measure to zero if and only if

$$\int \frac{|f_n|}{1 + |f_n|} \, d\mu \to 0 \qquad \text{as } n \to \infty.$$

3.2.12. Let g be a measurable function such that $|\int g \cdot f \, d\mu| < \infty$ for every integrable function f. Show that g is bounded a.e., whenever μ is σ-finite. (This result is false if μ is not σ-finite.)

3.2.13. Find a nonnegative Lebesgue-integrable function f such that for any real number a and positive integers n and k,

$$m\big(\{x: f(x) \geq k\} \cap (a, a + 1/n)\big) > 0.$$

3.2.14. Let f be a bounded Lebesgue-measurable function on $[0, 1]$. Show that

$$\left[\int_0^1 f(x) \, dx\right]^2 \leq \int_0^1 f^2(x) \, dx,$$

using the inequality $b \leq (1 + b^2)/2$.

3.2.15. Show that the difference set $D(E) = \{x - y: x, y \in E\}$ of a Lebesgue-measurable set E with $\mu(E) > 0$ contains an open interval. {Hint: Let $f(x) = \int_a^b \chi_{E_1}(y)\chi_{E_1}(x + y) \, dy$, where $E_1 = E \cap [a, b]$ and $\mu(E_1) > 0$. Then $f(0) > 0$ and f is continuous at 0.}

3.2.16. Let $\mu(X) < \infty$ and $f\colon X \times [0, 1] \to R$. Suppose that (i) for each $t \in [0, 1]$, $f(x, t)$ is integrable, and (ii) the partial derivative $\partial f(x, t)/\partial t$ exists and is uniformly bounded for all $x \in X$. Show that $\int f(x, t)\, d\mu$ is differentiable on (0, 1) and

$$\frac{d}{dt}\int f(x, t)\, d\mu = \int \frac{\partial f(x, t)}{\partial t}\, d\mu.$$

3.2.17. Prove the following version of Egoroff's Theorem: Let (f_n) be a sequence of measurable functions converging a.e. to a measurable function f. Suppose $|f_n| \leq g$ for all n, where g is an integrable function. Show that given $\varepsilon > 0$, there exists $A \in \mathscr{A}$ with $\mu(A) < \varepsilon$ and $f_n \to f$ uniformly on $X - A$. Note that $\mu(X)$ is not necessarily finite here. [Hint: Let $E_k = \{x\colon g(x) \geq 1/k\}$. Then $\mu(E_k) < \infty$, $E_k \subset E_{k+1}$. By Theorem 3.1, there exists $B_k \subset E_k$, $\mu(B_k) < \varepsilon \cdot 2^{-k}$ and $f_n \to f$ uniformly on $E_k - B_k$. Then $A = \bigcup_{k=1}^{\infty} B_k$ is the desired set.] (The reader can now do Problem 2.4.9.)

3.2.18. Prove the Riemann–Lebesgue theorem: Show that

$$\lim_{n\to\infty}\int_{-\infty}^{\infty} f(x)\cos nx\, dx = \lim_{n\to\infty}\int_{-\infty}^{\infty} f(x)\sin nx\, dx = 0$$

for every Lebesgue-integrable function f. [Hint: First prove the theorem when f is a step function. Then use Problem 3.2.3 above and Problem 2.3.9 (Chapter 2), (i)–(c).]

3.2.19. Given any Lebesgue-integrable function f and $\varepsilon > 0$, show that there exists an infinitely differentiable bounded function g vanishing outside a finite interval such that $\int_{-\infty}^{\infty} |f - g|\, d\mu < \varepsilon$. [Hint: First, by Problem 3.2.3, there is a simple function f_1 vanishing outside an interval $[a, b]$ such that $\int_{-\infty}^{\infty} |f - f_1|\, d\mu < \varepsilon/3$. By Problem 2.3.9 there exists a step function f_2 such that $\int_{-\infty}^{\infty} |f_1 - f_2|\, d\mu < \varepsilon/3$. Therefore f can be assumed to be a step function.]

3.2.20. (*J. Gillis*). Suppose for each $\lambda \in \Lambda$, an infinite set, there is a Lebesgue-measurable subset $A_\lambda \subset [0, 1]$ such that $m(A_\lambda) \geq \beta > 0$. Show that:

(a) Given $\varepsilon > 0$ and any positive integer p, there exist $\lambda_1, \lambda_2, \ldots, \lambda_p$ in Λ such that $m(\bigcap_{i=1}^{p} A_{\lambda_i}) \geq \beta^p - \varepsilon$. [Hint for $p = 2$, use Problem 3.2.14 to show

$$\left[\sum_{i=1}^{n} m(A_i)\right]^2 \leq \sum_{i=1}^{n} m(A_i) + 2\sum_{i<j\leq n} m(A_i \cap A_j).]$$

(b) If Λ is uncountable, then there is an uncountable subset $\Lambda_1 \subset \Lambda$ such that for any $\lambda, \lambda' \in \Lambda_1$, $m(A_\lambda \cap A_{\lambda'}) \geq \beta - \varepsilon$. [Hint: The set $\{A_\lambda : \lambda \in \Lambda\}$ has a condensation point C with respect to the pseudometric d, where $d(A, B) = m(A \triangle B)$, i.e., every open set containing C contains uncountably many of the A_λ.]

★ **3.2.21.** *An Extension of the Riemann–Lebesgue Theorem* (Kestelman). Suppose f is a Lebesgue-integrable function on $(0, \infty)$, and, for each positive λ, $I(\lambda)$ is a subinterval of $(0, \infty)$. Then $\lim_{\lambda\to\infty} \int_{I(\lambda)} f(t) \cos \lambda t \, dt = 0$. This result is false if we assume only that $I(\lambda)$ is a finite union of intervals. [Hint: Note that

$$s_\lambda = \int_{a_\lambda}^{b_\lambda} f(t) \cos \lambda t \, dt = -\int_{a_\lambda - \pi/\lambda}^{b_\lambda - \pi/\lambda} f\left(t + \frac{\pi}{\lambda}\right) \cos \lambda t \, dt,$$

and so

$$2 \mid s_\lambda \mid \leq \int_{a_\lambda}^{b_\lambda} \left| f(t) - f\left(t + \frac{\pi}{\lambda}\right)\right| dt + \int_{a_\lambda}^{a_\lambda + \pi/\lambda} \mid f(t) \mid dt + \int_{b_\lambda}^{b_\lambda + \pi/\lambda} \mid f(t) \mid dt.]$$

3.2.22. Suppose (f_n) is a sequence of continuous real-valued functions on $[0, 1]$ such that $f_1(x) \geq f_2(x) \geq \cdots \geq 0$ for all $x \in [0, 1]$. Suppose also that the only *continuous* function f such that $f_n(x) \geq f(x) \geq 0$ for all $x \in [0, 1]$ and all n is the zero function. Show by example that $\int_0^1 f_n(x) \, dx$ need *not* have limit 0 as $n \to \infty$. [Hint: Let K be a Cantor set $\subset [0, 1]$ with $m(K) > 0$, m being the Lebesgue measure. Then since $\chi_K(x)$ is upper semicontinuous, there exists a sequence of continuous functions f_n with $f_n(x) \geq f_{n+1}(x)$ for all n, $\lim_{n\to\infty} f_n(x) = \chi_K(x)$, and $\lim_{n\to\infty} \int_0^1 f_n(x) \, dx = m(K) > 0$.]

3.2.23. *Limit of the Derivatives as the Derivative of the Limit* (*without Uniform Convergence*). Suppose f_n is a sequence of continuously differentiable functions on (a, b) such that $\lim_{n\to\infty} f_n(x) = f(x)$ and $\lim_{n\to\infty} f_n'(x) = \Phi(x)$. Then if Φ and f' are continuous, prove that $f' = \Phi$. {Hint: Let $[c, d] \subset (a, b)$ and let $A_n = \{x \in [c, d] : \mid f_k'(x) - \Phi(x) \mid \leq 1 \text{ for all } k \geq n\}$. Since $[c, d]$ is of the second category, A_n is dense in some subinterval of $[c, d]$ for some n, and hence (f_n') is uniformly bounded on some $[s, t] \subset [c, d]$. Now use the Dominated Convergence Theorem on $[s, x]$, $s < x < t$.}

3.2.24. *Integration of Complex-Valued Functions.* Let u and v be the real and imaginary parts respectively of a complex-valued function f. Then f is called measurable (integrable) if both u and v are measurable (integrable). If f is integrable, we write: $\int f \, d\mu = \int u \, d\mu + i \int v \, d\mu$. Show that the integrable complex-valued functions form a vector space over the complex numbers. Also show that a measurable complex-valued function f is integrable if and only if $|f|$ is integrable, and then $|\int f\,d\mu| \leq \int |f|\,d\mu$. [Hint: $R(\overline{\int f\,d\mu} \cdot f) \leq |\int f\,d\mu| \cdot |f|$ and then integrate both sides. Here $R(g)$ is the real part of g and the bar denotes the conjugate.]

3.3. Lebesgue–Stieltjes Measure and the Riemann–Stieltjes Integral

In this section we will study the Riemann–Stieltjes integral from the point of view of measure and integration theory presented thus far. We will show that a bounded function f defined on $[a, b]$, $-\infty < a < b < \infty$ is Riemann–Stieltjes integrable with respect to a right-continuous monotonic increasing function g on $[a, b]$ if and only if f is cotinuous almost everywhere with respect to the Lebesgue–Stieltjes measure induced by g. In particular, we will show that every Riemann-integrable function is continuous almost everywhere with respect to the Lebesgue measure (hence Riemann-integrable functions are Lebesgue integrable), and the integrals are equal.

Let us recall our discussion of Lebesgue–Stieltjes measures that we presented briefly at the end of Section 2.3 in Chapter 2. There we found that the Lebesgue–Stieltjes measure μ induced by a monotonic increasing right-continuous function f is a complete measure on a σ-algebra containing all the Borel sets of R such that $\mu((a, b]) = f(b) - f(a)$. Actually, there the measure μ was obtained by considering the outer measure μ^* induced by

$$\mu^*(A) = \inf\left\{ \sum_{i=1}^{\infty} f(b_i) - f(a_i):\ A \subset \bigcup_{i=1}^{\infty} (a_i, b_i] \right\} \tag{3.22}$$

and considering the completion of its restriction on all the Borel sets, which are μ^*-measurable. Since the Borel sets are μ^*-measurable, μ^* is a metric outer measure by Theorem 2.3 of Chapter 2. It should be noticed that the condition $\mu((a, b]) = f(b) - f(a)$ requires the function f to be right continuous since $\lim_{h_n \to 0+} \mu((a, b + h_n]) = \mu((a, b])$. A natural question then is: What happens when we consider equation (3.22) for an *arbitrary* monotonic increasing function f? Of course, μ^* will be a metric outer measure as before. [This can be checked directly by the reader, or the reader can follow the proof of part (ii) of Proposition 3.9 below.] Therefore μ^* will again give us a measure μ on the class of all Borel sets by Theorem 2.3; but in this case the reader can easily verify that $\mu((a, b]) \leq f(b) - \lim_{h \to 0+} f(a + h)$. Should we then call the completion of μ the Lebesgue–Stieltjes measure induced by f? This is possible. However, a more satisfactory definition can be obtained by considering open intervals instead of half-open intervals as follows:

$$\mu^*(A) = \inf\left\{ \sum_{i=1}^{\infty} f(b_i) - f(a_i) \colon A \subset \bigcup_{i=1}^{\infty} (a_i, b_i) \right\}. \tag{3.23}$$

In what follows we will show that the outer measure in equation (3.23) induced by f (not necessarily right continuous) coincides with the outer measure in equation (3.23) induced by the right-continuous monotonic increasing function f_2 defined by

$$f_2(x) = \lim_{h\to 0+} f(x+h) \equiv f(x+).$$

In the case of a right-continuous monotonic increasing function, equations (3.22) and (3.23) induce the same outer measure. (See Remark 3.23.)

Proposition 3.9. Let f be a monotonic increasing function and μ^* be as in equation (3.23). Then

(i) $\mu^*(\{a\}) = f(a+) - f(a-)$;
(ii) μ^* is a metric outer measure;
(iii) $\mu^*([a, b]) = f(b+) - f(a-)$,
$\mu^*((a, b]) = f(b+) - f(a+)$,
$\mu^*([a, b)) = f(b-) - f(a-)$,
$\mu^*((a, b)) = f(b-) - f(a+)$;
(iv) $\mu^*(A) = \inf\{\mu^*(0) \colon A \subset 0 \text{ open}\}$;
(v) the outer measure induced as in equation (3.23) by f_2 or f_1, where $f_2(x) = f(x+)$ and $f_1(x) = f(x-)$, coincides with μ^*. ∎

Proof. We omit the proof of (i). For (ii) let $A, B \subset R$ and $d(A, B) > 0$. Then there is a positive integer m such that $d(A, B) > 1/m$. We must prove that for any $\varepsilon > 0$

$$\mu^*(A) + \mu^*(B) \leq \mu^*(A \cup B) + 2\varepsilon. \tag{3.24}$$

We may and do assume that $\mu^*(A \cup B) < \infty$. Then there exist open intervals (a_i, b_i) such that $A \cup B \subset \bigcup_{i=1}^{\infty}(a_i, b_i)$ and

$$\sum_{i=1}^{\infty} f(b_i) - f(a_i) < \mu^*(A \cup B) + \varepsilon. \tag{3.25}$$

We know that the points of discontinuity of a monotonic increasing function are at most countable, and therefore the points of continuity of f are dense if every open interval. Therefore, for each i we can find subintervals (c_{ij}, d_{ij}), $1 \le j \le n_i$, such that

$$|d_{ij} - c_{ij}| < 1/m \qquad \text{for each } i, j, \tag{3.26}$$

$$(a_i, b_i) \subset \bigcup_{j=1}^{n_i} (c_{ij}, d_{ij}) \qquad \text{for each } i, \tag{3.27}$$

$$f(b_i) - f(a_i) + \varepsilon/2^i \ge \sum_{j=1}^{n_i} f(d_{ij}) - f(c_{ij}) \qquad \text{for each } i. \tag{3.28}$$

[To obtain equations (3.26)–(3.28), we can argue briefly as follows: Let $0 < 2s_i < 1/m$ and k be a positive integer such that $b_i - a_i < ks_i$. Choose $c_{i1} = a_i$. Let d_{i1} be a point of continuity of f such that $c_{i1} < d_{i1} \le b_i$ and $s_i < d_{i1} - c_{i1} < 2s_i$. Choose c_{i2} such that $c_{i1} < c_{i2} < d_{i1}$, $s_i < c_{i2} - c_{i1}$ and $f(d_{i1}) - f(c_{i2}) < \varepsilon/(k2^i)$. In this way, we construct the intervals (c_{ij}, d_{ij}). If p is the smallest positive integer such that $b_i - c_{ip} < 2s_i$, then we take $d_{ip} = b_i$.] Now since

$$A \cup B \subset \bigcup_{i=1}^{\infty} \bigcup_{j=1}^{n_i} (c_{ij}, d_{ij})$$

and

$$|d_{ij} - c_{ij}| < 1/m \qquad \text{for all } i, j,$$

each subinterval (c_{ij}, d_{ij}) can intersect only one of A and B, not both. Therefore, some of these intervals will cover A while the rest will cover B. Since $\sum_{i=1}^{\infty} \sum_{j=1}^{n_i} f(d_{ij}) - f(c_{ij})$ is a convergent series of nonnegative terms, every rearrangement will also converge to the same sum. It is now clear from equation (3.25) that

$$\begin{aligned} \mu^*(A) + \mu^*(B) &\le \sum_{i=1}^{\infty} \sum_{j=1}^{n_i} f(d_{ij}) - f(c_{ij}) \\ &\le \sum_{i=1}^{\infty} f(b_i) - f(a_i) + \varepsilon \\ &< \mu^*(A \cup B) + 2\varepsilon. \end{aligned}$$

Since $\varepsilon > 0$ is arbitrary, $\mu^*(A \cup B) = \mu^*(A) + \mu^*(B)$. This proves (ii).

To prove (iii), we notice that $[a, b] \subset (a - 1/n, b + 1/n)$ for each n, and therefore

$$\begin{aligned} \mu^*([a, b]) &\le \lim_{n\to\infty} f(b + 1/n) - f(a - 1/n) \\ &= f(b+) - f(a-). \end{aligned}$$

For the converse inequality, let $[a, b] \subset \bigcup_{i=1}^{\infty}(a_i, b_i)$. Then since $[a, b]$ is compact, there exists N such that $[a, b] \subset \bigcup_{i=1}^{N}(a_i, b_i)$. It is easy to see that

$$f(b+) - f(a-) \leq \sum_{i=1}^{N} f(b_i) - f(a_i).$$

This proves that $f(b+) - f(a-) = \mu^*([a, b])$. The rest of the proof of this proposition is left to the reader. ∎

Remark 3.23. For right-continuous, monotonic increasing functions f, equations (3.22) and (3.23) above yield the *same* outer measure μ^*.

Proposition 3.9 now leads to the following definitions:

Definition 3.9. Let f be a monotonic increasing function and μ^* be the outer measure induced by f as in equation (3.23). Then the completion μ_f of the restriction of μ^* on the Borel sets is called the *Lebesgue–Stieltjes measure* induced by f. (If f is defined on a finite closed interval $[a, b]$ only, then μ_f is defined as before, restricting μ^* to subsets of $[a, b]$.) ∎

Definition 3.10. Let g be any μ_f-measurable function on R, that is, $g^{-1}(B)$ belongs to the domain of μ_f for every Borel subset B of R. Then $\int g\, d\mu_f$, when it exists, is called the *Lebesgue–Stieltjes integral* of g with respect to the monotonic increasing function f. ∎

We now consider briefly the classical Riemann–Stieltjes integral via measure theory, and discuss its connections with the Lebesgue–Stieltjes integral defined above. In order to get nicer and less complicated results in this context, we need to assume that the function f is right continuous.

Let g be a *bounded* real-valued function on the closed and bounded interval $[a, b]$, and let f be a *right-continuous, monotonic increasing* function on $[a, b]$.

For a partition P_n: $a = x_0 < x_1 < \cdots < x_n = b$, let us define

$$M_i = \sup\{g(x): x_{i-1} \leq x \leq x_i,\ 1 \leq i \leq n\}, \tag{3.29}$$

$$m_i = \inf\{g(x): x_{i-1} \leq x \leq x_i,\ 1 \leq i \leq n\}, \tag{3.30}$$

$$g_n = \sum_{i=1}^{n} M_i \chi_{(x_{i-1}, x_i]}, \tag{3.31}$$

$$h_n = \sum_{i=1}^{n} m_i \chi_{(x_{i-1}, x_i]}, \tag{3.32}$$

$$\text{the upper sum } U(P_n) = \sum_{i=1}^{n} M_i[f(x_i) - f(x_{i-1})]$$
$$= \int g_n \, d\mu_f \tag{3.33}$$

$$\text{the lower sum } L(P_n) = \sum_{i=1}^{n} m_i[f(x_i) - f(x_{i-1})]$$
$$= \int h_n \, d\mu_f. \tag{3.34}$$

Note that the notation x_i above should really be $x_{i(n)}$; however, we have decided to use x_i for notational simplicity.

Let us consider a sequence of partitions $P_1, P_2, \ldots$ of $[a, b]$ such that $P_{k+1} \supset P_k$ for each k, and the norm of P_k (the length of the largest subinterval of P_k) tends to zero as $k \to \infty$. Then we have

$$g_n \geq g_{n+1} \geq \cdots \geq g \geq \cdots \geq h_{n+1} \geq h_n. \tag{3.35}$$

Suppose $s(x) = \lim_{n\to\infty} g_n(x)$ and $t(x) = \lim_{n\to\infty} h_n(x)$. Then by the Dominated Convergence Theorem we have

$$\int s \, d\mu_f = \lim_{n\to\infty} \int g_n \, d\mu_f \geq \lim_{n\to\infty} \int h_n \, d\mu_f = \int t \, d\mu_f. \tag{3.36}$$

We are now in a position to define the Riemann–Stieltjes integral properly.

Definition 3.11. If $\int s \, d\mu_f$ and $\int t \, d\mu_f$ above are equal and have a common value α that is independent of the particular sequence of partitions P_k chosen, then g is called *Riemann–Stieltjes integrable* with respect to f on $[a, b]$; and we write

$$R \int_a^b g \, df = \alpha$$

as the Riemann–Stieltjes integral of g with respect to f on $[a, b]$. ∎

The next theorem will now give a necessary and sufficient condition of Riemann–Stieltjes integrability in terms of μ_f.

Theorem 3.5. Let g be a bounded real-valued function on $[a, b]$ and f be a right continuous monotonic increasing function on $[a, b]$. Then g is Riemann–Stieltjes integrable with respect to f if and only if g is continuous a.e. (μ_f). Furthermore, if $R \int_a^b g \, df$ exists, then g is integrable with respect to μ_f, and $R \int_a^b g \, df = \int g \, d\mu_f$. ∎

Proof. Suppose $R \int_a^b g\, df$ exists. Consider a sequence of partitions P_n, $P_{n+1} \supset P_n$ and the corresponding functions g_n and h_n, as in (3.35). Then from (3.36), $\int [s(x) - t(x)]\, d\mu_f(x) = 0$. By Proposition 3.8(iii), $s(x) = t(x) = g(x)$ a.e. (μ_f). Let $x \in [a, b]$, $x \notin \bigcup_{n=1}^{\infty} P_n$ and $s(x) = t(x) = g(x)$. Then given $\varepsilon > 0$, there exists a positive integer n such that $g_n(x) - h_n(x) < \varepsilon$. Since x is an interior point of one of the intervals $(x_{i-1}, x_i]$, where $x_i \in P_n$, $0 \leq i \leq n$, there exists $\delta > 0$ such that $(x - \delta, x + \delta) \subset (x_{i-1}, x_i]$ for some i. Then it is clear that

$$|x - y| < \delta \Rightarrow |g(x) - g(y)| \leq g_n(x) - h_n(x) < \varepsilon.$$

It follows that g is continuous at x. Thus the set of discontinuities of g is a subset of $\bigcup_{n=1}^{\infty}(P_n \cup E)$, where $\mu_f(E) = 0$. Now considering a different sequence of partitions whose intervals will contain the points of P_n as interior points and noting that $\mu_f(\{a\}) = 0$, it follows that g is continuous a.e. (μ_f).

Conversely, suppose that g is continuous a.e. (μ_f). Let x $(\neq a)$ be a point of continuity of g. Then given $\varepsilon > 0$, there exists a $\delta > 0$ such that $\sup g(y) - \inf g(y) < \varepsilon$ for $y \in (x - \delta, x + \delta)$. Let $P_n(\subset P_{n+1})$ be a sequence of partitions of $[a, b]$ such that the norm of P_n tends to 0 as $n \to \infty$. Then for some positive integer N, there exist x_{i-1}, x_i in P_N such that $x \in (x_{i-1}, x_i] \subset (x - \delta, x + \delta)$. It follows that $s(x) - t(x) \leq g_N(x) - h_N(x) < \varepsilon$. Since ε is arbitrary, we have $s = t = g$ a.e. (μ_f). It is clear that g is μ_f-measurable (since s is) and $\int g\, d\mu_f$ exists (since g is bounded). The rest is clear. ∎

Corollary 3.4. A bounded function g on $[a, b]$ is Riemann integrable if and only if g is continuous a.e. with respect to the Lebesgue measure on $[a, b]$. Moreover, every Riemann-integrable function is Lebesgue integrable, and the integrals, when they exist, are equal. ∎

Proof. In Theorem 3.5, if $f(x) = x$, then μ_f becomes the Lebesgue measure on $[a, b]$ and $(R) \int_a^b g\, df$ becomes the standard Riemann integral $\int_a^b g(x)\, dx$. The corollary now follows easily. ∎

We note that the definition for Riemann–Stieltjes integral used above is equivalent to the standard definition in terms of the equality of the upper and lower integrals

$$\overline{\int_a^b} g\, df \quad \text{and} \quad \underline{\int_a^b} g\, df$$

when f is a continuous function.

Let us consider now a slightly more restricted definition originally considered by Young. Suppose f is a monotonic increasing function on $[a, b]$. If independent of the choice of partitions $a = x_1 < x_2 < \cdots < x_{n-1} < x_n = b$ and the choice of points $y_i \in [x_i, x_{i+1}]$, the sums

$$\sum_{i=1}^{n-1} g(y_i)[f(x_{i+1}) - f(x_i)]$$

converge to a unique and finite limit $\int_a^b g\,df$ as the length of the greatest subinterval approaches zero (and $n \to \infty$), then g is said to be Riemann–Stieltjes integrable with respect to f on $[a, b]$.

Now we state the following theorem of Young characterizing the above Riemann–Stieltjes integrability of g with respect to f in terms of the variation of f at the points of discontinuity of g. We omit the proof; for the details, the serious reader can consult page 133 of Young's paper.†

Theorem 3.6. (*Young*). In order that a bounded function g be Riemann–Stieltjes integrable with respect to a monotonic increasing function f on $[a, b]$, it is necessary and sufficient that for every $\varepsilon > 0$ it be possible to include the set of discontinuities of g in a set of intervals $(I_j)_{j=1}^{\infty}$ such that $\sum_{j=1}^{\infty} w(f, I_j) < \varepsilon$, where $w(f, I_j) = \sup_{x \in I_j} f(x) - \inf_{x \in I_j} f(x)$. ∎

Problems

{In Problems 3.3.1–3.3.5, f is an absolutely continuous monotonic increasing function. Assume that f is absolutely continuous on $[a, b]$ if and only if f' is Lebesgue integrable on each $[c, d] \subset [a, b]$ and $\int_c^d f'(x)\,dx = f(d) - f(c)$. This will be discussed at length in the next chapter.}

3.3.1. Show that $(R) \int_a^b df = \int_a^b f'(x)\,dx$.

3.3.2. If E is Lebesgue measurable and $m(E) = 0$ (m = Lebesgue measure), then show that E is Lebesgue Stieltjes measurable and $\mu_f(E) = 0$. Conversely, if $\mu_f(E) = 0$, then $f'(x) = 0$ a.e. (Lebesgue measure) on E.

3.3.3. If E is any Lebesgue-measurable set $\subset [a, b]$, then show that $\mu_f(E) = \int_E f'(x)\,dx$.

3.3.4. If E is any μ_f-measurable set, then show that $E - \{x : f'(x) = 0\}$ is Lebesgue measurable.

† W. H. Young, Integration with respect to a function of bounded variation, *Proc. London Math. Soc.* **13**(2), 109–150 (1914).

3.3.5. Show that a function g is μ_f-measurable if and only if $g \cdot f'$ is Lebesgue measurable.

3.3.6. Prove the following theorem: Let f be an absolutely continuous monotonic increasing function on $[a, b]$. Suppose that *either* g is integrable with respect to μ_f *or* $g \cdot f'$ is Lebesgue integrable. Then both integrals exist and

$$\int_a^b g\, d\mu_f = \int_a^b g(x) f'(x)\, dx.$$

3.3.7. Let f be a continuous monotonic increasing function on $[a, b]$ with $f(a) = c$, $f(b) = d$. Then show that for any nonnegative Borel-measurable function g on $[c, d]$,

$$\int_a^b g \circ f\, d\mu_f = \int_c^d g(y)\, dy.$$

(Hint: First consider the case when g is a simple function.)

3.3.8. Prove that a real-valued function defined on $[a, b]$ is Riemann integrable if it has a finite limit at each point of $[a, b]$. [Hint: Let $f = g + h$, where $g(x) = \lim_{y \to x} f(y)$; show that the sets $\{x: h(x) \geq 1/n\}$ and $\{x: h(x) \leq -1/n\}$ are all finite so that $h(x) = 0$, a.e. and therefore f is continuous a.e.]

3.3.9. *Integration by Parts.* Let f and g be right-continuous, monotonic increasing functions on R with no common discontinuity point in $(a, b]$. Then prove that

$$\int_{(a,b]} f\, d\mu_g + \int_{(a,b]} g\, d\mu_f = f(b)g(b) - f(a)g(a).$$

3.3.10. *Another Necessary and Sufficient Condition for Riemann Integrability* (La Vita). Prove that a bounded real-valued function f on $[a, b]$ is Riemann integrable if and only if f has a finite right-hand limit a.e. {Hint: Suppose f has a finite right-hand limit at x and the oscillation of f at x, i.e.,

$$\lim_{\delta \to 0+} \Big[\sup_{|x-y|<\delta} f(y) - \inf_{|x-y|<\delta} f(y) \Big]$$

is greater than $1/n$. Then there is an interval $(x, x + \delta)$ where the oscillation of f is less than or equal to $1/n$. It follows that the set of discontinuity points of f where f has a finite right-hand limit is at most countable.}

3.3.11. *A Bounded Derivative Which Is Not Riemann Integrable* (Goffman). Let $H \subset [0, 1]$ be the union of pairwise disjoint open intervals (I_n) (forming the complement of a Cantor set) of total length $\frac{1}{2}$. Let $J_n \subset I_n$ be a closed subinterval in the center of I_n such that $l(J_n) = [l(I_n)]^2$, $l =$ length.

Define f $(0 \leq f \leq 1)$ on each J_n to be continuous, 1 at the center and 0 at the end points. Let $f = 0$ in the complement of the J_n. Then f is not Riemann integrable, but $f(x) = F'(x)$, $F(x) = \int_0^x f(t)\,dt$. (Hint: If $x \notin H$, then $l([x, y] \cap J_n) \leq 16\{l([x, y] \cap I_n)\}^2$. Use this to show that $F'(x) = 0$ for $x \notin H$.)

3.4. Product Measures and Fubini's Theorem

Lebesgue theory in $R^n (n > 1)$ is in most respects completely analogous to the theory in the one-dimensional case. Nevertheless the evaluation of an integral in R^n is usually accomplished by a succession of n one-dimensional integrals. This method of evaluation is based on a result known as Fubini's Theorem, named after G. C. Fubini (1879–1943). It is one of the most useful results in analysis. The theory of Fourier transforms, which has numerous applications in analysis and many applied areas, owes much to this theorem. Fubini's Theorem is applicable often in various contexts in probability theory. Its presentation requires some information regarding product measures, one of the principal concerns of this section. Further considerations of product measures in a topological setting are given in Chapter 7, Section 7.5.

First, we consider several examples illustrating a few iterated integrals.

Examples

3.4. Let $I^2 = (0, 1) \times (0, 1)$ and

$$f(x, y) = \frac{x^2 - y^2}{(x^2 + y^2)^2}, \qquad (x, y) \in I^2.$$

Then

$$\int_0^1 f(x, y)\,dy = \frac{1}{1 + x^2}$$

and therefore

$$\int_0^1 \left[\int_0^1 f(x, y)\,dy\right] dx = \frac{\pi}{4}.$$

But

$$\int_0^1 \left[\int_0^1 f(x, y)\,dx\right] dy = -\frac{\pi}{4}.$$

Notice that for $0 < x < 1$

$$\int_0^x f(x, y)\,dy = \frac{1}{2x}.$$

It follows that

$$\int_0^1 | f(x, y) | \, dy \geq \frac{1}{2x}$$

and

$$\int_0^1 \left[\int_0^1 | f(x, y) | \, dy \right] dx \geq \frac{1}{2} \int_0^1 \frac{dx}{x} = \infty.$$

This will mean, as will be apparent later, that f is not Lebesgue integrable on I^2. Notice that the iterated integrals are unequal in this example.

3.5. Let $I_1^2 = [0, 1] \times [0, 2]$ and $f(x, y) = x^2 + xy$. In this case

$$\int_0^2 \left[\int_0^1 f(x, y) \, dx \right] dy = \frac{5}{3}$$

and

$$\int_0^1 \left[\int_0^2 f(x, y) \, dy \right] dx = \frac{5}{3}.$$

Note that here $f(x, y)$ is a nonnegative bounded continuous function on I_1^2 and hence integrable with respect to the Lebesgue measure on I_1^2.

3.6. Let $f(x, y) = xy/(x^2 + y^2)^2$, whenever $(x, y) \neq (0, 0)$. Then

$$\int_0^\infty f(x, y) \, dy = \frac{1}{2x}$$

and

$$\int_{-\infty}^0 f(x, y) \, dy = -\frac{1}{2x}.$$

Therefore,

$$\int_{-\infty}^\infty \left[\int_{-\infty}^\infty f(x, y) \, dy \right] dx = 0.$$

Similarly,

$$\int_{-\infty}^\infty \left[\int_{-\infty}^\infty f(x, y) \, dx \right] dy = 0.$$

But

$$\int_{-\infty}^\infty | f(x, y) | \, dy = \frac{1}{|x|},$$

and therefore

$$\int_{-\infty}^\infty \left[\int_{\infty}^\infty | f(x, y) | \, dy \right] dx = \int_{-\infty}^\infty \frac{1}{|x|} \, dx = \infty.$$

In this case also, as in Example 3.4, f is not Lebesgue integrable.

3.7. Let $I = [0, 1] \times [0, 1]$ and μ_1, μ_2 be the Lebesgue and the counting measures, respectively, on $[0, 1]$. Let $f(x, y) = \chi_\triangle(x, y)$, where

$$\triangle = \{(x, x)\colon 0 \le x \le 1\}.$$

Then

$$\int_0^1 \left[\int_0^1 f(x, y) \mu_1(dx) \right] \mu_2(dy) = 0,$$

and

$$\int_0^1 \left[\int_0^1 f(x, y) \mu_2(dy) \right] \mu_1(dx) = 1.$$

Note that in this case f is nonnegative and the iterated integrals are unequal.

The above four examples clearly show that the iterated integrals of many "nice" functions need not always be equal. One way to find a useful criterion for the equality of the iterated integrals is through the introduction of product measures. First we need the following lemma.

Lemma 3.3. Let $\mathscr{F}$ be a *semialgebra* of subsets of X (that is, a collection $\mathscr{F}$ closed with respect to finite intersections and such that the complement of any set in $\mathscr{F}$ is a finite disjoint union of sets in $\mathscr{F}$), and let μ be a nonnegative set function defined on $\mathscr{F} \cup \{\varnothing\}$ with $\mu(\varnothing) = 0$. Then μ has a unique extension to a measure on the algebra generated by $\mathscr{F} \cup \{\varnothing\}$, provided that the following two conditions are satisfied:

(i) If $A \in \mathscr{F}$ and $A = \bigcup_{i=1}^n A_i$, where A_i's are a finite disjoint collection in $\mathscr{F}$, then

$$\mu(A) = \sum_{i=1}^n \mu(A_i).$$

(ii) If $A \in \mathscr{F}$ and $A = \bigcup_{i=1}^\infty A_i$, where A_i's are an at most countable collection of disjoint sets in $\mathscr{F}$, then

$$\mu(A) \le \sum_{i=1}^\infty \mu(A_i). \quad \blacksquare$$

Proof. First, the class $\mathscr{A}$ consisting of the empty set and all finite disjoint unions of sets in $\mathscr{F}$ is the algebra generated by $\mathscr{F} \cup \{\varnothing\}$. Second, the condition (i) implies that if a set A is the union of each of two finite disjoint collections $\{A_1, A_2, \ldots, A_n\}$ and $\{B_1, B_2, \ldots, B_m\}$ in $\mathscr{F}$, then

$$\sum_{i=1}^n \mu(A_i) = \sum_{j=1}^m \mu(B_j).$$

Defining $\lambda(A) = \sum_{i=1}^{n} \mu(A_i)$, the reader can verify through the condition (ii) that λ is countably additive and the unique extension to $\mathscr{A}$. ∎

We are now ready to introduce the product measure. One way to do this is by using the Carathéodory Extension Theorem. We will present this method, although a second method of obtaining the product measure is inherent in Theorem 3.9.

Let $(X, \mathscr{A}, \mu)$ and $(Y, \mathscr{B}, \nu)$ be two measure spaces and $X \times Y$ be the Cartesian product of X and Y. If $A \subset X$, $B \subset Y$, then $A \times B$ is called a rectangle (a measurable rectangle when $A \in \mathscr{A}$ and $B \in \mathscr{B}$). The reader can easily verify that the class $\mathscr{R}$ of all measurable rectangles forms a semi-algebra.

We define the set function λ on $\mathscr{R}$ by $\lambda(A \times B) = \mu(A) \cdot \nu(B)$. We have in mind the idea of extending λ to a measure on the algebra generated by $\mathscr{R}$ and then easily to a complete measure on a σ-algebra containing $\mathscr{R}$ by using the Extension Theorem (Theorem 2.4). To do this the following lemma is necessary.

Lemma 3.4. Let $(A_i \times B_i)_{i=1}^{\infty}$ be an at most countable disjoint collection of measurable rectangles whose union is a measurable rectangle $A \times B$. Then

$$\lambda(A \times B) = \sum_{i=1}^{\infty} \lambda(A_i \times B_i).$$ ∎

Proof. Since $A \times B = \bigcup_{i=1}^{\infty} (A_i \times B_i)$ and these rectangles are pairwise disjoint,

$$\chi_A(x)\chi_B(y) = \chi_{A \times B}(x, y) = \sum_{i=1}^{\infty} \chi_{A_i \times B_i}(x, y) = \sum_{i=1}^{\infty} \chi_{A_i}(x)\chi_{B_i}(y).$$

Integrating with respect to ν, we have: $\chi_A(x)\nu(B) = \sum_{i=1}^{\infty} \chi_{A_i}(x)\nu(B_i)$. The lemma now follows by integrating with respect to μ. ∎

Now by Lemmas 3.3 and 3.4 above and the Extension Theorem (Theorem 2.4, Chapter 2), λ can be extended to a complete measure on a σ-algebra $\mathscr{M}$ containing $\mathscr{R}$, called the product of μ and ν and denoted by $\mu \times \nu$.

Remark 3.24. $\mu \times \nu$ is defined on a σ-algebra containing $\mathscr{A} \times \mathscr{B}$, the smallest σ-algebra containing $\mathscr{R}$. The measure space $(X \times Y, \mathscr{A} \times \mathscr{B}, \mu \times \nu)$ can be incomplete even if μ and ν are complete. For example, if $A \subset X$, $A \notin \mathscr{A}$ and $\nu(B) = 0$ with B nonempty, then $A \times B \notin \mathscr{A} \times \mathscr{B}$; but $A \times B \subset X \times B$ and $\mu \times \nu(X \times B) = 0$. Furthermore, when μ and ν are both σ-finite, $\mu \times \nu$ is the unique extension of λ on $\mathscr{R}$ to a measure on $\mathscr{A} \times \mathscr{B}$. The motivation behind considering $\mu\times\nu$ on $\mathscr{M}$ rather than on $\mathscr{A}\times\mathscr{B}$ is that $\mu\times\nu$ on $\mathscr{M}$ is exactly the Lebesgue measure on R^2 when $\mu=\nu$ is the Lebesgue measure on R.

The next few lemmas will lead to Fubini's Theorem, the main result of this section, and will also describe the structure of the sets in $\mathscr{M}$.

First, a notation: For $E\subset X\times Y$, let

$$E_x=\{y\in Y\colon (x,y)\in E\}$$

and

$$E^y=\{x\in X\colon (x,y)\in E\}.$$

Then it is easy to verify the following:

(i) $(E^c)_x=(E_x)^c$

(ii) $(\cup E_\alpha)_x=\cup(E_\alpha)_x$, for any collection $\{E_\alpha\}$.

Lemma 3.5. If $E\in\mathscr{A}\times\mathscr{B}$, then

(i) $E_x\in\mathscr{B}$ for all $x\in X$,

and

(ii) $E^y\in\mathscr{A}$ for all $y\in Y$. ∎

Proof. We will prove only (i). Let

$$\mathscr{F}=\{E\in\mathscr{A}\times\mathscr{B}\colon E_x\in\mathscr{B}\text{ for all } x\in X\}.$$

Then $\mathscr{F}$ contains $\mathscr{R}$. The reader can easily verify that $\mathscr{F}$ is a σ-algebra. Hence $\mathscr{F}\supset\mathscr{A}\times\mathscr{B}$. ∎

Lemma 3.6. Let f be an extended real-valued $\mathscr{A}\times\mathscr{B}$ measurable function on $X\times Y$. Then

(i) f_x, where $f_x(y)=f(x,y)$, is $\mathscr{B}$-measurable for all $x\in X$,

and

(ii) f^y, where $f^y(x)=f(x,y)$, is $\mathscr{A}$-measurable for all $y\in X$. ∎

Proof. This lemma follows easily from Lemma 3.5 when f is a simple function. Since there exist simple functions $f_n(x,y)$ such that $\lim_{n\to\infty}f_n(x,y)=f(x,y)$, the rest of the proof is clear. ∎

Lemma 3.7. Suppose that μ and ν are complete. Let E be a set in $\mathscr{R}_{\sigma\delta}$ (that is, of the form $\bigcap_{i=1}^{\infty}\bigcup_{j=1}^{\infty}E_{ij}$ where $E_{ij}\in\mathscr{R}$) with $\mu\times\nu(E)<\infty$. Then

(i) the function $\Gamma_E(x)=\nu(E_x)$ is $\mathscr{A}$-measurable,

(ii) the function $\Gamma^E(y)=\mu(E^y)$ is $\mathscr{B}$-measurable,

and

(iii) $\mu\times\nu(E)=\int\mu(E^y)\nu(dy)=\int\nu(E_x)\mu(dx)$. ∎

Proof. This lemma is clearly true for $E\in\mathscr{R}$. Now suppose $E\in\mathscr{R}_\sigma$. Then we can write $E=\bigcup_{j=1}^{\infty}E_j$, $E_j\in\mathscr{R}$ and $E_i\cap E_k=\varnothing$ $(i\neq k)$. Since $\nu(E_x)=\sum_{j=1}^{\infty}\nu\big((E_j)_x\big)$ and by the Monotone Convergence Theorem,

$$\int\nu(E_x)\mu(dx)=\sum_{j=1}^{\infty}\int\nu\big((E_j)_x\big)\mu(dx)$$

$$=\sum_{j=1}^{\infty}\mu\times\nu(E_j)=\mu\times\nu(E)$$

and similarly

$$\mu\times\nu(E)=\int\mu(E^y)\nu(dy),$$

the lemma is true for all $E\in\mathscr{R}_\sigma$.

Finally, let $E\in\mathscr{R}_{\sigma\delta}$ and $\mu\times\nu(E)<\infty$. Then $E=\bigcap_{j=1}^{\infty}E_j$, $E_j=\mathscr{R}_\sigma$, $E_j\supset E_{j+1}$, and $\mu\times\nu(E_1)<\infty$. Now $\mu\times\nu(E_1)=\int\nu\big((E_1)_x\big)\mu(dx)<\infty$, and therefore $\nu\big((E_1)_x\big)<\infty$ a.e. (μ). Hence we have

$$\nu(E_x)=\lim_{j\to\infty}\nu\big((E_j)_x\big)\text{ a.e. }(\mu).$$

Since μ is complete, Γ_E is measurable; and by the Dominated Convergence Theorem,

$$\int\nu(E_x)\mu(dx)=\lim_{j\to\infty}\int\nu\big((E_j)_x\big)\mu(dx)$$

$$=\lim_{j\to\infty}\mu\times\nu(E_j)$$

$$=\mu\times\nu(E)$$

and similarly

$$\mu\times\nu(E)=\int\mu(E^y)\nu(dy).$$ ∎

It may be noted that Lemma 3.7 uses the completeness of μ and ν. Nevertheless, completeness is not needed if μ and ν are assumed σ-finite, as the next lemma will demonstrate. First we give an example showing that Lemma 3.7 need *not* be valid for all $E \in \mathscr{R}_{\sigma\delta}$.

Example 3.8. Let $X = Y = [0, 1]$, $\mathscr{A} = \mathscr{B} =$ Lebesgue-measurable subsets of $[0, 1]$, and $\mu = \nu$ be the measure defined by

$$\mu\{x\} = \begin{cases} 2, & x \in A \\ 1, & x \notin A, \end{cases}$$

where A is a nonmeasurable subset of $[0, 1]$. Then μ and ν are complete, and the set $D = \{(x, y) \in X \times Y\colon\ x = y\} \in \mathscr{R}_{\sigma\delta}$, but $\nu(D_x) = \chi_A(x) + 1$, so Γ_D is nonmeasurable.

Lemma 3.8. The conclusion for Lemma 3.7 remains true (without the assumption of completeness) for *all* $E \in \mathscr{A} \times \mathscr{B}$, if μ and ν are both σ-finite. ∎

Proof. We will briefly outline the proof. Observe that as in Lemma 3.7 $\mathscr{R}_\sigma \subset \mathscr{F}$, the class of sets for which the lemma holds. The reader can verify that $\mathscr{F}$ is a monotone class by first assuming that μ and ν are both finite. This means that $\mathscr{F} \supset \mathscr{A} \times \mathscr{B}$. The σ-finite case then follows easily. ∎

Proposition 3.10. Let μ and ν be complete. Let $E \in \mathscr{M}$ and $\mu \times \nu(E) < \infty$. Then for almost all x, $E_x \in \mathscr{B}$ and $\Gamma_E(x)$ $[= \nu(E_x)]$ is defined and $\mathscr{A}$-measurable. In addition

$$\mu \times \nu(E) = \int \nu(E_x)\mu(dx).$$

The same is true for $\Gamma^E(y)$ and $\mu \times \nu(E) = \int \mu(E^y)\nu(dy)$. ∎

Proof. By the definition of $\mu \times \nu$, there exists $F \in \mathscr{R}_{\sigma\delta}$, $F \supset E$ and $\mu \times \nu(F - E) = 0$. Hence if $G = F - E$, then $\mu \times \nu(G) = 0$ and

$$\mu \times \nu(E) = \mu \times \nu(F) - \mu \times \nu(G).$$

Because of Lemma 3.7, it is enough to prove the proposition for G. To do this, let $H \in \mathscr{R}_{\sigma\delta}$, $G \subset H$, and $\mu \times \nu(H) = 0$. Then by Lemma 3.7

$$\mu \times \nu(H) = \int \nu(H_x)\mu(dx) = \int \mu(H^y)\nu(dy).$$

Hence $\nu(H_x) = 0$ a.e. (μ) and $\mu(H^y) = 0$ a.e. (ν). Since $G_x \subset H_x$, $G^y \subset H^y$, and μ, ν are complete, the proposition is clear. ∎

Theorem 3.7. *Fubini's Theorem.* Let μ and ν be complete and f be integrable with respect to $\mu\times\nu$ on $\mathscr{M}$. Then

(i) $\int f(x, y)\nu(dy)$ and $\int f(x, y)\mu(dx)$ are integrable functions of x and y, respectively, and

(ii) $\int f\, d(\mu\times\nu) = \iint f\, d\mu\, d\nu = \iint f\, d\nu\, d\mu$. ∎

Proof. By Proposition 3.10 this theorem is clearly true for $\chi_E(x, y)$, where $E \in \mathscr{M}$ and $\mu\times\nu(E) < \infty$. Suppose then that f is nonnegative. Since f is integrable, by Proposition 3.5 f is the limit of an increasing sequence of simple functions f_n, each vanishing outside a set of finite measure. The theorem then follows for this f by an application of the Monotone Convergence Theorem. Since for any function f, $f = f^+ - f^-$, where $f^+ = \sup\{f(x), 0\}$ and $f^- = \sup\{-f(x), 0\}$, the theorem follows easily. ∎

Theorem 3.8. *Tonelli's Theorem.* Let μ and ν be complete and σ-finite. Let f be a nonnegative $\mathscr{M}$-measurable function on $X\times Y$. Then

(i) For almost all x, $f_x(y) = f(x, y)$ is $\mathscr{B}$-measurable and for almost all y, $f^y(x) = f(x, y)$ is $\mathscr{A}$-measurable.

(ii) $\int f(x, y)\, d\nu(y)$ and $\int f(x, y)\, d\mu(x)$ are both measurable functions of x and y, respectively.

(iii) $\int f\, d(\mu\times\nu) = \iint f\, d\mu\, d\nu = \iint f\, d\nu\, d\mu$. ∎

Proof. Since μ and ν are σ-finite, $\mu\times\nu$ is σ-finite; and therefore by Proposition 3.5 every nonnegative f is the limit of an increasing sequence of simple functions, each vanishing outside a set of finite measure. The proof now follows as in Theorem 3.7. ∎

Remark 3.25. We note that the integrability condition in Theorem 3.7 and the nonnegativeness (as well as the σ-finiteness) condition in Theorem 3.8 are essential, as Examples 3.4 and 3.7 demonstrate. The completeness assumption for μ and ν has been necessary in the proofs of Fubini's and Tonelli's Theorems (Theorems 3.7 and 3.8). But these theorems above have been stated for $\mathscr{M}$-measurable functions. As we will see shortly in this section, the completeness assumption is unnecessary in the above two theorems if f is $\mathscr{A}\times\mathscr{B}$ measurable. (Recall that $\mathscr{M} \supset \mathscr{A}\times\mathscr{B}$. This inclusion may be proper.)

Product measures are very useful in probability theory. One simple instance in which they can arise in probability theory is the following. Suppose we make two observations, one resulting in a point x in X and the other in a point y in Y. Suppose also that the probability $P(x, B)$ that the second observation belongs to B given that the first observation is x, is a probability measure on $\mathscr{B}$ for each x in X, and is $\mathscr{A}$-measurable for each B in $\mathscr{B}$. Now if μ is a probability measure such that $\mu(A)$ is the probability that the first observation belongs to A, then intuitively the probability that the first observation results in a point of A and the second in a point of B should be given by $\int_A P(x, B)\mu(dx)$. The next theorem will show that this probability is actually a product measure in $\mathscr{A}\times\mathscr{B}$. This theorem is also a generalization of Lemma 3.8.

• **Theorem 3.9.** Let $(X, \mathscr{A}, \mu)$ be a σ-finite measure space and $(Y, \mathscr{B})$ be a measurable space. (Here $\mathscr{A}$ and $\mathscr{B}$ are σ-algebras.) Suppose that $P(x, B)$ is an extended real-valued function on $X\times\mathscr{B}$ such that (i) for each x in X, $P(x, B)$ is a measure on $\mathscr{B}$ and for each B in $\mathscr{B}$, $P(x, B)$ is an $\mathscr{A}$-measurable function; (ii) $Y = \bigcup_{n=1}^{\infty} B_n$ and for each x in X, $P(x, B_n) \leq k_n < \infty$. Then there is a unique σ-finite measure Q on $\mathscr{A}\times\mathscr{B}$ such that

$$\text{(i)}\quad Q(A\times B) = \int_A P(x, B)\mu(dx), \qquad A\in\mathscr{A},\ B\in\mathscr{B}$$

and

$$\text{(ii)}\quad Q(E) = \int P(x, E_x)\mu(dx), \qquad E\in\mathscr{A}\times\mathscr{B}. \qquad \blacksquare$$

Proof. With no loss of generality, we can assume that the sets B_n are pairwise disjoint. Then let $P_n(x, B) = P(x, B\cap B_n)$. If we can show the existence of a unique measure Q_n on $\mathscr{A}\times\mathscr{B}$ such that

$$Q_n(E) = \int P_n(x, E_x)\mu(dx), \qquad E\in\mathscr{A}\times\mathscr{B}$$

for each n, then the measure Q given by $Q(E) = \sum Q_n(E)$ will be the desired product measure of the theorem. Therefore, it is no loss of generality to assume that the measures $P(x, B)$ are finite.

We observe that for each $E\in\mathscr{A}\times\mathscr{B}$, the function $P(x, E_x)$ is $\mathscr{A}$-measurable. To see this, let

$$\mathscr{F} = \{E\in\mathscr{A}\times\mathscr{B}\colon P(x, E_x) \text{ in } \mathscr{A}\text{-measurable}\}.$$

Then the reader can easily verify that $\mathscr{F}$ contains finite disjoint union of measurable rectangles, and $\mathscr{F}$ is a monotone class. Thus, $\mathscr{F} = \mathscr{A}\times\mathscr{B}$.

Since $P(x, E_x)$ is $\mathscr{A}$-measurable, we can define

$$Q(E) = \int P(x, E_x)\mu(dx).$$

Now it follows easily that Q is a measure, and $Q(A \times B) = \int_A P(x, B)\mu(dx)$.

Finally for the uniqueness proof, we observe that another measure Q_0 satisfying the properties of Q has to be σ-finite, will coincide with Q on $\mathscr{R}$, and will, by the Carathéodory Extension Theorem, coincide with Q on $\mathscr{A} \times \mathscr{B}$. ∎

Now we will state the σ-finite version of the Fubini–Tonelli Theorem, the most important theorem in this section.

Theorem 3.10. *The Fubini–Tonelli Theorem* (*σ-finite version*). Let $(X, \mathscr{A}, \mu)$ and $(Y, \mathscr{B}, \nu)$ be two σ-finite measure spaces. Suppose f is an extended real-valued $\mathscr{A} \times \mathscr{B}$-measurable function on $X \times Y$ such that f is either nonnegative or integrable. Then the iterated integrals of f are defined and

$$\int f\, d(\mu \times \nu) = \int\int f\, d\mu\, d\nu = \int\int f\, d\nu\, d\mu. \quad ∎$$

Proof. Using Lemma 3.8, the theorem follows immediately for simple functions. The proof then follows as in Theorem 3.7 and 3.8. ∎

Remarks

3.26. Interestingly enough, Fubini's Theorem (Theorem 3.7) has a category analog, proved in 1932 by Kuratowski and Ulam, which can be stated as follows:

> If E is a plane set ($\subset R^2$) of first category, then E_x is a linear set ($\subset R$) of first category for all x outside a set of the first category. If E is a nowhere dense subset of the plane, then E_x is a nowhere dense linear subset for all x outside a set of first category.

Actually this theorem can be reduced to Fubini's Theorem using the following fact: For any set E of first category in the plane, there exists a product homeomorphism h of the plane onto itself such that $m_2(h(E)) = 0$, m_2 being the Lebesgue measure on R^2. For proofs of these facts the interested reader can consult *Measure and Category* by J. C. Oxtoby.†

† J. C. Oxtoby, *Measure and Category*, Springer Verlag, pp. 56–60 (1971).

3.27. The requirement that f be $\mathscr{A}\times\mathscr{B}$-measurable (or $\mathscr{M}$-measurable) is *crucial* in Fubini's Theorem as well as Tonelli's Theorem. Even measurability of all the sections f_x and f^y is not enough for the validity of the above theorems. For instance, let $(X, \mathscr{A}, \mu) = (Y, \mathscr{B}, \nu)$ be the Lebesgue measure space on [0, 1]. By the continuum hypothesis and Proposition 1.6 there is a one-to-one mapping Φ of [0, 1] onto a well-ordered set (the set of all ordinals $< \Omega$, the first uncountable ordinal) such that $\Phi(x)$ has at most countably many predecessors in this set for each $x \in [0, 1]$. Let $Q = \{(x, y) \in [0, 1]\times[0, 1]: \Phi(x) < \Phi(y)\}$. Then for $x \in [0, 1]$, Q_x is the complement of a countable set and for $y \in [0, 1]$, Q^y is countable. Let $f = \chi_Q$. Then

$$\int_0^1 \int_0^1 f\,dy\,dx = 1, \qquad \int_0^1 \int_0^1 f\,dx\,dy = 0.$$

3.28. Note that $\mu\times\nu$ need not be semifinite even when μ and ν are both semifinite. For instance, consider Example 3.7. There $\mu\times\nu(\triangle) = \infty$, since otherwise by Fubini's Theorem the iterated integrals will be equal. If $A \subset \triangle$, $A \in \mathscr{A}\times\mathscr{B}$ and $\mu\times\nu(A) < \infty$, then by Fubini's Theorem

$$\mu\times\nu(A) = \iint \chi_A(x, y)\,d\mu\,d\nu = 0.$$

This shows that $\mu\times\nu$ is not semifinite.

Actually semifiniteness of the product measure is inseparably connected with the validity of Tonelli's Theorem. *By the validity of Tonelli's Theorem we mean* the following: For every nonnegative extended real-valued $\mathscr{A}\times\mathscr{B}$-measurable function $f(x, y)$ whenever one of the iterated integrals $\iint f\,d\mu\,d\nu$ or $\iint f\,d\nu\,d\mu$ is well defined and finite, the other one is also; and $\int f\,d(\mu\times\nu) = \iint f\,d\mu\,d\nu = \iint f\,d\nu\,d\mu$. Tonelli's Theorem is, of course, valid when μ and ν are both σ-finite. But even when μ is an arbitrary measure, Tonelli's Theorem can be valid provided ν is the counting measure on a countable set.

• **Proposition 3.11.** Let μ, ν be complete and $\mu\times\nu$ be semifinite. Then Tonelli's Theorem is valid. ▮

Proof. Suppose $f(x, y)$ is a nonnegative $\mathscr{A}\times\mathscr{B}$-measurable function such that $\iint f(x, y)\,d\mu\,d\nu$ is well-defined and equal to some nonnegative number k. Then $\{(x, y): f(x, y) \neq 0\}$ has σ-finite $\mu\times\nu$ measure. If not, we can find a positive integer n such that $\mu\times\nu(A_n) = \infty$, where $A_n = \{(x, y): f(x, y) > 1/n\}$. Since $\mu\times\nu$ is semifinite, there exists $B \subset A_n$, $B \in \mathscr{A}\times\mathscr{B}$

and $2kn < \mu \times \nu(B) < \infty$. Then, by Fubini's Theorem,

$$2k < \frac{1}{n} \mu \times \nu(B) = \iint \frac{1}{n} \cdot \chi_B(x, y)\, d\mu\, d\nu$$
$$\leq \iint f(x, y)\, d\mu\, d\nu,$$

which is a contradiction. Therefore, the support of f (that is, $\{x : f(x) \neq 0\}$) is σ-finite, and f is the limit of an increasing sequence of simple functions each vanishing outside a set of finite measure. The proof now follows as in Theorem 3.7. ∎

Actually, much more than Proposition 3.11 is true.

• **Theorem 3.11.** Suppose μ and ν are complete measures. Then the following conditions are equivalent:

(a) $\mu \times \nu$ is semifinite on $\mathscr{A} \times \mathscr{B}$ (respectively, $\mathscr{M}$).

(b) μ and ν are semifinite and Tonelli's Theorem is valid for $\mathscr{A} \times \mathscr{B}$ (respectively, $\mathscr{M}$) measurable functions. ∎

The proof is omitted. The interested reader is referred to the paper on Tonelli's Theorem by A. Mukherjea.[†]

The validity of Tonelli's Theorem for nonatomic measures does imply the σ-finiteness of the measures in certain cases. This can be seen in what follows. First we need a result for nonatomic measures.

Definition 3.12. An atom in a measure space $(X, \mathscr{A}, \mu)$ is defined as a set $A \in \mathscr{A}$ such that $\mu(A) > 0$ and $A \supset B \in \mathscr{A}$ implies $\mu(B)$ or $\mu(A - B) = 0$. $(X, \mathscr{A}, \mu)$ is called nonatomic if it has no atoms. ∎

Lemma 3.9. Let $(X, \mathscr{A}, \mu)$ be a nonatomic measure space. Let $A \in \mathscr{A}$ and $0 < \mu(A) < \infty$. Then for each a with $0 < a < \mu(A)$, there exists $B \subset A$, $B \in \mathscr{A}$ such that $\mu(B) = a$. ∎

Proof. First notice that given $\varepsilon > 0$ there exists $B \subset A$ such that $0 < \mu(B) < \varepsilon$. This is because, by nonatomicity, there exists $B_1 \subset A$ such that either $0 < \mu(B_1) \leq \frac{1}{2}\mu(A)$ or $0 < \mu(A - B_1) \leq \frac{1}{2}\mu(A)$. By repeating this process n times, where $1/2^n \mu(A) < \varepsilon$, the existence of $B \subset A$ with $0 < \mu(B) < \varepsilon$ follows.

[†] A. Mukherjea, *Indiana Univ. Math. J.* **23**(8), 679–684 (1974).

Now let $\mathscr{F} = \{B \subset A: B \in \mathscr{A}, \mu(B) \leq a\}$. Then $\mathscr{F}$ is a partially ordered set under inclusion. We note that in a linearly ordered subset $\mathscr{F}_0$ of $\mathscr{F}$, identifying sets of equal measure, there can only be an at most countable number of distinct sets, say $(A_n)_{n=1}^{\infty}$; then $\bigcup_{n=1}^{\infty} A_n \in \mathscr{F}$ and is an upper bound of $\mathscr{F}_0$. By Zorn's Lemma, $\mathscr{F}$ has a maximal element, say E. Then $\mu(E) = a$. If not, then $\mu(E) < a$. If $0 < \varepsilon < a - \mu(E)$, then by the previous paragraph we can find $F \subset A - E$ such that $0 < \mu(F) < \varepsilon$. Then $E \cup F \in \mathscr{F}$ and $\mu(E) < \mu(E \cup F)$, contradicting the maximality of E. Hence $\mu(E) = a$. ∎

• **Lemma 3.10.** Suppose $(X, \mathscr{A}, \mu)$ is a nonatomic measure space and $(Y, \mathscr{B}, \nu)$ is an arbitrary measure space such that there is a one-to-one mapping $f: X \to Y$ mapping measurable sets into measurable sets. Then $D = \{(x, f(x)): x \in X\}$ is $\mathscr{M}$-measurable; also $A \subset D$ and $\mu \times \nu(A) < \infty$ imply $\mu \times \nu(A) = 0$. ∎

Proof. To prove that D is $\mathscr{M}$-measurable, it is sufficient to prove that for every $H \subset X \times Y$ with $\lambda^*(H) < \infty$ (recall that λ^* is the outer measure induced by λ and $\mu \times \nu$ is the restriction of λ^* on $\mathscr{M}$, the λ^*measurable sets—see the discussion preceding Remark 3.24), $\lambda^*(H) \geq \lambda^*(H \cap D) + \lambda^*(H \cap D^c)$. We will prove this by showing that $\lambda^*(H \cap D) = 0$.

Suppose $\lambda^*(H \cap D) = a$. If $a > 0$, then there exist $P_n \in \mathscr{A}$, $Q_n \in \mathscr{B}$ such that $H \cap D \subset \cup(P_n \times Q_n)$ and $\sum_{n=1}^{\infty} \mu(P_n)\nu(Q_n) < 3a/2$. By the nonatomic property of μ and by Lemma 3.9, there exist $R_n \subset P_n$ such that $\frac{1}{3}\mu(P_n) \leq \mu(R_n) \leq \frac{1}{2}\mu(P_n)$. Then $H \cap D \subset \cup\{R_n \times [Q_n \cap f(R_n)]\} \cup \{(P_n - R_n) \times [Q_n \cap f(P_n - R_n)]\}$. This means that

$$\begin{aligned} a &= \lambda^*(H \cap D) \\ &\leq \sum \mu(R_n)\nu(Q_n \cap f(R_n)) + \sum \mu(P_n - R_n)\nu(Q_n \cap f(P_n - R_n)) \\ &< \tfrac{2}{3} \sum \mu(P_n)\nu(Q_n \cap f(R_n)) + \tfrac{2}{3} \sum \mu(P_n)\nu(Q_n \cap f(P_n - R_n)) \\ &= \tfrac{2}{3} \sum \mu(P_n)\nu(Q_n) < a, \end{aligned}$$

which is a contradiction. Hence $\lambda^*(H \cap D) = 0$, and therefore D is $\mathscr{M}$-measurable. The argument above also shows that if $A \subset D$ and $\lambda^*(A) < \infty$, then $\lambda^*(A) = 0$. ∎

• **Theorem 3.12.** Suppose μ and ν are complete and nonatomic. Suppose also that there is a bijection f from X into Y taking and carrying back, measurable sets into measurable sets such that

(i) $\mu(A) = \infty$ implies $\nu(f(A)) > 0$

and

(ii) $\nu(B) = \infty$ implies $\mu(f^{-1}(B)) > 0$.

Then Tonelli's Theorem is valid for $\mathscr{M}$-measurable functions and points are measurable if and only if μ and ν are both σ-finite. ∎

Proof. If μ, ν are complete, nonatomic, and σ-finite, the points in X and Y are measurable. For if $x \in A$ and $0 < \mu(A) < \infty$, by the nonatomicity of μ, there exists a sequence of subsets $A_n \in \mathscr{A}$, $x \in A_n$ such that $A_{n+1} \subset A_n$ and $\mu(A_n) = (1/2^{n+1}) \cdot \mu(A)$. Hence $x \in \bigcap_{n=1}^{\infty} A_n$ ($= B$, say) and $\mu(B) = 0$. Since μ is complete, $\{x\} \in \mathscr{A}$.

Suppose now that Tonelli's Theorem is valid for $\mathscr{M}$-measurable functions and points are measurable. Then by Lemma 3.10, $D = \{(x, f(x)): x \in X\}$ is $\mathscr{M}$-measurable and $\iint \chi_D(x, y)\, d\mu\, d\nu$ is well defined and $= 0$. Hence $\mu \times \nu(D) = 0$. Then there exist $P_n \in \mathscr{A}$ and $Q_n \in \mathscr{B}$ such that $D \subset \cup(P_n \times Q_n)$ and $\sum \mu(P_n)\nu(Q_n) < \infty$. Let $E_n = P_n \cap f^{-1}(Q_n)$ and $F_n = Q_n \cap f(P_n)$. Then $f(E_n) = F_n$. Also $D \subset \cup(E_n \times F_n)$ and $\sum \mu(E_n)\nu(F_n) < \infty$. Clearly, $X \subset \cup E_n$. If μ is not σ-finite, there is one E_n such that $\mu(E_n) = \infty$; but then $\nu(F_n) > 0$ by hypothesis. This contradicts that the above sum is finite. Hence μ (and similarly ν) is σ-finite. ∎

• **Corollary 3.5.** Suppose $X = Y$, $\mathscr{A} = \mathscr{B}$, and $\mu = \nu$, where μ is complete and nonatomic. Then Tonelli's Theorem is valid for nonnegative $\mathscr{M}$-measurable functions, and singletons in X are in $\mathscr{A}$ if and only if μ is σ-finite. ∎

Problems

3.4.1. Let $X = Y = [0, 1]$, and $\mathscr{A} = \mathscr{B}$ is a σ-algebra containing the open sets. Show that $\mathscr{A} \times \mathscr{B}$ contains all the open sets of $X \times Y$.

3.4.2. Let f and g be integrable on $(X, \mathscr{A}, \mu)$ and $(Y, \mathscr{B}, \nu)$, respectively. Show that $f(x) \cdot g(y)$ is integrable on $(X \times Y, \mathscr{A} \times \mathscr{B}, \mu \times \nu)$ and

$$\int f(x)g(y)\, d(\mu \times \nu) = \int f\, d\mu \cdot \int g\, d\nu.$$

3.4.3. Show that

$$\lim_{n\to\infty} \int_0^n \frac{\sin x}{x}\, dx = \frac{\pi}{2}.$$

[Hint: Note that

$$\int_0^n \frac{\sin x}{x}\,dx = \int_0^n \left(\int_0^\infty e^{-xt}\,dt\right)\sin x\,dx.$$

and then use Fubini's Theorem.]

3.4.4. Show that completeness of the measures is essential in Theorem 3.7. [Hint: Consider $\mu = \nu =$ the Borel measure on R and the function $\chi_{A\times[0,1]}(x, y)$ where A is a non-Borel Lebesgue-measurable set with finite positive Lebesgue measure.]

3.4.5. Show that a nonnegative real-valued function f on R is Lebesgue measurable if and only if the set $E = \{(x, y): 0 \leq y \leq f(x)\}$ is product measurable. (Hint: Consider simple functions first and then pass to the limit, for the "only if" part. To prove the "if" part, observe that

$$\{(x, y): f(x) > c, y > 0\} = \bigcup_{n=1}^{\infty}\left\{(x, y): \left(x, \frac{y}{n} + c\right) \in E,\ y > 0\right\}.\Big)$$

Using the set E and also the set $D = \{(x, y): 0 \leq y < f(x)\}$, show that the graph of a measurable function is measurable. Note that the graph may be product measurable even for a nonmeasurable function.

3.4.6. Let f be Lebesgue integrable in Problem 3.4.5. Show that $\mu\times\mu(E) = \int f(x)\,dx$, where μ is the Lebesgue measure on R. Thus, the integral of f is the area under the curve $y = f(x)$.

3.4.7. Prove that a real-valued function f on R^2 is Borel measurable (i.e., $f^{-1}(H)$ is a Borel set for any open set H), if each section f_x is Borel measurable on R and each section f^y is continuous on R. [Hint: Consider the functions

$$\sum_{i=-n^2}^{i=n^2} f\left(\frac{i-1}{n}, y\right)\chi_{[(i-1)/n,\,i/n)}(x).\Big]$$

3.4.8. (a) Show that the set $\{(x, y): x - y \in E\}$ is $\mu\times\mu$ measurable on R^2, where μ is the Lebesgue measure on R and E is a Lebesgue-measurable set on R. (Hint: Consider first when E is open and then, when E is a G_δ-set or a set of measure zero.) Use this to show that $f(x - y)$ is $\mu\times\mu$ measurable for a Lebesgue-measurable function f.

(b) Let f and g be complex-valued Lebesgue-integrable functions on R. Use Tonelli's theorem to show that $f * g(x) = \int f(x - y)g(y)\,dy$ is a well-defined Lebesgue-integrable function on R.

(c) Let f and g be as in (b). We define the Fourier transform $\hat{f}$ of f by $\hat{f}(t) = \int e^{-itx}f(x)\,dx$. [Sometimes by analogy with the Fourier series and also for the purpose of getting simpler forms for certain results,

$$\hat{f}(t) = \frac{1}{2\pi} \int e^{-itx} f(x)\, dx.$$

defines the Fourier transform.] Show that

(i) $\hat{f}$ is a complex-valued continuous function on R vanishing at ∞;

(ii) $\widehat{f*g} = \hat{f} \cdot \hat{g}$;

(iii) $\bar{\hat{f}} = \widehat{f^*}$, $f^*(x) = \overline{f(-x)}$ (bar = conjugate);

(iv) the set of all Fourier transforms of complex-valued Lebesgue-integrable functions is dense (relative to uniform convergence) in the set of all complex-valued continuous functions vanishing at ∞;

(v) for any two finite Borel measures μ and ν on R, the equality $\int e^{itx}\, d\mu(x) = \int e^{itx}\, d\nu(x)$ for all real t implies that $\mu = \nu$;

(vi) $\hat{f} = \hat{g}$ implies hat $f = g$ a.e.

[Hint: (iv) Use the Stone–Weierstrass theorem and see also Remark 1.92. (v) Note that $\int \hat{f}(-t)\, d\mu(t) = \int \hat{f}(-t)\, d\nu(t)$. Hence using (iv), $\int f\, d\mu = \int f\, d\nu$ for all f in $C_0(R)$. Use Problem 2.3.12 to conclude that $\mu = \nu$. (vi) Suppose first that f and g are real-valued. Write $f = f^+ - f^-$ and $g = g^+ - g^-$ (usual notation). Let

$$\mu(A) = \int_A (f^+ + g^-)\, dx \quad \text{and} \quad \nu(A) = \int_A (g^+ + f^-)\, dx.$$

Then $\int e^{itx}\, d\mu(x) = \int e^{itx}\, d\nu(x)$. Use (v).]

3.4.9. Let X be an uncountable set and $\mathscr{A}$ be the smallest σ-algebra of subsets of $X \times X$ containing all sets of the form $A \times B$, where $A \subset X$, $B \subset X$. When does $\mathscr{A}$ contain all subsets of $X \times X$? (Hint: (1) Assume card $X > c$. Let $D = \{(x, y)\colon x = y\} \subset X \times X$. Suppose $D \in \mathscr{A}$. Then there is a class $\mathscr{F}$ consisting of an at most countable number of rectangles $(A_i \times B_i)$ such that $D \in \sigma(\mathscr{F})$. Let the A_i's generate the σ-algebra $\mathscr{B}$. Then by Proposition 1.10 of Chapter 1, card $\mathscr{B} \leq c$. But for each $y \in X$, $D^y \in \mathscr{B}$, which is a contradiction. (2) Assume card $X = c$. By assuming the continuum hypothesis, identify X with $[0, \Omega)$ and $[0, 1]$, where Ω is the first uncountable ordinal. Observe the following:

(i) For any function f from $[0, \Omega)$ (or from one of its subsets) to $[0, 1]$, the graph of $f \in \mathscr{A}$.

(ii) For $B \subset C_1 = \{(x, y) \in [0, \Omega) \times [0, \Omega)\colon y \leq x\}$ or $B \subset C_2 = \{(x, y) \in [0, \Omega) \times [0, \Omega)\colon y > x\}$, $B \in \mathscr{A}$.

Therefore $\mathscr{A}$ consists of all subsets of $X \times X$.)

The first case of this result is due to P. R. Halmos, and the second one is due to B. V. Rao.

3.4.10. *An Example of a Set of Positive Product Measure Not Containing Any Measurable Rectangle of Positive Product Measure.* Let m be the Lebesgue measure on R and C be a Cantor set $\subset[0, 1]$ with $m(C) > 0$, as in Problem 2.3.7(b) of Chapter 2. Let $S = \{(x, y): x - y \in C\} \subset [0, 1]\times[0, 1]$. Show that S is a compact subset of R^2, $m\times m(S) > 0$, and S is the desired set. [Hint: If $A\times B \subset S$, then $A - B \subset C$; since $m(A) > 0$ and $m(B) > 0$ imply $A - B$ has nonempty interior (see Problem 3.2.15), $m\times m(A\times B) = 0$.]

3.4.11. *Integration by Parts Using Fubini's Theorem.* Let f and g be two monotonic increasing, bounded, right-continuous functions on $[a, b]$. Let $f^*(x) = f(x) - f(x-)$ and $g^*(x) = g(x) - g(x-)$, and let (d_i) be the set of common points of discontinuity of f and g. Prove that

$$\int_{(a,b]} f\, d\mu_g + \int_{(a,b]} g\, d\mu_f = f(b)g(b) - f(a)g(a) + \sum_{a<d_i\leq b} f^*(d_i)g^*(d_i).$$

[Hint: Let $A = \{(x, y): a < x \leq b,\ y \leq x\}$, $B = \{(x, y): a < x \leq b,\ y \leq a\}$, and $C = \{(x, y): y \leq x \leq b,\ a < y \leq b\}$, so that $A = B \cup C$ and $B \cap C = \varnothing$. By Fubini's Theorem,

$$\int_{(a,b]} f\, d\mu_g = \int_A \int d\mu_f\, d\mu_g = \int_B \int d\mu_g\, d\mu_f + \int_C \int d\mu_g\, d\mu_f.\Big]$$

3.4.12. *An Application of the Baire Category Theorem* (splitting of a family of sets of positive measure). Let $(X, \mathscr{A}, \mu)$ be a measure space and $\mathscr{F}$ be a family of sets in $\mathscr{A}$ with positive measure. Then $A \in \mathscr{A}$ is said to *split* $\mathscr{F}$ if $0 < \mu(A \cap B) < \mu(B)$ for every $B \in \mathscr{F}$. Prove the following result due to R. B. Kirk: Suppose μ is nonatomic and $\mathscr{F}$ countable. Then there exists $A \in \mathscr{A}$ which splits $\mathscr{F}$. In particular, *there exists a Lebesgue-measurable set $A \subset R$ which splits all open sets with respect to the Lebesgue measure.* [Hint: With no loss of generality, let $\mu(X) < \infty$. Let $d(A, B) = \mu(A \triangle B)$. Then $(\mathscr{A}, d)$ is a complete pseudometric space. If $\mathscr{F} = (B_n)$, let $\mathscr{F}_n = \{A \in \mathscr{A}: \mu(A \cap B_n) = 0 \text{ or } \mu(A \cap B_n) = \mu(B_n)\}$. Then $\mathscr{F}_n$ is closed. Also by the nonatomicity of μ (through Lemma 3.9), $\mathscr{F}_n$ is nowhere dense. By the Baire Category Theorem, there exists $A \in \bigcap_{n=1}^{\infty}(\mathscr{A} - \mathscr{F}_n)$. This A splits $\mathscr{F}$.]

3.4.13. *Another Proof of the Ulam Theorem Using Fubini's Theorem.* Let μ be a measure defined on the class 2^X of all subsets of $X = [0, \Omega)$, Ω being the first uncountable ordinal such that $\mu(\{x\}) = 0$ for every singleton $\{x\}$. Then μ is the zero measure. [Hint: By Problem 3.4.9, the product σ-

algebra $2^X \times 2^X = 2^{X \times X}$. Let $X_1 = \{(\alpha, \beta): \alpha \leq \beta, \alpha \text{ and } \beta \text{ in } X\}$ and $X_2 = \{(\alpha, \beta): \alpha > \beta, \alpha \text{ and } \beta \text{ in } X\}$. By Fubini's Theorem, $\mu \times \mu(X_1) = \mu \times \mu(X_2) = 0$.]

3.4.14. *Uniqueness of Translation-Invariant Measures*: Let μ and ν be two σ-finite translation-invariant measures defined on the Borel sets of R^n such that for *some* Borel set $A \subset R^n$ and some constant c, $0 < \mu(A) = c \cdot \nu(A) < \infty$. Then $\mu = c \cdot \nu$. Can R^n be replaced by any separable metric space with a vector space structure where the addition is continuous? [Hint: Apply Theorem 3.10 to the iterated integral

$$\iint \chi_B(x)\chi_A(x + y)\mu(dx)\nu(dy).]$$

3.4.15. *Interchange of the Order of Partial Differentiation*: Let f be a continuous real function defined on an open rectangle V of R^2 such that (i) $\partial f/\partial x$ exists and is continuous on V; (ii) for some a, $(d/dy)[f(a, y)]$ exists at all points (a, y) in V; and (iii) $(\partial/\partial y)(\partial f/\partial x)$ exists and is continuous on V. Then $\partial f/\partial y$ and $(\partial/\partial x)(\partial f/\partial y)$ exist on V; moreover, $(\partial/\partial x)(\partial f/\partial y) = (\partial/\partial y)(\partial f/\partial x)$ on V. [Hint: Use Fubini's Theorem and the fundamental theorem of calculus to show that

$$f(x, y) - f(a, y) - f(x, b) + f(a, b) = \int_b^y \int_a^x \frac{\partial}{\partial y}\left(\frac{\partial f}{\partial x}\right) dx\, dy.$$

Notice that for each x, the inner integral defines a continuous function of y.]

3.4.16. *Measures Which Coincide on the Translates of a Set.* Suppose μ and ν are any two Borel measures on R such that $\mu(R) = \nu(R) = 1$. Let B be a Borel set with finite positive Lebesgue measure. Suppose that $\hat{\chi}_B(t) = \int e^{-itx}\chi_B(x)\, dx$ and the set $\{t: \hat{\chi}_B(t) \neq 0\}$ is dense in R. Also, let $\mu(B + t) = \nu(B + t)$ for all t in R. Show that $\mu = \nu$. [Hint: If $f(x) = \int \chi_B(y - x)\, d\mu(y)$, then $\hat{f}(t) = \hat{\mu}(t) \cdot \hat{\chi}_B(-t)$, where $\hat{\mu}(t) = \int e^{-itx}\, d\mu(x)$. It follows that $\hat{\mu}(t) = \hat{\nu}(t)$. Now use Problem 3.4.8c (v).] This result remains true without the hypothesis on $\hat{\chi}_B(t)$, and is due to N. A. Sapogov.

3.4.17. Let $p \geq 1$ and $f(x, y)$ be any product-measurable function on $[0, 1] \times [0, 1]$. Let us write

$$\| g \| = \left(\int_0^1 | g(y) |^p\, dy\right)^{1/p}.$$

It will be proven in Chapter 5 that

(i) $\| g + h \| \leq \| g \| + \| h \|$

(ii) $\int_0^1 | g(y) |\, dy \leq \| g \|$.

Use these facts to prove that

$$\left\| \int | f(x, y) | \, dy \right\| \leq \int \| f(x, y) \| \, dx.$$

(Hint: Assume that f is bounded. Note that $L_p = \{g: \| g \| < \infty\}$ is a separable pseudometric space with the metric $d(g, h) = \| g - h \|$. Then there exist functions $f_n(x, y) = \sum_i \chi_{A_i}(x) g_i(y)$, $g_i \in L_p$ such that for almost all x, $\lim_{n\to\infty} \| f - f_n \| = 0$. Note that the inequality in the problem holds for f_n and by (ii) above,

$$\lim_{n\to\infty} \left\| \int | f(x, y) | \, dy - \int | f_n(x, y) | \, dy \right\| = 0.\Big]$$

4

Differentiation

There are two theorems from the theory of Riemann integration which indicate the sense in which integration and differentiation are inverse processes. They read as follows:

(1) If f is Riemann integrable on $[a, b]$ and F is the function given by the equation

$$F(x) = \int_a^x f(t)\,dt \qquad \text{for } x \in [a, b]$$

then F is continuous on $[a, b]$; also if f is continuous at x_0 in $[a, b]$, then F is differentiable at x_0 and $F'(x_0) = f(x_0)$.

(2) If f is differentiable on $[a, b]$ and f' is Riemann integrable on $[a, b]$, then

$$\int_a^b f'(x)\,dx = f(b) - f(a).$$

In this chapter we shall first explore the analogs of these theorems in the theory of Lebesgue integration of real-valued functions on an interval $[a, b]$. In doing so, we shall study absolutely continuous functions and functions of bounded variation and prove Lebesgue's well-known result that every monotone increasing function is differentiable almost everywhere. This exploration will lead us to more abstract notions of indefinite integrals and absolute continuity, and in particular we shall prove the famous Radon–Nikodyn Theorem. This outstanding result has many uses in measure theory and probability theory, and in a later chapter we shall use it in the study of conjugate spaces. At the close of this chapter we shall study an application of this key theorem in changing the variable in Lebesgue integration.

4.1. Differentiation of Real-Valued Functions

In order to define the derivative of a real function f on R, we first define four quantities known as the *derivates* of f at a point a.

Definition 4.1. Let $a \in R$ and let δ be a real number such that $\delta > 0$. If f is defined in the interval $[a, a + \delta)$, the *upper right derivate* $D^+f(a)$ and the *lower right derivate* $D_+f(a)$ of f at a are the extended real numbers given by

$$D^+f(a) = \limsup_{h \to 0^+} \frac{f(a+h) - f(a)}{h}$$

and

$$D_+f(a) = \liminf_{h \to 0^+} \frac{f(a+h) - f(a)}{h}.$$

If f is defined in the interval $(a - \delta, a]$, the *upper left derivate* $D^-f(a)$ and the *lower left derivate* $D_-f(a)$ of f at a are the extended real numbers given by

$$D^-f(a) = \limsup_{h \to 0^+} \frac{f(a) - f(a-h)}{h},$$

and

$$D_-f(a) = \liminf_{h \to 0^+} \frac{f(a) - f(a-h)}{h}.$$

If $D^+f(a) = D_+f(a)$, then the common value is called the *right derivative* $f_+'(a)$ of f at a, and f is said to be *right differentiable* at a if this common value is finite. In a similar fashion, the *left derivative* $f_-'(a)$ is defined as well as *left differentiability*. If $f_+'(a)$ and $f_-'(a)$ exist and are equal, their common value $f'(a)$ is the *derivative* of f at a, and f is *differentiable* at a if this common value $f'(a)$ is finite. ∎

It is clear that $D_+f(a) \leq D^+f(a)$ and $D_-f(a) \leq D^-f(a)$.

Proposition 4.1. Let f be a monotone increasing function on a bounded or unbounded interval I. The four derivate functions of f are measurable functions on I (defined except perhaps at the endpoints of I). ∎

Proof. We will show that the function D^+f whose value at a in I (a not an endpoint of I) is $D^+f(a)$ is a measurable function. We may assume for convenience that I is open. The proofs that the other derivates are measurable are similar.

By definition

$$D^+f(a) = \limsup_{h\to 0^+} \frac{f(a+h)-f(a)}{h}.$$

For each positive integer n the function f_n given by

$$f_n(a) = \sup_{\substack{0<h<1/n \\ a+h\in I}} \frac{f(a+h)-f(a)}{h} \tag{4.1}$$

is a well-defined extended real-valued function on I. Also

$$D^+f(a) = \lim_{n\to\infty} f_n(a)$$

so that our proof will be complete if we can show that for all values of n, f_n is a measurable function.

Consider the function $\hat{f}_n$ defined on the interval I by

$$\hat{f}_n(a) = \sup_{\substack{0<h_i<1/n \\ a+h_i\in I \\ h_i \text{ rational}}} \frac{f(a+h_i)-f(a)}{h_i}.$$

Since $\hat{f}_n$ is thus the pointwise supremum of a sequence (g_i) of measurable functions, where g_i is given by

$$g_i(a) = \begin{cases} \dfrac{f(a+h_i)-f(a)}{h_i} & \text{if } a+h_i \in I \\ 0 & \text{otherwise,} \end{cases}$$

each $\hat{f}_n$ is measurable. The proof is therefore reduced to showing that $f_n = \hat{f}_n$.

Clearly, $\hat{f}_n \le f_n$ for all n. Now let n be fixed and ε be an arbitrary positive number. Suppose first that $f_n(a) < \infty$. From equation (4.1), there exists an element h_0 [or $h_0(a)$] on $(0, 1/n)$ such that $a + h_0 \in I$ and

$$f_n(a) - \frac{\varepsilon}{2} < \frac{f(a+h_0)-f(a)}{h_0}.$$

Let r be a rational number in $(h_0, 1/n)$. Since f is nondecreasing,

$$f_n(a) - \frac{\varepsilon}{2} < \frac{f(a+h_0)-f(a)}{h_0} \le \frac{f(a+r)-f(a)}{h_0}. \tag{4.2}$$

Multiplying the inequalities (4.2) by h_0/r, we obtain the inequality

$$f_n(a) + \left(\frac{h_0}{r} - 1\right) f_n(a) - \frac{\varepsilon h_0}{2r} < \frac{f(a+r)-f(a)}{r}. \tag{4.3}$$

Choose r very close to h_0 in $(h_0, 1/n)$; then (4.3) becomes

$$f_n(a) - \varepsilon < \frac{f(a+r) - f(a)}{r}$$

when r is chosen so that $(h_0/r - 1)f_n(a) > -\varepsilon/2$. Thus $f_n(a) \leq \hat{f}_n(a)$ when $\hat{f}_n(a) < \infty$. It is easy to see that $\hat{f}_n(a) = \infty$ whenever $f_n(a) = \infty$. Hence $f_n \leq \hat{f}_n$. ∎

Proposition 4.2. Let (a, b) be an open interval of R and f a real-valued function defined on (a, b). The set of all points in (a, b), where $f_-'(x)$ and $f_+'(x)$ both exist (although possibly infinite) and are not equal, is at most countable. ∎

Proof. Let $A = \{x \in (a, b): f_-'(x) \text{ and } f_+'(x) \text{ exist but } f_-'(x) < f_+(x)\}$. For each x in A choose a rational number r_x such that $f_-'(x) < r_x < f_+'(x)$. Also choose rational numbers s_x and t_x such that $a < s_x < x$ and $x < t_x < b$ and such that

$$r_x < \frac{f(y) - f(x)}{y - x} \qquad \text{for all } y \text{ in } (x, t_x) \tag{4.4}$$

and

$$\frac{f(y) - f(x)}{y - x} < r_x \qquad \text{for all } y \text{ in } (s_x, x). \tag{4.5}$$

The inequalities (4.4) and (4.5) together give

$$f(y) - f(x) > r_x(y - x) \tag{4.6}$$

for all $y \neq x$ in (s_x, t_x). Define $\varphi: A \to Q \times Q \times Q$ by

$$\varphi(x) = (s_x, r_x, t_x).$$

Then φ is a one-to-one function since if $\varphi(x) = \varphi(z)$ for x and z in A, then the intervals (s_x, t_x) and (s_z, t_z) are identical and x and z are both in this interval. If $x \neq z$, then (4.6) gives us

$$f(z) - f(x) > r_x(z - x)$$

and

$$f(x) - f(z) > r_z(x - z),$$

whereas the addition of these inequalities establishes that

$$0 > 0.$$

Thus φ is a one-to-one function and A must be at most countable. The proof that the set

$$B = \{x \in (a, b): f_-'(x) \text{ and } f_+'(x) \text{ exist but } f_-'(x) > f_+'(x)\}$$

is at most countable is similar. ∎

It is easy to construct a function continuous on an interval $[a, b]$ but which fails to have a derivative at any given finite number of points in $[a, b]$. Perhaps contrary to what we would imagine, however, continuous functions can be found whose derivatives do not exist at points of a dense subset of the interval (see Problem 4.1.3). Furthermore, if a is a positive odd integer and $0 < b < 1$, the function f given by the series

$$f(x) = \sum_{n=0}^{\infty} b^n \cos(a^n \pi x)$$

is a continuous function; but if $ab > 1 + \frac{3}{2}\pi$ the function f is not differentiable at *any* point. This example is due to Weierstrass (see [26]). Another example of an everywhere continuous but nowhere differentiable function on R is given in problem 4.1.7. A careful examination of these functions reveals that they are continuous but not monotonic on *any* interval. That this should be the case is evident from the theorem we prove below that a real-valued monotonic function on an interval is differentiable almost everywhere. To prove this we use the theorem below of Vitali. In this theorem we need the following terminology.

Definition 4.2. Let $E \subset R$. A family $\mathscr{F}$ of intervals, each having positive length, is said to be a *Vitali covering* of E if for each $\varepsilon > 0$ and any $x \in E$ there is an interval $I \in \mathscr{F}$ such that $x \in I$ and $l(I) < \varepsilon$. ∎

Example. Let $E = [a, b]$. The collection of $I_{n,i} = [r_n - 1/i, r_n + 1/i]$ for $n, i = 1, 2, 3, \ldots$, where r_n is an enumeration of the rationals in E, is a Vitali covering of E.

Theorem 4.1. *Vitali's Covering Theorem.*† Let E be an arbitrary subset of R and let $\mathscr{F}$ be a family of closed intervals of positive length forming a Vitali covering of E. Let μ be any measure on R defined on a σ-algebra containing all intervals in R such that $\mu^*(E) < \infty$, where μ^* is the outer measure induced by μ. Then there exists a sequence (I_n) of disjoint intervals in $\mathscr{F}$ such that

$$\lim_{m \to \infty} \mu^*\left(E - \bigcup_{n=1}^{m} I_n\right) = \mu^*\left(E - \bigcup_{n=1}^{\infty} I_n\right) = 0.$$ ∎

† This general form of Vitali's Theorem is due to Miguel de Guzman.

(When μ is the Lebesgue measure, this theorem holds even when the intervals in $\mathscr{F}$ are not closed since the endpoints of an interval have Lebesgue measure zero.)

Proof. The proof is given in the following three steps.

Step I. Let A be any subset of R. Suppose that for each $x \in A$, there corresponds a finite interval (of positive length) I_x containing x. Then there exists a sequence (x_i) in A such that $A \subset \cup I_{x_i}$.

Because of Remark 1.46, there is no loss of generality in assuming that each x in A is either the left or right endpoint of the corresponding I_x. It is sufficient to give the argument when each x in A is a left endpoint of the corresponding I_x.

For each integer n and positive integer k, let us write

$$A_k = \{x \in A\colon \operatorname{diam} I_x > 1/k\} \quad \text{and} \quad a_{nk} = \inf\{x\colon\ x \in A_k \cap [n/k, (n+1)/k]\}.$$

Let $(x_{nk}^j) \subset A_k$ and $x_{nk}^j \to a_{nk}+$. Then we have

$$A_k \cap [n/k, (n+1)/k] \subset \bigcup_j I_{x_{nk}^j}.$$

By taking the union over all n and k, the assertion of this step follows.

Step. II. From any finite collection of intervals in R, one can select two disjoint subcollections of disjoint intervals such that the union of the intervals in the subcollections contains the intervals in the given collection.

To prove this step, there is no loss of generality in assuming that the given collection is $\{I_1, I_2, \ldots, I_n\}$ such that for each i, $I_i \cap (\bigcup_{j \neq i} I_j)^c$ is nonempty and contains some point a_i. We can also assume (after renaming the intervals if necessary) that the a_i are arranged in increasing order. Then $\{I_1, I_3, I_5, \ldots\}$ and $\{I_2, I_4, I_6, \ldots\}$ are subcollections of disjoint intervals meeting the requirements of this step.

Step III. In this step, we complete the proof of the theorem. By Step I, there is a sequence (J_i) of intervals in $\mathscr{F}$ such that $E \subset \cup J_i$. Then using Problem 2.3.2, we have

$$\infty > \mu^*(E) = \mu^*\left(E \cap \bigcup_{i=1}^{\infty} J_i\right) = \lim_{n\to\infty} \mu^*\left(E \cap \bigcup_{i=1}^{n} J_i\right)$$

and we can choose N so that

$$\mu^*\left(E \cap \bigcup_{i=1}^{N} J_i\right) > \frac{3}{4} \cdot \mu^*(E).$$

By Step II, we can choose from the collection (J_i), $1 \leq i \leq N$, a subcollection of disjoint intervals (I_i), $1 \leq i \leq n_1$, such that

$$\mu^*\left(E \cap \bigcup_{i=1}^{n_1} I_i\right) > \frac{3}{8} \cdot \mu^*(E). \tag{4.7}$$

Now let B be a μ^*-measurable set such that $B \supset E$ and $\mu^*(B) = \mu^*(E)$. Then we have

$$\mu\left(B - \bigcup_{i=1}^{n_1} I_i\right) = \mu(B) - \mu\left(B \cap \bigcup_{i=1}^{n_1} I_i\right) \leq \mu(B) - \mu^*\left(E \cap \bigcup_{i=1}^{n_1} I_i\right)$$

and therefore,

$$\mu^*\left(E - \bigcup_{i=1}^{n_1} I_i\right) \leq \left(1 - \frac{3}{8}\right)\mu^*(E),$$

by (4.7) and above. Since $\bigcup_{i=1}^{n_1} I_i$ is closed, for each x in the set $E_1 = E - \bigcup_{i=1}^{n_1} I_i$ we can take an interval from $\mathscr{F}$ and disjoint from $\bigcup_{i=1}^{n_1} I_i$, and then repeat the above procedure with E_1 instead of E, obtaining disjoint intervals (I_i), $n_1 + 1 \leq i \leq n_2$, from $\mathscr{F}$ such that

$$\mu^*\left(E - \bigcup_{i=1}^{n_2} I_i\right) = \mu^*\left(E_1 - \bigcup_{i=n_1+1}^{n_2} I_i\right) < \frac{5}{8} \cdot \mu^*(E_1) < \left(\frac{5}{8}\right)^2 \cdot \mu^*(E).$$

By continuing this process, the theorem follows. ∎

The Vitali Covering Theorem is now used in the proof of the following theorem of Lebesgue.

Theorem 4.2. If f is a nondecreasing function on an interval I (bounded or unbounded), then f is differentiable almost everywhere. ∎

Proof. Recall that f is differentiable at a point a in the interior of I if $D^+f(a) = D^-f(a) = D_+f(a) = D_-f(a) \neq \pm\infty$. The idea of the proof is to show that the set of points for which the derivates are not all equal is a set of measure zero and then establish that the common value is also finite almost everywhere. The proof is given for the case when $I = (c, d)$, a bounded and open interval—the extensions being left to the reader.

We shall show that $E = \{x \in I\colon D^+f(x) > D_+f(x)\}$ is a null set. The proof that $\{x \in I\colon D_-f(x) < D^-f(x)\}$ is a null set is similar and omitted. By Proposition 4.2, these facts establish that $D^+f(x) = D_+f(x) = D_-f(x) = D^-f(x)$ almost everywhere.

For every pair of positive rationals u and v, let

$$E_{uv} = \{x \in E\colon D_+f(x) < u < v < D^+f(x)\}.$$

Clearly $E = \bigcup_{u<v} E_{uv}$ and this is a countable union with u and v from the positive rationals. It suffices to show that $m(E_{uv}) = 0$ for all rational u and v with $0 < u < v$.

Let us assume the contrary: There exist rationals u and v with $0 < u < v$ and $m(E_{uv}) = \alpha > 0$. Letting $\varepsilon > 0$ be arbitrary, choose an open set O containing E_{uv} such that $m(0) < \alpha + \varepsilon$. For each $x \in E_{uv}$ there exists an arbitrarily small interval $[x, x + h]$ contained in $O \cap I$ such that $f(x + h) - f(x) < uh$. The family of all such intervals forms a Vitali covering of E_{uv} and so there exists a finite pairwise disjoint subfamily $\{[x_i, x_i + h_i]: i = 1, 2, \ldots, N\}$ such that

$$m\left(E_{uv} - \bigcup_{i=1}^{N} [x_i, x_i + h_i]\right) < \varepsilon. \tag{4.8}$$

Clearly $m(E_{uv} - \bigcup_{i=1}^{N}(x_i, x_i + h_i))$ is also less than ε, and since each $(x_i, x_i + h_i) \subset O$, with $A = \bigcup_{i=1}^{N}(x_i, x_i + h_i)$, we have

$$m(A) = \sum_{i=1}^{N} h_i \leq m(O) < \alpha + \varepsilon.$$

Thus the inequalities $f(x_i + h_i) - f(x_i) \leq uh_i$ imply

$$\sum_{i=1}^{N} f(x_i + h_i) - f(x_i) < u \sum_{i=1}^{N} h_i < u(\alpha + \varepsilon).$$

Likewise for each $y \in E_{uv} \cap A$, there is an arbitrarily small interval $[y, y + k] \subset A$ such that

$$f(y + k) - f(y) > vk. \tag{4.9}$$

The family of all such intervals forms a Vitali covering of $E_{uv} \cap A$ so that there exists a finite, pairwise disjoint subfamily $\{[y_i, y_i + k_i]: i = 1, \ldots, M\}$ of such intervals such that

$$m\left(E_{uv} \cap A - \bigcup_{i=1}^{M} [y_i, y_i + k_i]\right) < \varepsilon. \tag{4.10}$$

Hence using inequalities (4.8) and (4.10)

$$\begin{aligned} \alpha = m(E_{uv}) &\leq m(E_{uv} \cap A) + m(E_{uv} \cap A^c) \\ &\leq m\left((E_{uv} \cap A) - \bigcup_{i=1}^{M} [y_i, y_i + k_i]\right) + m\left(\bigcup_{i=1}^{M} [y_i, y_i + k_i]\right) + m(E_{uv} \cap A^c) \\ &< \varepsilon + \sum_{i=1}^{M} k_i + \varepsilon = 2\varepsilon + \sum_{i=1}^{M} k_i. \end{aligned} \tag{4.11}$$

By inequalities (4.9) and (4.11)

$$v(\alpha - 2\varepsilon) < v\left(\sum_{i=1}^{M} k_i\right) < \sum_{i=1}^{M} f(y_i + k_i) - f(y_i).$$

Since

$$\sum_{i=1}^{M} [y_i, y_i + k_i] \subset \bigcup_{j=1}^{N} [x_j, x_j + h_j]$$

and f is nondecreasing, we have

$$v(\alpha - 2\varepsilon) < \sum_{i=1}^{M} f(y_i + k_i) - f(y_i) \leq \sum_{j=1}^{N} f(x_j + h_j) - f(x_j) < u(\alpha + \varepsilon).$$

Therefore $v\alpha < u\alpha +$ (an arbitrarily small positive number), so that $v \leq u$. This contradicts our original condition on v and u so that $m(E_{uv}) = 0$.

It remains to show that the set F of all x in $I = (c, d)$, where $f'(x) = \infty$, has measure zero. It is sufficient to show that for each $\varepsilon > 0$, $F_\varepsilon = F \cap (c + \varepsilon, d - \varepsilon)$ has measure zero. If β is an arbitrary positive number, then for each $x \in F_\varepsilon$ there is an arbitrarily small positive number h such that $[x, x + h] \subset (c + \varepsilon, d - \varepsilon)$ and

$$f(x + h) - f(x) > \beta h. \tag{4.12}$$

By Vitali's Theorem (Theorem 4.1), there exists a pairwise disjoint, countable family $\{[x_n, x_n + h_n]\}$ of such intervals such that $m(F_\varepsilon - \bigcup_n [x_n, x_n + h_n]) = 0$. Hence using inequality (4.12)

$$\beta m(F_\varepsilon) \leq \beta \sum_{n=1}^{\infty} h_n < \sum_{n=1}^{\infty} f(x_n + h_n) - f(x_n) \leq f(d - \varepsilon) - f(c + \varepsilon).$$

Since β is arbitrary, $m(F_\varepsilon)$ must be zero. ∎

(One could also show $m(F) = 0$ by using Proposition 4.4 of the next section since the integrability of the function $f'(x)$ implies it is finite almost everywhere.)

The next result characterizes all functions on an interval $[a, b]$ that can be expressed as the difference of two nondecreasing functions. They are the functions described in the next definition.

Definition 4.3. Let f be a real-valued function defined on an interval $I = [a, b]$. For each partition P: $a = x_0 < x_1 < x_2 < \cdots < x_n = b$ of $[a, b]$, let

$$V_a^b(f, P) = \sum_{i=1}^{n} | f(x_i) - f(x_{i-1}) |.$$

The extended real number $V_a^b f$ given by

$$V_a^b f = \sup\{V_a^b(f, P) \colon P \text{ is a partition of } [a, b]\}$$

is called the *total variation* of f. If $V_a^b f < \infty$, then f is said to be a function of *bounded variation* on $[a, b]$. ∎

Remarks

4.1. The equality $V_a^c f + V_c^b f = V_a^b f$ holds for all $c \in [a, b]$ when $V_d^d f$ is defined to be zero for all d.

4.2. $v(x) = V_a^x f$ is a nondecreasing function of x on $[a, b]$.

Proposition 4.3. f is a function of bounded variation on $[a, b]$ if and only if it can be written as the difference of two nondecreasing functions g and h. ∎

Proof. The sufficiency of this statement is easily verified. We here establish the converse.

Assume $V_a^b f < \infty$. Define real-valued functions g and h on $[a, b]$ by the rules

$$g(x) = V_a^x f \quad \text{and} \quad h(x) = V_a^x f - f(x),$$

where, as noted above, $V_a^a f$ is defined to be zero. By Remark 4.2, $g(x)$ is a nondecreasing function of x. Also $V_a^x f - f(x)$ is nondecreasing. To see this, suppose $x_1 < x_2$. Then

$$[V_a^{x_1} f - f(x_2)] - [V_a^{x_2} f - f(x_1)] = V_{x_1}^{x_2} f - [f(x_2) - f(x_1)] \geq 0.$$

This establishes the converse. ∎

Corollary 4.1. A real-valued function f on $[a, b]$ of bounded variation has a finite derivative $f'(x)$ a.e. ∎

Problems

4.1.1. Let f be the function given by

$$f(x) = \begin{cases} ax \sin^2 \dfrac{1}{x} + bx \cos^2 \dfrac{1}{x} & \text{if } x > 0 \\ 0 & \text{if } x = 0 \\ cx \sin^2 \dfrac{1}{x} + dx \cos^2 \dfrac{1}{x} & \text{if } x < 0. \end{cases}$$

where $a < b$ and $c < d$. Find the four derivates of f at 0.

4.1.2. Let

$$f(x) = \begin{cases} 0 & \text{if } x = 0 \\ x^\alpha \sin 1/x^\beta & \text{if } 0 < x \leq 1. \end{cases}$$

(a) If $0 < \beta < \alpha$ show that f is absolutely continuous.† {Hint: Observe that for every $\varepsilon > 0$, $f(1) - f(\varepsilon) = \int_\varepsilon^1 f'(x)\,dx$ and then show that $f'(x) \in L_1[0, 1]$.}

(b) If $0 < \alpha \leq \beta$ show that f is not of bounded variation.

4.1.3. Let $\{a_1, a_2, \ldots\}$ be an enumeration of the rationals in $[0, 1]$. Prove that the function f defined on $[0, 1]$ by

$$f(x) = \sum_{n=1}^{\infty} \frac{|x - a_n|}{3^n}$$

is continuous on $[0, 1]$, but not differentiable at each rational a_n by showing $f_+'(a_n) \neq f_-'(a_n)$.

4.1.4. Let f be a continuous function on $[a, b]$. Show that the length of the curve $C = \{(x, y)\colon y = f(x) \text{ for } a \leq x \leq b\}$, which is given by

$$L = \sup\left\{ \sum_{i=1}^{n} [(x_i - x_{i-1})^2 + (f(x_i) - f(x_{i-1}))^2]^{1/2} : \right.$$
$$\left. a = x_0 < x_1 < \cdots < x_n = b \right\},$$

is finite if and only if $f(x)$ is of bounded variation on $[a, b]$.

4.1.5. State and prove the Vitali Covering Theorem for the case when $m^*(E) = \infty$, m being the Lebesgue measure.

4.1.6. If f is continuous on $[a, b]$ and one of its derivates is nonnegative in the interval (a, b), show that $f(a) \leq f(b)$. [Hint: Say $D^+f \geq 0$. Suppose that $f(b) - f(a) < -\varepsilon(b - a)$ and let $\varphi(x) = f(x) - f(a) + \varepsilon(x - a)$. Let ξ be the largest value in (a, b) such that $\varphi(\xi) = 0$. Then $D^+\varphi(\xi) \leq 0$.]

4.1.7. B. L. Van der Waerden gave the following example of an everywhere nondifferentiable but continuous function. The reader should verify the succeeding statements. For each x let $f_n(x)$ denote the distance from x to the nearest number that can be written in the form $m/10^n$, where m is an integer. The function f given by $f = \sum_{n=1}^{\infty} f_n$ is then continuous since each

† Note that we define "absolute continuity" in the next section; at this point it suffices to show $f(1) - f(0) = \int_0^1 f'(x)\,dx$.

f_n is continuous and the series converges uniformly. To show f is not differentiable, write any x in (0, 1) as a decimal $x = .a_1a_2a_3 \cdots a_q \cdots$. If the qth term is 4 or 9, let $x_q = x - 10^{-q}$; otherwise let $x_q = x + 10^{-q}$. The sequence (x_q) thus obtained converges to x but

$$f_n(x_q) - f_n(x) = \begin{cases} \pm (x_q - x) & \text{if } n < q \\ 0 & \text{if } n \geq q \end{cases}$$

so that $f(x_q) - f(x) = p(x_q - x)$, where p is an integer that is odd or even with $q - 1$.

4.1.8. This result is due to Fubini. Let $S = f_1 + f_2 + \cdots$ be a pointwise convergent series of monotonic functions—either all increasing or all decreasing—defined on $[a, b]$. Prove S is differentiable a.e. and $S'(x) = f_1'(x) + f_2'(x) + \cdots$ a.e. [Hint: Assuming each f_i is increasing we may also assume $f_i(a) = 0$ for all i. Set $S_n = \sum_{i=1}^n f_i$. Then $S_n \to S$, S is differentiable a.e., and $S_n'(x) \leq S_{n+1}'(x) \leq S'(x)$. Since the S_n' form an increasing sequence, S_n' converges, and to show $S_n' \to S'$ it suffices to show $S' - S_{n_k}' \to 0$ for a suitably chosen subsequence (n_k) of positive integers. Choose n_k so that $\sum_{k=1}^\infty S(b) - S_{n_k}(b) < \infty$. Then the series $\sum_{k=1}^\infty S - S_{n_k}$ converges everywhere and hence so does the series formed by $S' - S_{n_k}'$. Hence $S'(x) - S_{n_k}'(x) \to 0$.]

4.1.9. Let $E(\subset R)$ be the union of an arbitrary family of intervals, each being open, closed, half-open, or half-closed. Show that E is Lebesgue measurable. [Hint: Use a Vitali covering.]

4.1.10. Give an example of a continuous function that is not of bounded variation in any interval.

4.1.11. Show that if $f(x)$ is not of bounded variation on [0, 1], then there is $x \in [0, 1]$ such that f is not of bounded variation in any open interval containing x.

4.1.12. Construct a Lebesgue-measurable set in R^2 that is not a Borel set. [Hint: Let E be a subset of the diagonal, which is a (one-dimensional) Lebesgue-measurable set but not a Borel set. To each point of E associate a square with this point as a corner. The union of these squares is Lebesgue measurable (by Vitali's Theorem) but not a Borel set.]

4.1.13. *A Continuous, Strictly Increasing Function With Derivative* 0 *A.E.* Let (I_n) be the intervals [0, 1], [0, $\frac{1}{2}$], [$\frac{1}{2}$, 1], [0, $\frac{1}{4}$], . . . , [0, $\frac{1}{8}$], etc., and let f_n be the Cantor function defined on I_n. Let $f(x) = \sum_{n=1}^\infty (1/2^n) f_n(x)$. Prove that f is the desired function. [Use Problem 4.1.8.] Another example will be given in the text in Section 4.2.

4.2. Integration Versus Differentiation I: Absolutely Continuous Functions

As stated in the introduction to this chapter, we wish to examine the analogs of statements (1) and (2) in the theory of Lebesgue integration. Our first proposition is a beginning of this exploration.

Proposition 4.4. Suppose f is a nondecreasing function on $[a, b]$. Then f' is integrable on $[a, b]$ and in fact

$$\int_a^b f'(x)\,dx \le f(b) - f(a). \qquad \blacksquare$$

Proof. We can extend the domain of definition of f to R by defining $f(x) = f(a)$ for $x \le a$ and $f(x) = f(b)$ for $x \ge b$. By Proposition 4.1 and Theorem 4.2 f is differentiable a.e. and f' is a measurable function. The functions

$$g_n(x) = n[f(x + 1/n) - f(x)] \qquad \text{for } n = 1, 2, \ldots$$

are nonnegative and $f'(x) = \lim_n g_n(x)$ for almost all x. Using Fatou's Lemma

$$\begin{aligned}
\int_a^b f'(x)\,dx &= \int_a^b \liminf_n g_n(x) \le \liminf_n \int_a^b g_n(x)\\
&= \liminf_n \left[n \int_a^b \left\{f\left(x + \frac{1}{n}\right) - f(x)\right\} dx\right]\\
&= \liminf_n n\left[\int_{a+1/n}^{b+1/n} f(x)\,dx - \int_a^b f(x)\,dx\right]\\
&= \liminf_n n\left[\int_{a+1/n}^{b+1/n} f(x)\,dx + \int_b^{a+1/n} f(x)\,dx - \int_b^{a+1/n} f(x)\,dx\right.\\
&\qquad \left. - \int_a^b f(x)\,dx\right]\\
&= \liminf_n n\left[\int_b^{b+1/n} f(x)\,dx - \int_a^{a+1/n} f(x)\,dx\right]\\
&= \liminf_n n\left[\frac{1}{n} f(b) - \int_a^{a+1/n} f(x)\,dx\right]\\
&\le f(b) - f(a). \qquad \blacksquare
\end{aligned}$$

Can the inequality of Proposition 4.4 be "improved" to an equality? As the following examples show, the answer is no—unless some additional restrictions are placed on f.

Examples

4.1. Let k: $[0, 1] \to [0, 1]$ be the Cantor ternary function. Since k is a constant on each interval contained in the complement of the Cantor set, $k'(x) = 0$, a.e. Thus

$$0 = \int_0^1 k'(x)\,dx \neq k(1) - k(0) = 1.$$

• **4.2.** We now present an example of a function that is *strictly* increasing and continuous with $f' = 0$ a.e. so that on any interval

$$0 = \int_a^b f'(x)\,dx < f(b) - f(a).$$

Let t_0 be a fixed but arbitrary point in $(0, 1)$. Let f be the function on R that is the pointwise limit of the nondecreasing sequence (f_n) of functions constructed by induction as follows:

$$f_0(x) = x,$$

$$f_n(k2^{-n}) = \begin{cases} \dfrac{1-t_0}{2} f_{n-1}((k-1)2^{-n}) + \dfrac{1+t_0}{2} f_{n-1}((k+1)2^{-n} & \text{if } k \text{ is odd} \\ f_{n-1}(k2^{-n}) & \text{if } k \text{ is even,} \end{cases}$$

where $k = 0, \pm 1, \pm 2, \ldots$, and make f_n linear in the intervals $[(k-1)2^{-n}, k2^{-n}]$. Figure 2 shows the first three functions f_n for $t_0 = \frac{1}{2}$. By induction it

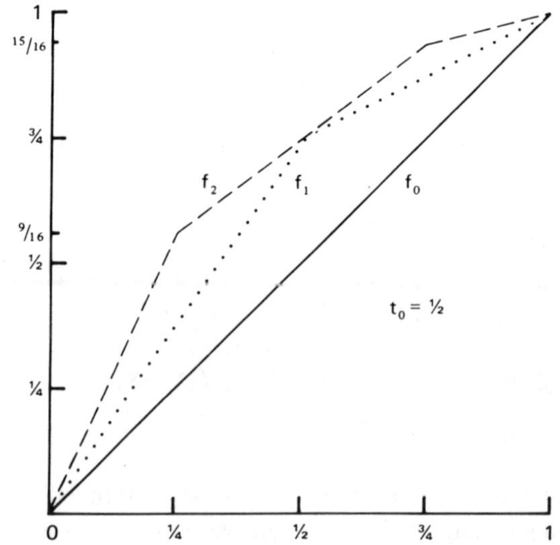

Fig. 2

can be shown that each f_n is nondecreasing and also $f_{n-1} \leq f_n$ for each $n \geq 1$. We now show that the limit function f [which exists since $f_n(x) \leq f_n(m) \leq m$ for $x \leq m$] is continuous, strictly increasing, and has derivative equal to zero a.e.

Proof. Let x be any real number and let a_n be the largest number $k2^{-n}$ satisfying $k2^{-n} < x$. Let $b_n = a_n + 2^{-n}$. Then $x \in (a_n, b_n]$ and $x = \lim_n a_n = \lim_n b_n$. Observing that for integers $n, s \geq 0$

$$f_n(a_n) = f_{n+s}(a_n) = f(a_n)$$

and

$$f_n(b_n) = f_{n+s}(b_n) = f(b_n),$$

it can be shown that

$$f(b_n) - f(a_n) = \frac{1 \pm t_0}{2} [f(b_{n-1}) - f(a_{n-1})] \qquad \text{for } n \geq 1$$

and (4.13)

$$f(b_0) - f(a_0) = 1$$

depending on whether k is odd or even. For example if $a_n = k2^{-n}$ and k is odd, then $a_{n-1} = (k-1)2^{-n}$, $b_{n-1} = (k+1)2^{-n} = b_n$ and thus

$$\begin{aligned} f(b_n) - f(a_n) &= f_{n-1}(b_{n-1}) - f_n(a_n) \\ &= f_{n-1}(b_{n-1}) - \frac{1-t_0}{2} f_{n-1}((k-1)2^{-n}) \\ &\quad - \frac{1+t_0}{2} f_{n-1}((k+1)2^{-n}) \\ &= f_{n-1}(b_{n-1}) - \frac{1-t_0}{2} f_{n-1}(a_{n-1}) - \frac{1+t_0}{2} f_{n-1}(b_{n-1}) \\ &= \frac{1-t_0}{2} [f(b_{n-1}) - f(a_{n-1})]. \end{aligned}$$

From equation (4.13) we thus obtain for $n \geq 1$

$$f(b_n) - f(a_n) = \prod_{i=1}^{n} \frac{1 + e_i t_0}{2} > 0, \tag{4.14}$$

where $e_i = \pm 1$. Then we have

$$f(x) - f(a_n) \leq f(b_n) - f(a_n) \leq \left(\frac{1+t_0}{2}\right)^n$$

so that $\lim_{n\to\infty} f(a_n) = f(x)$ and f is left continuous at x. Similarly, by taking b_n to be the smallest number $(k+1)2^{-n}$ larger than x, we can show that f is right continuous at x.

The function f is also strictly increasing. Let (a, b) be any interval and pick x in (a, b). Choose n such that $(a_n, b_n) \subset (a, b)$. By (4.14), $f(b_n) - f(a_n) > 0$.

Finally $f'(x) = 0$ a.e. Let x be a point at which f is differentiable. Then by (4.14)

$$f'(x) = \lim_{n\to\infty} \frac{f(b_n) - f(a_n)}{b_n - a_n} = \lim_{n\to\infty} \prod_{i=1}^{n} (1 + e_i t_0).$$

However,

$$\lim_{n\to\infty} \prod_{i=1}^{n} (1 + e_i t_0) = \lim_{n\to\infty} \exp \sum_{i=1}^{n} \ln(1 + e_i t_0) = 0$$

since $\sum_{i=1}^{\infty} \ln(1 + e_i t) = -\infty$ as the terms $\ln(1 + e_i t_0)$ are bounded away from zero. ∎

Our goal is to characterize the class of functions for which the inequality in Proposition 4.4 becomes an equality. More precisely we wish to find necessary and sufficient conditions on any function f defined on a closed interval $[a, b]$ to ensure that for all $x \in [a, b]$

$$f(x) - f(a) = \int_a^x f'(t)\, dt.$$

Let us first list some necessary properties possessed by any function G defined on $[a, b]$ by

$$G(x) = \int_a^x g(t)\, dt \tag{4.15}$$

where g is an integrable function on $[a, b]$. These are listed as Remarks 4.3–4.5 and Proposition 4.5.

Remarks

4.3. The function G defined in equation (4.15) is continuous on $[a, b]$.

4.4. The function G defined in equation (4.15) is a function of bounded variation on $[a, b]$ and thus differentiable a.e. Indeed, G is the difference of two nondecreasing functions:

$$G(x) = \int_a^x [g(t) \vee 0]\, dt - \int_a^x [-g(t)] \vee 0\, dt.$$

4.5. The function G defined in equation (4.15) is the zero function if and only if $g(t) = 0$ a.e.

Proof. It is clear that if $g(t) = 0$ a.e., then $G(x) = 0$ for all x in $[a, b]$. Conversely, suppose $g(t) > 0$ on a set E, necessarily measurable, of positive measure. Then by Problem 2.3.9 there is a closed set $F \subset E$ of positive measure. The set $H = (a, b) - F$ is open and can be written as a disjoint union of open intervals $\bigcup_n (a_n, b_n)$. Therefore

$$0 = \int_a^b g(t)\, dt = \int_F g(t)\, dt + \int_H g(t)\, dt = \int_F g(t)\, dt + \sum_{n=1}^{\infty} \int_{a_n}^{b_n} g(t)\, dt.$$

However, for each n,

$$\int_{a_n}^{b_n} g(t)\, dt = \int_a^{b_n} g(t)\, dt - \int_a^{a_n} g(t)\, dt = G(b_n) - G(a_n) = 0.$$

Therefore $\int_F g(t)\, dt = 0$; but since $g > 0$ on F and $m(F) > 0$, $\int_E g > 0$. This is a contradiction and E must have measure zero. ∎

The fourth necessary property possessed by G of equation (4.15) is the analog of Theorem (1) listed in the introduction to this chapter and is important enough to warrant a separate proposition.

Proposition 4.5. If g is integrable over $[a, b]$, then

$$G(x) = \int_a^x g(t)\, dt$$

is differentiable a.e. and $G' = g$ a.e. ∎

Proof. By Remark 4.4 G is differentiable a.e. It remains to show that $G' = g$ a.e. For this proof we may assume that g is nonnegative. We divide the proof into two cases.

Case 1. Suppose that g is bounded; that is, $|g| \leq K$ on $[a, b]$. Let

$$f_n(x) = n[G(x + 1/n) - G(x)] \qquad \text{for } x \in [a, b], \tag{4.16}$$

where $G(y)$ is defined to be $G(b)$ for $y \geq b$. Defining $g(y)$ to be $g(b)$ for $y \geq b$, then

$$f_n(x) = n \int_x^{x+1/n} g(t)\, dt \tag{4.17}$$

so that $|f_n(x)| \leq K$ for all x in $[a, b]$. Since $\lim_n f_n(x) = G'(x)$ a.e., the Dominated Convergence Theorem (Theorem 3.3) implies that G' is integrable and

$$\int_a^x G'(t)\,dt = \lim_n \int_a^x f_n(t)\,dt = \lim_n n \int_a^x \left[G\left(t + \frac{1}{n}\right) - G(t)\right] dt$$
$$= \lim_n n\left[\int_x^{x+1/n} G(t)\,dt - \int_a^{a+1/n} G(t)\,dt\right]. \tag{4.18}$$

Since G is continuous on $[x, x + 1/n]$, the mean value theorem for integrals implies that there exist δ and δ' such that $0 \leq \delta, \delta' \leq 1$ and

$$\int_x^{x+1/n} G(t)\,dt = \frac{1}{n}\, G\left(x + \frac{\delta}{n}\right)$$

and (4.19)

$$\int_a^{a+1/n} G(t)\,dt = \frac{1}{n}\, G\left(a + \frac{\delta'}{n}\right).$$

Putting equation (4.19) into equation (4.18) we obtain

$$\int_a^x G'(t)\,dt = G(x) - G(a) = \int_a^x g(t)\,dt.$$

Since $\int_a^x [G'(t) - g(t)]\,dt = 0$ for all x, we infer from Remark 4.5 that $G'(t) = g(t)$ a.e.

Case 2. If g is not bounded, let

$$g_n(x) = \begin{cases} g(x) & \text{if } g(x) \leq n \\ n & \text{if } g(x) > n. \end{cases}$$

By the Lebesgue Convergence Theorem,

$$\int_a^b g(t)\,dt = \lim_n \int_a^b g_n(t)\,dt.$$

By Case 1, since

$$G(x) = \int_a^x [g(t) - g_n(t)]\,dt + \int_a^x g_n(t)\,dt,$$

$$G'(x) = \frac{d}{dx}\int_a^x [g(t) - g_n(t)]\,dt + g_n(x) \text{ a.e.}$$

However, since $[g - g_n] \geq 0$, $\int_a^x [g - g_n]\,dt$ is an increasing function of x so that its derivative is nonnegative. Therefore $G'(x) \geq g_n(x)$ a.e. for all n.

Therefore $g(x) = \lim_n g_n(x) \le G'(x)$ a.e. Using Proposition 4.4 we obtain

$$G(b) - G(a) = \int_a^b g(t)\,dt \le \int_a^b G'(t)\,dt \le G(b) - G(a)$$

so that $g(t) = G'(t)$ a.e. ∎

Clearly, the fact that a function f is continuous, differentiable a.e., and of bounded variation on an interval $[a, b]$ is not sufficient to guarantee that

$$f(x) - f(a) = \int_a^x f'(t)\,dt \qquad \text{for all } x \text{ in } [a, b]. \tag{4.20}$$

Indeed, the Cantor ternary function is continuous and nondecreasing and yet equation (4.20) is not satisfied. The Cantor ternary function (or the function in Example 4.2) is peculiar in that it is a nonconstant function with derivative equal to zero a.e. What can we conclude about such functions? The answer is given in the following proposition.

Proposition 4.6. Suppose $f'(x) = 0$ a.e. on $[a, b]$ and $f(x)$ is not a constant function on $[a, b]$. Then there exists some $\varepsilon > 0$ such that for any $\delta > 0$ there is a finite disjoint collection of intervals

$$\{(x_1, y_1), (x_2, y_2), \ldots, (x_n, y_n)\}$$

with $\sum_{i=1}^n y_i - x_i < \delta$ but $\sum_{i=1}^n | f(y_i) - f(x_i) | > \varepsilon$. ∎

Proof. Since $f(x)$ is not a constant function, there exists a c in $(a, b]$ such that $f(a) \ne f(c)$. Let $E_c = [a, c) \cap \{x: f'(x) = 0\}$. For each x in E_c and any positive number γ, there are arbitrarily small intervals $[x, x + h] \subset [a, c]$ such that

$$| f(x + h) - f(x) | < \gamma h.$$

For a fixed γ, the set of all these intervals forms a Vitali covering of E_c. Hence by the Vitali theorem, for any $\delta > 0$ there is a finite disjoint collection

$$\{[x_1, x_1 + h_1), \ldots, [x_n, x_n + h_n)\}$$

such that $m\big(E_c - \bigcup_{i=1}^n [x_i, x_i + h_i)\big) = m\big([a, c] - \bigcup_{i=1}^n [x_i, x_i + h_i)\big) < \delta$. Rename the intervals so that

$$a = x_0 \le x_1 < x_1 + h_1 < x_2 < x_2 + h_2 < \cdots < x_n + h_n \le x_{n+1} = c.$$

Then with $h_0 = 0$ and $\varepsilon > 0$ chosen so that $2\varepsilon < |f(c) - f(a)|$,

$$\begin{aligned} 2\varepsilon < |f(c) - f(a)| &\leq \sum_{i=0}^{n} |f(x_{i+1}) - f(x_i + h_i)| \\ &+ \sum_{i=1}^{n} |f(x_i + h_i) - f(x_i)| \\ &\leq \sum_{i=0}^{n} |f(x_{i+1}) - f(x_i + h_i)| + \gamma \sum_{i=1}^{n} h_i \\ &\leq \sum_{i=0}^{n} |f(x_{i+1}) - f(x_i + h_i)| + \gamma(b - a). \end{aligned}$$

Since γ is arbitrary, pick γ small enough so that $\gamma(b - a) < \varepsilon$. Then

$$\varepsilon < \sum_{i=0}^{n} |f(x_{i+1}) - f(x_i + h_i)|$$

but

$$\sum_{i=1}^{n} [x_{i+1} - (x_i + h_i)] = m\Big(E_c - \bigcup_{i=1}^{n} [x_i, x_i + h_i]\Big) < \delta. \qquad \blacksquare$$

The negation of the conclusion of the preceding proposition is the property enjoyed by an important class of functions called *absolutely continuous* functions.

Definition 4.4. A function f on $[a, b]$ is said to be *absolutely continuous* if for each $\varepsilon > 0$ there exists a $\delta > 0$ such that whenever

$$\{(x_1, y_1), (x_2, y_2), \ldots, (x_n, y_n)\}$$

is a finite collection of nonoverlapping subintervals of $[a, b]$ with

$$\sum_{i=1}^{n} [y_i - x_i] < \delta$$

then

$$\sum_{i=1}^{n} |f(y_i) - f(x_i)| < \varepsilon. \qquad \blacksquare$$

Remark 4.6. The word "nonoverlapping" is important in Definition 4.4, and removing it results in what may be termed the definition of a "strongly" absolutely continuous function. It can be shown that a necessary and sufficient condition that function f be "strongly" absolutely continuous on $[a, b]$ is that it satisfies the Lipschitz condition: There exists a constant K

such that $| f(x) - f(y) | < K | x - y |$ for all x and y in $[a, b]$. (See Problem 4.2.10.) In contrast, although every function satisfying the Lipschitz condition on $[a, b]$ is absolutely continuous, that the converse is false is evidenced by the function $f(x) = x^{1/2}$ on $[0, 1]$.

Is the function G on $[a, b]$ given by the indefinite integral

$$G(x) = \int_a^x g(t)\, dt$$

of an integrable function g an absolutely continuous function? The reader can readily verify using Problem 3.2.1 in Chapter 3 that the answer, which we record in the following lemma, is in the affirmative.

Lemma 4.1. The indefinite integral G of an integrable function g on $[a, b]$ is absolutely continuous. ∎

We will now establish that the absolute continuity of a function G on an interval $[a, b]$ is a sufficient condition for G to be defined as an indefinite integral of an integrable function g. According to what we have already shown, G must then be necessarily continuous and of bounded variation on $[a, b]$ with $G' = g$ a.e. However, below we establish independently the fact that any absolutely continuous function is of bounded variation.

Lemma 4.2. If f is an absolutely continuous function on $[a, b]$, then f is a function of bounded variation on $[a, b]$. ∎

Proof. Let $\delta > 0$ correspond to the value $\varepsilon = 1$ in the definition of absolute continuity. Choose an integer n such that $n > (b - a)/\delta$. Subdivide $[a, b]$ into adjacent subintervals with the points

$$a < x_0 < x_1 < x_2 < \cdots < x_n = b,$$

where $x_{i+1} - x_i = (b - a)/n < \delta$ for $i = 1, 2, \ldots, n - 1$. Then for each i, $V_{x_i}^{x_{i+1}} < 1$ since any finite collection of nonoverlapping subintervals of $[x_i, x_{i+1}]$ will have total length less than δ. Hence

$$V_a^b f = \sum_{i=0}^{n-1} V_{x_i}^{x_{i+1}} < n.$$ ∎

Corollary 4.2. An absolutely continuous function on $[a, b]$ is differentiable a.e. on $[a, b]$. ∎

The converse of Lemma 4.2 is false, as the Cantor ternary function illustrates. (See Problem 4.2.4.) Problem 4.2.9 gives a sufficient condition for the converse to hold. In addition, the next result gives a necessary and sufficient condition for functions of bounded variation to be absolutely continuous.

Proposition 4.7. If f is of bounded variation on $[a, b]$, then f is absolutely continuous on $[a, b]$ if and only if the total variation function v, given by $v(x) = V_a{}^x f$, is absolutely continuous on $[a, b]$. ∎

Proof. Assume f is absolutely continuous. If $\varepsilon > 0$ is arbitrary, there exists a $\delta > 0$ such that, for any finite collection

$$\mathscr{C} = \{(x_1, y_1), (x_2, y_2), \ldots, (x_n, y_n)\}$$

of nonoverlapping subintervals of $[a, b]$ with $\sum_i (y_i - x_i) < \delta$ we have $\sum_i | f(y_i) - f(x_i) | < \varepsilon$. For each i, let

$$P_i\colon\ x_i = a_{i0} < a_{i1} < a_{i2} < \cdots < a_{im_i} = y_i$$

be a partition of $[x_i, y_i]$. Since

$$\sum_{i=1}^{n} \sum_{j=1}^{m_i} [a_{ij} - a_{i,j-1}] = \sum_{i=1}^{n} [y_i - x_i] < \delta,$$

then

$$\sum_{i=1}^{n} \sum_{j=1}^{m_i} | f(a_{ij}) - f(a_{i,j-1}) | < \varepsilon.$$

Fixing the collection $\mathscr{C}$ but varying the partitions P_i of each $[x_i, y_i]$, we obtain, upon taking the supremum over all such partitions P_i,

$$\sum_{i=1}^{n} V_{x_i}^{y_i} f < \varepsilon$$

or

$$\sum_{i=1}^{n} v(y_i) - v(x_i) < \varepsilon.$$

This proves v is absolutely continuous.

Conversely, since $| f(x_i) - f(x_{i-1}) | \le v(x_i) - v(x_{i-1})$ for $a \le x_{i-1} < x_i < b$, the absolute continuity of v implies that of f. ∎

If G is an indefinite integral as in equation (4.15), we know from Lemmas 4.1 and 4.2 that $V_a{}^b G < \infty$. In fact we have a formula for $V_a{}^b G$.

Proposition 4.8. If

$$G(x) = \int_a^x g(t)\,dt \text{ on } [a, b],$$

where g is an integrable function on $[a, b]$, then

$$V_a^b G = \int_a^b |\, g(t) \,|\, dt. \qquad \blacksquare$$

Proof. For any partition P: $a = x_0 \leq x_1 \leq x_2 \leq \cdots \leq x_n = b$ we have

$$\sum_{i=1}^n |\, G(x_i) - G(x_{i-1}) \,| = \sum_{i=1}^n \left| \int_{x_{i-1}}^{x_i} g(t)\,dt \right| \leq \int_a^b |\, g(t) \,|\, dt$$

so that $V_a^b G \leq \int_a^b |\, g(t) \,|\, dt$.

To prove the opposite inequality, we need the observation that if s is a step function such as $\sum_{i=1}^n c_i \chi_{[x_{i-1}, x_i)}$, where $a = x_0 < x_1 < x_2 < \cdots < x_n = b$ and each $c_i = -1$, 0, or 1, then

$$\begin{aligned} \int_a^b sg\,dt &= \sum_{i=1}^n c_i \int_{x_{i-1}}^{x_i} g\,dt \leq \sum_{i=1}^n \left| \int_{x_{i-1}}^{x_i} g\,dt \right| \\ &= \sum_{i=1}^n |\, G(x_i) - G(x_{i-1}) \,| \leq V_a^b G. \end{aligned} \tag{4.21}$$

For each n, there is a step function σ_n such that $m\{x\colon |\, g(x) - \sigma_n(x) \,| \geq 1/2^n\} < 1/2^n$. Hence the sequence σ_n converges to g outside the null set

$$\bigcap_{n=1}^\infty \bigcup_{m=n}^\infty \left\{x\colon |\, g(x) - \sigma_m(x) \,| \geq \frac{1}{2^m}\right\};$$

that is, $\sigma_n \to g$ a.e. on $[a, b]$. Let s_n be the step function given by

$$s_n(x) = \begin{cases} 1 & \text{if } \sigma_n(x) > 0 \\ 0 & \text{if } \sigma_n(x) = 0 \\ -1 & \text{if } \sigma_n(x) < 0. \end{cases}$$

Then by equation (4.21), $\int_a^b s_n g\,dt \leq V_a^b G$. However, $s_n g$ converges to $|\, g \,|$ a.e. on $[a, b]$ and $|\, s_n g \,| \leq |\, g \,|$ so that the Dominated Convergence Theorem gives

$$\int_a^b |\, g \,|\, dt = \lim_n \int_a^b s_n g\,dt \leq V_a^b G. \qquad \blacksquare$$

By examining the contrapositive of Proposition 4.6 we have also proved the next result.

Lemma 4.3. If f is absolutely continuous on $[a, b]$ and $f' = 0$ a.e., then f is a constant function. ∎

This leads us to our goal.

Theorem 4.3. If G is a real-valued function on $[a, b]$, the following statements are equivalent:

(i) G is an absolutely continuous function on $[a, b]$.

(ii) G is defined by

$$G(x) = \int_a^x g(t)\,dt + G(a),$$

where g is an integrable function on $[a, b]$.

(iii) G is differentiable a.e. on $[a, b]$ and G is defined by

$$G(x) = \int_a^x G'(t)\,dt + G(a). \quad ∎$$

Proof. Referring back to Proposition 4.5 and Lemma 4.1, we see that all that remains to be proved is that (i) implies (ii).

We know by Corollary 4.2 that G' exists a.e. and by Proposition 4.4 that G' is integrable. Define F on $[a, b]$ by

$$F(x) = \int_a^x G'(t)\,dt - G(x). \tag{4.22}$$

The function F is differentiable a.e. on $[a, b]$ and

$$F'(x) = G'(x) - G'(x) = 0 \text{ a.e.}$$

Since F is also absolutely continuous, by Lemma 4.3 F is a constant function. Thus by equation (4.22)

$$G(x) = \int_a^x G'(t)\,dt + G(a). \quad ∎$$

Our final result of this section generalizes Theorem 4.3 to functions defined on all of R. If f is defined on R, define $V_{-\infty}^{\infty} f$ to be $\lim_{a\to\infty} V_{-a}^{a} f$. If $V_{-\infty}^{\infty} f$ is finite, f is said to be of *bounded variation on R*. A function f with domain R is *absolutely continuous on R* if for each $\varepsilon > 0$ there is a $\delta > 0$ such that $\sum |f(y_i) - f(x_i)| < \varepsilon$ for every finite collection $\{(x_1, y_1), (x_2, y_2),$

$\ldots, (x_n, y_n)\}$ of nonoverlapping intervals with $\sum^n(y_i - x_i) < \delta$. Clearly a function that is absolutely continuous on R is absolutely continuous on every closed interval of R, but the converse is not true, as the function $f(x) = x^2$ illustrates.

Theorem 4.4. A function F on R has the form

$$F(x) = \int_{-\infty}^{x} f(t)\, dt \tag{4.23}$$

for some integrable function f on R if and only if F is absolutely continuous on R, of bounded variation on R, and $\lim_{x\to-\infty} F(x) = 0$. ∎

Proof. Suppose F has the form of equation (4.23). That F is absolutely continuous on R follows from Problem 3.2.1. Also, since on $[-a, a]$ for $a > 0$

$$F(x) - F(-a) = \int_{-\infty}^{x} f(t)\, dt - \int_{-\infty}^{-a} f(t)\, dt = \int_{-a}^{x} f(t)\, dt,$$

using Proposition 4.8 we have

$$V_{-\infty}^{\infty} F = \lim_{a\to\infty} V_{-a}^{a} F = \lim_{a\to\infty} \int_{-a}^{a} |f(t)|\, dt \leq \int_{-\infty}^{\infty} |f(t)|\, dt < \infty.$$

Using the Lebesgue Convergence Theorem with the functions $f_n = |f| \cdot \chi_{(-\infty,-n)}$, we see that the limit of $F(x)$ as $x \to -\infty$ is zero.

Conversely, if F is absolutely continuous on R then for all $a > 0$,

$$F(x) = \int_{-a}^{x} F'(t)\, dt + F(-a) \qquad \text{for } x \in [-a, a].$$

Let $a \to \infty$. Then

$$F(x) = \lim_{a\to\infty} \int_{-a}^{x} F'(t)\, dt$$

for any real x since $F(-a) \to 0$. However, using the Monotone Convergence Theorem and Proposition 4.8

$$\begin{aligned}
\int_{-\infty}^{\infty} |F'(t)|\, dt &= \int_{-\infty}^{\infty} \lim_{n\to\infty} |F'(t)|\, \chi_{(-n,n)}\, dt \\
&= \lim_{n\to\infty} \int_{-\infty}^{\infty} |F'(t)|\, \chi_{(-n,n)}\, dt \\
&= \lim_{n\to\infty} \int_{-n}^{n} |F'(t)|\, dt = \lim_{n\to\infty} V_{-n}^{n} F = V_{-\infty}^{\infty} F < \infty.
\end{aligned}$$

Therefore $|F'(t)|$ and hence $F'(t)$ is integrable. Also since

$$F(x) = \lim_{n\to\infty} \int_{-n}^{x} F'(t)\,dt = \lim_{n\to\infty} \int_{-\infty}^{x} F'(t)\chi_{(-n,\infty)}\,dt,$$

the Lebesgue Convergence Theorem implies that

$$F(x) = \int_{-\infty}^{x} F'(t)\,dt. \qquad \blacksquare$$

Problems

4.2.1. Let f and g be absolutely continuous functions on $[a, b]$. Prove

(a) $f \pm g$ are absolutely continuous,

and

(b) $f \cdot g$ is absolutely continuous.

4.2.2. Let f be defined on $[0, 1]$. Then show that

(a) If f is differentiable on $[0, 1]$ with a bounded derivative, then f is absolutely continuous.

(b) f need not be absolutely continuous on $[0, 1]$ even if f is so on $[\varepsilon, 1]$ for every $\varepsilon > 0$ and continuous at 0.

(c) f is necessarily absolutely continuous on $[0, 1]$ if f is so on $[\varepsilon, 1]$ for every $\varepsilon > 0$, continuous at 0, and of bounded variation on $[0, 1]$. {Hint: if f is of bounded variation on $[0, 1]$, then

$$\int_0^1 f'(x)\,dx = \lim_{\varepsilon_n\to 0} \int_{\varepsilon_n}^{1} f'(x)\,dx, \quad 0 < \varepsilon_n.\}$$

4.2.3. Show

$$\int_a^b |f'(t)|\,dt = V_a^b f < \infty$$

if and only if f is absolutely continuous.

4.2.4. Prove directly from the definition of absolute continuity that the Cantor ternary function is not absolutely continuous.

4.2.5. This problem gives integration by parts formulas.

(a) If f and g are integrable functions on $[a, b]$, let

$$F(x) = \int_a^x f(t)\,dt$$

and

$$G(x) = \int_a^x g(t)\,dt.$$

Prove

$$\int_a^b G(t)f(t)\,dt + \int_a^b g(t)F(t)\,dt = F(b)G(a) - F(a)G(a).$$

(b) If f and g are absolutely continuous functions on $[a, b]$, prove

$$\int_a^b f(t)g'(t)\,dt + \int_a^b f'(t)g(t)\,dt = f(b)g(b) - f(a)g(a).$$

(c) Let f and g be functions on R each of which is absolutely continuous on R, of bounded variation on R, and vanishes as $x \to -\infty$. Prove

$$\int_R f(t)g'(t)\,dt + \int_R f'(t)g(t)\,dt = \int_R f'(t)\,dt \cdot \int_R g'(t)\,dt$$
$$= \lim_{x\to\infty} f(x) \cdot \lim_{x\to\infty} g(x).$$

4.2.6. Show that any nondecreasing function f on $[a, b]$ can be written as $g + h$, where g and h are nondecreasing, g is absolutely continuous, and h is singular, i.e., $h' = 0$ a.e.

4.2.7. (a) If f is a function defined on the set $[a, b]$ and E is any subset of $[a, b]$ on which f' exists and $|f'| \leq K$, show $m^*f(E) \leq Km^*(E)$. (Hint: Let

$$E_n = \{x \in E\colon |y - x| < 1/n \Rightarrow |f(y) - f(x)| \leq (K + \varepsilon)|y - x|\}.$$

Then $m^*(E_n)$ increases to $m^*(E)$. Consider intervals I_{nj} of length $1/n$ such that $E_n \subset \bigcup_{j=1}^{\infty} I_{nj}$ and $\sum_{j=1}^{\infty} m(I_{nj}) < m^*(E) + \varepsilon$.)

(b) Let f be defined and measurable on $[a, b]$ and let E be any set on which f is differentiable. Then

$$m^*(f(E)) \leq \int_E |f'(x)|\,dx.$$

[Hint: Suppose first $|f'(x)| < N$ (an integer) on E. Let $E_k{}^n = \{x \in E\colon (k-1)/2^n \leq |f'(x)| < k/2^n\}$ for $k = 1, 2, \ldots, N\cdot 2^n$ and $n = 1, 2, \ldots$. Show for each n that

$$m^*(f(E)) \leq \sum_{k=1}^{N\cdot 2^n} \frac{k}{2^n} m(E_k{}^n) = \sum_{k=1}^{N\cdot 2^n} \frac{k-1}{2^n} m(E_k{}^n) + \frac{1}{2^n} \sum_{k=1}^{N\cdot 2^n} m(E_k{}^n).\Big]$$

The next two problems give another characterization of absolutely continuous functions.

4.2.8. (a) An absolutely continuous function maps null sets into null sets. (b) Prove that an absolutely continuous function maps measurable sets into measurable sets.

4.2.9. Prove that if f is continuous and of bounded variation on $[a, b]$ and if f maps sets of measure zero into sets of measure zero, then f is absolutely continuous on $[a, b]$. [Hint: If $\{(a_k, b_k): k = 1, 2, \ldots, n\}$ is any collection of nonoverlapping subintervals of $[a, b]$, let $E_k = \{x \in (a_k, b_k): f \text{ is differentiable at } x\}$. Show $m(f(a_k, b_k)) = m(f(E_k))$ and using Problem 4.2.7 show

$$\sum^{n} |f(b_k) - f(a_k)| \leq \sum_{i=1}^{n} \int_{a_k}^{b_k} |f'(x)| \, dx.\Big]$$

4.2.10. Verify Remark 4.6; that is, show that a function is "strongly" absolutely continuous on $[a, b]$ if and only if it satisfies the Lipschitz condition on $[a, b]$. [In establishing the necessity of this result the following hint may be useful. If there does not exist an $\varepsilon > 0$ and a positive K such that $|f(x) - f(y)| \leq K|x - y|$ whenever $|x - y| < \varepsilon$, then, given any positive integer n, there exist x_n and y_n in $[a, b]$ such that $|x_n - y_n| < 1/n$ and $|f(y_n) - f(x_n)| > n|x_n - y_n|$. Use the strong absolute continuity of f on each $[x_n, y_n]$ to obtain a contradiction for some n.]

4.2.11. Show that $v'(x) = |f'(x)|$ a.e. if $v(x)$ is the total variation on $[a, x]$ of a function f of bounded variation on $[a, b]$. {Hint: Choose a partition P_n for $[a, b]$ such that $\sum |f(x_k) - f(x_{k-1})| > v(b) - 1/2^n$. In each segment $x_{k-1} \leq x \leq x_k$ of P_n, let $f_n(x) = f(x) + c$ or $-f(x) + c$ according as $f(x_k) - f(x_{k-1}) \geq 0$ or ≤ 0, where c is chosen so that $f_n(a) = 0$ and the values of f_n at x_k agree. Then the function $v(x) - f_n(x)$ is increasing and $\sum_{n=1}^{\infty}[v(x) - f_n(x)] < \infty$. By Problem 4.1.8 of this chapter, $v'(x) - f_n'(x) \to 0$ a.e. as $n \to \infty$. Since $f_n'(x) = \pm f'(x)$ a.e., $v'(x) = |f'(x)|$ a.e.}

4.2.12. If f is of bounded variation on $[a, b]$, then show that its total variation $v(x)$ satisfies $v(b) \geq \int_a^b |f'(x)| \, dx$. (Compare with Problem 4.2.3.) Also show that $m^*(v(E)) \geq m(g(E))$, where

$$E = \{x \in [a, b]: v'(x) \text{ exists}\}$$

and $v = g + h$, where g is absolutely continuous and nondecreasing, and h is singular and nondecreasing. Finally, show that $m^*(v(E)) = \int_a^b v'(x) \, dx$.

4.2.13. Suppose that f is a continuous function defined on $[a, b]$ and that $N(y)$, called the *Banach indicatrix*, is the number of solutions x of the equation $y = f(x)$. Show that $v(x)$, the total variation of f, satisfies

$$v(b) = \int_{-\infty}^{\infty} N(y) \, dx.$$

{Hint: Let $P_1 \subset P_2 \subset \cdots$ be a sequence of partitions of $[a, b]$ such that $|P_n| \to 0$ as $n \to \infty$. If $N_n(y) = \sum_k \chi_{E_{n,k}}(y)$, where $E_{n,k} = f([x_{k-1}^{(n)}, x_k^{(n)}])$ and $[(x_{k-1}^{(n)}, x_k^{(n)}]$ is the kth segment in P_n, then show that $N_n(y) \to N(y)$ a.e. and apply the Monotone Convergence Theorem.}

4.2.14. Give an example of a real-valued, absolutely continuous function on [0, 1] that is monotone on no interval. {Hint: Using Cantor sets of positive Lebesgue measure, construct a Lebesgue-measurable set $A \subset [0, 1]$ such that $m(A \cap I) > 0$ and $m(A^c \cap [0, 1]) > 0$ for any interval I. Then consider $f(x) = \int_0^x [\chi_A(y) - \chi_{A^c}(y)]\, dy$.}

At this point the reader should be aware of the fact that there exists a real-valued, everywhere differentiable function f defined on R such that f is monotone on no subinterval of R and f' is bounded. For a proof of this, the reader is referred to the paper of Katznelson and Stromberg.[†]

4.2.15. Extend the result in Problem 4.2.12 as follows: If f and v are as in Problem 4.2.12 and A is a Lebesgue-measurable subset of $[a, b]$, then

$$m^*\big(v(A)\big) \geq \int_A |f'(x)|\, dx.$$

Here the equality holds for every $A \subset [a, b]$ if and only if f is absolutely continuous. [Hint: Suppose first that f is absolutely continuous; then v is also so. It is easy to show that for open sets G, $m^*\big(v(G)\big) = \int_G v'(x)\, dx$. By approximation through open sets prove this also for A, and then use Problems 4.2.6 and 4.2.11.] This result is due to D. E. Varberg.

★ **4.2.16.** If f is monotonically increasing or absolutely continuous on $[a, b]$ and A is any Lebesgue-measurable subset for which $m\big(f(A)\big) = 0$, then show that $f'(x) = 0$ a.e. on A. Is this result true for functions of bounded variation? (For related questions, the reader is referred to Varberg.[‡])

4.2.17. If f is a real-valued continuous function on $[a, b]$ and f' exists for all but an at most countable set of points and if f' is Lebesgue integrable, then prove that f is absolutely continuous. (Hint: Let $E = \{x \in [a, b]: f'(x) \text{ exists}\}$. Then

$$\begin{aligned}\sum_{k=1}^n |f(b_k) - f(a_k)| &\leq \sum_{k=1}^n mf([a_k, b_k]) \\ &= \sum_{k=1}^n mf([a_k, b_k] \cap E) \leq \sum_{k=1}^n \int_{[a_k, b_k]} |f'(x)|\, dx\end{aligned}$$

by Problem 4.2.7(b).)

[†] Y. Katznelson and K. Stromberg, *Amer. Math. Monthly* **81**(4), 349–354 (1974).

[‡] D. E. Varberg, *Amer. Math. Monthly* **72**, 831–841 (1965).

4.2.18. Let f be absolutely continuous on $[c, d]$ and g be absolutely continuous on $[a, b]$, where $[c, d] = g([a, b])$. Prove that $f \circ g$ is absolutely continuous on $[a, b]$ if and only if $f \circ g$ is of bounded variation on $[a, b]$. Note that if $f(x) = x^{1/2}$ and $g(x) = x^2 \mid \sin(1/x) \mid$ for $x > 0$, $g(x) = 0$ for $x = 0$, then f and g are both absolutely continuous on $[0, 1]$ even though $f \circ g$ is not.

★ **4.2.19.** *Periods of Measurable Functions.* A real-valued function f defined on R has period t if $f(x + t) = f(x)$ for all x. For a periodic function, *either* there exists the smallest positive period t_0 and all periods are of the form nt_0, where n is any integer, *or* the set of the periods is dense. The characteristic function of the rationals has any rational number as its period. Prove the following assertions:

(i) If t is a positive period of a measurable function f and $\int_0^t \mid f(x) \mid dx < \infty$, then the limit

$$L(f) = \lim_{x \to \infty} \frac{1}{2x} \int_{-x}^{x} f(s)\, ds$$

exists and equals $(1/t) \int_0^t f(s)\, ds$.

(ii) If a measurable function has a dense set of periods, then it is constant a.e.

{Hint: Let $g = f/(1 + |f|)$ and $G(x) = \int_0^x g(s)\, ds$. Then by (i),

$$L(g) = \frac{[G(x + t_n) - G(x)]}{t_n},$$

where (t_n) are periods of f, $t_n \to 0+$ as $n \to \infty$. Hence $g(x) = G'(x) = L(g)$ a.e..} This result is due to A. Lomnicki.

★ **4.2.20.** *The Stone–Čech Compactification and Nonmeasurable Functions* (W. Sierpinski). Let $\beta(Z^+)$ be the Stone–Čech compactification of the positive integers, and let $w \in \beta(Z^+) - Z^+$. Let

$$\Phi(x) = \begin{cases} 0, & \text{if } 2^k x \text{ is an integer for some } k > 0 \\ 0, & \text{if } w \in \overline{\{n: x_n = 0\}}, \text{ where } x_n = 0 \text{ or } 1 \text{ and } x = \sum_{n=1}^{\infty} x_n/2^n \\ 1, & \text{otherwise.} \end{cases}$$

Then Φ is *not* Lebesgue measurable. {Hint: The idea of this hint is due to Z. Semadeni. If $A, B (\subset Z^+)$ differ by a finite set, then $w \in \bar{A} \Leftrightarrow w \in \bar{B}$; therefore any dyadic x (i.e., $2^k \cdot x$ is an integer for some k) is a period of Φ. By Problem 4.2.19 (ii), Φ is constant a.e. Now note that if $D \subset Z^+$, then $\bar{D} \cap \overline{(Z^+ - D)} = \varnothing$; the reason is that the continuous map $f\colon Z^+ \to [0, 1]$ with $f(D) = \{0\}$ and $f(Z^+ - D) = \{1\}$ has an extension $F\colon \beta(Z^+) \to [0, 1]$

so that $\bar{D} \subset F^{-1}(0)$ and $\overline{Z^+ - D} \subset F^{-1}(1)$. Since $1 - x = 1 - \sum_{n=1}^{\infty}(x_n/2^n) = \sum_{n=1}^{\infty} (1 - x_n)/2^n$, it follows that $\Phi(1 - x) = 1 - \Phi(x)$ a.e. or $\Phi(x) = \frac{1}{2}$ a.e. This contradicts the definition of Φ.}

4.3. Integration Versus Differentiation II: Absolutely Continuous Measures, Signed Measures, the Radon-Nikodym Theorem

In the preceding section we characterized indefinite integrals of integrable functions f on $[a, b]$ as absolutely continuous functions. We shall now generalize and extend the notion of an indefinite integral to an arbitrary measure space $(X, \mathscr{B}, \mu)$. If f is an integrable function on X with respect to μ, for each E in $\mathscr{B}$ let us consider the real number $\nu(E)$ given by the "indefinite integral"

$$\nu(E) = \int_E f \, d\mu. \tag{4.24}$$

What are some properties of the real-valued set function ν defined on $\mathscr{B}$ by equation (4.24)? Clearly $\nu(\varnothing) = 0$ and $\nu(\bigcup_{i=1}^{\infty} E_i) = \sum_{i=1}^{\infty} \nu(E_i)$ for any sequence (E_i) of pairwise disjoint, measurable sets in $\mathscr{B}$. However, since ν is not necessarily nonnegative it falls short of being a measure. Nevertheless, it does fulfill the requirements of being a "signed measure."

Definition 4.5. Let $(X, \mathscr{B})$ be a measurable space. An extended real-valued function ν on $\mathscr{B}$ is called a *signed measure* provided the following conditions hold:

(i) $\nu(\varnothing) = 0$.

(ii) ν assumes at most one of the values $+\infty$ and $-\infty$.

(iii) for all pairwise disjoint sequences (E_i) of measurable sets in $\mathscr{B}$, $\nu(\bigcup_{i=1}^{\infty} E_i) = \sum_{i=1}^{\infty} \nu(E_i)$, where equality means that the series on the right converges absolutely if $\nu(\bigcup_{i=1}^{\infty} E_i)$ is finite and properly diverges otherwise. ▌

Observe that the signed measure defined by equation (4.24) has the following property: If $\mu(E) = 0$, then $\nu(E) = 0$. If ν is any signed measure on $(X, \mathscr{B})$ and μ is any measure on $(X, \mathscr{B})$ such that $\mu(E) = 0$ implies $\nu(E) = 0$ for all E in $\mathscr{B}$, then we say ν is "absolutely continuous" with respect to μ and write $\nu << \mu$. This concept will be precisely defined below.

Our goal in this section is to show that if ν is any signed measure on

$(X, \mathscr{B})$ absolutely continuous with respect to a certain type of measure μ on $(X, \mathscr{B})$, then ν has an integral representation as in equation (4.24).

As a beginning let us look at a particular situation. Let f be a bounded distribution function—a bounded real-valued nondecreasing function on R continuous from the right—and let μ_f be the unique corresponding Borel measure such that for all a and b in R

$$\mu_f(a, b] = f(b) - f(a).$$

Since f is differentiable a.e. and f' is thus a nonnegative measurable function, we can define a measure ν on the class of Borel sets by

$$\nu(E) = \int_E f'(x)\, dx. \tag{4.25}$$

In particular, for the Borel set $(a, b]$,

$$\nu(a, b] = \int_a^b f'(x)\, dx \leq f(b) - f(a).$$

Thus ν is a Borel measure and the question of whether $\nu = \mu_f$ is equivalent to the question of whether

$$\int_a^b f'(x)\, dx = f(b) - f(a) \tag{4.26}$$

for all intervals $(a, b]$. By Theorems 4.3 and 4.4, a necessary and sufficient condition for equation (4.26) to be satisfied is that f be absolutely continuous on R. We have almost proved the necessity condition in the next proposition.

Proposition 4.9. Let μ_f be the Borel measure corresponding to a bounded distribution function f on R. The function f is absolutely continuous on R if and only if $\mu_f << m$, that is, whenever E is a Borel set with $m(E) = 0$ then $\mu_f(E) = 0$. ∎

Proof. If f is absolutely continuous, then $\mu_f = \nu$, where ν is defined in equation (4.25). Hence $\mu_f << m$.

Conversely, suppose $\mu_f << m$. The measure μ_f then satisfies the condition that given $\varepsilon > 0$, there exists a $\delta > 0$ such that for every Borel-measurable set with $m(E) < \delta$, $\mu_f(E) < \varepsilon$. Indeed, if not then there exists some $\varepsilon_0 > 0$ and a sequence E_n of measurable sets such that for each n

$$m(E_n) < 1/2^n \qquad \text{but} \qquad \mu_f(E_n) \geq \varepsilon_0.$$

Let $E_0 = \bigcap_{k=1}^{\infty}(\bigcup_{n=k}^{\infty} E_n)$. Then

$$m(E_0) \leq \sum_{n=k}^{\infty} m(E_n) < \frac{1}{2^{k-1}} \qquad \text{for all } k$$

so that $m(E_0) = 0$. However since $\mu_f(\bigcup_{n=1}^{\infty} E_n) < \infty$, using Proposition 2.2 we obtain

$$\mu_f(E_0) = \lim_{k\to\infty} \mu_f\left(\bigcup_{n=k}^{\infty} E_n\right) \geq \lim_{k\to\infty} \mu_f(E_k) \geq \varepsilon_0 > 0,$$

which is a contradiction to the fact that $\mu_f << m$. Therefore given any collection $(a_1, b_1), (a_2, b_2). \ldots, (a_n, b_n)$ of nonoverlapping intervals with

$$\sum_{i=1}^{n} m(a_i, b_i] = \sum_{i=1}^{n} b_i - a_i < \delta,$$

then

$$\mu_f\left(\bigcup_{i=1}^{n} (a_i, b_i]\right) = \sum_{i=1}^{n} \mu_f(a_i, b_i] = \sum_{i=1}^{n} |f(b_i) - f(a_i)| < \varepsilon.$$

Thus f is absolutely continuous. ∎

It is important to observe that f must be a bounded function in Proposition 4.9. For example if f is the function

$$f(x) = \begin{cases} 0 & \text{if } x < 0 \\ x^2 & \text{if } x \geq 0, \end{cases}$$

then

$$\mu_f(E) = \int_{E\cap[0,\infty)} 2x\, dx \qquad \text{for all Borel sets } E$$

so that $\mu_f << m$ but f is not absolutely continuous on R.

Let us note the following conclusions from the preceding discussion. If μ_f is the Borel measure corresponding to the bounded distribution function f, then $\mu_f << m$ if and only if f is absolutely continuous on R. However, a necessary and sufficient condition for

$$\int_{(a,b]} f'(x)\, dx = f(b) - f(a)$$

for all intervals $(a, b]$ is the absolute continuity of f on R. Thus we may conclude from the uniqueness of μ_f that if $\mu_f << m$ then

$$\mu_f(E) = \int_E f'(x)\, dx \tag{4.27}$$

for all Borel sets E. Conversely, if equation (4.27) is satisfied for all Borel sets E, then $\mu_f << m$.

We can generalize this discussion in the following manner. Suppose f is a bounded variation function on R ($V_{-\infty}^{\infty} f < \infty$) that is continuous from the right. Then f can be written as the difference $f_1 - f_2$ of two bounded monotonic functions each continuous from the right ($f_1 = V_{-\infty}^{x} f = \lim_{a\to\infty} V_{-a}^{x} f$ and $f_2 = V_{-\infty}^{x} f - f$). Let μ_1 and μ_2 be the unique finite Borel measure corresponding to f_1 and f_2, respectively, such that

$$\mu_i(a, b] = f_i(b) - f_i(a), \qquad i = 1, 2$$

for all intervals $(a, b]$. Let ν_1 and ν_2 be the measures given by

$$\nu_i(E) = \int_E f_i'(x)\, dx, \qquad i = 1, 2$$

for each Borel set E. Using Proposition 4.7, as above we can verify that f is absolutely continuous if and only if f_1 and f_2 are absolutely continuous, which is true if and only if $\nu_i = \mu_i$ for $i = 1$ and 2. Consequently, the absolute continuity of f implies that ν given by

$$\nu(E) = \nu_1(E) - \nu_2(E) = \int_E f'(x)\, dx$$

is a well-defined signed measure equal to μ_f. Clearly $\nu << m$ so that $\mu_f << m$ when f is absolutely continuous.

Analogous to the argument given in Proposition 4.9, the reader can show that $\mu_f << m$ needs f to be absolutely continuous. We may conclude as before then that $\mu_f << m$ if and only if

$$\mu_f(E) = \int_E f'\, dx,$$

where f is the distribution corresponding to μ_f.

Having seen the implication of absolute continuity of a distribution function or a bounded variation function on the associated Borel measure, it does seem appropriate to say that μ_f is absolutely continuous with respect to m if $m(E) = 0$ implies $\mu_f(E) = 0$. Let us formalize this concept.

Definition 4.6. Let $(X, \mathscr{B}, \mu)$ be a measure space and ν a signed measure on $\mathscr{B}$. ν is *absolutely continuous* with respect to μ, written $\nu << \mu$, if whenever E is a set in $\mathscr{B}$ with $\mu(E) = 0$, then $\nu(E) = 0$. ∎

Remark 4.7. In the case that $|\nu(E)| < \infty$ whenever $\mu(E) < \infty$, $\nu << \mu$ if and only if for each $\varepsilon > 0$, there is a $\delta > 0$ such that whenever $\mu(E) < \delta$ then $|\nu(E)| < \varepsilon$. See Problem 4.3.1.

Later in this section we will further discuss the concept of absolute continuity of measures when we characterize signed measures absolutely continuous with respect to a measure. First we must "decompose" any signed measure as the difference of two measures.

Let ν be a signed measure on $(X, \mathscr{B})$. A set P in $\mathscr{B}$ is called *positive* with respect to ν if for every measurable set E in $\mathscr{B}$, $\nu(P \cap E) \geq 0$. A set N in $\mathscr{B}$ is called *negative* if for every measurable set E, $\nu(N \cap E) \leq 0$. A set that is both positive and negative with respect to ν is called *a null set.*

Remarks

4.8. A measurable set is a null set if and only if every measurable subset of it has ν-measure zero. However, a set of ν-measure zero is not necessarily a null set unless ν is a measure. For example, if ν is the signed measure on the Lebesgue measurable subsets of R given by

$$\nu(E) = \int_E f(x)\, dx,$$

where $f(x) = x$ for $-1 \leq x \leq 1$ and 0 otherwise, then $[-1, 1]$ is a set of ν-measure zero but is not a null set.

4.9. Every measurable subset of a positive (negative) set is itself positive (negative). A set of positive (negative) measure is not necessarily a positive (negative) set. A countable union of positive (negative) sets is positive (negative).

Theorem 4.5. If ν is a signed measure on $(X, \mathscr{B})$, then there exist two disjoint sets P and N in X such that $X = P \cup N$, P is positive, and N is negative. ∎

This theorem is known as the *Hahn Decomposition Theorem* and the sets P and N are said to form a *Hahn decomposition* of X with respect to ν.

To prove Theorem 4.5 we make use of the following lemma.

Lemma 4.4. If ν is a signed measure and E is a measurable set such that $0 < \nu(E) < \infty$, then E contains a positive set A with $\nu(A) > 0$. ∎

Proof. If E itself is not positive it contains sets of negative measure. Let k_1 be the smallest positive integer such that there is a measurable subset E_1 of E with $\nu(E_1) < -1/k_1$. Since

$$0 < 1/k_1 + \nu(E) < \nu(E) - \nu(E_1) = \nu(E - E_1) < \infty,$$

the argument just used for E applies to $E - E_1$. Thus by an inductive argument we obtain a sequence (k_i) of positive integers such that for each i, k_i is the smallest positive integer for which there is a measurable set E_i such that

$$E_i \subset E - \left[\bigcup_{j=1}^{i-1} E_j \right]$$

and

$$\nu(E_i) < -1/k_i$$

Setting

$$A = E - \left[\bigcup_{i=1}^{\infty} E_i \right],$$

we have

$$E = A \cup \left[\bigcup_{i=1}^{\infty} E_i \right].$$

Since this is a disjoint union,

$$\nu(E) = \nu(A) + \sum_{i=1}^{\infty} \nu(E_i) \tag{4.28}$$

and hence the series $\sum_{i=1}^{\infty} \nu(E_i)$ must converge since $\nu(E)$ is finite. Thus $\sum_{i=1}^{\infty} 1/k_i$ converges and $k_i \to \infty$. Also since the series in equation (4.28) converges to a negative number with $\nu(E) > 0$, necessarily $\nu(A) > 0$.

A is indeed a positive set. For each $\varepsilon > 0$, A contains no measurable subsets of measure less than $-\varepsilon$. If $\varepsilon > 0$ choose i so large that $(k_i - 1)^{-1} < \varepsilon$. Since

$$A \subset E - \left[\bigcup_{j=1}^{i} E_j \right],$$

A contains no measurable sets of measure less than $-(k_i - 1)^{-1}$ or less than $-\varepsilon$. ∎

Proof of Theorem 4.5. Since ν assumes at most one of the values of $+\infty$ or $-\infty$ we may assume that $+\infty$ is the value never attained. Let $\alpha = \sup \nu(A)$ for all positive sets A. Since $\varnothing$ is a positive set, $\alpha \geq 0$. Let (P_i) be a sequence of positive sets such that

$$\alpha = \lim_{i \to \infty} \nu(P_i).$$

By Remark 4.9, $P = \cup P_i$ is positive so that $\nu(P) \leq \alpha$. However, for each i,

$P - P_i \subset P$ so that $\nu(P - P_i) \geq 0$ and

$$\nu(P) = \nu(P_i) + \nu(P - P_i) \geq \nu(P_i).$$

Hence $\nu(P) = \alpha$.

The proof will be complete if we can show that $N = X - P$ is a negative set. Suppose N contains a subset E with $\nu(E) > 0$. By Lemma 4.4, E then contains a positive set A with $\nu(A) > 0$. Since $A \cap P = \emptyset$, $A \cup P$ is positive and

$$\nu(A \cup P) = \nu(A) + \nu(P) > \nu(P) = \alpha,$$

which contradicts the definition of α. Hence every subset E of N has nonpositive measure and N is negative. ∎

The Hahn decomposition of a measurable space with respect to a signed measure ν need not be unique. However, it can easily be shown that if $X = P_1 \cup N_1$ and $X = P_2 \cup N_2$ are two Hahn decompositions of X, then

$$\nu(P_1 \triangle P_2) = 0 \qquad \text{and} \qquad \nu(N_1 \triangle N_2) = 0.$$

Therefore if E is any measurable set in $\mathscr{B}$,

$$\nu(E \cap P_1) = \nu\big(E \cap (P_1 \cup P_2)\big) = \nu(E \cap P_2)$$

and

$$\nu(E \cap N_1) = \nu\big(E \cap (N_1 \cup N_2)\big) = \nu(E \cap N_2).$$

Definition 4.7. Two measures μ_1 and μ_2 on $(X, \mathscr{B})$ are said to be *mutually singular*, written $\mu_1 \perp \mu_2$, if there are disjoint measurable sets A and B in X with $X = A \cup B$ such that $\mu_1(A) = \mu_2(B) = 0$. ∎

The concepts of mutual singularity and absolute continuity are in a sense opposite concepts. If $\mu_1 \perp \mu_2$ then each nonzero value of μ_1 is taken on a set where μ_2 is zero. If in addition $\mu_1 << \mu_2$, then μ_1 must be the zero measure.

Proposition 4.10. Let ν be a signed measure on $(X, \mathscr{B})$. Then there is a unique pair of mutually singular measures ν^+ and ν^- such that $\nu = \nu^+ - \nu^-$. Moreover, if ν_1 and ν_2 are measures such that $\nu = \nu_1 - \nu_2$, then $\nu^+(E) \leq \nu_1(E)$ and $\nu^-(E) \leq \nu_2(E)$ for all E in $\mathscr{B}$. ∎

Proof. For E in $\mathscr{B}$ define ν^+ and ν^- by

$$\nu^+(E) = \nu(E \cap P) \qquad \text{and} \qquad \nu^-(E) = -\nu(E \cap N),$$

where P and N form a Hahn decomposition of X. The uniqueness of ν^+ and ν^- is left as Problem 4.3.2 as is the proof of the last statement (Problem 4.3.3). ∎

The measures ν^+ and ν^- defined in the proof of Proposition 4.10 are called the *positive variation* and *negative variation* of ν, respectively. The measure $|\nu|$ given by

$$|\nu|(E) = \nu^+(E) + \nu^-(E)$$

is called the *total variation* of ν. The representation of ν as $\nu^+ - \nu^-$ is called the *Jordan decomposition* of ν.

Example 4.3. Let $(X, \mathscr{B}, \mu)$ be a measure space and let ν be defined by

$$\nu(E) = \int_E f\, d\mu,$$

where f is an integrable function on X. Writing f as $f^+ - f^-$,

$$\nu(E) = \int_E f^+\, d\mu - \int_E f^-\, d\mu.$$

Is $\nu^+(E) = \int_E f^+\, d\mu$ and $\nu^-(E) = \int_E f^-\, d\mu$ for each E in $\mathscr{B}$? Let P be the measurable set $\{x \in X\colon f(x) > 0\}$. Clearly P is a positive set and P^c is negative. By definition

$$\nu^+(E) = \nu(E \cap P) = \int_{E\cap P} f\, d\mu = \int_E f^+\, d\mu$$

and

$$\nu^-(E) = -\nu(E \cap P^c) = -\int_{E\cap P^c} f\, d\mu = \int_E f^-\, d\mu.$$

Thus the Jordan decomposition of ν is generated by the decomposition of the measurable function f into its positive and negative parts.

The Jordan decomposition of a signed measure ν into the difference of two measures (nondecreasing set functions) is reminiscent of the decomposition of a function of bounded variation, in particular an absolutely continuous function, as the difference of two nondecreasing functions. In fact we have the following equality analogous to the definition of the total variation of a function.

Proposition 4.11. Let ν be a signed measure on $(X, \mathscr{B})$. Then

$$|\nu|(E) = \sup\left\{\sum_{k=1}^{n} |\nu(E_k)| : E = \bigcup_{k=1}^{n} E_k,\ E_k \in \mathscr{B},\ E_k \cap E_j = \varnothing \text{ if } k \neq j\right\}. \blacksquare$$

The proof of this proposition is left as Problem 4.3.9.

We now state the main result of this section—a realization of the goal of this section.

Theorem 4.6. *Radon–Nikodym Theorem.* Let μ be a σ-finite measure and ν a signed measure on $(X, \mathscr{B})$. Assume ν is absolutely continuous with respect to μ. Then there is a μ-measurable function f_0 such that for each set E in $\mathscr{B}$ we have

$$\nu(E) = \int_E f_0 \, d\mu.$$

The function f_0 is unique in the sense that if g is any measurable function with this property, then $g = f_0$ a.e. $[\mu]$. ∎

In the proof of this outstanding result we need the following lemma.

Lemma 4.5. Let $(X, \mathscr{B}, \mu)$ be a measure space and let ν be a measure on $\mathscr{B}$. Assume both μ and ν are finite. Let $\mathscr{C}$ denote the collection of all μ-integrable nonnegative functions f such that

$$\nu(E) \geq \int_E f \, d\mu \tag{4.29}$$

for all E in $\mathscr{B}$. Then there is a function f_0 in $\mathscr{C}$ such that

$$\int f_0 \, d\mu = \sup_{f \in \mathscr{C}} \int f \, d\mu. \quad \blacksquare \tag{4.30}$$

Proof. Let α be the finite real number which is the supremum in equation (4.30) and let (f_n) be a sequence of functions in $\mathscr{C}$ such that $\int f_n \, du \to \alpha$. Let (g_n) be the nondecreasing sequence of μ-integrable functions given by

$$g_n(x) = \max\{f_1(x), \ldots, f_n(x)\}.$$

Then each g_n is in $\mathscr{C}$ since for any set E in $\mathscr{B}$

$$\int_E g_n \, d\mu = \sum_{i=1}^{n} \int_{E_i} g_n \, d\mu \leq \sum_{i=1}^{n} \nu(E_i) = \nu(E),$$

where E_i are disjoint measurable sets such that $E = \bigcup_{i=1}^{n} E_i$ and $g_n = f_i$ on E_i. Since for each n

$$\int g_n \, d\mu \leq \nu(X) < \infty,$$

the Monotone Convergence Theorem guarantees the existence of *a* μ-integrable function f_0 such that $g_n \uparrow f_0$ and

$$\int_E f_0 \, d\mu = \lim_n \int_E g_n \, d\mu \leq \nu(E)$$

for all E in $\mathscr{B}$. Since the g_n are in $\mathscr{C}$ we have the following inequalities:

$$\alpha = \lim_n \int f_n \, d\mu \leq \lim_n \int g_n \, d\mu = \int f_0 \, d\mu \leq \alpha,$$

which establishes the result. ∎

Proof of Theorem 4.6. We will divide the proof of the existence of the function f_0 into several parts, each part being a generalization of the preceding part.

(i) Assume first μ and ν are finite measures. By Lemma 4.5, there is a nonnegative μ-integrable function f_0 such that

$$\nu(E) \geq \int_E f_0 \, d\mu \tag{4.31}$$

for all E in $\mathscr{B}$ and $\int f_0 \, d\mu \geq \int f \, d\mu$ for any nonnegative function f such that

$$\nu(E) \geq \int_E f \, d\mu$$

for all E in $\mathscr{B}$. We wish to establish that inequality (4.31) is an equality.

Let ν_1 be the measure given by

$$\nu_1(E) = \nu(E) - \int_E f_0 \, d\mu.$$

Clearly $\nu_1(X) < \infty$, and since $\nu << \mu$, also $\nu_1 << \mu$. If ν_1 is not the zero measure, then $\nu_1(X) > 0$. Hence for some constant $k > 0$

$$\mu(X) - k\nu_1(X) < 0.$$

Let P and N be a Hahn decomposition for the signed measure $\mu - k\nu_1$. We assert that $\mu(N) > 0$. Indeed if $\mu(N) = 0$, then $\nu_1(N) = 0$ since

$\nu_1 << \mu$. However, then

$$0 \leq (\mu - k\nu_1)(P \cap X) = \mu(P \cap X) - k\nu_1(P \cap X) = \mu(X) - k\nu_1(X) < 0.$$

Since this is clearly false, we must have $\mu(N) > 0$.

Define h on X by $h(x) = 1/k$ if $x \in N$ and $h(x) = 0$ if $x \notin N$. If $E \in \mathscr{B}$,

$$\int_E h\, d\mu = \frac{1}{k}\mu(N \cap E) \leq \nu_1(N \cap E) \leq \nu_1(E) = \nu(E) - \int_E f_0\, d\mu.$$

Thus

$$\int_E (h + f_0)\, d\mu \leq \nu(E).$$

Since $h > 0$ on N, $h + f_0 > f_0$ on a set of positive μ-measure. Hence using inequality (4.31) we obtain

$$\int f_0\, d\mu < \int h + f_0\, d\mu,$$

a contradiction. Hence our original assumption that ν_1 is not the zero measure is false, so that (4.31) is an equality.

(ii) Assume μ is a finite measure and ν is a σ-finite measure. Let $X = \cup X_n$, X_n pairwise disjoint with finite ν-measure. By part (i) for each positive integer n, there exists a nonnegative measurable function f_n such that

$$\nu_n(E) = \int_E f_n\, d\mu$$

for all E in $\mathscr{B}$, where ν_n is the finite measure $\nu_n(E) = \nu(E \cap X_n)$. Thus

$$\nu(E) = \sum_n \nu(E \cap X_n) = \int_E f_0\, d\mu,$$

where $f_0 = \sum f_n$.

(iii) Assume now that μ is a finite measure and ν is an arbitrary measure. Let $\mathscr{C}$ be the nonempty class of measurable subsets of X that can be written as a countable union of measurable sets with finite ν-measure. Let $\alpha = \sup\{\mu(A)\colon A \in \mathscr{C}\}$; there exists a sequence (A_i) of sets from $\mathscr{C}$ such that $\mu(A_i) \to \alpha$. Let $A = \bigcup_{i=1}^{\infty} A_i$, a set in $\mathscr{C}$. Clearly $\alpha = \mu(A)$, and by (ii) there exists a nonnegative measurable function f_1 such that for all sets E in $\mathscr{B}$

$$\nu_A(E) = \int_E f_1\, d\mu, \tag{4.32}$$

where ν_A is the σ-finite measure given by $\nu_A(E) = \nu(A \cap E)$. Now if $\mu(E \cap A^c) = 0$, then $\nu(E \cap A^c) = 0$ since $\nu << \mu$. However, if $\mu(E \cap A^c) > 0$, then $\nu(E \cap A^c) = \infty$. Indeed, if $\nu(E \cap A^c) < \infty$, then $E \cap A^c \in \mathscr{C}$ and

$$\alpha = \mu(A) < \mu(A) + \mu(E \cap A^c) = \mu(A \cup (E \cap A^c)) \leq \alpha,$$

which is a contradiction. Therefore no matter what the value of $\mu(E \cap A^c)$,

$$\nu(E \cap A^c) = \int_{E \cap A^c} \infty \, d\mu.$$

Defining f_0 on X as f_1 on A and ∞ on A^c, equation (4.32) gives

$$\nu(E) = \nu(A \cap E) + \nu(A^c \cap E) = \nu_A(E) + \nu(A^c \cap E) = \int_E f_0 \, d\mu$$

for all E in $\mathscr{B}$.

(iv) Assume now that μ is a σ-finite measure and ν is an arbitrary measure. Then $X = \cup X_n$, where the X_n are pairwise disjoint and $\mu(X_n) < \infty$ for each n. For each n there exists a nonnegative measurable function f_n such that for all sets E in $\mathscr{B}$ with $E \subset X_n$,

$$\nu(E) = \int_E f_n \, d\mu_n,$$

where μ_n is the finite measure given by $\mu_n(A) = \mu(A \cap X_n)$. Let $f_0 = \sum_{n=1}^{\infty} f_n$. Then f_0 is a nonnegative measurable function and

$$\nu(E) = \sum_n \nu(E \cap X_n) = \sum_n \int_{E \cap X} f_n \, d\mu_n = \sum_n \int_E f_n \, d\mu = \int_E f_0 \, d\mu$$

for all E in $\mathscr{B}$.

(v) Assume now finally that μ is a σ-finite measure and ν is an arbitrary signed measure. By Proposition 4.10 we can write ν as $\nu^+ - \nu^-$, where ν^+ and ν^- are measures—one being finite. By part (iv), there exist nonnegative functions f_1 and f_2 such that

$$\nu^+(E) = \int_E f_1 \, d\mu \quad \text{and} \quad \nu^-(E) = \int_E f_2 \, d\mu$$

for all E in $\mathscr{B}$. Letting $f_0 = f_1 - f_2$ (since one of f_1 or f_2 is integrable, one may be assumed to be real valued and hence $f_1 - f_2$ is well defined), we have

$$\nu(E) = \int_E f_0 \, d\mu$$

for all E in $\mathscr{B}$. The existence of f_0 is established.

The uniqueness of the function f remains to be established. Suppose f_0 and g are both measurable functions such that for all E in $\mathscr{B}$

$$\nu(E) = \int_E f_0 \, d\mu \quad \text{and} \quad \nu(E) = \int_E g \, d\mu.$$

In particular since μ is σ-finite and for each E in $\mathscr{B}$

$$\int_E f_0 \, d\mu \leq \int_E g \, d\mu$$

we know from Proposition 3.8 that $f_0 \leq g$ a.e. $[\mu]$. Similarly $g \leq f_0$ a.e. $[\mu]$. ∎

Remark 4.10. The assumption that μ be σ-finite is vital and cannot be removed. As an example, let μ be a counting measure on the Lebesgue-measurable subsets of $[0, 1]$ and ν the Lebesgue measure on this same measure space. No function can be found satisfying Theorem 4.6.

We have proved in particular that if ν is a measure absolutely continuous with respect to a σ-finite measure μ, then there is a nonnegative measurable function f_0 such that for any measurable set

$$\nu(E) = \int \chi_E \cdot f_0 \, d\mu.$$

If $\varphi = \sum a_i \chi_{E_i}$ is a simple function, the additivity of the integral gives

$$\int \varphi \, d\nu = \int \varphi \cdot f_0 \, d\mu.$$

Since any nonnegative measurable function f is the pointwise limit of a nondecreasing sequence (φ_n) of simple functions, the Monotone Convergence Theorem implies that

$$\int_E f \, d\nu = \int_E f \cdot f_0 \, d\mu$$

for all measurable E. Finally if f is integrable with respect to ν, it can be written as the difference of two nonnegative ν-integrable functions $f_1 - f_2$ so that $f_1 f_0$ and $f_2 f_0$ are each μ-integrable and

$$\int_E f \, d\nu = \int_E f_1 \, d\nu - \int_E f_2 \, d\nu = \int_E f_1 f_0 \, d\mu - \int_E f_2 f_0 \, d\mu = \int_E f f_0 \, d\mu$$

for all measurable sets E.

Thus we have proved the following corollary to the Radon–Nikodym Theorem, which is a strengthening of the result for the case when ν is a measure.

Corollary 4.3. Suppose μ and ν are measures on $(X, \mathscr{B})$ with μ σ-finite and $\nu << \mu$. Then there exists a unique nonnegative μ-measurable function f_0 such that (i) for each nonnegative measurable function f on X

$$\int_E f\,d\nu = \int_E ff_0\,d\mu$$

and (ii) for each ν-integrable function f, ff_0 is μ-integrable and in fact

$$\int_E f\,d\nu = \int_E ff_0\,d\mu$$

for all E in $\mathscr{B}$. ∎

The function f_0 whose existence and "uniqueness" is established in the Radon–Nikodym Theorem is appropriately called the *Radon–Nikodym derivative* of ν with respect to μ and is denoted by $d\nu/d\mu$. Observe that the Radon–Nikodym derivative of the Borel measure μ_f (generated by a bounded absolutely continuous distribution function f on R) with respect to Lebesgue measure m is the usual derivative f'.

It is interesting that the Radon–Nikodym derivatives satisfy rules that are analogous to some familiar rules for derivatives of real-valued functions of a real variable, primarily the following.

(i) If $\nu_1 << \mu$ and $\nu_2 << \mu$ with μ σ-finite and a and b reals,

$$\frac{d(a\nu_1 + b\nu_2)}{d\mu} = a\,\frac{d\nu_1}{d\mu} + b\,\frac{d\nu_2}{d\mu} \quad \text{a.e. } [\mu].$$

(ii) If $\nu << \mu$ and $\mu << \lambda$ (with μ and λ σ-finite measures), then $\nu << \lambda$ and

$$\frac{d\nu}{d\lambda} = \frac{d\nu}{d\mu} \cdot \frac{d\mu}{d\lambda} \quad \text{a.e. } [\lambda].$$

(iii) If $\nu << \mu$ and $\mu << \nu$ (ν and μ both σ-finite), then

$$\left(\frac{d\nu}{d\mu}\right)^{-1} = \frac{d\mu}{d\nu} \quad \text{a.e. } [\mu].$$

The proofs of these statements are trivial and left as exercises (Problem 4.3.6).

As a consequence of the Radon–Nikodym Theorem we can prove the following decomposition theorem for a σ-finite measure.

Theorem 4.7. *Lebesgue Decomposition Theorem.* Let $(X, \mathscr{B}, \mu)$ be a σ-finite measure space and ν a σ-finite measure defined on $\mathscr{B}$. Then there exist unique measures ν_1 and ν_2 with $\nu_1 \perp \mu$ and $\nu_2 << \mu$ such that $\nu = \nu_1 + \nu_2$. ∎

Proof. The measure $\lambda = \mu + \nu$ is also σ-finite and clearly $\mu << \lambda$ and $\nu << \lambda$. Hence the Radon–Nikodym Theorem asserts the existence of nonnegative measurable functions f_0 and g_0 such that for all measurable sets E,

$$\mu(E) = \int_E f_0 \, d\lambda \quad \text{and} \quad \nu(E) = \int_E g_0 \, d\lambda.$$

Let $U = \{x: f_0(x) > 0\}$ and $V = \{x: f_0(x) = 0\}$. Then $X = U \cup V$, $U \cap V = \varnothing$, and $\mu(V) = 0$. Defining ν_1 by

$$\nu_1(E) = \nu(E \cap V),$$

$\nu_1(U) = 0$ and so $\nu_1 \perp \mu$. Defining ν_2 by

$$\nu_2(E) = \nu(E \cap U) = \int_{E \cap U} g_0 \, d\lambda,$$

we have $\nu = \nu_1 + \nu_2$. It remains to show that $\nu_2 << \mu$. If $\mu(E) = 0$, then

$$0 = \int_E f_0 \, d\lambda,$$

so that $f_0 = 0$ a.e. $[\lambda]$ on E. Since $f_0 > 0$ on U, $\lambda(U \cap E) = 0$. Hence

$$\nu(U \cap E) = \int_{U \cap E} g_0 \, d\lambda = 0,$$

so that $\nu_2(E) = 0$. Uniqueness is left for the reader to prove. ∎

Example 4.4. Let f be a bounded distribution function and μ_f the corresponding Borel measure. As we have already seen in Proposition 4.9 $\mu_f << m$ if and only if f is absolutely continuous. In that case

$$\mu_f(E) = \int_E f' \, d\mu$$

(f' being the derivative of f). We assert now that $\mu_f \perp m$ if and only if $f' = 0$ a.e.

To prove this assertion, suppose first that μ_f and m are not mutually singular. Since μ_f and m are σ-finite measures, Theorem 4.7 gives two measures ν_1 and ν_2 such that

$$\mu_f = \nu_1 + \nu_2, \ \nu_1 \perp m, \text{ and } \nu_2 << m.$$

Now ν_2 is a Borel measure so that to it corresponds a distribution function g_2 given by

$$g_2(x) = \begin{cases} -\nu_2(x, 0] & \text{if } x < 0 \\ 0 & \text{if } x = 0 \\ \nu_2(0, x] & \text{if } x > 0. \end{cases}$$

Since $\nu_2 << m$, g_2 is absolutely continuous and

$$\nu_2(E) = \int_E g_2'(t)\, dt$$

for all Borel sets E. Since ν_2 cannot be zero (otherwise $\mu_f \perp m$), g_2' is positive on a Borel set E such that $m(E) > 0$. If distribution function g_1 corresponds to ν_1, then defining g as $g_1 + g_2$, we have $g' = g_1' + g_2' \geq g_2'$, which is positive on E. Since g is the distribution function of μ_f up to a constant, $f' = g' > 0$. Hence $f' = 0$ implies $\mu_f \perp m$.

Conversely, suppose $f' > 0$ on a Borel set A of measure greater than zero. Since f' is Lebesgue measurable, we can define λ by

$$\lambda(E) = \int_E f'(t)\, dt.$$

Clearly $\lambda << m$ and $\lambda(A) > 0$. Also

$$\lambda(a, b] = \int_a^b f'(t)\, dt \leq f(b) - f(a) = \mu_f(a, b]$$

for all $a < b$ in R and thus for all Borel-measurable sets E, $\lambda(E) \leq \mu_f(E)$. Thus we have found a nonzero σ-finite measure λ such that $\lambda \leq \mu_f$ and $\lambda << m$.

This rules out the possibility that $\mu_f \perp m$. Indeed, if $\mu_f \perp m$, then since $\lambda \leq \mu_f$, $\lambda \perp m$; but $\lambda << m$, which implies that $\lambda = 0$. This is a contradiction.

Using this example, the reader should readily see the analogy between the Lebesgue Decomposition Theorem and Problem 4.2.6.

Finally we indicate that the Lebesgue Decomposition Theorem can be proved easily without using the Radon–Nikodym Theorem even when

ν is not σ-finite. Note that if $\mu(X) < \infty$, $\mathscr{F} = \{E: \nu(E) = 0\}$, and $s = \sup\{\mu(E): E \in \mathscr{F}\}$, then there exists $B \in \mathscr{F}$ such that $\mu(B) = s$. Defining $\nu_1(A) = \mu(E - A)$ and $\nu_2(A) = \mu(E \cap A)$, it can be easily verified that ν_1 and ν_2 meet the requirements of the theorem. It is easy to extend this argument when μ is σ-finite.

Problems

4.3.1. (a) Suppose ν is a signed measure and μ a measure on $(X, \mathscr{B})$ such that $|\nu(E)| < \infty$ whenever $\mu(E) < \infty$. Prove $\nu << \mu$ if and only if for each $\varepsilon > 0$, there is a $\delta > 0$ such that whenever $\mu(E) < \delta$ then $|\nu(E)| < \varepsilon$.

(b) Show that statement (a) is not necessarily valid if the condition that $\mu(E) < \infty$ implies $|\nu(E)| < \infty$ does not hold. [Hint: Let X = positive integers, $\mu(E) = \sum_{n \in E} 2^{-n}$ and $\nu(E) = \sum_{n \in E} 2^n$.]

4.3.2. Show that the Jordan decomposition of a signed measure is unique. (Hint: Show that any such decomposition yields a Hahn decomposition.)

4.3.3. (a) Let ν be a signed measure on $(X, \mathscr{B})$. Prove for all E in $\mathscr{B}$ that $\nu^+(E) = \sup\{\nu(F): F \in \mathscr{B}, F \subset E\}$, and $\nu^-(E) = -\inf\{\nu(F): F \in \mathscr{B}, F \subset E\}$, where $\nu = \nu^+ - \nu^-$ is the Jordan decomposition of ν.

(b) If $\nu = \nu_1 - \nu_2$, where ν_1 and ν_2 are measures, show that $\nu^+ << \nu_1$ and $\nu^- << \nu_2$.

4.3.4. Let μ be a measure and ν a signed measure on the measurable space $(X, \mathscr{B})$. Let $\nu = \nu^+ - \nu^-$ be the Jordan decomposition of ν. Prove that the following are equivalent:

(i) $\nu << \mu$,

(ii) $\nu^+ << \mu$ and $\nu^- << \mu$,

(iii) $|\nu| << \mu$.

4.3.5. Show that if E is any measurable set, then

$$-\nu^-(E) \leq \nu(E) \leq \nu^+(E) \quad \text{and} \quad |\nu(E)| \leq |\nu|(E).$$

4.3.6. Prove the formulas for differentiation stated in the text before Theorem 4.7.

4.3.7. Let $(X, \mathscr{B}, \mu)$ be an arbitrary measure space and let ν be the measure on $\mathscr{B}$ given by $\nu(A) = 0$ if $\mu(A) = 0$ and $\nu(A) = \infty$ if $\mu(A) > 0$. Prove ν is a measure, $\nu << \mu$, and find f_0 such that $\nu(E) = \int_E f_0\, d\mu$ for all E in $\mathscr{B}$.

4.3.8. Let $\mathscr{B} = \{A \subset R\colon \text{either } A \text{ or } A^c \text{ is countable}\}$. Prove the following:

(i) $\mathscr{B}$ is a σ-algebra.

(ii) The set functions μ and ν on $\mathscr{B}$ given, respectively, by

$$\mu(A) = \text{cardinality of } A$$

and

$$\nu(A) = \begin{cases} 0 & \text{if } A \text{ is countable} \\ \infty & \text{if } A \text{ is not countable} \end{cases}$$

are measures.

(iii) $\nu << \mu$ but the Radon–Nikodym Theorem is not satisfied.

4.3.9. Prove Proposition 4.11.

4.3.10. Let k be the Cantor ternary function on [0, 1] and extend k to R by defining $k(x) = 1$ if $x \geq 1$ and $k(x) = 0$ if $x \leq 0$. Let μ_k be the corresponding Borel measure on the Borel sets of R. If m is the Lebesgue measure on the Borel sets, show $\mu_k \perp m$.

4.3.11. If μ is a signed measure and f is a measurable function such that f is $|\mu|$-integrable, define

$$\int f\,d\mu = \int f\,d\mu^+ - \int f\,d\mu^-.$$

If μ is a finite signed measure, show that

$$|\mu|(E) = \sup\left\{\left|\int_E f\,d\mu\right| : |f| \leq 1\right\}.$$

4.3.12. Let $(X, \mathscr{A}, \mu)$ be a σ-finite measure space. Prove that there is a unique decomposition $\mu = \mu_1 + \mu_2 + \mu_3$, where $\mu_1 << \mu$, μ_2 is *purely atomic* [i.e., $\mu_2(B) > 0$ implies that there exist $C \subset B$ such that C is an atom of μ_2], and $\mu_3 \perp \mu$ with $\mu_3(\{x\}) = 0$ for every x. [Hint: By Theorem 4.7, $\mu = \mu_0 + \mu_1$, $\mu_1 << \mu$, and $\mu_0 \perp \mu$. Write $\mu_2(E) = \mu_0(E \cap A)$ and $\mu_3(E) = \mu_0(E - A)$, where $A = \{x\colon \mu_0(\{x\}) \neq 0\}$.]

★ **4.3.13.** Let $(X, \mathscr{A})$ and $(Y, \mathscr{B})$ be two measurable spaces. Let μ_1, μ_2 be two nonzero measures on $\mathscr{A}$ and ν_1, ν_2 two nonzero measures on $\mathscr{B}$. Prove that

(i) $\mu_1 \times \nu_1 << \mu_2 \times \nu_2$ if and only if $\mu_1 << \mu_2$ and $\nu_1 << \nu_2$;

(ii) $\mu_1 \times \nu_1 \perp \mu_2 \times \nu_2$ if and only if $\mu_1 \perp \mu_2$ or $\nu_1 \perp \nu_2$.

★ **4.3.14.** *An Extension of the Lebesgue Decomposition Theorem* (R. A. Johnson). Let μ and ν be measures on $(X, \mathscr{A})$. ν is called *$\mathscr{A}$-singular* with respect to μ, denoted by $\nu S \mu$, if given $B \in \mathscr{A}$ there exists $A \in \mathscr{A}$ with $A \subset B$, $\nu(B) = \nu(A)$, and $\mu(A) = 0$. Prove the following assertions:

(i) Suppose $\nu S \mu$ and ν is σ-finite. Then $\mu S \nu$ and $\nu \perp \mu$.

(ii) If $\nu S \mu$ and $\lambda << \mu$, then $\nu S \lambda$.

(iii) If $\nu S \mu$ and $\nu << \mu$, then $\nu = 0$.

(iv) There exist $\nu_0 << \mu$, $\nu_1 S \mu$, and $\nu_0 S \nu_1$ such that $\nu = \nu_0 + \nu_1$, where ν_1 is always unique. If ν is σ-finite, then ν_0 is also unique.

[Hint: Let $\mathscr{A}_1 = \{E \in \mathscr{A}: \mu(E) = 0\}$. If $\nu_1(A) = \sup\{\nu(A \cap E): E \in \mathscr{A}_1\}$, then ν_1 is a measure on $\mathscr{A}$ and $\nu_1 S \mu$. Let $\mathscr{A}_2 = \{E \in \mathscr{A}: \nu_1(E) = 0\}$. If $\nu_0(A) = \sup\{\nu(A \cap E): E \in \mathscr{A}_2\}$, then ν_0 is a measure on $\mathscr{A}$ and $\nu_0 << \mu$. Show that $\nu = \nu_0 + \nu_1$.]

★ **4.3.15.** Prove the following result due to R. A. Johnson: Let $(X, \mathscr{A})$ be a measurable space. Then the following hold.

(i) If μ_1 and μ_2 are *purely atomic* (see Problem 4.3.12) measures on $\mathscr{A}$, then so is $\mu_1 + \mu_2$.

(ii) If $\mu = \mu_1 + \mu_2$ and $\mu_1 S \mu_2$, then μ_1 is purely atomic or nonatomic according to whether μ is so. (See Problem 4.3.14.)

(iii) If $\mu << \mu_1 + \mu_2$, $\mu_2 S \mu_1$, μ is purely atomic, and μ_2 is nonatomic, then $\mu_2 S \mu$. [Hint: Use part (iv) of Problem 4.3.14 to write $\mu_2 = \lambda_1 + \lambda_2$, $\lambda_1 << \mu$, $\lambda_2 S \mu$, and $\lambda_1 S \lambda_2$. Now show that $\lambda_1 = 0$.]

(iv) For any measure μ on $\mathscr{A}$, there exist measures μ_1 and μ_2 such that $\mu = \mu_1 + \mu_2$, where μ_1 is purely atomic and μ_2 is nonatomic. μ_1 and μ_2 may be chosen such that $\mu_1 S \mu_2$ and $\mu_2 S \mu_1$, and under these conditions the decomposition is unique. [Hint: Let $\mathscr{A}_0$ be the family of sets that are countable unions of atoms of μ. Let $\mu_1(E) = \sup\{\mu(E \cap A): A \in \mathscr{A}_0\}$ and $\mu_2(E) = \sup\{\mu(E \cap A): \mu_1(A) = 0\}$. Then $\mu = \mu_1 + \mu_2$ is the desired decomposition.]

4.3.16. *An Application of the Baire-Category Theorem* (Vitali–Hahn–Saks). Let (μ_n) be a sequence of finite measures in a measure space $(X, \mathscr{A}, \mu_0)$ such that for each measurable set E, $\lim_{n\to\infty}\mu_n(E) = \mu(E)$ exists. Suppose that for each n, μ_n is absolutely continuous with respect to the measure μ_0. Then (i) the sequence (μ_n) is uniformly absolutely continuous with respect to μ_0 (i.e., given $\varepsilon > 0$, there exists $\delta > 0$ such that $\mu_0(E) < \delta \Rightarrow \mu_n(E) < \varepsilon$ for each n), and (ii) if μ_0 is finite, then μ is a measure and absolutely continuous with respect to μ_0. (Hint: $\mathscr{A}$ is a complete pseudometric space with the metric d defined by $d(A, B) = \arctan \mu_0(A \triangle B)$. Then $\mathscr{A} = \bigcup_{k=1}^{\infty} \mathscr{A}_k$, where $\mathscr{A}_k = \{E \in \mathscr{A}: |\mu_n(E) - \mu_m(E)| \leq \varepsilon$ for $n, m \geq k\}$. By the Baire-Category Theorem, there exist $a > 0$, $A \in \mathscr{A}$ and a positive integer N such that $\mu_0(A \triangle E) < a$ implies that $|\mu_n(E) - \mu_m(E)| \leq \varepsilon$ for $n, m \geq N$.)

4.3.17. *The Setwise Limit of a Sequence of Finite Measures on a σ-Algebra Is a Measure* (Nikodym). Let (μ_n) be a sequence of finite measures in a measure space such that for each measurable set E, $\lim_{n\to\infty}\mu_n(E) = \mu(E)$ exists. Then μ is a measure. (Compare Problems 2.1.5 and 2.2.7.) [Hint: Write $\nu_n(E) = \mu_n(E)/\mu_n(X)$ and let $\nu = \sum(1/2^n)\nu_n$. Now use part (ii) of Problem 4.3.16.]

• 4.4. Change of Variables in Integration

In this section, we shall give an application of the Radon–Nikodym Theorem by proving a general theorem that may by thought of as a change-of-variable formula. We shall first prove a change-of-variable formula for integration of real-valued functions of real variables by use of some results on differentiation of composite functions. Specifically, if g: $[a, b] \to [c, d]$ is a function differentiable a.e. on $[a, b]$ and f is a real-valued integrable function on $[c, d]$, we wish to analyze when the formula

$$\int_{g(\alpha)}^{g(\beta)} f(x)\,dx = \int_{\alpha}^{\beta} f(g(s))g'(s)\,ds \tag{4.33}$$

is valid for all α and β in $[a, b]$.

Let us first examine a few examples.

Examples

4.5. Let k be the Cantor ternary function on [0, 1] and let $f(x) = x$ on [0, 1]. Since

$$\frac{1}{2} = \int_{k(0)}^{k(1)} f(x)\,dx \neq \int_0^1 k(x)k'(x)\,dx = 0,$$

the change-of-variable formula (4.33) does not hold.

4.6. Let $g(x) = x\sin(1/x)$ if $x \neq 0$ and $g(0) = 0$ be defined on [0, 1] and let $f(x) = x$. Then g is not absolutely continuous (Problem 4.1.2) on [0, 1], but direct computation shows that the change-of-variable formula (4.33) holds.

4.7. Let $g(x) = x^6\sin^3(1/x)$ if $x \neq 0$ and $g(0) = 0$. Let $f(x) = x^{-2/3}/3$ if $x \neq 0$ and $f(0) = 0$. Then f is integrable on [0, 1], and using Problems 4.1.2 and 4.2.1, g is seen to be absolutely continuous on [0, 1]. However, the change-of-variable formula (4.33) does not hold.

These examples show in particular that the absolute continuity of g is in general neither sufficient nor necessary for equation (4.33) to hold. Theorem 4.8 below gives a necessary and sufficient condition for equation (4.33) to hold. To prove this result we need the following results (see also [58]).

Proposition 4.12. If g is differentiable on a set E with $g' = 0$ a.e. on E, then $m(g(E)) = 0$. Conversely, if g has a derivative (finite or infinite) on a set E with $m(g(E)) = 0$, then $g' = 0$ a.e. on E. ∎

Proof. Using problem 4.2.7(a), if $E_K = \{x \in E: K - 1 < |g'(x)| \leq K\}$,

$$m^*(g(E)) \leq \sum_{K=0}^{\infty} m^*(g(E_K)) \leq \sum_{K=0}^{\infty} K m^*(E_K) = 0. \tag{4.34}$$

Conversely, let $B = \{x \in E: |g'(x)| > 0\}$. Then $B = \bigcup_{n=1}^{\infty} B_n$, where

$$B_n = \{x \in B: |g(t) - g(x)| \geq |t - x|/n \text{ for } |t - x| < 1/n\}.$$

Fix n and let $A = I \cap B_n$, where I is any interval of length less than $1/n$. To show $m(B) = 0$, we will show $m(A) = 0$. Since $m(g(A)) = 0$, for any $\varepsilon > 0$ there exists a sequence (I_K) of intervals such that $g(A) \subset \bigcup_{K=1}^{\infty} I_K$ and $\sum_{K=1}^{\infty} m(I_K) < \varepsilon$. Let $A_K = g^{-1}(I_K) \cap A$. Since $A \subset \bigcup_{K=1}^{\infty} A_K$ and $A_K \subset I \cap B_n$, we have

$$m^*(A) \leq \sum_{K=1}^{\infty} m^*(A_K) \leq \sum_{k=1}^{\infty} \sup_{s,t \in A_K} |s - t| \leq \sum_{k=1} n \sup_{s,t \in A_K} |g(s) - g(t)|.$$

However, $\sup_{s,t \in A_K} |g(s) - g(t)| \leq m(I_K)$ since $g(A_K) \subset I_K$ so that $m^*(A) \leq \sum_{k=1}^{\infty} n m(I_K) \leq n\varepsilon$. Since ε is arbitrary, $m(A) = 0$. ∎

The function f of Example 4.2 on the interval [0, 1] is an example of a *strictly* increasing function with $f' = 0$ a.e. Letting $F = f^{-1}$, then $(F \circ f)(x) = x$ for all x in [0, 1] so that $(F \circ f)' = 1$. However, $f' = 0$ so that the "chain rule" $(F' \circ f)(x) f'(x) = (F \circ f)'(x)$ does not hold for any x. The next result gives a condition to ensure that the "chain rule" holds.

Proposition 4.13. Suppose F, g, and $F \circ g$ are differentiable almost everywhere on their domains $[c, d]$ and $[a, b]$, where $g([a, b]) \subset [c, d]$. If $m(F(E)) = 0$ whenever $m(E) = 0$, then

$$(F \circ g)' = (F' \circ g) g' \tag{4.35}$$

holds almost everywhere on $[a, b]$. ∎

Proof. Let $Z = \{y \in [c, d]: F \text{ is not differentiable at } y\}$, $E = g^{-1}(Z)$ and $G = [a, b] - E$. For x in G we can write

$$\Delta F = F(g(x + \Delta x)) - F(g(x)) = [F'(g(x)) + \varepsilon]\, \Delta g,$$

where $\Delta g = g(x + \Delta x) - g(x)$ and $\varepsilon \to 0$ as $\Delta g \to 0$. Dividing both sides of this equation by Δx and letting $\Delta x \to 0$ we obtain

$$(F \circ g)'(x) = F'\big(g(x)\big)g'(x)$$

if g is differentiable at x, that is, almost everywhere on G.

Clearly $m\big(g(E)\big) = 0$. By our hypothesis, $mF\big(g(E)\big) = 0$. Using Proposition 4.12, we see that g' and $(F \circ g)'$ equal zero a.e. on E, so that equation (4.35) holds a.e. on $[a, b]$. ∎

Using Problem 4.2.8(a) we have immediately the following corollary.

Corollary 4.4. If g and $F \circ g$ are differentiable a.e. on $[a, b]$ and F is absolutely continuous on $[c, d]$, then equation (4.35) is valid almost everywhere on $[a, b]$. ∎

Theorem 4.8. Suppose that g is differentiable almost everywhere on $[a, b]$ and that f is integrable on $[c, d]$, where $g([a, b]) \subset [c, d]$. If $F(x) = \int_c^x f(t)\, dt$, then the absolute continuity of $F \circ g$ is a necessary and sufficient condition that

$$\int_{g(\alpha)}^{g(\beta)} f(x)\, dx = \int_\alpha^\beta f\big(g(s)\big)g'(s)\, ds \tag{4.36}$$

for all α and β in $[a, b]$. ∎

Proof. If the formula (4.36) holds, then for any x in $[a, b]$,

$$F \circ g(x) - F \circ g(a) = \int_{g(a)}^{g(x)} f(t)\, dt = \int_a^x f\big(g(s)\big)g'(s)\, ds,$$

so that by Theorem 4.3, $F \circ g$ is absolutely continuous.

Conversely, if $F \circ g$ is absolutely continuous, then its derivative, which is $(F' \circ g)g'$ or $(f \circ g)g'$ a.e. by Proposition 4.13, is integrable and

$$\int_{g(\alpha)}^{g(\beta)} f(x)\, dx = F\big(g(\beta)\big) - F\big(g(\alpha)\big) = \int_\alpha^\beta f\big(g(s)\big)g'(s)\, ds. \quad ∎$$

Problems 4.4.2–4.4.4 below give some important corollaries of Theorem 4.8.

We will now use the Radon–Nikodym Theorem to prove an analog of Theorem 4.8 in arbitrary measure spaces. Given two measurable spaces $(X, \mathscr{B})$ and $(Y, \mathscr{A})$, a *measurable transformation* is a surjective function $\tau: X \to Y$ such that for each E in $\mathscr{A}$, $\tau^{-1}(E)$ is in $\mathscr{B}$. Thus a real-valued function f on $(X, \mathscr{B})$ is measurable if and only if f is a measurable transformation from $(X, \mathscr{B})$ to R with the class of Borel sets.

A function $\tau\colon X \to Y$ assigns in an obvious manner a function f on X to every function g on Y by the formula $f(x) = g(\tau(x))$. A measurable transformation $\tau\colon (X, \mathscr{B}) \to (Y, \mathscr{A})$ also assigns in an obvious manner a measure μ_τ on $\mathscr{A}$ to each measure μ on $\mathscr{B}$ by the formula $\mu_\tau(E) = \mu(\tau^{-1}(E))$.

Proposition 4.14. If $\tau\colon (X, \mathscr{B}) \to (Y, \mathscr{A})$ is a measurable transformation, μ is a measure on $\mathscr{B}$, and if g is a measurable function on Y, then

$$\int_Y g\, d\mu_\tau = \int_X g \circ \tau\, d\mu \tag{4.37}$$

in the sense that if either integral exists they both do and are equal. ∎

Proof. Note first that $g \circ \tau$ is measurable, since for any real number α

$$\{x\colon\, g \circ \tau(x) > \alpha\} = \tau^{-1}\{y\colon\, g(y) > \alpha\}.$$

If $g = \chi_E$, where $E \in \mathscr{A}$, then $g \circ \tau = \chi_{\tau^{-1}(E)}$. Clearly the proposition is valid for such functions g and thereby for simple functions defined on Y and measurable with respect to $\mathscr{A}$. The assertion of the proposition now follows by writing any nonnegative measurable function as the limit of a nondecreasing sequence of simple functions, invoking the Monotone Convergence Theorem, and then writing any measurable function as the difference of two nonnegative measurable functions. ∎

In the special case that $g = \chi_E \cdot h$ for $E \in \mathscr{A}$ and some measurable function h with respect to $\mathscr{A}$, equation (4.37) becomes

$$\int_E h\, d\mu_\tau = \int_{\tau^{-1}(E)} h \circ \tau\, d\mu.$$

This brings us to our goal—a formula for integration by substitution.

Theorem 4.9. Suppose $\tau\colon (X, \mathscr{B}) \to (Y, \mathscr{A})$ is a measurable transformation and μ and ν are measures on $(X, \mathscr{B})$ and $(Y, \mathscr{A})$, respectively, such that $\nu(\tau(E)) = 0$ whenever E is a measurable set in $\mathscr{B}$ with $\mu(E) = 0$. Suppose also that μ_τ is a σ-finite measure on Y. Then there is a nonnegative measurable function t_0 on X so that for a ν-integrable function f on Y, $(f \circ \tau)t_0$ is μ-integrable and

$$\int_E f(y)\, d\nu = \int_{\tau^{-1}(E)} (f \circ \tau) \cdot t_0\, d\mu \tag{4.38}$$

for each measurable set E in $\mathscr{A}$. ∎

Proof. Suppose $E \in \mathscr{A}$ and $\mu_\tau(E) = \mu(\tau^{-1}(E)) = 0$. By hypothesis, $\nu(E) = \nu[\tau(\tau^{-1}(E))] = 0$ so that $\nu << \mu_\tau$. By Corollary 4.3 there exists a nonnegative measurable function f_0 on X such that for each ν-integrable function f on Y, ff_0 is μ_τ-integrable and

$$\int_Y f(y)\, d\nu = \int_Y f(y) f_0(y)\, d\mu_\tau. \tag{4.39}$$

By Proposition 4.14

$$\int_Y f(y) f_0(y)\, d\mu_\tau = \int_X f(\tau(x)) f_0(\tau(x))\, d\mu. \tag{4.40}$$

Combining equations (4.39) and (4.40) we have for each ν-integrable function f

$$\int_Y f(y)\, d\nu = \int_X f(\tau(x)) t_0(x)\, d\mu,$$

where $t_0 = f_0 \circ \tau$. In the special case that f is replaced by $f \cdot \chi_E$ for some measurable set E in Y, we get equation (4.38). ∎

If τ is a monotonic, absolutely continuous function with domain $[a, b]$ and range $[\alpha, \beta]$, the function t_0 is equal to $|\tau'|$ a.e. on $[a, b]$. (See Problem 4.4.5.) Thus if f is Lebesgue integrable on $[\alpha, \beta]$, we have that $(f \circ \tau)|\tau'|$ is Lebesgue integrable on $[a, b]$ and

$$\int_\alpha^\beta f(y)\, dy = \int_a^b f \circ \tau(x) \,|\, \tau'(x) \,|\, dx.$$

In general the function t_0 plays the role of the absolute value of the Jacobian of a nonsingular transformation τ from R^n to R^n. The interested reader may consult [70] for a discussion of this situation.

Problems

4.4.1. Suppose F and g are real-valued functions on the closed intervals $[a, b]$ and $[c, d]$, respectively, with $g([a, b]) \subset [c, d]$. If F and g have finite derivatives a.e. on their respective domains and g' is zero at most on a null set, then prove that $F \circ g$ is differentiable a.e. and $(F \circ g)' = (F' \circ g)g'$ a.e.

In problems 4.4.2–4.4.4 function g is defined on $[a, b]$ and f is defined on $[c, d]$ with $g([a, b]) \subset [c, d]$.

4.4.2. If g is monotonic and absolutely continuous and f is integrable prove that $(f \circ g)g'$ is integrable and the change-of-variable-formula (4.33) holds.

4.4.3. If g is absolutely continuous and f is bounded and measurable, then prove that $(f \circ g)g'$ is integrable and the change-of-variable formula (4.33) holds.

4.4.4. If g is absolutely continuous, and f and $(f \circ g)g'$ are integrable, then prove that the change-of-variable formula (4.33) holds. (Hint: Use Problem 4.4.3 and the Dominated Convergence Theorem.)

4.4.5. Prove that if τ is a monotonic, absolutely continuous function with domain $[a, b]$ and range $[\alpha, \beta]$, then the function t_0 in Theorem 4.9 is equal to $| \tau' |$ a.e. on $[a, b]$. [Hint: Show $\tau(x) = \int_a^x t_0(s)\, ds + \tau(a)$ for each x when τ is increasing.]

5

Banach Spaces

Integral equations occur in a natural way in numerous physical problems and have attracted the attention of many mathematicians including Volterra, Fredholm, Hilbert, Schmidt, F. Riesz, and others. The works of Volterra and Fredholm on integral equations emphasized the usefulness of the techniques of the integral operators. Soon it was realized that many problems in analysis could be attacked with greater ease if placed under a suitably chosen axiomatic framework. Axioms closely related to those of a Banach space were introduced by Bennett.[†] Using the axioms of a Banach space, F. Riesz[‡] extended much of the Fredholm theory of integral equations. In 1922 using similar sets of axioms for such spaces, Banach,[§] Wiener,[||] and Hahn[¶] all independently published papers. But it was Banach who continued making extensive and fundamental contributions in the development of the theory of these spaces, now well known as Banach spaces. Banach-space techniques are widely known now and applied in numerous physical and abstract problems. For example, using the Hahn–Banach Theorem (asserting the existence of nontrivial continuous linear real-valued functions on Banach spaces), one can show the existence of a translation-invariant, finitely additive measure on the class of all bounded subsets of the reals such that the measure of an interval is its length.

The purpose of this chapter is to present some of the basic properties and principles of Banach spaces. In Sections 5.1 and 5.2 we introduce the

† A. A. Bennett, *Proc. Nat. Acad. Sci. U.S.A.* **2**, 592–598 (1916).

‡ F. Riesz, *Acta Math.* **41**, 71–98 (1918).

§ S. Banach, *Fund. Math.* **3**, 133–181 (1922).

|| N. Wiener, *Bull. Soc. Math. France* **50**, 119–134 (1922).

¶ H. Hahn, *Monatsh. Math. Phys.* **32**, 3–88 (1922).

basic concepts and definitions and the L_p spaces. In Sections 5.3 and 5.4 we present what are acknowledged as the four most important theorems in Banach spaces—the Hahn–Banach Theorem, the Open Mapping Theorem, the Closed Graph Theorem, and the Principle of Uniform Boundedness. In Section 5.5 we introduce the reflexive spaces, derive representation theorems for the duals of various important Banach spaces, and present in detail the interplay between reflexivity and weak topology. In Section 5.6 we introduce compact operators, present the classical Fredholm alternative theory, and discuss spectral concepts for such operators.

Throughout this chapter, F will denote either the real numbers R or the complex numbers C.

5.1. Basic Concepts and Definitions

We begin with several fundamental definitions.

Definition 5.1. A *linear space* X over a field F is an Abelian group under addition $(+)$, together with a scalar multiplication from $F \times X$ into X such that

(i) $\alpha(x + y) = \alpha x + \alpha y$,
(ii) $(\alpha + \beta)x = \alpha x + \beta x$,
(iii) $(\alpha\beta)x = \alpha(\beta x)$,

and

(iv) $1x = x$

for all α, $\beta \in F$, x and $y \in X$. (Here 1 denotes the multiplicative identity in F and 0 will denote the additive identity in X.) ∎

Definition 5.2. A linear space X over a field F is called a *normed linear space* if to each $x \in X$ is associated a nonnegative real number $\| x \|$, called the norm of x such that

(i) $\| x \| = 0$ if and only if $x = 0$;
(ii) $\| \alpha x \| = | \alpha | \, \| x \|$ for all $\alpha \in F$;
(iii) $\| x + y \| \leq \| x \| + \| y \|$ for all $x, y \in X$.

Defining $d(x, y) = \| x - y \|$, it can easily be verified that d defines a metric in X. A normed linear space X is called a *Banach* space if it is complete in this metric. The topology induced by d will usually be referred to as the topology in X. ∎

Examples

5.1. $R^n = \{(a_1, a_2, \ldots, a_n): a_i \in R, 1 \leq i \leq n\}$ is a Banach space over R if we define addition and scalar multiplication in the natural way and

$$\| (a_1, a_2, \ldots, a_n) \| = \sum_{i=1}^{n} | a_i |.$$

5.2. $C_1[0, 1]$, the usual linear space of complex-valued continuous functions over $[0, 1]$, is a Banach space over F if we define

$$\| f \| = \sup_{0 \leq x \leq 1} | f(x) |.$$

(See Problem 5.1.5.) This norm will be referred to as the uniform or "sup" norm.

5.3. The set of all polynomials on $[0, 1]$ as a subspace of $C[0, 1]$ in Example 5.2 is a normed linear space, but not a Banach space. The reason is that nonpolynomial continuous functions can be uniformly approximated by polynomials on $[0, 1]$, by Weierstrass' theorem.

5.4. The space l_p of all sequences $\{x = (x_1, x_2, \ldots): x_i \in F, \sum | x_i |^p < \infty\}$ is a Banach space under natural addition and scalar multiplication, and norm defined by $\| x \|_p = (\sum_{i=1}^{\infty} | x_i |^p)^{1/p}$, where p is a real number with $1 \leq p < \infty$. The l_p spaces are special cases of the important L_p spaces (studied later in this chapter), where the set $\{1, 2, \ldots\}$ is considered a measure space with each point in this set having measure 1. The proof that l_p spaces are Banach spaces is not trivial and will not be presented at this time. This will follow later from the corresponding fact about L_p spaces (Section 5.2).

Definition 5.3. Suppose X is a linear space over F, and $\| \cdot \|_1$ and $\| \cdot \|_2$ are two norms defined on it. Then these norms are called *equivalent* if there exist positive numbers a and b such that

$$a \cdot \| x \|_1 \leq \| x \|_2 \leq b \cdot \| x \|_1$$

for all $x \in X$. ∎

Equivalence of norms is clearly reflexive, symmetric, and transitive. In an infinite-dimensional normed linear space, there are always two norms that are not equivalent (Problem 5.1.12). But the situation is much nicer in the finite-dimensional case, as the following theorem shows.

Theorem 5.1. In a finite-dimensional linear space X, all norms are equivalent. ∎

Proof. Let us define for

$$x = (x_1, x_2, \ldots, x_n) \in X, \quad \| x \| = \sup_{1 \leq i \leq n} | x_i |.$$

Then $\| \cdot \|$ defines a norm in X, in which X is complete (see Problem 5.1.2). The theorem will be proved by showing that any norm $\| \cdot \|_1$ on X is equivalent to this sup-norm $\| \cdot \|$ on X.

Let $\{z_1, z_2, \ldots, z_n\}$ be a basis in X. Then since $x = \sum_{i=1}^n x_i z_i$, we have

$$\| x \|_1 \leq \sum_{i=1}^n | x_i | \; \| z_i \|_1 \leq \| x \| \cdot \sum_{i=1}^n \| z_i \|_1 .$$

If $b = \sum_{i=1}^n \| z_i \|_1$, then $b > 0$ and

$$\| x \|_1 \leq b \cdot \| x \| \qquad \text{for all } x \in X.$$

To find $a > 0$ such that $a \| x \| \leq \| x \|_1$ for all $x \in X$, we will use induction. Clearly when $n = 1$,

$$\| x \|_1 = | x_1 | \; \| z_1 \|_1 = \| x \| \cdot \| z_1 \|_1$$

and therefore, $\| z_1 \|_1$ can be taken as a. Suppose the theorem is true for all spaces of dimension less than or equal to $n - 1$, $n > 1$. Let X_i be the subspace of X spanned by $\{z_1, \ldots, z_{i-1}, z_{i+1}, \ldots, z_n\}$. Then $\dim X_i = n - 1$. In X_i, the norms $\| \cdot \|$ and $\| \cdot \|_1$ are equivalent by induction-hypothesis. Since X_i is complete in $\| \cdot \|$, it is also complete in $\| \cdot \|_1$ and therefore a closed subspace of $(X, \| \cdot \|_1)$. If $z_i + X_i = \{z_i + y\colon y \in X_i\}$, then $z_i + X_i$ is also a closed subset of $(X, \| \cdot \|_1)$ (see Problem 5.1.1). If $0 \in z_i + X_i$, then

$$0 = z_i + \sum_{j=1}^n \alpha_j z_j,$$

for suitable scalars $\alpha_j \in F$, with $\alpha_i = 0$. But this contradicts the linear independence of the basis vectors $\{z_1, z_2, \ldots, z_n\}$. Hence $0 \notin z_i + X_i$, and therefore there exists a positive number d_i such that

$$\| z_i + y \|_1 > d_i \qquad \text{for all } y \in X_i.$$

This means that, for scalars $\alpha_j \in F$,

$$\left\| \sum_{j=1}^n \alpha_j z_j \right\|_1 \geq | \alpha_i | \, d_i .$$

We can repeat the process for each i, $1 \leq i \leq n$, and for each i find positive numbers $d_i > 0$. Then

$$\left\| \sum_{i=1}^{n} \alpha_i z_i \right\|_1 \geq \sup_{1 \leq i \leq n} | \alpha_i | \cdot \min_{1 \leq i \leq n} | d_i |$$

which means that

$$a \| x \| \leq \| x \|_1 \qquad \text{for all } x \in X,$$

where $a = \min_{1 \leq i \leq n} | d_i | > 0$. ∎

Corollary 5.1. Every finite-dimensional normed linear space is complete. ∎

Corollary 5.2. Let $(X_1, \| \cdot \|_1)$ and $(X_2, \| \cdot \|_2)$ be any two finite-dimensional normed linear spaces of the same dimension over F. Then they are topologically isomorphic, i.e., there is a mapping from one onto the other, which is an algebraic isomorphism as well as a topological homeomorphism. ∎

The proof is an easy consequence of Theorem 5.1 and is left to the reader. Next, we present a result due to Riesz, often useful in proving various results in the theory of normed linear spaces, besides being of independent geometric interest.

Proposition 5.1. Let Y be a proper closed linear subspace of a normed linear space X over F. Let $0 < a < 1$. Then there exists some $x_a \in X$ such that $\| x_a \| = 1$ and $\inf_{y \in Y} \| x_a - y \| \geq a$. ∎

Proof. Let $x \in X - Y$ and $d = \inf_{y \in Y} \| x - y \|$. Then $d > 0$, since Y is closed. Now there exists $y_0 \in Y$ such that $0 < \| x - y_0 \| < d/a$. Let $x_a = [x - y_0]/\| x - y_0 \|$. Now the reader can easily check that x_a satisfies the requirements of the propositon. ∎

Note that in the above proposition, the proper subspace Y has to be necessarily *closed*. For instance, if X is $C[0, 1]$ with the uniform norm and Y is the subspace of all polynomials on [0, 1], then $\bar{Y} = X$ and therefore the proposition fails to work in this case.

Also, one cannot generally take $a = 1$ in Proposition 5.1. For example, let X be the real-valued continuous functions on [0, 1] which vanish at 0, with the uniform norm, and $Y = \{ f \in X: \int_0^1 f(x)\, dx = 0 \}$. Then Y is a closed proper subspace of X. Suppose there exists $h \in X - Y$ such that

$\inf_{f\in Y}\| h - f \| \geq 1$, where for $g \in X$, $\| g \| = \sup_{0\leq x\leq 1}| g(x) |$. If $g \in X - Y$ and $a(g) = [\int_0^1 h(x)\, dx]/[\int_0^1 g(x)\, dx]$, then

$$\int_0^1 [h(x) - a(g) \cdot g(x)]\, dx = 0 \quad \text{or} \quad h - a(g) \cdot g \in Y;$$

therefore, $\| h - [h - a(g) \cdot g] \| = \| a(g) \cdot g \| \geq 1$. Let $g_n(x) = x^{1/n}$, $1 \leq n < \infty$. Then $g_n \in X - Y$ and hence $\| a(g_n) \cdot g_n \| \geq 1$. But $a(g_n) = [(n+1)/n] \int_0^1 h(x)\, dx$. Hence, since $\| g_n \| = 1$, we have

$$\left| \int_0^1 h(x)\, dx \right| \geq \frac{n}{n+1}$$

for each positive integer n. This means that $| \int_0^1 h(x)\, dx | \geq 1$. But since $h(0) = 0$ and $\| h \| = 1$, $| \int_0^1 h(x)\, dx | < 1$, which is a contradiction.

Now we show how Proposition 5.1 can help us understand the notion of compactness in a normed linear space. We know that a set in R^n or in any finite-dimensional normed linear space is compact if and only if it is closed and bounded. But if X is an infinite-dimensional normed linear space, then by Proposition 5.1 we can find (x_n) with $\| x_n \| = 1$, $1 \leq n < \infty$ and for each n, $\| x_n - x_i \| \geq \frac{1}{2}$, for $1 \leq i < n$; clearly these x_n's cannot have a limit point and therefore the closed unit ball in X is not compact. Thus we have the following theorem.

Theorem 5.2. Let X be a normed linear space over F. Then its closed unit ball is compact if and only if X is finite dimensional. ∎

Definition 5.4. A series $\sum_{k=1}^{\infty} x_k$ in a normed linear space X over F is called *summable* if $\| \sum_{k=1}^{n} x_k - x \| \to 0$ as $n \to \infty$ for some $x \in X$. For a summable series, we write $\sum_{k=1}^{\infty} x_k = \lim_{n\to\infty} \sum_{k=1}^{n} x_k$. The series $\sum_{k=1}^{\infty} x_k$ is called absolutely summable if $\sum_{k=1}^{\infty} \| x_k \| < \infty$. ∎

We know that an absolutely summable series of real numbers is summable. This is a consequence of the completeness of the real numbers. In fact, we have the following theorem, which is often useful in establishing the completeness of a normed linear space.

Theorem 5.3. Every absolutely summable series in a normed linear space X is summable if and only if X is complete. ∎

Proof. For the "if" part, let X be complete and for each positive integer n let x_n be an element of X such that $\sum_{n=1}^{\infty} \| x_n \| < \infty$. Let

$y_k = \sum_{n=1}^{k} x_n$. Then

$$\| y_{k+p} - y_k \| = \left\| \sum_{n=k+1}^{k+p} x_n \right\| \leq \sum_{n=k+1}^{k+p} \| x_n \|,$$

which converges to zero as $k \to \infty$. Hence the sequence $(y_k)_{k=1}^{\infty}$ is Cauchy in X. Therefore since X is complete, there exists $x \in X$ such that $x = \lim_{k\to\infty} \sum_{n=1}^{k} x_n$. This proves the "if" part.

For the "only if" part, suppose every absolutely summable series in X is summable. Let (x_n) be a Cauchy sequence in X. For each positive integer k, there is a positive integer n_k such that $\| x_n - x_m \| < 1/2^k$ for all n and m greater than or equal to n_k. We choose $n_{k+1} > n_k$. Let $y_1 = x_{n_1}$ and $y_{k+1} = x_{n_{k+1}} - x_{n_k}$, $k \geq 1$. Then $\sum_{k=1}^{\infty} \| y_k \| < \infty$. Therefore, there exists $y \in X$ such that

$$y = \lim_{m\to\infty} \sum_{k=1}^{m} y_k = \lim_{m\to\infty} x_{n_m}.$$

Since (x_n) is Cauchy, $\lim_{n\to\infty} x_n$ is also y. ∎

Finally, in this section we introduce the concept of quotient spaces. Quotient spaces are often useful in tackling certain problems in normed linear spaces, as will be seen in the later sections of this chapter. See also Problem 5.1.10.

Suppose M is a closed subspace of a normed linear space X over a field F. Let $[x] = \{y \in X\colon y \sim x\}$, where $x \in X$ and $y \sim x$ if and only if $y - x \in M$. Clearly, "$\sim$" defines an equivalence relation on X. Let X/M denote the set of all equivalence classes. For $\alpha \in F$ and $[x]$, $[y]$ in X/M, we define

$$[x] + [y] = [x + y] \quad \text{and} \quad \alpha[x] = [\alpha x].$$

These operations are well defined and make X/M a vector space. Let $\| [x] \|_1 = \inf_{y\in M} \| x - y \|$. One can easily check that $(X/M, \| \cdot \|_1)$ is a normed linear space, usually called the *quotient of X by M*. If Φ is the natural map from X onto X/M defined by $\Phi(x) = [x]$, then Φ is a continuous, linear, and open mapping. The linearity of Φ is trivial. The continuity of Φ follows from the fact that $\| [x_n - x] \|_1 \leq \| x_n - x \|$. Also, if $V = \{y \in X\colon \| y - x \| < r\}$, then $\Phi(V) = \{[y] \in X/M\colon \| [y] - [x] \|_1 < r\}$. This implies that Φ is open.

In many cases, several properties of the quotient space X/M are strongly related to similar properties of the space X. Problem 5.1.9 and the next proposition will illustrate two of them.

Proposition 5.2. Let M be a closed linear subspace of a normed linear space X. Then X is complete if and only if M and X/M are complete. ▮

Proof. For the "if" part, let M and X/M be complete. Let $(x_n)_{n=1}^{\infty}$ be a Cauchy sequence in X. Then $([x_n])_{n=1}^{\infty}$ is Cauchy in X/M and therefore there exists $[y] \in X/M$ such that $[x_n] \to [y]$ in X/M as $n \to \infty$. This means that $\inf_{z \in M} \|x_n - y - z\| \to 0$ as $n \to \infty$. Hence there exists a subsequence (n_k) of positive integers and a sequence (z_k) in M such that $x_{n_k} - y - z_k \to 0$ as $k \to \infty$. This means that the sequence $(z_k)_{k=1}^{\infty}$ is Cauchy in M and therefore there exists $z \in M$ such that $z_k \to z$ as $k \to \infty$ so that $x_n \to y + z$ as $n \to \infty$. The "if" part is proved.

For the "only if" part, we will use Theorem 5.3. Let X be complete. Then M, being a closed subspace, is also complete. To prove the completeness of X/M, let $\sum_{n=1}^{\infty} \| [x_n] \|_1 < \infty$. We are finished if we can show that there exists $[y] \in X/M$ such that $[y] = \lim_{k\to\infty} \sum_{n=1}^{k} [x_n]$. Let $y_n \in M$ be chosen such that $\| x_n + y_n \| \leq \| [x_n] \|_1 + 1/2^n$, for each positive integer n. Then $\sum_{n=1}^{\infty} \| x_n + y_n \| < \infty$. Since X is complete, there exists $y \in X$ such that $y = \lim_{k\to\infty} \sum_{n=1}^{k} x_n + y_n$. Since the natural map Φ is continuous and linear,

$$[y] = \lim_{k\to\infty} \sum_{n=1}^{k} [x_n + y_n] = \lim_{k\to\infty} \sum_{n=1}^{k} [x_n]. \qquad ▮$$

Problems

5.1.1. Let X be a normed linear space over a field F. Let $y \in X$ and $\alpha \in F$, $\alpha \neq 0$. Show that the mappings $x \to x + y$ and $x \to \alpha \cdot x$ are homeomorphisms of X onto itself.

5.1.2. Let X be a linear space (of dimension n) over a field F. Let $\{z_1, z_2, \ldots, z_n\}$ be a basis of X. If $x = \sum_{i=1}^{n} \alpha_i \cdot z_i$, $\alpha_i \in F$ and $\| x \| = \sup_{1 \leq i \leq n} | \alpha_i |$, then show that $(X, \| \cdot \|)$ is a Banach space.

5.1.3. Let X be a normed linear space. If $S_x(r) = \{y \in X\colon \| y - x \| < r\}$, then show that $\overline{S_x(r)} = \{y \subset X\colon \| y - x \| \leq r\}$.

5.1.4. Let A and B be two subsets of a normed linear space X. Let

$$A + B = \{x + y\colon x \in A,\ y \in B\}.$$

Show that (a) $A + B$ is open whenever either A or B is open; and (b) $A + B$ is closed whenever A is compact and B is closed. (Note that $A + B$ need not be closed even if A and B are both closed.)

5.1.5. Prove that $C_1[0, 1]$, the linear space of complex-valued continuous functions on $[0, 1]$, is a Banach space under the uniform norm.

5.1.6. Show that a finite-dimensional subspace of an infinite-dimensional normed linear space X is nowhere dense in X.

5.1.7. Use Problem 5.1.6 and the Baire Category Theorem to prove that an infinite-dimensional Banach space cannot have a countable basis.

5.1.8. Let X be a normed linear space and f be the mapping defined by $f(x) = rx/(1 + \| x \|)$. Show that f is a homeomorphism from X onto $\{x\colon \| x \| < r\}$. [Hint: $f^{-1}(y) = y/(r - \| y \|)$.]

5.1.9. Let M be a closed subspace of a normed linear space X. Show that X is separable if and only if M and X/M are both separable.

5.1.10. Show that the sum of two closed subspaces of a normed linear space is closed whenever one of the subspaces is finite dimensional. [Hint: Let A be a closed subspace and B be a finite-dimensional subspace of a normed linear space X. If Φ is the natural map from X onto X/A, then $\Phi^{-1}(\Phi(B)) = A + B$.]

5.1.11. Show that a normed linear space whose closed unit ball is totally bounded is finite dimensional.

5.1.12. Let X be an infinite-dimensional normed linear space over F. Let (x_n) be an infinite set contained in H, a Hamel basis for X. If $x \in X$ and $x = \sum a_\alpha x_\alpha$, $x_\alpha \in H$ and $a_\alpha \in F$, then define $\| x \|_1 = \sup | a_\alpha |$ and $\| x \|_2 = \sum | a_\alpha |$. Show that these two norms are not equivalent. [Hint: Take $y_n = (1/n)(x_1 + x_2 + \cdots + x_n)$. Then $\| y_n \|_1 \to 0$ as $n \to \infty$; $\| y_n \|_2 = 1$ for each n.]

5.2. The L_p Spaces

The subject of this section is the L_p spaces,† an important class of Banach spaces.

Let $(X, \mathscr{A}, \mu)$ be a measure space. If p is a positive real number, a measurable real or complex-valued function f (i.e., the real part as well as the imaginary part of f is measurable) is said to be in L_p if $| f |^p$ is integrable. Since for any two scalars α and β in F

$$| \alpha f + \beta g |^p \leq 2^p(| \alpha |^p | f |^p + | \beta |^p | g |^p),$$

it follows that L_p is a linear space over F. For a function $f \in L_p$, we define

$$\| f \|_p = \left(\int | f |^p \, d\mu \right)^{1/p}.$$

† The reflexivity of these spaces is studied in Section 5.5.

If $p < 1$, the function $\| \cdot \|_p$ generally does not have the triangle property. For example, if $p = \frac{1}{2}$ and the measure space is the Lebesgue measure space on [0, 1], then for $f = 4\chi_{[0,1/2]}$, $g = 4\chi_{[1/2,1]}$ we have

$$\| f + g \|_{1/2} = 4, \qquad \| f \|_{1/2} + \| g \|_{1/2} = 2.$$

But for $0 < p < 1$ and $f, g \in L_p$, if we define

$$d(f, g) = \int | f - g |^p \, d\mu,$$

then (L_p, d) becomes a metric linear space (i.e., a linear space with a metric topology where vector addition and scalar multiplication are continuous mappings). This is due to the fact that

$$x^p + y^p \geq (x + y)^p, \qquad 0 < p < 1, \qquad x \geq 0, \qquad y \geq 0.$$

M. M. Day† showed that even this consideration of L_p (as a metric linear space for $0 < p < 1$) fails to be useful in a certain sense. He proved that in most measure spaces any continuous linear functional (i.e., a scalar-valued function) on the L_p spaces, $0 < p < 1$, is identically zero. Many important results in functional analysis are based on the existence of nontrivial continuous linear functionals (this will be discussed in the next section) and hence cannot be valid for such spaces.

The function $\| \cdot \|_p$, $p \geq 1$, however, is more interesting, and it will be shown in this section that $(L_p, \| \cdot \|_p)$, $p \geq 1$, is a Banach space.

Identifying all measurable functions that are equal almost everywhere, we have

$$\| f \|_p = 0 \text{ if and only if } f = 0$$

Also $\| \alpha f \|_p = | \alpha | \, \| f \|_p$, for $\alpha \in R$. The next four results will prove that L_p is indeed a Banach space under $\| \cdot \|_p$, $p \geq 1$.

Lemma 5.1. Let a and b be nonnegative reals and p, q be reals with $p > 1$, $q > 1$ and $1/p + 1/q = 1$. Then

$$ab \leq a^p/p + b^q/q$$

with equality if and only if $a^p = b^q$. ∎

† M. M. Day, *Bull. Amer. Math. Soc.* **46**, 816–823 (1940).

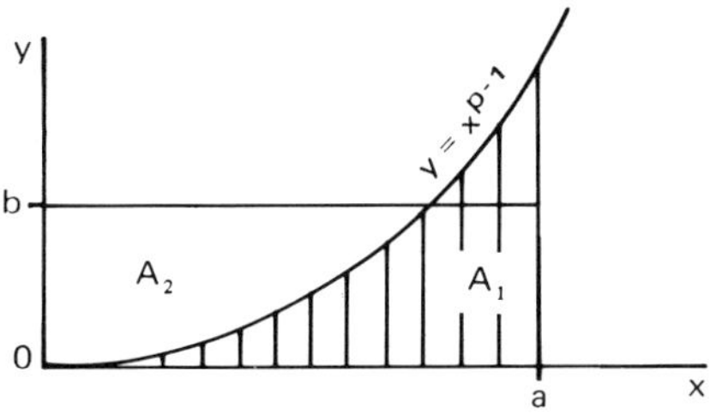

Fig. 3

Proof. Suppose $a > b$. The function $y = x^{p-1}$ is sketched in Figure 3. Clearly, from the picture, $a \cdot b \leq A_1 + A_2$, where

$$A_1 = \int_0^a x^{p-1}\,dx = \frac{a^p}{p}$$

and

$$A_2 = \int_0^b x\,dy = \int_0^b y^{q-1}\,dy = \frac{b^q}{q}.$$

Hence $a \cdot b \leq a^p/p + b^q/q$. The case $a \leq b$ is taken care of similarly. Also, from the picture it is clear that $ab = A_1 + A_2$ if and only if $a^{p-1} = b$, i.e., $a^p = b^q$. ∎

Proposition 5.3. *Hölder Inequality.* Let p and q be real numbers with $p > 1$, $q > 1$ and $1/p + 1/q = 1$. If $f \in L_p$, $g \in L_q$, then $f \cdot g \in L_1$ and $\int |f \cdot g|\,d\mu \leq \|f\|_p \cdot \|g\|_q$, with equality if and only if $\alpha |f|^p = \beta |g|^q$ a.e. for some constants α and β, not both zero. ∎

Proof. Let $f_1 = f/\|f\|_p$ and $g_1 = g/\|g\|_q$. (The proposition is trivial when either $\|f\|_p$ or $\|g\|_q = 0$; so we may assume $\|f\|_p > 0$ and $\|g\|_q > 0$.) Now $\|f_1\|_p = \|g_1\|_q = 1$. Taking $a = |f_1|$, $b = |g_1|$ in Lemma 5.1, we have

$$|f_1 g_1| \leq \frac{|f_1|^p}{p} + \frac{|g_1|^q}{q}.$$

Therefore,

$$\int |f_1 g_1|\,d\mu \leq \frac{1}{p} \cdot \|f_1\|_p{}^p + \frac{1}{q} \cdot \|g_1\|_q{}^q,$$

which means

$$\int |fg|\,d\mu \leq \|f\|_p \|g\|_q.$$

The rest is left to the reader. ∎

Proposition 5.4. *Minkowski Inequality.* If f and g are in L_p, $p \geq 1$, then

$$\| f + g \|_p \leq \| f \|_p + \| g \|_p. \qquad \blacksquare$$

Proof. The case $p = 1$ is trivial. Hence we assume $p > 1$. Then

$$\int | f + g |^p \, d\mu \leq \int | f + g |^{p-1} | f | \, d\mu + \int | f + g |^{p-1} | g | \, d\mu.$$

By the Hölder Inequality (taking $q = p/(p - 1)$),

$$\int | f + g |^{p-1} | f | \, d\mu \leq \| f \|_p \cdot \| \, | f + g |^{p-1} \|_q$$

and

$$\int | f + g |^{p-1} | g | \, d\mu \leq \| g \|_p \, \| \, | f + g |^{p-1} \|_q.$$

Since

$$\| \, | f + g |^{p-1} \|_q = (\| f + g \|_p)^{p-1},$$

the proposition follows from the last two inequalities. $\blacksquare$

Theorem 5.4. (*Riesz–Fischer*) For $p \geq 1$, $(L_p, \| \cdot \|_p)$ is a Banach space. $\blacksquare$

Proof. We only need to show that $(L_p, \| \cdot \|_p)$ is complete. We will use Theorem 5.3. Let $(f_n)_{n=1}^{\infty}$ be a sequence in L_p with $\sum_{n=1}^{\infty} \| f_n \|_p = a < \infty$. We must show that $\sum_{n=1}^{\infty} f_n$ is summable in L_p.

Set $g_n(x) = \sum_{k=1}^{n} | f_k(x) |$. Hence, by the Minkowski Inequality,

$$\| g_n \|_p \leq \sum_{k=1}^{n} \| f_k \|_p \leq a$$

or

$$\int g_n{}^p \, d\mu \leq a^p.$$

Let $\lim_{n \to \infty} g_n(x) = g(x)$. By Fatou's Lemma,

$$\int g^p \, d\mu \leq a^p.$$

This means that $g(x)$ is finite a.e. Now since $\sum_{k=1}^{\infty} | f_k(x) | < \infty$ a.e., there is a function $h(x)$ with $|h(x)|$ finite a.e. and $\sum_{k=1}^{\infty} f_k(x) = h(x)$. Since

$$\left| \sum_{k=1}^{n} f_k(x) - h(x) \right|^p \le 2^p[g(x)]^p,$$

by the Lebesgue Convergence Theorem

$$\left\| \sum_{k=1}^{n} f_k(x) - h(x) \right\|_p \to 0 \qquad \text{as } n \to \infty.$$

This completes the proof. ∎

So far we have considered p to be a real number ≥ 1. We now take $p = \infty$ and denote by L_∞ the space of all measurable functions that are bounded except possibly on a set of measure zero. As before, we identify functions in L_∞ that are equal a.e. and L_∞ becomes a complete normed linear space if we define

$$\|f\|_\infty = \operatorname{ess\,sup}|f|, \qquad \text{where } \operatorname{ess\,sup}|f| = \inf\{M: \mu\{x: |f(x)| > M\} = 0\}.$$

(See Problem 5.2.1.) The Holder Inequality, as well as the Riesz–Fischer theorem, remains true when $p = \infty$ and $q = 1$.

When the measure space X is the set of positive integers with each integer having measure 1, the L_p spaces are usually called the l_p spaces. Thus, we define the l_p space $(1 \le p < \infty)$ as the class of all sequences $x = (x_n)_{n=1}^{\infty}$, $x_n \in F$ with $\sum_{n=1}^{\infty} |x_n|^p < \infty$. Then l_p, like L_p, is a Banach space under the norm $\|x\|_p = (\sum |x_n|^p)^{1/p}$. Also, we define the l_∞ space as the class of all bounded sequences $x = (x_n)_{n=1}^{\infty}$, $x_n \in F$. Again l_∞, like L_∞, is a Banach space under the norm $\|x\|_\infty = \sup_n |x_n|$.

Finally, in this section we study an important theorem on the convergence of certain arithmetic means of L_1 functions.

The Individual Ergodic Theorem

Many interesting problems in mathematics and physics arise naturally in the context of L_p, $p \ge 1$ spaces. The problem of the convergence of certain arithmetic means is one such problem in ergodic theory. To be more specific, we need the following concept.

Let $(X, \mathscr{A}, \mu)$ be a measure space. Then a mapping $\Phi: X \to X$ is called *measure preserving* if

(i) Φ is bijective;

(ii) $A \in \mathscr{A} \Leftrightarrow \Phi(A) \in \mathscr{A}$;

(iii) $\mu(\Phi(A)) = \mu(A), \qquad A \in \mathscr{A}$.

A simple example of a measure-preserving mapping is $\Phi(x) = x + t$, where the measure is the Lebesgue measure on R. The reader can find nontrivial examples of such mappings in [21].

Let T be a linear operator on the linear space of measurable functions defined by

$$T(f)(x) = f(\Phi(x)),$$

where Φ is a measure-preserving mapping on the measure space $(X, \mathscr{A}, \mu)$. It can be verified easily that T is one-to-one and onto.

An important concern of ergodic theory is to find when the averages

$$\frac{1}{n} \sum_{k=0}^{n-1} T^k(f)$$

converge for various classes of measurable functions f. The individual ergodic theorem solves this problem for L_1 functions with respect to pointwise convergence. This theorem was first proven by G. D. Birkhoff in 1931. We present below F. Riesz' proof of this theorem.

Theorem 5.5. *The Individual Ergodic Theorem.* Let $(X, \mathscr{A}, \mu)$ be a finite measure space and Φ a measure-preserving mapping on X. Then for $f \in L_1$, the sequence

$$T_n(f)(x) = \frac{1}{n} \sum_{k=0}^{n-1} f(\Phi^k(x))$$

converges pointwise a.e. ∎

Proof. The proof will be given in two steps. It is no loss of generality to assume that f is real valued.

Step I. In this step, we show that

$$\int_E f \, d\mu \geq 0,$$

where

$$E = \left\{x: \sum_{k=0}^{n} f(\Phi^k(x)) \geq 0 \text{ for some } n = 0, 1, 2, \ldots\right\}.$$

To show this, we consider the set

$$E_m = \left\{x: \sum_{k=0}^{n} f(\Phi^k(x)) \geq 0 \text{ for some } n \leq m\right\}$$

Then it is enough to show that

$$\int_{E_m} f\,d\mu \geq 0, \qquad m = 1, 2, \ldots .$$

Let m be a fixed positive integer and n be any given positive integer. For $k \leq n$, we write

$$F_k = \{x: f(\Phi^k(x)) + \cdots + f(\Phi^{k+i-1}(x)) \geq 0 \qquad \text{for some } i \leq m\}.$$

For $k > n$, we write

$$F_k = \{x: f(\Phi^k(x)) + \cdots + f(\Phi^{k+i-1}(x)) \geq 0 \qquad \text{for some } i \text{ such that } k + i - 1 \leq n + m - 1\}.$$

Then it is clear that for each x

$$\sum_{k=0}^{n+m-1} \chi_{F_k}(x) f(\Phi^k(x)) \geq 0.$$

The reason is that, in the above summation, if a particular term (say, the kth term) is negative, then there is a positive integer $i(k)$ such that $k + i(k) - 1 \leq n + m - 1$ and

$$\sum_{j=k}^{k+i(k)-1} \chi_{F_j}(x) f(\Phi^j(x)) \geq 0.$$

Thus the above summation is the sum of a number of nonnegative subsums. Since Φ is bijective, $F_0 = E_m$ and $\Phi(F_k) = F_{k-1}$, $k \leq n - 1$. This means that

$$n \int_{E_m} f\,d\mu + \sum_{k=n}^{n+m-1} \int_{F_k} f \circ \Phi^k\,d\mu \geq 0.$$

It follows that

$$n \int_{E_m} f\,d\mu + m \int |f|\,d\mu \geq 0.$$

Since this inequality is valid for all positive integers n, it follows that $\int_{E_m} f\,d\mu \geq 0$. Thus we have proven step I.

Step II. Let r, s be rationals and

$$E_{rs} = \left\{x: \liminf_{n\to\infty} T_n(f)(x) < r < s < \limsup_{n\to\infty} T_n(f)(x)\right\}.$$

The theorem will be proven if $\mu(E_{rs}) = 0$ for all rationals r and s. It is clear that $\Phi(E_{rs}) = E_{rs}$. Therefore, replacing X by E_{rs} and applying step I, we have

$$\int_{E_{rs}} (f - s)\, d\mu \geq 0, \qquad \int_{E_{rs}} (r - f)\, d\mu \geq 0.$$

This means that $\int_{E_{rs}}(r - s)\, d\mu \geq 0$, proving that $\mu(E_{rs}) = 0$. ∎

Problems

5.2.1. Prove that L_∞ is a Banach space under the norm $\| f \|_\infty = \operatorname{ess\,sup} | f |$.

5.2.2. Let $f \in L_p$, $1 \leq p < \infty$. Then prove the following.

(a) There exists a sequence $f_n \in L_p$, each vanishing outside a set of finite measure, $| f_n | \leq | f |$ for all n and $\| f_n - f \|_p \to 0$ as $n \to \infty$.

(b) Given $\varepsilon > 0$, there exists $g \in L_p$ such that $| g |$ is bounded, $| g | \leq | f |$ and $\| g - f \|_p < \varepsilon$.

(c) Given ε positive, there exists a simple function g in L_p such that $| g | \leq | f | + \varepsilon$ and $\| g - f \|_p \leq \varepsilon$.

5.2.3. In a finite-measure space, every L_p function is in L_q whenever $1 \leq q \leq p \leq \infty$. Prove this and then show that this is false in an infinite-measure space.

5.2.4. Show that in any measure space for $f \in L_1 \cap L_\infty$, the limit of $\| f \|_p$ is $\| f \|_\infty$ as $p \to \infty$. (Hint: $\int | f |^{p+1}\, d\mu \leq \| f \|_\infty^p \int | f |\, d\mu$.)

5.2.5. Let (f_n) be a sequence of L_p $(1 \leq p < \infty)$ functions, each vanishing outside a set E_n, where (E_n) is a sequence of pairwise disjoint measurable sets. Show that $f = \sum_{n=1}^\infty f_n$ is in L_p if and only if $\sum_{n=1}^\infty \| f_n \|_p{}^p < \infty$, and in this case $\| f - \sum_{n=1}^k f_n \|_p \to 0$ as $k \to \infty$.

5.2.6. Let g be a real-valued integrable function in a finite-measure space such that for some constant M and all simple functions h, $| \int gh\, d\mu | \leq M \| h \|_p$, where $1 \leq p < \infty$. Show that g is in L_q, where $1/p + 1/q = 1$. [Hint: Let $p > 1$ and (g_n) be a sequence of nonnegative simple functions which increase to $| g |^q$. Set $h_n = (g_n)^{1/p} \cdot g/| g |$ when $g \neq 0$, and $= 0$ when $g = 0$. Then $g_n \leq h_n g$ and $\int g_n\, d\mu \leq M^q$.]

5.2.7. Let $1 \leq p < \infty$ and let (f_n) be a sequence in L_p converging a.e. to a function f in L_p. Show that $\| f_n - f \|_p \to 0$ as $n \to \infty$ if and only if $\| f_n \|_p \to \| f \|_p$ as $n \to \infty$. (Hint: The "if" part is nontrivial. Use Theorem 3.4.)

5.2.8. Let $f \in L_p$ and (f_n) be a sequence in L_p, where $1 < p < \infty$, and $f_n \to f$ a.e. Suppose also that for some constant M, $\| f_n \|_p \leq M$ for all

n. Then show that for each function $g \in L_q$, $1/p + 1/q = 1$,

$$\int f \cdot g \, d\mu = \lim_{n\to\infty} \int f_n \cdot g \, d\mu.$$

This result is false for $p = 1$. Give an example. [Hint: For this problem use the Hölder inequality and reduce the problem to one in a finite-measure space, then use Egoroff's Theorem.]

5.2.9. Let $1 \leq p \leq \infty$ and $1/p + 1/q = 1$. Let μ be a σ-finite measure and let g be a measurable function such that $f \cdot g \in L_1$ for every $f \in L_p$. Prove that $g \in L_q$.

5.2.10. Let $\| f_n - f \|_p \to 0$ as $n \to \infty$, $1 \leq p < \infty$, where (f_n) is a sequence in L_p and $f \in L_p$. Suppose (g_n) is a sequence of measurable functions such that $| g_n | \leq M$ for some constant M, for all n and $g_n \to g$ a.e. Show that $\| g_n f_n - gf \|_p \to 0$ as $n \to \infty$.

5.2.11. *A Generalization of the Lebesgue Convergence Theorem* (W. R. Wade). Let $(X, \mathscr{A}, \mu)$ be a complete finite-measure space, and let (f_n) be a sequence of measurable functions such that for some r with $0 < r \leq \infty$, $\| f_n \|_r \leq M$ for all n. If $\lim_{n\to\infty} f_n(x) = f(x)$ and $0 < p < r$, then $\lim_{n\to\infty} \| f_n - f \|_p = 0$. This result is false if $p = r$ or if $\mu(X) = \infty$. (Hint: Let $0 < r < \infty$. Use Fatou's Lemma to show that f^p is integrable for $0 < p < r$; then use Egoroff's Theorem and the Hölder Inequality.)

5.2.12. *An Infinite-Measure Version of Problem 5.2.11* (W. R. Wade). Let $(X, \mathscr{A}, \mu)$ be a complete measure space. Suppose as in Problem 5.2.11 that $0 < r \leq \infty$, $\| f_n \|_r \leq M$ for all n, and $\lim_{n\to\infty} f_n(x) = f(x)$ a.e. If $0 < p < r$ and $1/q + p/r = 1$, then $\lim_{n\to\infty} \| f_n \cdot g - f \cdot g \|_p = 0$ for every $g \in L_q$.

5.2.13. *A Stone-Weierstrass Theorem for L_2-Valued Real Functions.* Let $(X, \mathscr{A}, \mu)$ be a finite-measure space with $\mu(X) = 1$, and let $x: T \to L_2(\mu)$ be a mapping where

(i) T is a compact set in some topological space

and

(ii) $\lim_{s\to t} \int | x_t - x_s |^2 \, d\mu = 0$, $x_t \equiv x(t)$.

Such a map is called *mean-square-continuous* (m.s.c.). A family of uniformly bounded [i.e., $| x_t(\cdot) | \leq M$ for all t] m.s.c. maps $[x_t(\cdot)]$ is called an *algebra* if it is closed under pointwise multiplication and finite linear combinations. An algebra $\mathscr{W}$ is said to separate points of T if, given $t_1, t_2 \in T$, there exists $x_t(\cdot) \in \mathscr{W}$ such that $| x_{t_1}(\cdot) - x_{t_2}(\cdot) | = 1$ a.e. Prove the following result: Let $\mathscr{W}$ be an algebra of uniformly bounded m.s.c. maps on T containing all bounded constant maps [i.e., $x_t(\cdot) = a(\cdot)$ for some $a(\cdot) \in L_2$,

$|a(\cdot)| \leq M]$ and separating points of T. Then given any m.s.c. map $y_t(\cdot)$, there exists a sequence $x_t^{(n)}(\cdot) \in \mathscr{W}$ converging in mean square to $y_t(\cdot)$ uniformly on T. (Hint: The proof can follow the same lines as used in that of Theorem 1.26.)

5.2.14. *Necessary and Sufficient Conditions for Convergence in L_p.* Let (f_n) be a sequence in $L_p(X, \mathscr{A}, \mu)$, $1 \leq p < \infty$ and $\lim_{n\to\infty} f_n = f$ a.e., where f is a real-valued measurable function. Prove that the following are equivalent:

(a) $f \in L_p$ and $\lim_{n\to\infty} \| f_n - f \|_p = 0$.

(b) Whenever (A_n) is a decreasing sequence of sets in $\mathscr{A}$ with $\mu(\bigcap_{n=1}^{\infty} A_n) = 0$, then $\lim_{k\to\infty} \int_{A_k} | f_n |^p \, d\mu = 0$ uniformly in n.

(c) (i) For each $\varepsilon > 0$, there exists $A_\varepsilon \in \mathscr{A}$ such that $\mu(A_\varepsilon) < \infty$ and $\int_{X-A_\varepsilon} | f_n |^p \, d\mu < \varepsilon$ for all n; and

(ii) for each $\varepsilon > 0$, there exists a $\delta > 0$ such that $\mu(A) < \delta$ implies $\int_A | f_n |^p \, d\mu < \varepsilon$ for all n.

(d) The condition (i) in (c) holds and $\lim_{k\to\infty} \int_{A_{nk}} | f_n |^p \, d\mu = 0$ uniformly in n, where $A_{nk} = \{x\colon | f_n(x) |^p \geq k\}$.

[Hint: Show $(a) \Leftrightarrow (c)$; then show $(a) \Rightarrow (b) \Rightarrow (d) \Rightarrow (c)$.]

5.2.15. *Convolution of Integrable Functions.* Let $f \in L_1$ and $g \in L_p$, $1 \leq p \leq \infty$ (with respect to the Lebesgue measure on R). Show that

(i) the convolution $f * g$ of f and g, defined by

$$f * g(x) = \int f(x - y) g(y) \, dy,$$

is in L_p;

(ii) $\| f * g \|_p \leq \| f \|_1 \cdot \| g \|_p$;

(iii) $f * g = g * f$ a.e.

Also show that for $f_1, f_2 \in L_1$ and $g \in L_p$, $1 \leq p \leq \infty$,

(iv) $f_1 * (f_2 * g) = (f_1 * f_2) * g$;

(v) $(f_1 + f_2) * g = f_1 * g + f_2 * g$.

[Hint: Use Fubini's Theorem; see also Problem 3.4.8.]

5.3. Bounded Linear Functionals and the Hahn–Banach Theorem

The aim of this section is to prove the famous Hahn–Banach Theorem. This theorem is one of the most fundamental theorems in functional analysis. It yields the existence of nontrivial continuous linear functionals on a normed linear space, a basic fact necessary for the development of a large portion of functional analysis. Moreover, it is an indispensible tool in the proofs of many important theorems in analysis. (For example, see Proposition 5.8.)

This theorem was first proved by Hahn in 1927 for a normed linear space over the reals, and then by Banach in 1929 for a real linear space (in the absence of any topology). The complex version of this theorem is due to Bohnenblust and Sobczyk in 1938 and, independently, to Soukhomlinoff, also in 1938.

We start this section by introducting the concept of a bounded linear operator.

Definition 5.5. Let X and Y be vector spaces over the same scalar field. Then a mapping T from X into Y is called a *linear operator* if for all $x_1, x_2 \in X$ and scalars α, β,

$$T(\alpha x_1 + \beta x_2) = \alpha T(x_1) + \beta T(x_2). \qquad \blacksquare$$

Examples

5.5. Let X be the real-valued continuous functions defined on $[0, 1]$, under the uniform norm and Y be the reals. Let

$$T(f) = \int_0^1 f(x)\,dx, \qquad f \in X.$$

Then T is a *continuous linear operator* from X into Y.

5.6. Let X be the class of real-valued continuously differentiable functions on $[0, 1]$ and Y be the class of real-valued continuous functions on $[0, 1]$, both under the uniform norm. Let

$$T(f) = \frac{df}{dx}, \qquad f \in X.$$

Then T is a *linear* operator. But T is *not continuous*, since the sequence $x^n/n \to 0$ in X, but the sequence $T(x^n/n) = x^{n-1}$ does not converge to 0 in Y.

5.7. Let X be a n-dimensional normed linear space over F and let Y be *any* normed linear space over F. Let

$$T\left(\sum_{i=1}^{n} \alpha_i x_i\right) = \sum_{i=1}^{n} \alpha_i y_i,$$

where α_i's are scalars, $\{x_1, x_2, \ldots, x_n\}$ is a basis of X, and $y_1, y_2, \ldots, y_n$ are arbitrarily chosen, but fixed elements of Y. Then T is a linear operator.

But T is also *continuous*, since for $x = \sum_{i=1}^{n} \alpha_i x_i$,

$$\| T(x) \| \leq \sum_{i=1}^{n} \| y_i \| \cdot \sup_{1 \leq i \leq n} | \alpha_i |$$
$$\leq \sum_{i=1}^{n} \| y_i \| \cdot K \cdot \| x \|,$$

where K is a constant such that

$$\sup_{1 \leq i \leq n} | \alpha_i | \leq K \left\| \sum_{i=1}^{n} \alpha_i x_i \right\|$$

for all n-tuples $\{\alpha_1, \alpha_2, \ldots, \alpha_n\}$. Such a K can always be found since in a finite-dimensional space any norm is equivalent to the sup-norm. (See Theorem 5.1.) Note that this example shows that a linear operator from a finite-dimensional normed linear space into any normed linear space is continuous.

5.8. Let T be a mapping from l_1 into F defined by

$$T(x) = \sum_{i=1}^{\infty} x_i, \qquad x = (x_i)_{i=1}^{\infty} \in l_1.$$

Then T is *linear*, but *not continuous* if we define a new norm in l_1 by considering it as a subspace of l_∞. T is not continuous since $T(z_n) = 1$ for all n, where

$$z_n = \Big(\underbrace{\frac{1}{n}, \frac{1}{n}, \ldots, \frac{1}{n}}_{n \text{ terms}}, 0, 0, \ldots\Big)$$

and $\| z_n \|_\infty \to 0$ as $n \to \infty$.

Definition 5.6. A linear operator T from a normed linear space X into a normed linear space Y is called *bounded* if there is a positive constant M such that

$$\| T(x) \| \leq M \| x \| \qquad \text{for all } x \in X.$$ ∎

Proposition 5.5. Let T be a linear operator from a normed linear space X into a normed linear space Y. Then the following are equivalent:

(a) T is continuous at a point.
(b) T is uniformly continuous on X.
(c) T is bounded. ∎

Proof. (a) ⇒ (b). Suppose T is continuous at a point x_0. Then given $\varepsilon > 0$, there exists $\delta > 0$ such that $\| x - x_0 \| < \delta \Rightarrow \| Tx - Tx_0 \| < \varepsilon$.

Now let y and z be elements in X with $\| y - z \| < \delta$. Then $\| (y - z + x_0) - x_0 \| < \delta$ and therefore $\| T(y - z + x_0) - T(x_0) \| < \varepsilon$, which means that $\| T(y) - T(z) \| < \varepsilon$, by the linearity of T. Hence (b) follows.

(b) $\Rightarrow$ (c). Suppose T is uniformly continuous on X and not bounded. Hence for each positive integer n there exists $x_n \in X$ such that $\| T(x_n) \| > n \cdot \| x_n \|$. This means that $\| T(x_n/n \| x_n \|) \| > 1$. But this contradicts the continuity of T at the origin since $\| x_n/n \| x_n \| \| \to 0$ as $n \to \infty$.

(c) $\Rightarrow$ (a) Boundedness of T trivially implies the continuity of T at the origin. ∎

The bounded linear operators from a normed linear space X into a normed linear space Y, denoted by $L(X, Y)$, form a vector space where addition of vectors and scalar multiplication of vectors are defined by

$$(T_1 + T_2)(x) = T_1(x) + T_2(x), \qquad (\alpha T)(x) = \alpha \cdot T(x).$$

Let us define on this vector space

$$\| T \| = \sup_{x \neq 0} \frac{\| T(x) \|}{\| x \|}$$

Equivalently,

$$\| T \| = \sup_{\| x \| \leq 1} \| T(x) \| = \sup_{\| x \| = 1} \| T(x) \| = \sup_{\| x \| < 1} \| T(x) \|$$

This defines a norm on $L(X, Y)$ and $L(X, Y)$ becomes a normed linear space. The completeness of $L(X, Y)$ in this norm depends upon that of Y. More precisely, we have the following proposition.

Proposition 5.6. $L(X, Y)$ is a Banach space if and only if Y is complete. (We assume, of course, that $X \neq \{0\}$.) ∎

Proof. For the "if" part, let (T_n) be a Cauchy sequence in $L(X,Y)$. Then for each $x \in X$, $\| T_n(x) - T_m(x) \| \leq \| T_n - T_m \| \, \| x \|$, so that $\lim_{n\to\infty} T_n(x)$ exists in Y, Y being complete. Let us define $T(x) = \lim_{n\to\infty} T_n(x)$. Then T is a linear operator from X into Y. We wish to show that T is bounded and $\| T_n - T \|$ converges to 0 as $n \to \infty$. To do this, let $\varepsilon > 0$. There exists N such that for $n, m \geq N$, we have $\| T_n - T_m \| < \varepsilon$ or $\| T_n \| \leq \| T_N \| + \varepsilon$. Therefore $\| T(x) \| = \lim_{n\to\infty} \| T_n(x) \| \leq (\| T_N \| + \varepsilon) \cdot \| x \|$. Hence T is bounded. Now for $x \in X$ and $n \geq N$,

$$\begin{aligned} \| T_n(x) - T(x) \| &= \lim_{m\to\infty} \| T_n(x) - T_m(x) \| \\ &\leq \lim_{m\to\infty} \| T_n - T_m \| \cdot \| x \| < \varepsilon \cdot \| x \|. \end{aligned}$$

Therefore,

$$\| T_n - T \| = \sup_{\|x\| \leq 1} \| T_n(x) - T(x) \| < \varepsilon$$

if $n \geq N$. The "if" part of the proof follows. The proof of the "only if" part is an application of the Hahn–Banach theorem and is left to the reader as a problem with hint. See Problem 5.3.4. ∎

Before we present the Hahn–Banach theorem we state a useful proposition providing a criterion for the existence of a bounded inverse of a bounded linear operator.

Proposition 5.7. Suppose T is a linear operator from a normed linear space X into a normed linear space Y. Then the inverse mapping T^{-1} exists and is a bounded linear operator from $T(X)$ into X if and only if there is some $k > 0$ such that $k \| x \| \leq \| T(x) \|$, for all $x \in X$. ∎

The proof of this proposition is routine and left to the reader.

Definition 5.7. When Y is the scalar field F, which is a Banach space over itself under the absolute-value norm, the elements of $L(X, Y)$ are called the *bounded linear functionals* on X. The class $L(X, F)$ is denoted by X^*. Hence X^* is also a Banach space called the dual of X. ∎

Example 5.9. Suppose X is a n-dimensional normed linear space under the "sup"-norm over the real numbers R. Let T be a bounded linear functional on X. Then $\| T \| = \sup_{\|x\|=1} | T(x) |$. To compute $\| T \|$, let $\{x_1, x_2, \ldots, x_n\}$ be a basis of X and $T(x_i) = r_i$, $1 \leq i \leq n$. Now for $x = \sum_{i=1}^n a_i x_i \in X$,

$$| T(x) | \leq \sup_{1 \leq i \leq n} | a_i | \cdot \sum_{i=1}^n | r_i |.$$

Again, if we define $b_i = 1$ if $r_i > 0$, and $b_i = -1$ if $r_i \leq 0$, then for $x = \sum_{i=1}^n b_i x_i$,

$$\| x \| = \sup_{1 \leq i \leq n} | b_i | = 1$$

and

$$| T(x) | = \sum_{i=1}^n b_i r_i = \sum_{i=1}^n | r_i |.$$

From this equality and the above inequality, it follows that

$$\| T \| = \sum_{i=1}^n | r_i |.$$

We now state and prove the main theorem of this section, which will show, among other things, the nontriviality of X^*, the dual (or adjoint) of a normed linear space X.

Theorem 5.6. (*Hahn–Banach*) Suppose X is a linear space over the *reals* R. Let S be a subspace of X and p be a real-valued function on X with the following properties:

(i) $p(x + y) \leq p(x) + p(y)$;

(ii) $p(\alpha x) = \alpha p(x), \qquad$ if $\alpha \geq 0$.

If f is a linear functional on S (i.e., a linear mapping from S into R) such that $f(s) \leq p(s)$ for all $s \in S$, then there exists a linear functional F on X such that $F(x) \leq p(x)$ for all $x \in X$ and $F(s) = f(s)$ for all $s \in S$. ▮

Proof. Let $\mathscr{F}$ be the set of all linear functionals g defined on a subspace of X containing S, such that $g(s) = f(s)$ for all $s \in S$ and $g(x) \leq p(x)$, whenever g is defined. Clearly $f \in \mathscr{F}$. We partially order $\mathscr{F}$ by requiring $g \leq h$ if and only if h is a linear extension of g. By the Hausdorff Maximal Principle, there is a maximal chain in $\mathscr{F}_0$. We define a functional F by setting

$$\text{domain } F = \bigcup_{g \in F_0} \text{domain } g$$

and

$$F(x) = g(x), \qquad \text{if } x \in \text{domain } g.$$

It is easy to check that domain F is a subspace of X and F is a linear extension of f. Moreover, F is a maximal extension, since if G is a proper extension of F, then $\mathscr{F}_0 \cup \{G\}$ will be a chain in $\mathscr{F}$, contradicting the maximality of $\mathscr{F}_0$.

We are finished if we can show that domain $F = X$. This will be shown by showing that any $g \in \mathscr{F}$, with its domain a proper subspace of X, has a proper extension.

Let $y \in X -$ domain g. We wish to extend g to the subspace spanned by y and domain g. Thus we need to define a functional h by

$$h(\alpha y + x) = \alpha h(y) + g(x),$$

$x \in$ domain g and α, a scalar where $h(y)$ is chosen so that $h(\alpha y + x) \leq p(\alpha y + x)$. In particular, we need to choose $h(y)$ so that

$$h(y) \leq p(y + x) - g(x)$$

and

$$-h(y) \leq p(-y - x) + g(x)$$

or

$$-p(-y - z) - g(z) \leq h(y) \leq p(y + x) - g(x),$$

where x and z are elements from domain g. Since for $x, z \in$ domain g,

$$p(y + x) - g(x) + p(-y - z) + g(z) \geq p(x - z) - g(x - z) \geq 0,$$
$$\sup_z [-p(-y - z) - g(z)] \leq \inf_x [p(y + x) - g(x)].$$

Hence we choose $h(y) = \inf\{p(y + x) - g(x)\colon x \in \text{domain } g\}$. Now the theorem will be proved if we only check that $h(\alpha y + x) \leq p(\alpha y + x)$. To do this, let $\alpha > 0$. Then

$$\begin{aligned} h(\alpha y + x) &= \alpha h(y + x/\alpha) = \alpha h(y) + g(x) \\ &\leq \alpha p(y + x/\alpha) - \alpha g(x/\alpha) + g(x) \\ &= p(\alpha y + x). \end{aligned}$$

The cases $\alpha = 0$ and $\alpha < 0$ can similarly be taken care of. ∎

Theorem 5.7. *Hahn–Banach Theorem (Complex Version).* Let X be a linear space over the complex numbers, S a linear subspace, and p, a real-valued function on X such that $p(x + y) \leq p(x) + p(y)$ and $p(\alpha x) = |\alpha| p(x)$. Let f be a linear functional on S such that $|f(s)| \leq p(s)$ for all $s \in S$. Then there is a linear functional F defined on X such that $F(s) = f(s)$ for all $s \in S$ and $|F(x)| \leq p(x)$ for all $x \in X$. ∎

Proof. Let us define the mappings g and h on S, by taking $g(s)$ to be the real part of $f(s)$ and $h(s)$ its imaginary part. Then g and h are linear in the real sense, by considering X (and hence S) as a vector space over the reals.

Now $f = g + ih$. Since for $s \in S$

$$g(is) + ih(is) = f(is) = if(s) = ig(s) - h(s),$$

we have $h(s) = -g(is)$. Since $g(s) \leq |f(s)| \leq p(s)$, by Theorem 5.6 we can extend g to a linear functional G on X considered as a real vector space such that $G(x) \leq p(x)$ for all $x \in X$. Let $F(x) = G(x) - iG(ix)$. Then for $s \in S$, $F(s) = g(s) + ih(s) = f(s)$. Also $F(ix) = G(ix) - iG(-x) = iF(x)$. It is now easy to check that F is linear in the complex sense. Finally for $x \in X$, if α is a complex number with $|\alpha| = 1$ and $\alpha F(x) = |F(x)|$, then $|F(x)| = F(\alpha x) = G(\alpha x) \leq p(\alpha x) = |\alpha| p(x) = p(x)$. ∎

In the rest of the section, we will consider some of the important consequences and a few applications of the Hahn–Banach theorem.

Theorem 5.8. *Hahn–Banach Theorem* (*Normed Linear Space Version*). Let X be a normed linear space over a field F and let Y be a subspace of X. If $y^* \in Y^*$, then there exists some $x^* \in X^*$ such that $\| y^* \| = \| x^* \|$ and $y^*(x) = x^*(x)$ for all $x \in Y$. ∎

Proof. Let us define $p(x) = \| y^* \| \cdot \| x \|$, $x \in X$. Then $p(x + y) \leq p(x) + p(y)$ and $p(\alpha x) = |\alpha| p(x)$. Also $|y^*(x)| \leq p(x)$, since $\| y^* \| = \sup_{x \neq 0} |y^*(x)| / \| x \|$, for $x \in Y$. Hence by Theorem 5.7, there exists a linear functional x^* on X such that (i) $x^*(x) = y^*(x)$ for all $x \in Y$, and (ii) $|x^*(x)| \leq p(x)$ for all $x \in X$. This means that $|x^*(x)| \leq \| y^* \| \; \| x \|$, $x \in X$ or $\| x^* \| \leq \| y^* \|$. But then (i) implies that $\| x^* \| = \| y^* \|$. ∎

Corollary 5.3. Let Y be a subspace of a normed linear space X over a field F. Suppose $x \in X$ and $\inf_{y \in Y} \| x - y \| = d > 0$. Then there exists $x^* \in X^*$ such that $\| x^* \| = 1$, $x^*(x) = d$, and $x^*(y) = 0$ for all $y \in Y$. ∎

Proof. Let Z be the subspace spanned by Y and x. We define a linear functional z^* on Z by

$$z^*(\alpha x + y) = \alpha d, \quad y \in Y \text{ and } \alpha \in F.$$

Then $z^*(y) = 0$ for all $y \in Y$ and $z^*(x) = d$. To show that $z^* \in Z^*$, for $\alpha \neq 0$, $\alpha \in F$ and $y \in Y$

$$\| \alpha x + y \| = |\alpha| \; \| x + y/\alpha \| \geq |\alpha| \, d$$

or $|z^*(\alpha x + y)| \leq \| \alpha x + y \|$ for all y in Y. Hence $\| z^* \| \leq 1$. But there exists a sequence $y_k \in Y$ such that $\| x - y_k \| \to d$ as $k \to \infty$. Therefore given $\varepsilon > 0$, $|z^*(\alpha x - \alpha y_k)| = |\alpha| d \geq \| \alpha x - \alpha y_k \| - \varepsilon$ for sufficiently large k. This means that $\| z^* \| \geq 1$. Hence $\| z^* \| = 1$. The corollary now follows by extending z^* by Theorem 5.8. ∎

The next corollary shows that there are sufficiently many bounded linear functionals on a normed linear space to separate points of the space.

Corollary 5.4. Given $x \in X$ $(\neq \{0\})$, a normed linear space over F, there exists $x^* \in X^*$ such that $\| x^* \| = 1$ and $x^*(x) = \| x \|$. In particular, if $x \neq y$, there exists $x^* \in X^*$ such that $x^*(x) - x^*(y) = \| x - y \| \neq 0$. ∎

Proof. The proof follows immediately from Corollary 5.3, by taking $Y = \{0\}$. ∎

Corollary 5.5. For any x in a normed linear space X over F,

$$\| x \| = \sup_{\substack{x^* \in X^* \\ \|x^*\|=1}} | x^*(x) |. \qquad ∎$$

We leave the proof to the reader.

Corollary 5.4 above is an important consequence of the Hahn–Banach Theorem. There are many applications of this outstanding theorem. One interesting application is the existence of a finitely additive measure, defined on the class of *all* bounded subsets of R, which is translation invariant and an extension of the Lebesgue measure. We will present this application in this section.

To do this, we first need an extension of the Hahn–Banach Theorem.

Theorem 5.9. (*An Extension of the Hahn–Banach Theorem*) Let p be a real-valued function of the linear space X over the reals such that $p(x + y) \leq p(x) + p(y)$ and $p(\alpha x) = \alpha p(x)$, if $\alpha \geq 0$. Suppose f is a linear functional on a subspace S such that $f(s) \leq p(s)$ for all $s \in S$. Suppose also that $\mathscr{T}$ is an Abelian semigroup of linear operators on X (that is $T_1, T_2 \in \mathscr{T}$ implies $T_1T_2 = T_2T_1 \in \mathscr{T}$) such that if $T \in \mathscr{T}$, then $p\big(T(x)\big) \leq p(x)$ for all $x \in X$ and $f\big(T(s)\big) = f(s)$ for all $s \in S$. Then there is an extension F of f to a linear functional on X such that $F(x) \leq p(x)$ and $F\big(T(x)\big) = F(x)$ for all $x \in X$. [Note that this theorem yields the Hahn–Banach Theorem (real version) when $\mathscr{T}$ consists of the identity operator alone. The complex version of this theorem can also be formulated like that of the Hahn–Banach Theorem and proved.] ∎

Proof. The proof will be an application of the Hahn–Banach Theorem. First we need to choose a new subadditive function (like p). Let us define

$$q(x) = \inf(1/n)p\big(T_1(x) + \cdots + T_n(x)\big), \qquad x \in X,$$

where the infimum is taken over all possible finite subsets $\{T_1, T_2, \ldots, T_n\} \in \mathscr{T}$. To show that this infimum is a real number, we have $q(x) \leq p(x)$. Also since

$$0 = \frac{1}{n}p(0) \leq \frac{1}{n}p\big(T_1(x) + \cdots + T_n(x)\big) + \frac{1}{n}p\big(-T_1(x) - \cdots - T_n(x)\big)$$

and

$$\frac{1}{n} p(-T_1(x) - \cdots - T_n(x)) \le p(-x),$$

we have

$$\frac{1}{n} p(T_1(x) + \cdots + T_n(x)) \ge -p(-x)$$

and therefore

$$-p(-x) \le q(x) \le p(x).$$

Moreover, $q(\alpha x) = \alpha q(x)$, $\alpha \ge 0$. Now we wish to show that $q(x+y) \le q(x) + q(y)$. So let $x, y \in X$ and $\varepsilon > 0$. Then there exist $\{T_1, T_2, \ldots, T_n\}$ and $\{S_1, S_2, \ldots, S_m\} \in \mathscr{F}$ such that

$$\frac{1}{n} p\{T_1(x) + \cdots + T_n(x)\} < q(x) + \frac{\varepsilon}{2}$$

and

$$\frac{1}{m} p\{S_1(y) + \cdots + S_m(y)\} < q(y) + \frac{\varepsilon}{2}.$$

Then

$$\begin{aligned} q(x+y) &\le \frac{1}{nm} p\Big(\sum_{i=1}^{n} \sum_{j=1}^{m} T_i S_j(x+y)\Big) \\ &\le \frac{1}{nm} p\Big[\sum_{j=1}^{m} S_j\Big(\sum_{i=1}^{n} T_i(x)\Big)\Big] + \frac{1}{nm} p\Big[\sum_{i=1}^{n} T_i\Big(\sum_{j=1}^{m} S_j(y)\Big)\Big] \\ &\le \frac{1}{n} p\Big(\sum_{i=1}^{n} T_i(x)\Big) + \frac{1}{m} p\Big(\sum_{j=1}^{m} S_j(y)\Big) \\ &< q(x) + q(y) + \varepsilon. \end{aligned}$$

Since ε is arbitrary, $q(x+y) \le q(x) + q(y)$. Also, since for $s \in S$

$$f(s) = \frac{1}{n} f(T_1(s) + \cdots + T_n(s)) \le \frac{1}{n} p(T_1(s) + \cdots + T_n(s)),$$

$f(s) \le q(s)$. Therefore, by Theorem 5.6, there exists a linear extension F of f to all of X such that $F(x) \le q(x) \le p(x)$ for all $x \in X$. The proof will be complete if we can show that $F(T(x)) = F(x)$, $x \in X$ and $T \in \mathscr{F}$. To do this, let $x \in X$, $T \in \mathscr{F}$, and n be any positive integer. Then

$$\begin{aligned} q(x - T(x)) &\le \frac{1}{n} p\Big(\sum_{i=1}^{n} T^i(x - T(x))\Big) \\ &= \frac{1}{n} p(T(x) - T^{n+1}(x)) \\ &\le \frac{1}{n} [p(x) + p(-x)]. \end{aligned}$$

Letting n approach ∞, $q(x - T(x)) \leq 0$. Since $F(x - T(x)) \leq q(x - T(x))$, we have $F(x) \leq F(T(x))$. Applying this to $-x$, we get $F(x) = F(T(x))$. ∎

We proved in Chapter 3 that it is impossible to define a translation-invariant countably additive measure on the class of all bounded subsets of R, such that the measure of an interval is its length. But interestingly enough, an application of Theorem 5.9 will show that there exists a finitely additive measure having the above properties.

Proposition 5.8. There is a finitely additive measure μ defined for all bounded subsets of R such that

(i) $\mu(A + t) = \mu(A)$, $A \subset R$ and $t \in R$;

(ii) if $A \subset R$ is Lebesgue measurable, then $\mu(A)$ is the Lebesgue measure of A. ∎

Proof. Let X be the linear space over R of all real bounded functions defined on $[0, 1)$, under the natural operations of pointwise addition and scalar multiplication. Let Y be the subspace of all bounded Lebesgue-measurable functions of $[0, 1)$. For $f \in X$, let us define

$$p(f) = \text{l.u.b. } \{f(x)\colon x \in [0, 1)\}.$$

Then

$$\text{(i)} \quad p(f_1 + f_2) \leq p(f_1) + p(f_2), \qquad f_1, f_2 \in X$$

and

$$\text{(ii)} \quad \text{for } \alpha \geq 0, \quad p(\alpha f) = \alpha p(f) \qquad \text{for all } f \in X.$$

Let Φ be the linear functional defined on Y by $\Phi(f) = \int_0^1 f(x)\, dx$. Let $\mathscr{T}$ be the Abelian semigroup of linear operators on X defined by

$$\mathscr{T} = \{T_t\colon 0 \leq t < 1 \quad \text{and} \quad T_t[f](x) = f(x \mathbin{\mathring{+}} t),$$

where

$$\begin{aligned} x \mathbin{\mathring{+}} t &= x + t, & &\text{if } x + t < 1 \\ &= x + t - 1, & &\text{if } x + t \geq 1\}. \end{aligned}$$

Now for $0 \leq t < 1$, an easy computation shows that for any $A \subset [0, 1)$,

$$x \mathbin{\mathring{+}} t \in A \Leftrightarrow x \in A + (1 - t).$$

Since the Lebesgue measure of a Lebesgue-measurable set A is the same as that of $A \mathbin{\mathring{+}} (1 - t)$, we have

$$\int_0^1 \chi_A(x)\, dx = \int_0^1 \chi_A(x \mathbin{\mathring{+}} t)\, dx.$$

Hence for $f \in Y$ and $0 \leq t < 1$, $\Phi(f) = \Phi(T_t[f])$. Also by the definition of p and T_t, we have

$$p(T_t[f]) \leq p(f) \qquad \text{for all } f \in X,\ 0 \leq t < 1.$$

Hence by Theorem 5.9, there exists a linear extension F of Φ defined on X such that $F(f) \leq p(f)$ and $F(T_t[f]) = F(f)$ for all $f \in X$.

For $A \subset [0, 1)$, let us define $\mu(A) = F(\chi_A)$. Then since F is linear, μ is *finitely additive*. Since $F(-\chi_A) \leq p(-\chi_A) = 0$, $-F(\chi_A) \leq 0$ or $\mu(A) \geq 0$. Also since for $0 \leq t < 1$, $\chi_A(x \mathbin{\mathring{+}} t) = \chi_{A \mathring{+} (1-t)}(x)$, we have

$$\mu(A) = F(\chi_A) = F\big(\chi_A(x \mathbin{\mathring{+}} t)\big) = \mu\big(A \mathbin{\mathring{+}} (1 - t)\big).$$

If $t > \frac{1}{2}$, then $1 - t < \frac{1}{2}$. Hence for $A \subset [0, \frac{1}{2})$,

$$\mu(A) = \mu(A + s) \text{ for all } s \in [0, \tfrac{1}{2}).$$

Now to extend μ to the class of all bounded subsets on R, consider any $B \subset [n/2, (n + 1)/2)$, where n is any integer. Then $B - (n/2) \subset [0, \frac{1}{2})$. It is the value $\mu(B - n/2)$ we will take as $\mu(B)$. Thus for any bounded set $A \subset [-m, m]$,

$$\mu(A) = \sum_{n=-2m}^{2m-1} \mu\left(A \cap \left[\frac{n}{2}, \frac{n+1}{2}\right)\right).$$

This μ will satisfy the requirements of the theorem. ∎

Before we close this section, we briefly discuss another useful concept —that of separability. A set A in a normed linear space X is called separable if there is a countable subset of A that is dense in A. $C[0, 1]$ is separable under the supremum norm. l_p for $1 \leq p < \infty$ is separable since the set $\{(\alpha_1, \alpha_2, \ldots, \alpha_k, 0, 0, \ldots)\}$ of all sequences where the α_i's are rational is countable and dense in l_p. Also the L_p space of Lebesgue-measurable functions $(1 \leq p < \infty)$ is separable since the polynomials with rational coefficients (being dense in $C[0, 1]$) are dense in L_p. But l_∞, as well as L_∞, is not separable. (See Problem 5.3.7.) Also the L_p $(1 \leq p \leq \infty)$ space over an arbitrary measure space is not, in general, separable. For example, if S is an uncountable set with the counting measure, then for $1 \leq p < \infty$

$$L_p = \left\{ f\colon S \to R\colon \sum_{s \in S} | f(s) |^p < \infty \right\};$$

and for $f \in L_p$, $\| f \|_p = [\sum_{s \in S} | f(s) |^p]^{1/p}$. If s_1 and s_2 are in S, then

$$\| \chi_{\{s_1\}}(s) - \chi_{\{s_2\}}(s) \|_p = 2^{1/p}.$$

This means that $L_p(S)$ is *not* separable.

It so happens that $l_1^* = l_\infty$ (equality means the existence of a linear isometry onto). This will be discussed in the next section. Therefore l_1^* is not separable, despite the separability of l_1. But the converse situation is different, and once again the Hahn–Banach Theorem helps us clarify the converse.

Proposition 5.9. If the dual X^* of a normed linear space X is separable, then X is also separable. ∎

Proof. Let $(x_k^*)_{k=1}^\infty$ be a countable dense set in X^*. Let $(x_k)_{k=1}^\infty$ be elements of X such that for all k, $\| x_k \| = 1$ and $| x_k^*(x_k) | \geq \frac{1}{2} \| x_k^* \|$. Let Y be the closed subspace spanned by the x_k's. Then Y is separable. We are finished if $Y = X$. If $Y \neq X$, by Corollary 5.3 there exists $x^* \in X^*$ such that $\| x^* \| > 0$ and $x^*(y) = 0$ for all $y \in Y$. Since $(x_k^*)_{k=1}^\infty$ is dense in X^*, there is a subsequence $(x_{k_i}^*)$ such that $\| x_{k_i}^* - x^* \| \to 0$ as $i \to \infty$. But

$$\begin{aligned} \| x_{k_i}^* - x^* \| &\geq | x_{k_i}^*(x_{k_i}) - x^*(x_{k_i}) | \\ &= | x_{k_i}^*(x_{k_i}) | \geq \tfrac{1}{2} \| x_{k_i}^* \|. \end{aligned}$$

Hence $\| x_{k_i}^* \| \to 0$ as $i \to \infty$. This means that $\| x^* \| = 0$, which is a contradiction. ∎

Problems

5.3.1. Let X be a finite-dimensional space with basis $\{x_1, x_2, \ldots, x_n\}$, such that $\| \sum_{i=1}^n \alpha_i x_i \| = \sup_{1 \leq i \leq n} | \alpha_i |$, $\alpha_i \in F$. Define the linear operator A from X into X by

$$A(x_i) = \sum_{j=1}^n \alpha_{ij} x_j, \qquad 1 \leq i \leq n.$$

Then find $\| A \|$.

5.3.2. Let X be as in Problem 5.3.1, with $\| \sum_{i=1}^n \alpha_i x_i \| = (\sum_{i=1}^n | \alpha_i |^2)^{1/2}$. Then if $f \in X^*$, find $\| f \|$.

5.3.3. Let X be a normed linear space over R and $f\colon X \to R$.

(a) Suppose that f is continuous and for all x, y in X, $f(x) + f(y) = f(x + y)$. Prove that $f \in X^*$.

(b) Suppose that f is linear, but discontinuous. Show that $f^{-1}(0)$ is dense in X. (Hint: Note that $Y = f^{-1}(0)$ is not closed since f is discontinuous. Let y be a limit point of Y but not in Y. Then for any x in X, $x = \{x - [f(x)/f(y)]y\} + [f(x)/f(y)]y$.)

5.3.4. Show that if $L(X, Y)$ is complete, where X and Y are any two normed linear spaces over F and $X \neq \{0\}$, then Y is complete. {Hint: Let (y_n) be a Cauchy sequence in Y. Then define $T_n \in L(X, Y)$ by $T_n(x) = x^*(x) \cdot y_n$, where $x^*(\neq 0) \in X^*$. Show that (T_n) is Cauchy in $L(X, Y)$ and hence $\lim_{n\to\infty} y_n = [1/x^*(x)] \lim_{n\to\infty} T_n(x)$, where $x^*(x) \neq 0$.}

5.3.5. Let $1 \leq p \leq \infty$ and $1/p + 1/q = 1$. For each $g \in L_q$, define $T_g \in L_p^*$ by $T_g(f) = \int fg \, d\mu$. Show that for $p > 1$, $\| T_g \| = \| g \|_q$, and that for $p = 1$ this equality holds for all $g \in L_\infty$ if and only if the measure is semifinite. [Hint: If $1 < p < \infty$ and $f = | g |^{q/p} \cdot \operatorname{sgn} g$,[†] then $T_g(f) = \| g \|_q \| f \|_p$.]

5.3.6. For any fixed $f(t) \in C[0, 1]$ (under the uniform norm), let Φ be the linear functional on $C[0, 1]$ defined by

$$\Phi(g(t)) = \int_0^1 f(t) \cdot g(t) \, dt.$$

Then show that Φ is bounded and find $\| \Phi \|$.

5.3.7. Prove that l_∞ as well as $L_\infty[0, 1]$ is not separable. [Hint: Let $x_k \in l_\infty$ and $x_k = (x_k^1, x_k^2, \ldots)$. Define $x = (\alpha_1, \alpha_2, \ldots)$ by $\alpha_k = 0$, if $| x_k^k | \geq 1$; $= 1 + | x_k^k |$, if $| x_k^k | < 1$. Then $x \in l_\infty$ and $\| x - x_k \|_\infty \geq 1$ for all k.]

5.3.8. If S is a linear subspace of a Banach space X, the *annihilator* S° of S is defined to be $S^\circ = \{x^* \in X^*: x^*(s) = 0 \text{ for all } s \in S\}$.[‡] If T is a subspace of X^*, then ${}^\circ T = \{x \in X: x^*(x) = 0 \text{ for all } x^* \in T\}$. Then show that

(i) S° is a closed subspace of X^*;
(ii) ${}^\circ(S^\circ) = S$;
(iii) if S is closed, then S^* is linearly isometric onto X^*/S° and S° is linearly isometric to $(X/S)^*$.

5.3.9. For $1 < p < \infty$, show that l_p^* is linearly isometric onto l_q, $1/p + 1/q = 1$. [Hint: Let

$$e_k = (\underbrace{0, 0, \ldots, 0, 1,}_{k} 0, 0, \ldots).$$

If $x^* \in l_p^*$, let $x^*(e_k) = b_k$. Show that $b = (b_1, b_2, \ldots) \in l_q$ and $\| x^* \| = \| b \|_q$.]

5.3.10. Show that l_1^* is linearly isometric onto l_∞.

† By definition, $\operatorname{sgn} \alpha = 0$ if $\alpha = 0$ and $\operatorname{sgn} \alpha = \alpha/| \alpha |$ if $\alpha \neq 0$.

‡ Although A° is also used to denote the interior of a set A in this text, the meaning of the symbol should be clear from the context.

5.3.11. Let c be the space of all sequences of complex numbers $(x_1, x_2, \ldots)$ such that $\lim_{n\to\infty} x_n$ exists with natural addition and scalar multiplication. For an element $x = (x_1, x_2, \ldots) \in c$, let $\| x \| = \sup_n | x_n |$. Then show that

(i) c is a Banach space;

(ii) c^* is linearly isometric to l_1.

[Hint: Let $e_k = (0, 0, \ldots, 0, 1, 0, \ldots)$ as in Problem 5.3.9 for $1 \leq k < \infty$, $e_0 = (1, 1, 1, \ldots)$, $x^*(e_k) = b_k$, $1 \leq k < \infty$, and $x^*(e_0) = m$, where $x^* \in c^*$. Show that $(d_1, d_2, \ldots) \in l_1$, where $d_1 = m - \sum_{k=1}^{\infty} b_k$, $d_{i+1} = b_i$, $1 \leq i < \infty$ and $\| x^* \| = \sum_{i=1}^{\infty} | d_i |$.]

5.3.12. Let c_0 be the subspace of c (see Problem 5.3.11 above) such that $\lim_{n\to\infty} x_n = 0$, if $(x_1, x_2, \ldots) \in c_0$. Show that $c_0{}^*$ is linearly isometric onto l_1.

5.3.13. Consider the linear space L_p, $0 < p < 1$, in the Lebesgue measure space on [0, 1], with the pseudometric

$$d(f, g) = \int_0^1 | f(t) - g(t) |^p \, dt.$$

Then show the following:

(i) (L_p, d), $0 < p < 1$, is a pseudometric linear space [that is, a vector space with metric (pseudo) topology where the vector addition and scalar multiplication are continuous].

(ii) f can be written as $g + h$, $d(g, 0) = d(h, 0) = \frac{1}{2} d(f, 0)$. {Hint: Let $g_x = f \cdot \chi_{[0,x]}$, $h_x = f \cdot \chi_{[x,1]}$. Then $d(g_x, 0) + d(h_x, 0) = d(f, 0)$. But $d(g_x, 0)$ is a continuous function from [0, 1] onto $[0, d(f, 0)]$. Use the Intermediate Value theorem.}

(iii) The only continuous linear functional on L_p is the zero functional. [Hint: Let $\Phi \in L_p{}^*$ with $\Phi(f) = 1$. Then, by part (ii), there exists $g \in L_p$ with $\Phi(g) \geq \frac{1}{2}$, $d(g, 0) = \frac{1}{2} d(f, 0)$. Let $g_1 = 2g$. Then $\Phi(g_1) \geq 1$, $d(g_1, 0) = 2^{p-1} d(f, 0)$. Continue the process to get $g_1, g_2, \ldots$ with $\Phi(g_n) \geq 1$ and $d(g_n, 0) = 2^{n(p-1)} d(f, 0)$.]

5.3.14. *The Volterra Fixed Point Theorem.* Let X be a Banach space and T be a bounded linear operator on X such that $\sum_{n=1}^{\infty} \| T^n \| < \infty$. Prove that the transformation S defined by

$$S(x) = y + T(x), \qquad y \text{ a fixed element in } X$$

has a unique fixed point given by

$$x_0 = y + \sum_{n=1}^{\infty} T^n(y).$$

[Hint: For any $z \in X$, consider $x_1 = z$ and $x_{n+1} = T(x_n)$; show that (x_n) is Cauchy in X and $\lim_{n\to\infty} x_n = 0$.]

5.3.15. *An Application of Problem 5.3.14.* Show that the Volterra equation

$$f(t) = g(t) + \int_0^t K(s, t) f(s)\, ds,$$

where $g \in C[0, 1]$ and $K \in C([0, 1] \times [0, 1])$ are given functions, has a unique solution $f \in C[0, 1]$. [Hint: Define T on the Banach space $C[0, 1]$ with "sup" norm by

$$T(f)(t) = \int_0^t K(s, t) f(s)\, ds.$$

Then

$$\| T \| \leq M = \sup_{0\leq s,t\leq 1} | K(s, t) | \quad \text{and} \quad \| T^n \| \leq M^n/n!.$$

Now use Problem 5.3.14.]

5.3.16. *Existence of Banach Limits* (An application of the Hahn–Banach Theorem). Banach limits are linear functionals F on the space l_∞ of bounded sequences of real numbers $(x_n)_{n=0}^\infty$ satisfying these conditions:

(i) $F[(x_n)] \geq 0$ if $x_n \geq 0$ for $0 \leq n < \infty$;

(ii) $F[(x_{n+1})] = F[(x_n)]$;

(iii) $F[(1)] = 1$, $(1) = (1, 1, 1, \ldots)$.

Prove the following assertions:

(1)

$$p[(x_n)] = \lim_{n\to\infty} \left(\sup_j \frac{1}{n} \sum_{i=0}^{n-1} x_{i+j} \right)$$

exists and is a sublinear functional on l_∞. [Hint: If

$$a_n = \sup_j \frac{1}{n} \sum_{i=0}^{n-1} x_{i+j},$$

then $a_{km} \leq a_m$ and $(r + km)a_{r+km} \leq ra_r + kma_m$ or $\lim_{k\to\infty} \sup a_{r+km} \leq a_m$ for $r = 1, 2, \ldots, m$. Hence $\lim\sup a_n \leq a_m$ for each m or $\lim a_n$ exists.]

(2) For $(x_n) \in c(\subset l_\infty)$, define

$$f[(x_n)] = \lim_{n\to\infty} x_n \leq p[(x_n)].$$

(3) By Theorem 5.6, there is an extension F of f to l_∞ and F is a Banach limit. {Hint: $| p[(x_{n+1}) - (x_n)] | = | \lim_n [\sup_j n^{-1}(x_{j+n} - x_j)] | \leq \lim_n 2n^{-1} \sup_j | x_j | = 0$; also $| p[(x_n) - (x_{n+1})] | = 0$.}

(4) The maximal value of Banach limits on (x_n) is $p(x_n)$. (Hint: First,

for any Banach limit L, $L[(x_n)] \leq p[(x_n)]$. To see this, by (i) $L[(\sup_j z_j - z_n)] \geq 0$, and therefore $\sup_j z_j \geq L[(z_n)]$; then if $z_n = (1/m)\sum_{i=0}^{m-1} x_{i+n}$, $L[(x_n)] = L[(z_n)] \leq \sup_j z_j$ for each m and therefore letting $m \to \infty$, $\leq p[(x_n)]$. Conversely, given (x_n), there exists a Banach limit L such that $L[(x_n)] = p[(x_n)]$. Indeed, if $(x_n) \notin c$, define L on c as f [in (2)], and as in the proof of Theorem 5.6, $L[(x_n)] = \inf_{(y_n)\in c}\{p[(x_n + y_n)] - \lim y_n\}$; then as in Theorem 5.6, L can be extended to l_∞ and $L[(x_n)] = p[(x_n)]$, since $p[(x_n + y_n)] = p[(x_n)] + \lim y_n$.)

(5) The minimal value of Banach limits on (x_n) is

$$\lim_{n\to\infty} \left(\inf_j \frac{1}{n} \sum_{i=0}^{n-1} x_{i+j} \right).$$

(6) A necessary and sufficient condition in order that all Banach limits on (x_n) agree and be equal is that $\lim_{n\to\infty}(1/n)\sum_{i=0}^{n-1} x_{i+j} = s$ uniformly in j. Hence on convergent sequences, Banach limits agree with limits of the sequences.

The formulation of the above results in the form they appear is due to L. Sucheston.

5.3.17. Let $(X, \mathscr{A}, \mu)$ be a finite measure space and let S be the linear space of all real-valued measurable functions (where two functions agreeing a.e. are equal) with the topology of convergence in measure {equivalently, with the metric $d(f, g) = \int [|f - g|/(1 + |f - g|)]\, d\mu$}. Prove the following assertions:

(i) If A is an atom in $\mathscr{A}$, then $\Phi_A(f) = \int_A f\, d\mu$ defines a continuous linear functional on S.

(ii) If Φ is a continuous linear functional on S, then $\Phi(f \cdot \chi_A) = 0$ for all $f \in S$ if A contains no atoms.

(iii) Let $\mathscr{F}$ be an (at most countable) collection of pairwise disjoint atoms of $\mathscr{A}$ and $T = \{\Phi_A : A \in \mathscr{F}\}$. Then the class S^* of all continuous linear functionals on S is precisely the linear span of T. [This result is due to T. K. Mukherjee and W. H. Summers.]

(iv) S^* contains a nonzero element if and only if $\mathscr{A}$ contains an atom.

5.3.18. *Another Application of the Hahn–Banach Theorem.* The classical moment problem can be stated as follows: Given a sequence of real numbers (a_n), when does there exist a real-valued function g of bounded variation on [0, 1] such that $\int_0^1 x^n\, dg(x) = a_n$, $n = 0, 1, 2, \ldots$? Show that this problem can be answered in an abstract setup as follows:

Let X be a normed linear space, $(x_\lambda)_{\lambda\in\Lambda}$ be elements in X, and $(a_\lambda)_{\lambda\in\Lambda}$ be scalars. Then the following are equivalent:

(i) There exists $x^* \in X^*$ such that $x^*(x_\lambda) = a_\lambda$ for each $\lambda \in \Lambda$.

(ii) There exists a positive number M such that

$$\left| \sum a_\lambda b_\lambda \right| \leq M \cdot \left\| \sum b_\lambda x_\lambda \right\|,$$

for every subset $(b_\lambda)_{\lambda \in \Lambda}$ of scalars with all but finitely many of the b_λ's zero.

[Hint: (ii) $\Rightarrow$ (i). Let Y be the subspace spanned by $(x_\lambda)_{\lambda \in \Lambda}$. Define $x^* \in Y^*$ by $x^*(\sum b_\lambda x_\lambda) = \sum b_\lambda a_\lambda$; show that x^* is well defined because of (ii). Then use the Hahn–Banach Theorem.]

5.4. The Open Mapping Theorem, the Closed Graph Theorem, and the Principle of Uniform Boundedness

In this section we will consider three basic principles of functional analysis, which rank in importance with the Hahn–Banach Theorem of the previous section. They provide the foundation for many far-reaching modern results in diverse disciplines of analysis such as ergodic theory, the theory of differential equations, integration theory, and so on. The first, called the Open Mapping Theorem, asserts that certain continuous linear mappings between Banach spaces map open sets into open sets. The second is called the Closed Graph Theorem and asserts that a linear map between two Banach spaces that has a closed graph is continuous. The third is the Principle of Uniform Boundedness, which asserts that a pointwise bounded family of continuous linear mappings from one Banach space to another is uniformly bounded. One form of the Open Mapping and the Closed Graph Theorems was first proved by Banach in 1929; a more general form is due to Schauder in 1930. The principle of uniform boundedness was proved for bounded linear functionals on a Banach space by Hahn in 1922, for continuous linear mappings between Banach spaces by Hildebrandt in 1923, and also by Banach and Steinhaus (in a more general case) in 1927. The Principle of Uniform boundedness is sometimes called the Banach–Steinhaus Theorem.

This section is based on the concept of a closed linear operator. We start with its definition.

Let X and Y be normed linear spaces over the same scalars. Then $X \times Y$ is the normed linear space of all ordered pairs (x, y), $x \in X$ and $y \in Y$, with the usual definitions of addition and scalar multiplication and norm defined by $\| (x, y) \| = \max\{\| x \|, \| y \|\}$.

Definition 5.8. Let $T: D \to Y$ be a linear operator, where D is a subspace of X. Then T is called *closed* if its *graph* $G_T = \{(x, T(x)): x \in D\}$ is a closed subspace of $X \times Y$. Equivalently, T is closed if and only if the following condition holds: Whenever $x_n \in D$, $x_n \to x \in X$, and $T(x_n) \to y \in Y$, then $x \in D$ and $y = T(x)$. ▮

Examples

5.10. *A Linear Operator That Is Closed But Not Continuous*: Let $X = Y = C[0, 1]$ and $D = \{x \in X: d[x(t)]/dt \in C[0, 1]\}$. Define $T: D \to Y$ by $T(x) = dx/dt$. Then if $x_n(t) = t^n$, $\| x_n(t) \| = 1$; but

$$\| T(x_n) \| = \| n \cdot t^{n-1} \| = n \to \infty.$$

Hence T is not bounded. The reader can easily check that T is closed. Note that D is not complete.

5.11. *A Linear Operator That Is Continuous But Not Closed*: Let D be a nonclosed subspace of any normed linear space X. Let $Y = X$ and $i: D \to Y$ be the identity map. Then i is clearly continuous but not closed.

Clearly we have the following proposition.

Proposition 5.10. Let $T: D \to Y$, $D \subset X$, be a continuous linear operator. Then if D is a closed subspace of X, T is closed. Conversely, if Y is complete and T is closed, then D is a closed subspace of X. ▮

We leave the proof to the reader.

Example 5.12. (*A Closed Linear Operator from a Banach Space into a Normed Linear Space That Is Not Continuous.*) Let X be any infinite-dimensional Banach space and let H be a Hamel basis† for X. We may and do assume that the elements in H have norm 1. Let Y be the vector space X with a new norm $\| \cdot \|_1$ given by

$$\left\| \sum_{i=1}^{n} \alpha_i x_i \right\|_1 = \sum_{i=1}^{n} | \alpha_i |, \qquad x_i \in H.$$

For each $x \in X$, $\| x \|_1 \geq \| x \|$. Let (y_n) be an infinite sequence in H. If

$$z_n = \sum_{k=1}^{n} \left(\frac{1}{k^2}\right) \cdot y_k,$$

then

$$\| z_n - z_m \|_1 = \sum_{k=n+1}^{m} \frac{1}{k^2} \to 0 \text{ as } n, m \to \infty.$$

† A linearly independent set which spans X.

Hence (z_n) is Cauchy in Y as well as in X. For any z in Y, $z = \sum_{i=1}^{p} \alpha_i h_i$ for some positive integer p with $h_i \in H$, for $i = 1, 2, \ldots, p$. Hence for $n \geq k$, $\| z_n - z \|_1 \geq 1/k^2$ if k is large enough for $y_k \neq h_i$, $i = 1, 2, \ldots, p$. This means that (z_n) does not converge to any z in Y, so that Y is not complete. Now if we consider the identity map $i: X \to Y$, then i is clearly closed, one-to-one, onto, and has a continuous inverse. However, i is *not* continuous, since in that case Y has to be complete.

Example 5.13. (*A Closed Linear Operator from a Normed Linear Space onto a Banach Space That Is Not an Open Map.*) Consider Example 5.12 and the identity map i: $Y \to X$. Then i is a closed linear map that is not open.

Now we state and prove the Open Mapping Theorem.

Theorem 5.10. [*The Open Mapping Theorem.*] Let X be a Banach space and Y be a normed linear space of the second category. If T is a closed linear operator from a linear subspace D in X onto Y, then T is an open map. ▮

Before we prove this theorem, we prove the following lemma.

Lemma 5.2. Let T be a closed linear operator from a Banach space X† into a normed linear space Y. Let

$$S_X(r) = \{x \in X: \| x \| \leq r\} \quad \text{and} \quad S_Y{}^\circ(r) = \{y \in Y: \| y \| < r\}.$$

If $S_Y{}^\circ(r) \subset \overline{TS_X(1)}$, then $S_Y{}^\circ(r) \subset TS_X(1)$. ▮

Proof. It is sufficient to prove that for each ε satisfying $0 < \varepsilon < 1$,

$$S_Y{}^\circ(r) \subset TS_X\left(\frac{1}{1-\varepsilon}\right). \quad \text{(Why?)}$$

Now let $0 < \varepsilon < 1$ and $\| y \| < r$, $y \in Y$. Then for each nonnegative integer,

$$S_Y{}^\circ(r \cdot \varepsilon^n) = \varepsilon^n S_Y{}^\circ(r) \subset \varepsilon^n \overline{TS_X(1)} = \overline{\varepsilon^n TS_X(1)} = \overline{TS_X(\varepsilon^n)}.$$

† Actually T can be defined only on a linear subspace D of X. In this case $S_X(r)$ in the proof is replaced by $S_D(r) = S_X(r) \cap D$.

Taking $n = 0$, we have $S_Y^\circ(r) \subset \overline{TS_X(1)}$. This means that there exists $x_0 \in S_X(1)$ such that $\| y - T(x_0) \| < r\varepsilon$. But if $n = 1$, we have $S_Y^\circ(r\varepsilon) \subset \overline{TS_X(\varepsilon)}$. Since $y - T(x_0) \in S_Y^\circ(r\varepsilon)$, there exists $x_1 \in S_X(\varepsilon)$ such that

$$\| y - T(x_0) - T(x_1) \| < r\varepsilon^2.$$

In this way, by induction, we obtain a sequence (x_n) such that $x_n \in S_X(\varepsilon^n)$ and

$$\| y - T(x_0) - T(x_1) - \cdots - T(x_n) \| < r \cdot \varepsilon^{n+1}.$$

Since $\sum_{n=0}^\infty \| x_n \| \leq 1/(1 - \varepsilon)$ and X is complete, $\sum_{n=0}^\infty x_n$ is summable. Let $\sum_{n=0}^k x_n \to x$. Since $T(\sum_{n=0}^k x_n) \to y$ and T is closed, $y = T(x)$. This proves the lemma. ∎

Proof of Theorem 5.10. It follows from Lemma 5.2 that the theorem will be proved if we can show that there exists $r > 0$ such that $S_Y^\circ(r) \subset \overline{TS_X(1)}$. [Note that $S_Y^\circ(r) \subset TS_X(1)$ implies that 0 is an interior point of $T(V)$, where V is any open neighborhood of 0. Then by the linearity of T, the openness of T follows.] Since $Y = \bigcup_{n=1}^\infty n \cdot TS_X(1)$ and Y is of the second category, there is a nonnegative integer k such that $\overline{k \cdot TS_X(1)} = \overline{TS_X(k)}$ contains a nonempty open set V. Hence $\overline{TS_X(1/2)}$ also contains a nonempty open set W [namely, $(1/2k)V$]. Since $W - W$ is an open set containing 0 and $W - W \subset \overline{TS_X(1)}$, the proof is complete. ∎

In connection with the "openness" property of a linear map, the following proposition is interesting.

Proposition 5.11. Let T be a bounded linear map from a normed linear space X *onto* a normed linear space Y. Then T is open if and only if there exists $k > 0$ such that for each $y \in Y$, there exists $x \in X$ with $T(x) = y$ and $\| x \| \leq k \cdot \| y \|$. ∎

We leave the proof to the reader.

Proposition 5.11 helps us obtain a partial converse of the Open Mapping Theorem.

Proposition 5.12. If T is a bounded linear operator from a Banach space X *onto* a normed linear space Y, then the "openness" of T implies the completeness of Y. ∎

Proof. Let (y_n) be a Cauchy sequence in Y. Then we can find a sequence (n_k) of positive integers such that $n_k < n_{k+1}$ and for each k, $\| y_{n_{k+1}} - y_{n_k} \|$

$< 1/2^k$. If T is open, by Proposition 5.11 there exists $N > 0$ such that we can find x_k with $T(x_k) = y_{n_{k+1}} - y_{n_k}$ and $\| x_k \| \leq N \cdot \| y_{n_{k+1}} - y_{n_k} \|$. Then $\sum_{k=1}^{\infty} \| x_k \| < \infty$. Since X is complete, there exists $x \in X$ such that $x = \lim_{n\to\infty} \sum_{k=1}^{n} x_k$. Since T is continuous and linear, $\sum_{k=1}^{n} T(x_k) \to T(x)$, which means that $y_{n_{k+1}} \to y_{n_1} + T(x)$. Since (y_n) is Cauchy, $y_n \to y_{n_1} + T(x)$. ∎

Now as applications of the Open Mapping Theorem, we present the next three propositions and also the Closed Graph Theorem.

Proposition 5.13. Let X be a vector space that is complete in each of the norms $\| \cdot \|$ and $\| \cdot \|_1$. Suppose there exists $k > 0$ such that

$$\| x \| \leq k \| x \|_1$$

for each $x \in X$. Then the norms are equivalent. ∎

Proof. If i is the identity map from $(X, \| \cdot \|_1)$ *onto* $(X, \| \cdot \|)$, then i is clearly closed and linear. By the Open Mapping Theorem, i is also open and so i^{-1} is continuous. ∎

Proposition 5.14. Let X and Y be Banach spaces. Then the set of all surjective maps in $L(X, Y)$ is open in $L(X, Y)$. ∎

Proof. Let T be a surjective map in $L(X, Y)$. Let $S \in L(X, Y)$ and $\| T - S \| < 1/2k$, where k is a real number with the property as stated in Proposition 5.11. We claim that S is surjective. To prove this, let $y \in Y$ and $\| y \| \leq 1$. Then by Proposition 5.11, there exists $x \in X$, $T(x) = y$ and $\| x \| \leq k$. Let $y_1 = T(x) - S(x)$. Then, $\| y_1 \| \leq \frac{1}{2}$ and there exists $x_1 \in X$, $T(x_1) = y_1$ and $\| x_1 \| \leq k/2$. Let $y_2 = T(x_1) - S(x_1)$. Then, $\| y_2 \| \leq 1/2^2$. Continuing inductively, we find x_n such that $T(x_n) = y_n$ and

$$y_{n+1} = T(x_n) - S(x_n), \qquad \| y_n \| \leq 1/2^n, \qquad \| x_n \| \leq k/2^n.$$

Then, $y = S(x) + S(x_1) + \cdots + S(x_n) + y_{n+1}$. Write $z = x + \sum_{n=1}^{\infty} x_n$. Then $S(z) = y$ and our claim is proven. ∎

Proposition 5.15. Let X be a separable Banach space. Then there exists a closed linear subspace L of the Banach space l_1 such that X is topologically isomorphic to the quotient space l_1/L. ∎

Proof. Let (x_n) be a dense sequence in the unit ball of X. We define the mapping T: $l_1 \to X$ by $T(y) = \sum a_n x_n$, $y = (a_n)_{n=1}^{\infty}$. Clearly, $\| T(y) \| \leq \| y \|_1$. Let $L = T^{-1}\{0\}$. Now the mapping τ defined by $\tau(y + L) = T(y)$

is a well-defined continuous linear one-to-one mapping from l_1/L into X. If T is surjective, then τ is surjective, in which case an application of the Open Mapping Theorem will finish the proof. So it suffices to show that T is surjective.

Suppose $x \in X$ and $\| x \| \leq 1$. By an induction argument, we can find a subsequence (x_{n_i}) such that $x = \sum_{i=0}^{\infty} x_{n_i}/2^i$. [Choose x_{n_0} such that $\| x - x_{n_0} \| < \frac{1}{2}$; then, choose $n_1 > n_0$ such that $\| 2(x - x_{n_0}) - x_{n_1} \| < \frac{1}{2}$ and so on.] Let $y = (a_k)$ be defined so that $a_k = 1/(2^i)$ if $k = n_i$, $= 0$ if $k \neq n_i$ for all i. Then $y \in l_1$ and $T(y) = x$. It follows that T is surjective. ▮

Theorem 5.11. *The Closed Graph Theorem.* A closed linear operator T mapping a Banach space X into a Banach space Y is continuous. ▮

Proof. Since T is closed, $G_T = \{(x, T(x)): x \in X\}$ is a closed subspace of the Banach space $X \times Y$. We define $P_1[(x, T(x))] = x$ and $P_2[(x, T(x))] = T(x)$. Then P_1 and P_2 are both continuous maps from G_T *onto* X and into Y, respectively. By the Open Mapping Theorem, P_1 is open. Since P_1 is one-to-one, P_1^{-1} is continuous and therefore, $P_2P_1^{-1} = T$ is continuous. ▮

Actually, we could state the Closed Graph Theorem in a more general form.

Theorem 5.12. A closed linear operator T mapping a normed linear space X of the second category into a Banach space Y is continuous. ▮

Proof. We will briefly outline the proof. Let $M = \{x \in X \mid T(x) = 0\}$. Then M is a closed subspace of X, since T is closed. If $X = M$, the theorem is trivial. Suppose $X \neq M$. Then the quotient space X/M is of the second category (see Problem 5.4.3). Define T_0: $X/M \to Y$ by $T_0(\Phi(x)) = T(x)$, where Φ is the natural map from X onto X/M. Then T_0 is well defined, one-to-one, closed, and linear. (See Problem 5.4.4.) Therefore, the mapping T_0^{-1}: $T(X)(\subset Y) \to X/M$ is again closed. By the Open Mapping Theorem, T_0^{-1} is open and therefore T_0 is continuous. This means that T is continuous. ▮

By means of quotient space arguments as used in the proof of Theorem 5.12, it is possible to give a proof of the Open Mapping Theorem for the case when D is closed, using Theorem 5.12 as a starting point. This means that the Open Mapping and the Closed Graph Theorems are two different forms of the same theorem.

Next, we present an incomplete normed linear space of the second category.

Example 5.14. Let X be the Lebesgue-integrable functions on $[0, 1]$ with the L_1 norm. Let (r_n) be the rationals in $(0, 1]$, and x_n be the characteristic function of $[0, r_n]$. Then A, the set containing the x_n's, is linearly independent and hence can be extended to a Hamel basis H of X. Since X is complete, H is uncountable. (See Problem 5.1.7.) Let (y_n) be a sequence of elements in $H - A$. Let Y_n be the subspace spanned by $H - \{y_n, y_{n+1}, \ldots\}$. Then since $X = \bigcup_{n=1}^{\infty} Y_n$, for some $n = p$, Y_p is *of the second category*. Also Y_p is not complete, since Y_p contains A and the span of A is dense in X (the step functions being dense in L_1).

Finally, we come to the Principle of Uniform Boundedness. We need first a lemma.

Lemma 5.3. Let S be a nonempty set and Y be a normed linear space. Let $B(S, Y) = \{f\colon S \to Y \mid \sup_{s\in S} \| f(s) \| < \infty\}$. If Y is complete and $\| f \| = \sup_{s\in S} \| f(s) \|$, then $B(S, Y)$ is a Banach space with this norm. ∎

The proof of this lemma is left as Problem 5.4.5.

Theorem 5.13. *The Principle of Uniform Boundedness.* Let S be a family of bounded linear operators from a Banach space X into a Banach† space Y. Suppose that for $x \in X$ there is a constant $M(x)$ such that $\sup_{T\in S} \| T(x) \| \leq M(x) < \infty$. Then there exists $M > 0$ such that

$$\sup\{\| T \| : T \in S\} < M. \qquad ∎$$

Proof. We define the linear operator A from X into $B(S, Y)$ by $A(x)[T] = T(x)$. Then A is well defined since $\sup_{T\in S} \| T(x) \| < \infty$. A is also closed. (Why?) By the Closed Graph Theorem, A is continuous. Therefore, $\sup_{\|x\|\leq 1} \| A(x) \| < M$, for some $M > 0$. But since $\| A(x) \| = \sup_{T\in S} \| T(x) \|$ we have $\sup_{\|x\|\leq 1} \sup_{T\in S} \| T(x) \| < M$ or $\sup_{T\in S} \| T \| < M$. ∎

Remarks on Theorem 5.13 and Some Applications

5.1. Theorem 5.13 need *not* be true if X is not complete. Let $X = \{(a_1, a_2, \ldots, a_n, 0, 0, 0, \ldots)\colon a_i \in R\}$. Then $X \subset l_2$. We define $T_n\colon X \to l_2$ linearly by

$$T_n(e_i) = \begin{cases} 0, & i \neq n \\ ne_n, & i = n \end{cases}$$

† Completeness of Y is not needed in this theorem. One can consider $B(S, \hat{Y})$, where $\hat{Y}$ is the completion of Y. See Proposition 5.17. Also it suffices to let X be a normed linear space of the second category.

where

$$e_i = (\underbrace{0, 0, \ldots, 0, 1}_{i}, 0, 0, \ldots).$$

Then for $x = \sum_{i=1}^{k} a_i e_i$, $T_n(x) = 0$ for $n > k$ and therefore $\sup_{1 \le n < \infty} \| T_n(x) \|$ is finite. But $\| T_n \| = n$ and $\sup_{1 \le n < \infty} \| T_n \| = \infty$.

5.2. From Theorem 5.13, it follows that for $S \subset X^*$, X a Banach space, $\sup_{x^* \in S} \| x^* \| < \infty$ whenever for each $x \in X$, $\sup_{x^* \in S} | x^*(x) |$ is finite.

5.3. If $S \subset X$ a normed linear space such that $\sup_{x \in S} | x^*(x) | < \infty$ for each $x^* \in X^*$, then $\sup_{x \in S} \| x \| < \infty$. This immediately follows by applying Theorem 5.13 to the family of operators $[J(x)]_{x \in S}$ from X^* into F (the scalar field), J being the natural map from X into X^{**}, that is $[J(x)](x^*) = x^*(x)$, $x^* \in X^*$.

5.4. *Analyticity of a Banach Space Valued Function of a Complex Variable.* Let X be a complex Banach space, G an open set in the complex plane, and f a function from G into X. Then f is called analytic on G if f is differentiable at each point $\lambda_0 \in G$, that is,

$$\left\| \frac{f(\lambda) - f(\lambda_0)}{\lambda - \lambda_0} - f'(\lambda_0) \right\| \to 0$$

as $\lambda \to \lambda_0$ for some function $f'\colon G \to X$. We will apply the principle of uniform boundedness (actually Remark 5.3 above) to show that a necessary and sufficient condition for f to be differentiable on G is that $x^*\big(f(\lambda)\big)$ be differentiable on G for each $x^* \in X^*$.

Since the "necessary" part is trivial, we show only the "sufficient" part. Let $x^*\big(f(\lambda)\big)$ be differentiable on G and $\lambda_0 \in G$. There exists $r > 0$ such that

$$| \lambda - \lambda_0 | < r \Rightarrow \lambda \in G.$$

By Cauchy's integral formula (in elementary complex analysis), we have

$$x^*\big(f(\lambda)\big) = \frac{1}{2\pi i} \int_C \frac{x^*\big(f(z)\big)}{z - \lambda}\, dz, \tag{5.1}$$

where $| \lambda - \lambda_0 | < r$ and C is the positively oriented circle $| z - \lambda_0 | = r$. It follows from equation (5.1) after a simple calculation that for $\lambda \ne \mu$, $| \lambda - \lambda_0 | \le \frac{1}{2} r$ and $| \mu - \lambda_0 | \le \frac{1}{2} r$ (and therefore, $| z - \lambda | \ge \frac{1}{2} r$ and $| z - \mu | \ge \frac{1}{2} r$ for $z \in C$),

$$x^*\left(\left[\frac{f(\lambda) - f(\lambda_0)}{\lambda - \lambda_0} - \frac{f(\mu) - f(\lambda_0)}{\mu - \lambda_0}\right] \Big/ \lambda - \mu\right)$$
$$= \frac{1}{2\pi i} \int_C \frac{x^*\big(f(z)\big)}{(z - \lambda)(z - \mu)(z - \lambda_0)}\, dz. \tag{5.2}$$

Since $M = \sup\{\| x^*(f(z)) \|: z \in C\} < \infty$, the absolute value of the left-hand side of equation (5.2) does not exceed $4M/r^2$. Applying Remark (5.3) above, we have

$$\left\| \frac{f(\lambda) - f(\lambda_0)}{\lambda - \lambda_0} - \frac{f(\mu) - f(\lambda_0)}{\mu - \lambda_0} \right\| \leq K \cdot | \lambda - \mu |$$

for some constant K, whenever $0 < | \lambda - \lambda_0 | \leq r/2$ and $0 < | \mu - \lambda_0 | \leq r/2$. Since X is complete, it follows that f is differentiable at λ_0.

5.5. *Divergence of the Fourier Series.* By the Fourier series of a function f in $L_1[-\pi, \pi]$, we mean the series

$$\sum_{k=-\infty}^{\infty} \hat{f}(k)e^{ikt},$$

where the Fourier transform $\hat{f}$ of f is defined by

$$\hat{f}(k) = \frac{1}{2\pi} \int_{-\pi}^{\pi} f(s)e^{-iks}\, ds, \qquad k \in Z.$$

Such series were invented originally to serve as tools to solve problems in heat conduction, the theory of oscillation, and various other fields. The many problems that arose to determine whether the Fourier series of f converges to f or whether f is determined by its Fourier series gave rise to an important branch of analysis, known as harmonic analysis. We will not consider the convergence of the Fourier series here. In the next chapter (Remark 6.4), we shall show that the Fourier series of f in $L_2[-\pi, \pi]$ converges to f in L_2-norm. Actually a much more nontrivial result holds, namely, that the Fourier series of such a function converges almost everywhere. This result was first conjectured by N. N. Lusin in 1951 and then proven in 1966 by L. Carleson [14]. Later on, in 1968, the result was extended to $L_p[-\pi, \pi]$, $1 < p \leq \infty$, by R. A. Hunt [29]. We will not go into the details of these results here. Here we will consider the divergence of the Fourier series to show another application of Theorem 5.13.

Theorem. There exists a continuous function f on $[-\pi, \pi]$ such that the Fourier series of f diverges at 0. The set of all such functions is of the second category in $C_1[-\pi, \pi]$. ∎

Proof. We write $S_n f(x) = \sum_{k=-n}^{n} \hat{f}(k)e^{ikx}$. Then we have, for any f in $C_1[-\pi, \pi]$,

$$S_n f(x) = \frac{1}{2\pi} \int_{-\pi}^{\pi} f(t)D_n(x - t)\, dt,$$

where

$$D_n(t) = \sum_{k=-n}^{n} e^{ikt} = \frac{\sin(n+\frac{1}{2})t}{\sin(\frac{1}{2}t)} \qquad \text{if } e^{it} \neq 1, = 2n+1 \text{ if } e^{it} = 1.$$

For each nonnegative integer n, we define

$$x_n^*(f) = \frac{1}{2\pi} \int_{-\pi}^{\pi} f(t)D_n(t)\,dt, \qquad f \in C[-\pi, \pi].$$

Then x_n^* is a bounded linear functional on $C_1[-\pi, \pi]$ with norm $(1/2\pi)\,\| D_n \|_1$. (The reader should verify this.) Now, suppose that the Fourier series of every $f \in C[-\pi, \pi]$ converges at $x = 0$. Then for each $f \in C_1[-\pi, \pi]$, $\sup_n | x_n^*(f) | < \infty$. By Theorem 5.13, it follows that

$$\sup_n \| x_n^* \| = \frac{1}{2\pi} \sup_n \| D_n \|_1 < \infty.$$

But we have

$$\| D_n \|_1 = \int_{-\pi}^{\pi} \left| \frac{\sin(n+\frac{1}{2})t}{\sin(\frac{1}{2}t)} \right| dt = 4 \int_0^{\pi/2} \left| \frac{\sin(2n+1)t}{\sin t} \right| dt$$
$$\geq 4 \int_0^{\pi/2} \left| \frac{\sin(2n+1)t}{t} \right| dt \geq \frac{8}{\pi} \sum_{k=0}^{2n} \frac{1}{k+1}.$$

Thus, $\sup_n \| D_n \|_1 = \infty$, a contradiction.

To complete the proof, we assume that the set $X \subset C_1[-\pi, \pi]$ of functions f such that $\sup_n | \sum_{k=-n}^{n} \hat{f}(k) | < \infty$ is of the second category. Noting that Theorem 5.13 holds even when X is a normed linear space of the second category, it follows as before that $\sup_n \| x_n^* \| < \infty$, which is a contradiction. The proof is complete. ∎

Before we close this section, we present a theorem on projection, an application of the Closed Graph Theorem. The reader will find projection operators extremely useful in Hilbert space theory.

Definition 5.9. A bounded linear operator P from a normed linear space X onto a subspace M of X is called a *projection* if $P^2 = P$. ∎

Remarks

5.6. If P is a projection and $P \neq 0$, then $\| P \| \geq 1$ since

$$\| P \| = \| P^2 \| \leq \| P \| \cdot \| P \|.$$

In fact, $\| P \|$ can be greater than 1. Let $X = R^2$ with $\| (x, y) \| = | x | + | y |$

and $M = \{(x, x) \mid x \in R\}$, a closed subspace of X. We define $P((x, y)) = (y, y)$. Then P is a projection onto M. But $P((0, 1)) = (1, 1)$ and therefore, $\| P \| > 1$.

5.7. There is always a projection from a normed linear space X onto any of its finite-dimensional subspaces M. To see this, let $x_1, x_2, \ldots, x_n$ be the basis of $M \subset X$. Define $x_i^* \in X^*$ by

$$x_i^*(x_j) = \begin{cases} 1, & i = j \\ 0, & i \neq j \end{cases}$$

and

$$x_i^*(x) = 0, \qquad x \notin M.$$

Let $P(x) = \sum_{i=1}^{n} x_i^*(x) x_i$. Then P is a projection.

Proposition 5.16. Let M be a closed subspace of a Banach space X. Then there exists a projection P from X onto M if and only if there is a closed subspace N of X such that $X = M + N$, $M \cap N = \{0\}$. ∎

Proof. For the "only if" part, let $N = P^{-1}(\{0\})$, where P is a projection of X onto M. Since $x = Px + (x - Px)$ and $x - Px \in N$, $X = M+N$. Also $z \in M \cap N$ implies $P(z) = 0$ and $z = P(x)$ for some $x \in X$, which means $z = 0$.

For the "if" part, let $X = M + N$, $M \cap N = \{0\}$. Then for each $x \in X$, there exist unique $m \in M$ and $n \in N$ such that $x = m + n$. We define $P(x) = m$. Then P is linear and $P^2 = P$. To show that P is bounded, it is sufficient to show that it is closed (because of the Closed Graph Theorem). Let $x_k \to x$ and $P(x_k) \to y$. Let $x_k = m_k + n_k$, $m_k \in M$ and $n_k \in N$. Also let $x = m + n$, $m \in M$ and $n \in N$. Then $P(x_k) = m_k \to y \in M$. Hence $n_k = (m_k + n_k) - m_k \to m + n - y$. This means that $m + n - y \in N$ or $m - y \in M \cap N = \{0\}$. Therefore $y = m = P(x)$ and P is closed. ∎

Problems

5.4.1. The notations are as in Lemma 5.2. Show that $S_y(r) \subset TS_X(1)$, if $S_Y(r) \subset TS_X(1/(1 - \varepsilon))$ for all $\varepsilon \in (0, 1)$.

5.4.2. Prove Proposition 5.11.

5.4.3. Let X be a normed linear space of the second category and M be a closed proper subspace of X. Show that X/M is of the second category.

5.4.4. Let T be a closed linear operator from a normed linear space X into a normed linear space Y. Let $M = T^{-1}\{0\}$. Show that there is a unique one-to-one, closed, and linear operator T_0: $X/M \to Y$ such that $T_0 \circ \Phi = T$, where Φ is the natural map from X onto X/M. Also show that T_0^{-1}: $T(X)(\subset Y) \to X/M$ is a closed linear operator.

5.4.5. Prove Lemma 5.3.

5.4.6. Let $T_n \in L(X, Y)$, where X is a Banach space and Y is any normed linear space. If for each $x \in X$, $T(x) = \lim_{n\to\infty} T_n(x)$, then show that $T \in L(X, Y)$.

5.4.7. Let A be a linear operator from X into Y such that $y^* \circ A$ is continuous for each $y^* \in Y^*$. Show that $A \in L(X, Y)$. (X and Y are normed linear spaces which are not necessarily complete.)

5.4.8. If A is a bounded linear operator from D (a dense subspace of a normed linear space X) into a Banach space Y, then show that there exists a unique bounded linear operator B from X into Y with $\| A \| = \| B \|$ and $B(x) = A(x)$, $x \in D$. [Note that completeness is essential here; for consider $D = Y =$ the polynomials on $[0, 1]$, $X = C[0, 1]$ and i: $D \to D$, the identity map.]

5.4.9. Let $A \in L(X, Y)$. The adjoint of A is an operator A^*: $Y^* \to X^*$ defined by $A^*y^*(x) = y^*(A(x))$, $x \in X$. Show the following:

(i) $\| A^* \| = \| A \|$.

(ii) $(\alpha A + \beta B)^* = \alpha A^* + \beta B^*$ for all $\alpha, \beta \in F$ and $A, B \in L(X, Y)$.

(iii) If $A \in L(X, Y)$, $B \in L(Y, Z)$, then $(BA)^* = A^*B^*$.

(iv) If $A \in L(X, Y)$, A is onto and A^{-1} exists and belongs to $L(Y, X)$, then $(A^{-1})^* = (A^*)^{-1}$.

Note that if $X = Y = l_p$, $1 \leq p < \infty$, then a bounded linear operator A on l_p can be represented by an infinite matrix (a_{ij}). If e_i is that element in l_p where the ith entry is 1 and every other entry is 0, and if $e_i^* \in l_p^*$ such that $e_i^*(e_j) = 1$ if $i = j$, $= 0$ if $i \neq j$, then $a_{ij} = e_i^*(A(e_j))$. If $A(x) = y$ and $x = \sum x_i e_i$, then $e_i^*(y) = \sum_j a_{ij} x_j$. Since l_p^* can be identified with l_q, where $pq = p + q$, the adjoint A^* can be easily verified to be represented by the transpose of the infinite matrix (a_{ij}) in the same sense as above.

In the next few problems and later on, the range of A is denoted by $R(A)$ or R_A.

5.4.10. Let $A \in L(X, Y)$, A be one-to-one and X, Y both Banach spaces. Then $R(A)$ is closed if and only if there exists $C > 0$ such that $\| x \| \leq C \| A(x) \|$ for each $x \in X$.

5.4.11. If X and Y are Banach spaces and $A \in L(X, Y)$, then $R(A)$ is closed in Y if and only if there exists $C > 0$ such that $\inf\{\| x - y \|: A(y) = 0\} \leq C \| A(x) \|$ for each $x \in X$.

5.4.12. Let $S \subset X$. Then S° (the *annihilator* of S) $= \{x^* \in X^*: x^*(y) = 0$ for each $y \in S\}$. $S^\circ = X^*$ when S is empty. Similarly, for $E \subset X^*$, ${}^\circ E = \{x \in X: x^*(x) - 0$ for each $x^* \in E\}$. ${}^\circ E = X$, when E is empty. Show the following:

(i) S° and ${}^\circ E$ are both closed subspaces.

(ii) For $S \subset X$, ${}^\circ(S^\circ)$ is the closed subspace spanned by S.

(iii) $[R(A)]^\circ = N(A^*)$, the null space of A^*

(iv) $\overline{R(A)} = {}^\circ[N(A^*)]$.

(v) ${}^\circ[R(A^*)] = N(A)$.

(vi) $R(A^*) \subset [N(A)]^\circ$.

5.4.13. Let $A \in L(X, Y)$, X and Y both Banach spaces. If $R(A)$ is closed in Y, then show that $R(A^*) = [N(A)]^\circ$ and hence is closed in X^*. {Hint: Use Problem 5.4.12 (vi) and the following. For $x^* \in [N(A)]^\circ$, define $f: R(A) \to F$ by $f(A(x)) = x^*(x)$. Show, by Problem 5.4.11, $|f(A(x))| \leq C \cdot \| x^* \| \cdot \| A(x) \|$, $C > 0$. Extend f to $y^* \in Y^*$ so that $A^*(y^*) = x^*$.]

5.4.14. Let $A \in L(X, Y)$. Then show that $R(A^*) = X^*$ if and only if A^{-1} exists and is continuous.

5.4.15. Let $A \in L(X, Y)$, Y complete and $R(A) = Y$. Then show that A^* has a continuous inverse.

5.4.16. Prove that $L_2[0, 1]$ is of the first category in $L_1[0, 1]$. (Hint: The identity map from L_2 into L_1 is continuous, but *not* onto. Use the Open Mapping Theorem.)

5.4.17. *Joint Continuity of a Separately Continuous Bilinear Function.* Suppose X and Y are Banach spaces and $T: X \times Y \to R$ is a mapping such that for each $x \in X$ and each $y \in Y$, the functions $y \to T(x, y)$ and $x \to T(x, y)$ are bounded linear functionals. Show that T is continuous on $X \times Y$. [Hint: Use the Principle of Uniform Boundedness.]

5.4.18. *Convex Functions and the Principle of Uniform Boundedness.* Let $\mathscr{F}$ be the smallest σ-algebra containing the open sets of a Banach space X. Then every real-valued convex function defined on X and measurable with respect to $\mathscr{F}$ is continuous. Using this result, the Principle of Uniform Boundedness can be proven as follows. The function $p(x) = \sup\{\| T_\lambda(x) \|: \lambda \in \Lambda\}$, where $\{T_\lambda: \lambda \in \Lambda\}$ is a family of bounded linear operators from X into a normed linear space Y such that $p(x)$ is a real-valued function on X, is a lower semicontinuous convex function and therefore continuous. Then there exists $\delta > 0$ such that $\| x \| \leq \delta \Rightarrow | p(x) - p(0) | \leq 1$. This means that for each $\lambda \in \Lambda$ $\| T_\lambda \| \leq 1/\delta$.

5.4.19. *Another Application of the Principle of Uniform Boundedness.* Let X be the Banach space (with "sup" norm) of periodic continuous functions f on R with period 2π. For $f \in X$, let $T_n f(t) = n(f(t + 1/n) - f(t))$. Then by Corollary 1.2 (Chapter 1), $\lim_{n\to\infty} T_n f(t) = f'(t)$ exists for all f in a dense subset of X. However, the set of nondifferentiable functions is of the second category in X. [Notice that $\| T_n \| = 2n$. If $D \subset X$ is the set of functions differentiable at 0 in R and D is of the second category, then the Principle of Uniform Boundedness applies to the sequence $x^* \circ T_n$, where $x^*(f) = f(0)$.]

5.4.20. *Equivalent Norms in* $C[0, 1]$. Any complete norm $\| \cdot \|$ in $C[0, 1]$, where $\lim_{n\to\infty} \| f_n - f \| = 0 \Rightarrow \lim_{n\to\infty} f_n(t) = f(t)$ for all $t \in [0, 1]$, is equivalent to the usual "sup" norm. {Here the Closed Graph Theorem can be used to show that the identity map from $C[0, 1]$ with "sup" norm into $(C[0, 1], \| \cdot \|)$ is continuous; then Proposition 5.13 applies.}

5.5. Reflexive Banach Spaces and the Weak Topology

In the theory of Banach spaces an often useful concept is that of reflexivity, which is based upon a characterization of a class of bounded linear functionals. The notion and properties of reflexive Banach spaces, along with some representation theorems for bounded linear functionals on certain well-known normed linear spaces, is the subject of this section. One reason for the study of such spaces is that a large number of useful results are obtainable in such spaces that are not true in general Banach spaces; and yet one comes across a wide class of such spaces in theory as well as in practice. For instance, the L_p $(1 < p < \infty)$ spaces† will be shown to belong to this class.

Among other things, we also show in this section the interplay between reflexivity and a basic convergence concept, the concept of weak convergence. The topics of weak convergence and weak topology are essential in functional analysis. They find an immense number of applications in various contexts in the theory of differential equations and in the calculus of variations.

Definition 5.10. The *natural map* J of a normed linear space X into its

† In this section, for convenience the L_p spaces are taken over the reals.

second conjugate space $X^{**}[= L(X^*, F)]$ is defined by

$$[J(x)](x^*) = x^*(x), \qquad x^* \in X^*.$$

If the range of J is all of X^{**}, then X is called *reflexive*. ∎

Remark 5.8. A reflexive normed linear space is complete. The reason is that

$$\| J(x) \| = \sup_{\|x^*\|=1} | x^*(x) | = \| x \|,$$

by Corollary 5.5; and therefore J is a linear isometry from X into X^{**}, which is complete. However, the existence of a linear isometry from X *onto* X^{**} does *not* guarantee the reflexivity of X. For an example demonstrating this fact, the serious reader is referred to R. C. James.†

Remark 5.9. A finite-dimensional normed linear space X is reflexive. The reason is that $\dim X = \dim X^* = \dim X^{**}$, and a one-to-one linear operator between finite-dimensional spaces of the same dimension is also *onto*.

Proposition 5.17. Every normed linear space X is a dense subspace of a Banach space. ∎

Proof. Let $\hat{X} = X \cup [\overline{J(X)} - J(X)]$, where J is the natural map and $\overline{J(X)} \subset X^{**}$. Let us define for $x, y \in \hat{X}$ and $\alpha, \beta \in F$

$$\begin{aligned} \alpha x + \beta y &= \alpha J(x) + \beta J(y), && \text{if } x, y \in X; \\ \alpha x + \beta y &= \alpha J(x) + \beta y, && \text{if } x \in X,\ y \in \hat{X} - X; \end{aligned}$$

and

$$\alpha x + \beta y = \alpha x + \beta y, \qquad \text{if } x, y \in \hat{X} - X.$$

(Here αx and $\alpha x + \beta y$ represent scalar multiplication and addition in X^{**}.) Also let

$$\| x \|_1 = \begin{cases} \| Jx \|, & \text{if } x \in X \\ \| x \|, & \text{if } x \in \hat{X} - X. \end{cases}$$

Then $(\hat{X}, \| \cdot \|_1)$ is a normed linear space. Let Φ be the mapping from $\hat{X}$ into $\overline{J(X)}$ defined by

$$\Phi(x) = \begin{cases} J(x), & \text{if } x \in X \\ x, & \text{if } x \in \hat{X} - X. \end{cases}$$

† R. C. James, A nonreflexive Banach space isometric with its second conjugate, *Proc. Nat. Acad. Sci. U.S.A.* **37**, 174–177 (1951).

Then Φ is a surjective linear isometry. Since $\overline{J(X)}$ is complete (being a closed subspace of X^{**}), $\hat{X}$ is also complete. The rest is clear. ∎

Before we get involved with the properties of reflexive spaces, we should consider some interesting examples of such spaces. To this end, we study first the conjugate (or dual) spaces of some important Banach spaces. The following theorem is one of the many important contributions of F. Riesz in this area and was proved by him in 1907 and 1909 for the Lebesgue measure space on [0, 1] when $1 < p < \infty$. The case $p = 1$ was treated later in 1919 by H. Steinhaus.

Theorem 5.14. *Riesz Representation Theorem.*† Let $(X, \mathscr{A}, \mu)$ be a σ-finite measure space and Φ be a bounded linear functional on L_p, $1 \leq p < \infty$. If $1/p + 1/q = 1$, then there is a *unique* element $g \in L_q$ such that

$$\Phi(f) = \int fg \, d\mu, \qquad f \in L_p$$

and

$$\| \Phi \| = \| g \|_q . \qquad ∎$$

Proof. Suppose first that μ is finite. Let $\nu(E) = \Phi(\chi_E)$, for $E \in \mathscr{A}$. Let $E_n \in \mathscr{A}$, $E_n \cap E_m = \varnothing \, (n \neq m)$. Then whenever $\alpha_n \in F$ with $| \alpha_n | = 1$, $\sum_{n=1}^{k} \alpha_n \chi_{E_n} \to \sum_{n=1}^{\infty} \alpha_n \chi_{E_n}$ in L_p. Therefore since Φ is continuous, we have

$$\begin{aligned}\sum_{n=1}^{\infty} | \nu(E_n) | &= \lim_{k\to\infty} \sum_{n=1}^{k} \Phi\big(\chi_{E_n} \cdot \operatorname{sgn} \Phi(\chi_{E_n})\big) \\ &= \Phi\Big(\sum_{n=1}^{\infty} \chi_{E_n} \cdot \operatorname{sgn} \Phi(\chi_{E_n})\Big) \\ &< \infty\end{aligned}$$

and similarly,

$$\sum_{n=1}^{\infty} \nu(E_n) = \Phi\Big(\sum_{n=1}^{\infty} \chi_{E_n}\Big) = \Phi(\chi_{\cup E_n}) = \nu\Big(\bigcup_{n=1}^{\infty} E_n\Big).$$

Hence ν is a signed measure and so by the Radon–Nikodym Theorem there is an integrable function g such that $\nu(E) = \int_E g \, d\mu$, $E \in \mathscr{A}$. If f is a simple function, we have

$$\left| \int fg \, d\mu \right| = | \Phi(f) | \leq \| \Phi \| \cdot \| f \|_p .$$

† In this theorem, $\Phi(f) = \int f\bar{g} \, d\mu$ when the scalars are complex numbers. The reader should verify this.

By Problem 5.2.6, $g \in L_q$. Let χ be the linear functional defined on L_p by

$$\chi(h) = \int hg \, d\mu, \qquad h \in L_p.$$

Then χ is clearly bounded (by the Hölder Inequality) and $\chi - \Phi$ is a bounded linear functional vanishing on the class of all simple functions that are dense in L_p by Problem 5.2.2. Hence $\chi - \Phi = 0$ and

$$\Phi(h) = \int hg \, d\mu \quad \text{for all } h \in L_p.$$

It follows easily that $\| \Phi \| = \| g \|_q$. The uniqueness part is trivial.

Now to prove the theorem when μ is σ-finite, let $X = \bigcup_{n=1}^{\infty} X_n$, $X_n \subset X_{n+1}$, and $\mu(X_n) < \infty$ for each n. Then there exists a sequence of functions (g_n) such that $g_n \in L_q$, $g_n(x) = 0$ for $x \notin X_n$, and

$$\Phi(f) = \int f g_n \, d\mu,$$

where $f \in L_p$ and $f(x) = 0$ for $x \notin X_n$. Also $\| g_n \|_q \leq \| \Phi \|$. Because of the uniqueness of the g_n's (except for changes on sets of measure zero) we can assume

$$g_{n+1}(x) = g_n(x) \quad \text{if} \quad x \in X_n.$$

Let us define

$$g(x) = g_n(x) \quad \text{if} \quad x \in X_n.$$

Then by the Monotone Convergence Theorem,

$$\int | g |^q \, d\mu = \lim_{n\to\infty} \int | g_n |^q \, d\mu \leq \| \Phi \|^q,$$

and so $g \in L_q$. Now by the Lebesgue Convergence Theorem, if $f \in L_p$, then

$$\begin{aligned} \int f \cdot g \, d\mu &= \lim_{n\to\infty} \int_{X_n} f \cdot g \, d\mu \\ &= \lim_{n\to\infty} \int f \cdot \chi_{X_n} \cdot g_n \, d\mu \\ &= \lim_{n\to\infty} \Phi(f \cdot \chi_{X_n}) \\ &= \Phi(f). \end{aligned}$$

The rest is left to the reader. ∎

In the theorem above, σ-finiteness is necessary when $p = 1$. (Problem 5.5.1). But for $p > 1$, σ-finiteness is not necessary, as the following corollary shows.

Corollary 5.6. Let $(X, \mathscr{A}, \mu)$ be any measure space and Φ be a bounded linear functional on L_p, $1 < p < \infty$. Then there is a unique element $g \in L_q$, $1/p + 1/q = 1$ such that

$$\Phi(f) = \int f \cdot g \, d\mu \quad \text{for all } f \in L_p$$

and $\| \Phi \| = \| g \|_q$. ∎

Proof. From Theorem 5.14 it follows that for $A \in \mathscr{A}$, A σ-finite, there exists a unique $g_A \in L_q$, vanishing outside A such that

$$\Phi(f) = \int f \cdot g_A \, d\mu$$

for every $f \in L_p$, vanishing outside A. Clearly, because of uniqueness $A \subset B$ implies $g_A = g_B$ a.e. on A. Let us define $\nu(E)$ for E σ-finite by

$$\nu(E) = \int | g_E |^q \, d\mu \qquad [= \| g_E \|_q^q \leq \| \Phi \|^q].$$

Then let (E_n) be an increasing sequence of σ-finite sets in $\mathscr{A}$ such that $\lim_{n\to\infty} \nu(E_n) = \sup\{\nu(E)\colon E \in \mathscr{A},\ E\ \sigma\text{-finite}\}$. Denote this supremum by s. Letting $C = \bigcup_{n=1}^{\infty} E_n$, then $\nu(C) = s$. Let us define

$$g(x) = \begin{cases} g_C(x), & x \in C \\ 0, & x \notin C. \end{cases}$$

Then $g \in L_q$. If D is σ-finite, $D \in \mathscr{A}$ and $C \subset D$, then $g_D = g_C$ a.e. on C and $\int | g_D |^q \, d\mu = \nu(D) \leq s = \int | g_C |^q \, d\mu$. Hence $g_D = g$ a.e. in X. If $f \in L_p$, then $N = \{x\colon f(x) \neq 0\}$ is σ-finite; let $E = N \cup C \supset C$. Then $g_E = g$ a.e., as above. Hence $\Phi(f) = \int f \cdot g_E \, d\mu = \int f \cdot g \, d\mu$. The rest is left to the reader. ∎

Corollary 5.7. For $1 < p < \infty$, L_p^* is linearly isometric onto L_q, $1/p + 1/q = 1$. ∎

The proof immediately follows from Corollary 5.6.

Corollary 5.8. For $1 \leq p < \infty$, l_p^* is linearly isometric to l_q, $1/p + 1/q = 1$. ∎

The proof follows immediately from Theorem 5.14 and Problem 5.3.5, since l_p, $1 \leq p < \infty$, is a special case of L_p in the σ-finite measure space of integers with each integer having measure 1.

Corollary 5.9. For $1 < p < \infty$, L_p is reflexive. ▮

Proof. Let J be the natural map from L_p into L_p^{**}. To show that J is surjective, let $x^{**} \in L_p^{**}$. Let χ be the map from L_q into L_p^*, $1/p + 1/q = 1$, defined by

$$[\chi(g)](f) = \int f \cdot g \, d\mu, \qquad f \in L_p, \; g \in L_q.$$

Then by Corollary 5.6 χ is a surjective linear isometry. Let us define the map

$$x^* = x^{**} \circ \chi.$$

Then $x^* \in L_q^*$. Hence by Corollary 5.6, there exists $h \in L_p$ such that $x^*(g) = \int gh \, d\mu$, $g \in L_q$. Let $y^* \in L_p^*$. Then there exists $g_0 \in L_q$ such that $\chi(g_0) = y^*$. Now clearly

$$x^{**}(y^*) = x^*(g_0) = \int g_0 h \, d\mu = [\chi(g_0)](h) = y^*(h) = [J(h)](y^*).$$

Hence J is surjective. ▮

Corollary 5.10. For $1 < p < \infty$, l_p is reflexive. However, l_1 is *not* reflexive. ▮

Proof. The first part follows from Corollary 5.9. For the second part suppose l_1 is reflexive. Then l_1^{**}, being homeomorphic to l_1, is separable and therefore, by Proposition 5.9, l_1^* (and hence l_∞) is separable, which is a contradiction. ▮

We have seen above that the dual of L_p, $1 \leq p < \infty$, is (linearly isometric to) L_q in a σ-finite measure space. Unfortunately, such a representation does not hold for the bounded linear functionals on L_∞. Consider the Lebesgue measure space on [0, 1]. Let Φ be the linear functional on $C[0, 1] \subset L_\infty$ defined by $\Phi(f) = f(0)$. Let Φ_0 be the extension of Φ (possible by the Hahn–Banach Theorem) to L_∞. Then $\| \Phi_0 \| \geq \| \Phi \| = 1$. Suppose there is a $g \in L_1$ such that $\Phi_0(f) = \int_0^1 f(t)g(t)\, dt$ for all $f \in L_\infty$. Let us define $f_n \in C[0, 1]$ by

$$f_n(t) = \begin{cases} 0, & 1/n \leq t \leq 1 \\ 1 - nt, & 0 \leq t \leq 1/n. \end{cases}$$

Then $\int_0^1 f_n(t)g(t)\,dt \to 0$ as $n \to \infty$, but $\Phi_0(f_n) = f_n(0) = 1$, which is a contradiction.

From the above results it is clear that there are plenty of infinite-dimensional reflexive spaces and therefore development of a theory for such spaces is in order and naturally will be very useful.

So far we have explored to some extent the class of all bounded linear functionals on a normed linear space. The concept of reflexivity is based upon characterizing a class of *bounded* linear functionals. Why is it that we do not consider with equal interest the class of all *linear* (not necessarily bounded) functionals on a normed linear space? The next result sheds some light on this question. First we need a definition.

Definition 5.11. Let X be a vector space over F, and let X' be the vector space (under natural operations) of all linear functionals on X—called the *algebraic dual* of X. Let J be the natural map from X into X'' (the algebraic dual of X') defined by

$$[J(x)](x') = x'(x), \qquad x' \in X'.$$

Then if J is onto, X is called *algebraically reflexive.* ∎

Proposition 5.18. A vector space X over F is algebraically reflexive if and only if X is finite-dimensional. ∎

Proof. We will only prove the "only if" part. Let H be the Hamel basis of X. Suppose X is infinite dimensional. Then H is infinite and let $H = \{x_i : i \in I\}$, I being an infinite indexed set.

Let us define $x_i' \in X'$ by

$$x_i'(x_j) = \begin{cases} 1, & i = j \\ 0, & i \neq j \end{cases}$$

and

$$x_i'(\sum \alpha_j x_j) = \sum \alpha_j x_i'(x_j).$$

Then $A' = \{x_i' : i \in I\}$ is linearly independent in X'. (Why?). Let H' be a Hamel basis of X' containing A'. Let us define $x'' \in X''$ such that

$$x''(x_i') = \beta_i, \qquad i \in I$$
$$x''(x') = 0, \qquad x' \in H' - A',$$

where infinitely many of the β_i's are nonzero. If J is onto, then there exists

$x \in X$ such that $[J(x)]x' = x'(x) = x''(x')$ for all $x' \in X'$. However $x_i'(x) = x''(x_i')$ for all $i \in I$. If $i \in I$ is such that the term x_i is missing in the unique representation of x as a linear combination of elements of H, and $x''(x_i') = \beta_i \neq 0$, then $x_i'(x) = 0 \neq \beta_i = x''(x_i')$, which is a contradiction. ∎

We see therefore from Proposition 5.18 that the concept of algebraic reflexivity is not very useful since the study of algebraically reflexive spaces is nothing more than the study of finite-dimensional spaces.

Now let us reconsider the space c (see Problem 5.3.11) a subspace of l_∞. By Problem 5.3.11, c^* is linearly isometric onto l_1 and therefore c^{**} is linearly isometric onto l_1^*. (Why?) If c is reflexive, then c^{**} is separable since c is separable. But this contradicts the nonseparability of l_1^* (or l_∞). Hence c as well as c^* (or l_1) is *not* reflexive. Here arises a natural question: Does the reflexivity of a Banach space X imply that of X^*? The following proposition answers this.

Proposition 5.19. A Banach space X is reflexive if and only if X^* is reflexive. ∎

Proof. Suppose X is reflexive. To prove that the natural map J: $X^* \to X^{***}$ is onto, let $x^{***} \in X^{***}$. We define

$$x^* = x^{***} \circ J_X,$$

where J_X is the natural map from X onto X^{**}. Then $x^* \in X^*$ and $J(x^*)[J_X(x)] = J_X(x)[x^*] = x^*(x) = x^{***}[J_X(x)]$. Since $J_X(X) = X^{**}$, $x^{***} = J(x^*)$ and J is onto.

To prove the converse, let X^* be reflexive. If X is not reflexive, then $J_X(X)$ is a closed proper subspace of X^{**}. By Corollary 5.3. there exists $x^{***} \in X^{***}$ such that $x^{***}(x^{**}) \neq 0$ for some $x^{**} \in X^{**} - J_X(X)$ and $x^{***}[J_X(x)] = 0$ for each $x \in X$. Since $J(X^*) = X^{***}$, there exists $x^* \in X^*$ such that $J(x^*) = x^{***}$. Therefore, $0 = J(x^*)[J_X(x)] = J_X(x)[x^*] = x^*(x)$ for each $x \in X$, which means that $x^* = 0$; and therefore $x^{***} = 0$, which is a contradiction. ∎

We will not try to determine the space l_∞^* in this book, since this determination is nontrivial; and we will not need it in our discussion. However, knowing that c is *not* reflexive, we can assert that l_∞ is also *not* reflexive. The following proposition makes it possible.

Proposition 5.20. Every closed subspace of a reflexive Banach space is reflexive. ∎

Proof. Let Y be a closed subspace of a reflexive Banach space X. Let $J_Y: Y \to Y^{**}$ and $J_X: X \to X^{**}$ be the natural maps. Given $y^{**} \in Y^{**}$, we define $x^{**} \in X^{**}$ by $x^{**}(x^*) = y^{**}(x^*|_Y)$. Since J_X is onto, there exists $x \in X$ such that $J_X(x) = x^{**}$. We claim that $x \in Y$ and $J_Y(x) = y^{**}$. If $x \notin Y$, then by Corollary 5.3, there exists $x^* \in X^*$ such that $x^*(x) \neq 0$ and $x^*(y) = 0$ for each $y \in Y$. Then $x^{**}(x^*) = y^{**}(x^*|_Y) = 0$ or $J_X(x)[x^*] = x^*(x) = 0$, which is a contradiction. Hence $x \in Y$. The rest is left to the reader. ∎

Next we consider a very important nonreflexive Banach space—the space $C[a, b]$ of (real valued) continuous functions under the uniform (sup) norm. It is clear that for any function g of bounded variation on $[a, b]$

$$\varphi(f) = \int_a^b f(t)\, dg(t)$$

defines a bounded linear functional φ on $C[a, b]$. Actually we will see in what follows that *every* bounded linear functional on $C[a, b]$ has the above form. This fact was first discovered by F. Riesz in 1909. Later, in 1937, it was extended by Banach to the case of a compact metric space (instead of $[a, b]$) and then by Kakutani in 1941 to the case of a compact Hausdorff space (instead of $[a, b]$). Kakutani considered signed measures instead of functions of bounded variation. (See Chapter 7.)

Let us denote by $\mathrm{BV}[a, b]$ the Banach space of real functions of bounded variation on $[a, b]$ with the norm

$$\| g \| = V(g) + | g(a) |,$$

where $V(g)$ denotes the total variation of g on $[a, b]$. (The reader should convince him- or herself that $\mathrm{BV}[a, b]$ is a Banach space—a fact that is neither difficult nor trivial to verify.) We will denote by $B[a, b]$ the Banach space of all real bounded functions on $[a, b]$ with the usual supremum norm.

Theorem 5.15.[†] (*Riesz*) For every bounded linear functional Φ on $C[a, b]$ there is a function g of bounded variation such that for each $f \in C[a, b]$,

$$\Phi(f) = \int_a^b f(t)\, dg(t)$$

and $\| \Phi \| = V(g)$. ∎

[†] The reader may note that a similar result holds in $C_1[a, b]$ when the scalars are complex numbers.

Proof. Since $C[a, b] \subset B[a, b]$, by the Hahn–Banach Theorem there exists a continuous linear functional Φ_0 on $B[a, b]$ extending Φ such that $\| \Phi_0 \| = \| \Phi \|$. We define for $s \in (a, b]$

$$z_s(t) = \begin{cases} 1, & a \leq t \leq s \\ 0, & s < t \leq b \end{cases}$$

$$z_a(t) \equiv 0.$$

Define $g(s) = \Phi_0(z_s)$. We claim that $V(g) \leq \| \Phi_0 \| < \infty$. Clearly if

$$a = t_0 < t_1 < t_2 < \cdots < t_{n-1} < t_n = b$$

and

$$k_i = \operatorname{sgn}[g(t_i) - g(t_{i-1})], \qquad 1 \leq i \leq n,$$

then we have

$$\begin{aligned} \sum_{i=1}^{n} | g(t_i) - g(t_{i-1}) | &= \sum_{i=1}^{n} k_i[\Phi_0(z_{t_i}) - \Phi_0(z_{t_{i-1}})] \\ &= \Phi_0\Big(\sum_{i=1}^{n} k_i[z_{t_i} - z_{t_{i-1}}]\Big) \\ &\leq \| \Phi_0 \|, \end{aligned}$$

since the function inside the parenthesis has norm 1. Hence $V(g) \leq \| \Phi_0 \|$. Now to complete the proof of the theorem, let $f \in C[a, b]$. For $a = t_0 < t_1 < \cdots < t_{n-1} < t_n = b$, let $h(t) = \sum_{i=1}^{n} f(t_i)[z_{t_i}(t) - z_{t_{i-1}}(t)]$. Then we have

$$\begin{aligned} | h(t) - f(t) | &= | f(t_i) - f(t) |, \qquad t_{i-1} < t \leq t_i \\ &= | f(t_1) - f(t) |, \qquad t = a \end{aligned}$$

and

$$\Phi_0(h) = \sum_{i=1}^{n} f(t_i)[g(t_i) - g(t_{i-1})].$$

It is clear now that in the limit when $n \to \infty$ and $\max_{1 \leq i \leq n} | t_i - t_{i-1} | \to 0$,

$$\Phi(f) = \Phi_0(f) = \int_a^b f(t)\, dg(t).$$

Since $| \int_a^b f(t)\, dg(t) | \leq \| f \| \, V(g)$, $\| \Phi \| \leq V(g)$. The proof is complete. ∎

The above theorem does not provide a one-to-one correspondence between bounded linear functionals Φ on $C[a, b]$ and functions of bounded variation on $[a, b]$, as the following lemma shows.

Lemma 5.4. Let $g \in \mathrm{BV}[a, b]$. Let h be defined by

$$h(t) = \begin{cases} g(t+0) - g(a), & a < t < b \\ g(b) - g(a), & t = b \\ 0, & t = a. \end{cases}$$

Then $h \in \mathrm{BV}[a, b]$ and for each $f \in C[a, b]$

$$\int_a^b f(t)\, dg(t) = \int_a^b f(t)\, dh(t)$$

and $V(h) \leq V(g)$. ∎

The proof of this lemma is left to the reader.

To provide a one-to-one correspondence between $C^*[a, b]$ and a suitable subspace of $\mathrm{BV}[a, b]$, we need the following definition.

Definition 5.12. A function $g \in \mathrm{BV}[a, b]$ is called *normalized* if $g(a) = 0$ and $g(t+0) = g(t)$, $a < t < b$. ∎

The collection of normalized functions of bounded variation on $[a, b]$ is denoted by $\mathrm{NBV}[a, b]$. The next two results will show that $\mathrm{NBV}[a, b]$ will provide us with the one-to-one correspondence desired above. We need another lemma.

Lemma 5.5. Let $g \in \mathrm{BV}[a, b]$ such that $\int_a^b f(t)\, dg(t) = 0$ whenever $f \in C[a, b]$. Then $g(a) = g(b)$ and for $a < t < b$, $g(t-0) = g(t+0) = g(a)$. ∎

Proof. Clearly if $f \equiv 1$, then $\int_a^b dg(t) = 0$ and therefore $g(a) = g(b)$. For $a \leq c < b$ and $0 < h < b - c$, let us define

$$f(t) = \begin{cases} 1, & a \leq t \leq c \\ 1 - (t-c)/h, & c \leq t \leq c+h \\ 0, & c+h \leq t \leq b. \end{cases}$$

Then $f \in C[a, b]$ and

$$0 = \int_a^b f(t)\, dg(t) = g(c) - g(a) + \int_c^{c+h} f(t)\, dg(t).$$

Integrating by parts and simplifying,

$$0 = -g(a) + \frac{1}{h} \int_c^{c+h} g(t)\, dt, \qquad 0 < h < b - c$$

and therefore by letting $h \to 0$, $g(c + 0) = g(a)$. Similarly, $g(c - 0) = g(a)$ for $a < c < b$. ∎

Theorem 5.16. The dual space of $C[a, b]$ is linearly isometric onto NBV$[a, b]$. ∎

Proof. For $\Phi \in C^*[a, b]$, by Theorem 5.15 we can find $g \in \text{BV}[a, b]$ such that for each $f \in C[a, b]$ we have

$$\Phi(f) = \int_a^b f(t)\, dg(t),$$

and $\| \Phi \| = V(g)$. Let h be as defined in Lemma 5.4. Then $h \in \text{NBV}[a, b]$, $V(h) \leq V(g)$ and for $f \in C[a, b]$

$$\Phi(f) = \int_a^b f(t)\, dh(t).$$

Then since $\| \Phi \| \leq V(h)$, $V(h) = \| \Phi \|$. If we define $T(\Phi) = h$, then T is a linear isometry from $C^*[a, b]$ *onto* NBV$[a, b]$. That T is well defined (that is, there is a unique $h \in \text{NBV}[a, b]$ with the above properties) is guaranteed by Lemma 5.5. ∎

Finally in this section, we wish to consider a new notion of convergence of a sequence of vectors in a normed linear space—called weak convergence. The reason for considering this topic is that a very important interplay exists between reflexivity and weak convergence, as will be clear in what follows.

Definition 5.13. A sequence (x_n) in a normed linear space X is said to converge *weakly* to $x \in X$ if for every $x^* \in X^*$, $x^*(x_n) \to x^*(x)$ as $n \to \infty$. ∎

Remarks

5.10. If $x_n \to x$ in X, then $x_n \xrightarrow{w} x$, i.e., x_n converges weakly to x.

5.11. The converse to Remark 5.10 need not be true. For instance, if $X = l_p$ $(1 < p < \infty)$ and $x_n = (0, 0, \ldots, 0, \underset{n}{1}, 0, \ldots)$, then $\| x_n \|_p = 1$ for each n while $x_n \xrightarrow{w} 0$. [Recall that for each $x^* \in X^*$ there is $b = (b_1, b_2, \ldots) \in l_q$, $1/p + 1/q = 1$ such that $x^*(x_n) = b_n$.]

5.12. If $x_n \xrightarrow{w} x$, then $\sup_{1 \leq n < \infty} \| x_n \| < \infty$. The reason is that if $J(x_n)[x^*] = x^*(x_n)$, then $[J(x_n)]_{n=1}^{\infty}$ becomes a pointwise bounded family of bounded linear operators from X^* into F; and therefore by the Principle of Uniform Boundedness $\sup_{1 \leq n < \infty} \| J(x_n) \| = \sup_{1 \leq n < \infty} \| x_n \| < \infty$.

5.13. If X is finite dimensional, then $x_n \xrightarrow{w} x$ if and only if $x_n \to x$. See Problem 5.5.3.

5.14. If a sequence (f_n) in $C[a, b]$ converges weakly to $f \in C[a, b]$, then the sequence is uniformly bounded and for $a \leq t \leq b$, $\lim_{n\to\infty} f_n(t) = f(t)$. This is because of Remark 5.12 above and because $y_t^*(f_n) \to y_t^*(f)$, where $y_t^* \in C^*[a, b]$ is defined by $y_t^*(h) = h(t)$.

5.15. If $x_n = (x_{n1}, x_{n2}, \ldots) \in l_p$, $z = (z_1, z_2, \ldots) \in l_p$ with $1 < p < \infty$, then $x_n \xrightarrow{w} z$ if and only if $\lim_{n\to\infty} x_{ni} = z_i$, $1 \leq i < \infty$, and $\sup_{1 \leq n < \infty} \times \| x_n \|_p < \infty$. See Problem 5.5.4.

5.16. A normed linear space X is called *weakly sequentially complete* if every weak Cauchy sequence $(x_n) \in X$ [that is, $(x^*(x_n))$ is Cauchy for each $x^* \in X^*$] converges weakly to some element $x \in X$. A reflexive Banach space X is weakly sequentially complete. To prove this, let (x_n) be a weak Cauchy sequence in X. Let $T(x^*) = \lim_{n\to\infty} x^*(x_n)$. Then T is linear; and by Remark 5.12 above, $\| T \| \leq \sup_{1 \leq n < \infty} \| x_n \| < \infty$ and therefore $T \in X^{**}$. If X is reflexive, there exists $x \in X$ such that $J(x) = T$, J being the usual natural map from X onto X^{**}. Hence $\lim_{n\to\infty} x^*(x_n) = x^*(x)$.

5.17. $C[0, 1]$ is *not* weakly sequentially complete and hence *not* reflexive. Consider $x_n(t) = (1 - t)^n$. Then $x_n(t)$ cannot converge pointwise to a continuous function and hence by Remark 5.14 cannot converge weakly. But for $y^* \in C^*[0, 1]$, there exists $g \in \mathrm{NBV}[0, 1]$ such that for $z_{mn} = x_m - x_n$

$$y^*(z_{mn}) = \int_0^1 z_{mn}(t)\, dg(t) = \int_0^{t_1} z_{mn}(t)\, dg^*(t) + \int_{t_1}^1 z_{mn}(t)\, dg(t),$$

where $g^*(t) = g(t + 0)$. The first integral can be made arbitrarily small by taking t_1 (>0) sufficiently close to 0, $g^*(t)$ being continuous from the right; and the second integral can be made arbitrarily small {since $x_n(t) \to 0$ uniformly in $[t_1, 1]$} by taking n and m sufficiently large. This means that (x_n) is weakly Cauchy in $C[0, 1]$.

5.18. In a reflexive Banach space X, each bounded (that is, norm bounded) sequence has a weakly convergent subsequence.

To prove this, let (x_n) be a sequence in X such that $\sup_{1 \leq n < \infty} \| x_n \| < \infty$. Consider the closed linear subspace Y spanned by the x_n's. Then Y is separable. Y is also reflexive by Proposition 5.20. Therefore Y^{**} is separable, and by Proposition 5.9 Y^* is also separable. Let $(y_m^*)_{m \in N}$ be dense in Y^*. Since the sequence $y_1^*(x_n)$ is bounded, there exists a subsequence $(x_{1,n})_{n \in N}$ such that $y_1^*(x_{1,n})$ converges as $n \to \infty$. Similarly, there exists a subsequence $(x_{2,n}) \subset (x_{1,n})$ such that $y_2^*(x_{2,n})$ converges as $n \to \infty$. Using induction we can find for each positive integer k a subsequence $(x_{k+1,n}) \subset (x_{k,n})$ such that for each i, $1 \leq i \leq k+1$, $y_i^*(x_{k+1,n})$ converges as $n \to \infty$. If $(x_{nn})_{n=1}^{\infty}$ is the diagonal sequence, then it is easy to show that for each i, $y_i^*(x_{nn})$ converges; and therefore for each $y^* \in Y^*$, $y^*(x_{nn})$ converges. We define

$$y^{**}(y^*) = \lim_{n \to \infty} y^*(x_{nn}).$$

Then $y^{**} \in Y^{**}$, since $\sup_{1 \leq n < \infty} \| x_n \| < \infty$. Since Y is reflexive, there exists $y \in Y$ such that $y^*(y) = y^{**}(y^*)$ whenever $y^* \in Y^*$. Hence $x_{nn} \xrightarrow{w} y$.

5.19. It is clear from the previous remark that in a *reflexive* Banach space if the weak convergence of a *sequence* implies its convergence, then its closed unit ball must be compact, and consequently the space must be finite dimensional. This is not true in a nonreflexive space, as the following remark shows.

5.20. In l_1, a nonreflexive space, the weak convergence of a sequence implies its (strong) convergence. If this were not the case, there would be an $\varepsilon > 0$ and a sequence of elements $x_n = (x_1{}^n, x_2{}^n, x_3{}^n, \ldots)$ in l_1 such that for each $(w_1, w_2, \ldots) \in l_\infty$, we have

$$\sum_{j=1}^{\infty} w_j x_j{}^n \to 0 \qquad \text{as } n \to \infty$$

and (5.3)

$$\sum_{j=1}^{\infty} | x_j{}^n | > \varepsilon, \qquad 1 \leq n < \infty.$$

Choosing the w_k's properly, it follows easily that for each k, $\lim_{n\to\infty} x_k{}^n = 0$. Let $m_0 = n_0 = 1$. Then the sequence (m_k, n_k) is defined inductively as follows: n_k is the smallest integer $n > n_{k-1}$ such that

$$\sum_{j=1}^{m_{k-1}} | x_j{}^{n_k} | < \frac{\varepsilon}{5}, \tag{5.4}$$

and m_k is the smallest integer $m > m_{k-1}$ such that

$$\sum_{j=m_k}^{\infty} | x_j{}^{n_k} | < \frac{\varepsilon}{5}. \tag{5.5}$$

We define $w = (w_1, w_2, \ldots)$ by

$$w_j = \begin{cases} \operatorname{sgn} x_j{}^{n_k}, & m_{k-1} \leq j < m_k \\ \operatorname{sgn} x_j{}^{n_{k+1}}, & m_k \leq j < m_{k+1}, \text{ etc.} \end{cases}$$

Then by inequalities (5.4) and (5.5) we have

$$\left| \sum_{j=1}^{\infty} w_j x_j^{n_k} - \sum_{j=1}^{\infty} | x_j^{n_k} | \right| \leq 2 \sum_{j=1}^{m_{k-1}} | x_j^{n_k} | + 2 \sum_{j=m_k}^{\infty} | x_j^{n_k} | < \frac{4\varepsilon}{5}.$$

Therefore it follows from (5.3) that for each k,

$$\left| \sum_{j=1}^{\infty} w_j x_j^{n_k} \right| \geq \frac{\varepsilon}{5},$$

which contradicts the first statement in (5.3).

From the above remarks the reader has some ideas about weak convergence in a normed linear space and its connection with reflexivity of the space. This connection will be more distinct if we study what is called the weak topology of a normed linear space X.

The *weak topology* of X is the weakest topology on X such that every element in X^* is continuous. Clearly a basis for the weak topology on X consists of sets of the form

$$\{x: | f_i(x) - f_i(x_0) | < \varepsilon, \qquad i = 1, 2, \ldots, n\},$$

where $x_0 \in X$, $\varepsilon > 0$, and $f_i \in X^*$. Then it is clear that the weak topology is contained in the metric topology of X so that every weakly closed set is strongly closed.

That the converse is not true is clear from Remark 5.11. However, the following proposition holds.

Proposition 5.21. A linear subspace Y is weakly closed if and only if it is strongly closed. ∎

Proof. Suppose Y is strongly closed and $y \notin Y$. Then $\inf_{z \in Y} \| y - z \| > \delta > 0$. Hence by Corollary 5.3. there is $x^* \in X^*$ such that $x^*(y) \neq 0$ and $x^*(z) = 0$ for each $z \in Y$. This means that $A = \{w \in X: | x^*(w) | > 0\}$ is an open set in the weak topology, but $A \cap Y$ is empty. Hence y is not a weak-closure point of Y. The rest is clear. ∎

In Remark 5.20 we have seen that in l_1 weak convergence of a sequence is equivalent to its strong convergence. However, the topology of a topological space is not determined by the concept of convergence of a sequence unless the space is first countable. We will see that in l_1, as well as in any infinite-dimensional normed linear space, the weak topology is *properly* contained in the strong topology. The following theorem demonstrates this.

Theorem 5.17. The weak topology of a normed linear space coincides with its strong topology if and only if the space is finite dimensional. ∎

Proof. We prove only the "only if" part. We will prove that if the open unit ball in X is weakly open, then X^* (and therefore X) is finite dimensional.

Suppose $S = \{x \in X: \| x \| < 1\}$ is weakly open. Then there exist $x_1^*, \ldots, x_n^*$ in X^* and positive real numbers $r_1, \ldots, r_n$ such that

$$\{x \in X: | x_i^*(x) | < r_i, \quad 1 \leq i \leq n\} \subset S.$$

If $\bigcap_{i=1}^n \{x \in X: x_i^*(x) = 0\}$ ($= A$, say) contains x_0, then for every real number r, $rx_0 \in A \subset S$. This implies that $x_0 = 0$, so $A = \{0\}$. We claim that $\{x_1^*, \ldots, x_n^*\}$ spans X^*. To prove this let $x^* \in X^*$. We define for $x \in X$, $T(x) = (x_1^*(x), \ldots, x_n^*(x)) \subset F^n$, where F is the scalar field. We also define h from $T(X)$ into F by $h(T(x)) = x^*(x)$. Then h is well defined since $T(x) = T(y)$ implies $x - y \in A = \{0\}$ or $x = y$. Since h is a linear functional on $T(X) \subset F^n$ and $T(X)$ [and therefore $T(X)^*$] is finite dimensional, we may assume that there exist $h_1, h_2, \ldots, h_m$ $(1 \leq m \leq n)$ in $T(X)^*$ such that $h_i(t_i, t_2, \ldots, t_n) = t_i$, $1 \leq i \leq m$ and $h = \sum_{i=1}^m a_i h_i$, $a_i \in F$. Then $x^* = \sum_{i=1}^m a_i x_i^*$. Hence X^* is finite dimensional. ∎

Before we close this section, we will present characterizations of a reflexive Banach space in terms of the weak compactness and also weak sequential compactness of its closed unit ball. We will do this by introducing another useful concept called the weak* topology, a topology of X^*.

We know that the weak topology in X^* is the weakest topology in X^* such that each element in X^{**} is continuous. However, this topology turns out to be less useful than the topology in X^* generated by the elements in $J(X)$, J being the natural map from X into X^{**}. This latter topology is called the weak* topology for X^* and is clearly weaker than its weak topology. A base for the weak* topology is given by the sets of the form

$$\{f \in X^*: | f(x_i) - f_0(x_i) | < \varepsilon, \quad i = 1, \ldots, n\},$$

where $x_1, x_2, \ldots, x_n \in X$, $\varepsilon > 0$, and $f_0 \in X^*$.

If X is reflexive, then $J(X) = X^{**}$ and therefore the weak topology for X^* and its weak* topology coincide. The usefulness of the weak* topology stems mainly from the following basic theorem. A sequential form of this important theorem for separable Banach spaces was proved by Banach in 1932. Alaoglu proved the theorem in the following general form in 1940.

Theorem 5.18. *The Banach–Alaoglu Theorem.* The closed unit ball in X^* is compact in its weak* topology. ▮

Proof. Let $S^* = \{f \in X^*: \| f \| \leq 1\}$. If $f \in S^*$, then $f(x) \in \{c \in F: | c | \leq \| x \|\} = I_x$, say. Then we can think of S^* as a subset of $P = \prod_{x \in X} I_x$, which is the set of all functions f on X with $f(x) \in I_x$, given the usual product topology. The topology which S^* inherits as a subset of P is the weak* topology of S^*. Since P is compact by Tychonoff's theorem (Theorem 1.5), S^* will be compact if it is closed as a subset of P.

Let f be a point of closure of S^* in P. Then $f: X \to F$ and $| f(x) | \leq \| x \|$. Now for $x, y \in X$ and $\alpha, \beta \in F$, the set

$$V = \{g \in P: | g(x) - f(x) | < \varepsilon, \qquad | g(y) - f(y) | < \varepsilon,$$

and

$$| g(\alpha x + \beta y) - f(\alpha x + \beta y) | < \varepsilon\}$$

is an open subset of P containing f and hence $V \cap S^* \neq \varnothing$. Since for $g \in V \cap S^*$, g is linear, $| f(\alpha x + \beta y) - \alpha f(x) - \beta f(y) | < \varepsilon (1 + | \alpha | + | \beta |)$. Since this inequality holds for every $\varepsilon > 0$, f is linear and therefore $f \in S^*$. ▮

If X is reflexive, then X and X^{**} can be identified, and therefore the weak topology on X can be regarded as the weak* topology of X^{**}. Hence by Theorem 5.18 the closed unit ball in X is weakly compact. The converse is also true. To prove this we need the following lemma.

Lemma 5.6. Let $S \subset X$ and $S^{**} \subset X^{**}$ be defined as follows:

$$S = \{x: \| x \| \leq 1\} \quad \text{and} \quad S^{**} = \{x^{**}: \| x^{**} \| \leq 1\}.$$

Then $J(S)$ is dense in S^{**} with the weak* topology of X^{**}, where J is the natural map from X into X^{**}. ▮

Proof. Let $x_0^{**} \in S^{**}$, $\varepsilon > 0$, and $x_1^*, \ldots, x_n^* \in X^*$. The lemma will be proved if we can find $x_0 \in S$ such that

$$| x_i^*(x_0) - x_0^{**}(x_i^*) | < \varepsilon, \qquad i = 1, 2, \ldots, n. \tag{5.6}$$

Let $Y = \cap \{x \in X: x_i^*(x) = 0, i = 1, 2, \ldots, n\}$. Then Y is a closed subspace of X. Let $Y^\circ = \{x^* \in X^*: x^*(y) = 0 \text{ for each } y \in Y\}$. Then by an argument similar to that used in the proof of Theorem 5.17 (see also Problem 5.5.7), Y° is the closed subspace of X^* spanned by $x_1^*, \ldots, x_n^*$. Consider the mapping $T: (X/Y)^* \to Y^\circ$ by $T(V^*) = x^*$, where $x^*(x) = V^*([x])$. T is well defined and an onto linear isometry. (The reader can easily verify this.) Hence $(X/Y)^*$ (and therefore X/Y) is finite dimensional. Define $V^{**} \in (X/Y)^{**}$ by $V^{**}(V^*) = x_0^{**} \circ T(V^*)$. Then $\| V^{**} \| \leq 1$ since T is an isometry. Since X/Y is reflexive (being finite dimensional), there exists $V \in X/Y$ such that $V^{**}(V^*) = V^*(V)$ for each $V^* \in (X/Y)^*$ and $\| V \| = \| V^{**} \| \leq 1$. Let us choose k such that $\sup_{1 \leq i \leq n} \| x_i^* \| < k$. Then there exists $x \in V$ such that

$$\| x \| \leq \| V \| + \varepsilon/k \leq 1 + \varepsilon/k,$$

and $V^*(V) = x^*(x)$, where $T(V^*) = x^*$. This means that for each $x^* \in Y^\circ$, $x^*(x) = x_0^{**}(x^*)$. If $x_0 = [k/(k + \varepsilon)] \cdot x$, then $\| x_0 \| \leq 1$; and for $x^* \in Y^\circ$,

$$\begin{aligned} | x^*(x_0) - x_0^{**}(x^*) | &= | x^*(x_0) - x^*(x) | \\ &\leq \| x^* \| \, \| x_0 - x \| < \varepsilon/k \cdot \| x^* \|. \end{aligned}$$

It follows that x_0 satisfies the inequalities (5.6). ∎

Theorem 5.19. *X is reflexive if and only if the closed unit ball in X is weakly compact.* ∎

Proof. The "only if" part follows easily from the Banach–Alaoglu Theorem. To prove the "if" part we see that the natural map J from X (with weak topology) into X^{**} (with its weak* topology) is a linear homeomorphism onto $J(X) \subset X^{**}$. Hence if the closed unit ball S of X is weakly compact, $J(S)$ is compact (and hence closed) in the weak* topology of X^{**}. By Lemma 5.6, $J(S) = S^{**} = \{x^{**} \in X^{**}: \| x^{**} \| \leq 1$. Since J is linear, $J(X) = X^{**}$. ∎

Our next theorem in this section is the well-known *Eberlein–Šmulian Theorem*, one of the most remarkable results in Banach spaces. In 1940, V. L. Šmulian proved that the weak countable compactness of the weak closure of a subset A of a Banach space X implies its weak sequential compactness. W. F. Eberlein, in 1947, proved that the subset A is weakly compact if and only if it is weakly closed and weakly sequentially compact. Note that these results are nontrivial since the weak topology need not be even first countable. It is relevant here only to mention (without proof) at this point that the weak topology of the closed unit ball of a Banach space X is a metric topology if and only if X^* is separable.

Definition 5.14. A set $A \subset X^*$, where X is a Banach space, is called *total* if whenever $x^*(x) = 0$ for every $x^* \in A$, then $x = 0$. ∎

If X is a separable Banach space, then X^* contains a *countable total* set; for if (x_n) is a dense sequence in $\{x: \| x \| = 1\}$ and $x_n^* \in X^*$ with $x_n^*(x_n) = \| x_n \| = 1$, then (x_n^*) is a *total set* and for all x in X, $\| x \| = \sup_n | x_n^*(x) |$.

The proof of the Eberlein–Šmulian Theorem that we present here is due to R. Whitley. First we need a lemma.

Lemma 5.7. Let X be a Banach space such that X^* contains a countable total set. Then the weak topology on a weakly compact subset of X is *metrizable.* ∎

Proof. Let (x_n^*) be total and $\| x_n^* \| = 1$. Let

$$d(x, y) = \sum_{n=1}^{\infty} \frac{1}{2^n} | x_n^*(x - y) |$$

and let A be a weakly compact subset of X. By Remark 5.3, A is bounded since $x^*(A)$ is compact for $x^* \in X^*$. Let i be the identity map from A (with weak topology) onto A (with the metric topology induced by d). i is clearly continuous and therefore a homeomorphism; for if B is a weakly closed subset of A, then B is weakly compact and $i(B)$ is compact in the metric d and therefore closed. ∎

Corollary 5.11. Let $A \subset X$, a Banach space. Let B, the weak closure of A, be compact in the weak topology. Then B is also sequentially compact in the weak topology. ∎

Proof. Let (a_n) be a sequence from B, and let $\overline{\text{sp}}(a_n)$ be the (norm) closure of the linear subspace spanned† by (a_n). By Proposition 5.21 $\overline{\text{sp}}(a_n)$ is weakly closed and therefore $B \cap \overline{\text{sp}}(a_n)$ is a weakly compact subset of the separable Banach space $\overline{\text{sp}}(a_n)$. By Lemma 5.7, the weak topology on $B \cap \overline{\text{sp}}(a_n)$ is metrizable and therefore sequentially compact by Proposition 1.22 of Chapter 1. The rest is clear. ∎

Theorem 5.20. *The Eberlein–Šmulian Theorem.* The following are equivalent for any subset A of a Banach space X:

† Sometimes the subspace spanned by a set E is also denoted by $[E]$.

(a) The weak closure of A is weakly compact.
(b) Any sequence in A has a weakly convergent (in X) subsequence.
(c) Every countable infinite subset of A has a weak-limit point in X. ▌

Proof. (a) ⇒ (b) by Corollary 5.11, and (b) ⇒ (c) trivially. We establish only (c) ⇒ (a). So we assume (c). Since $x^*(A)$ is a bounded set of scalars for each $x^* \in X^*$, by Remark 5.3 A is bounded. If J is the natural map of X into X^{**}, $J(A)$ is bounded; and therefore by Theorem 5.18, $w^*(J(A))$, the weak* closure of $J(A)$, is compact in the weak* topology of X^{**} (i.e., the topology induced by X^*). Since J is a homeomorphism from X (with the weak topology) onto $J(X)$ (with the weak* topology), it is sufficient to show that $w^*(J(A)) \subset J(X)$.

To show this, let $x^{**} \in w^*(J(A))$. We will use induction. Let $x_1^* \in X^*$ with $\| x_1^* \| = 1$. Now there is $a_1 \in A$ with $| (x^{**} - J(a_1))(x_1^*) | < 1$. Let E_2 be the finite-dimensional subspace spanned by x^{**} and $x^{**} - J(a_1)$. Since the surface of the closed unit ball in E_2 is compact, there are $(y_i^{**})_{i=2}^{n_2} \in E_2$ with $\| y_i^{**} \| = 1$ such that for any $y^{**} \in E_2$ and $\| y^{**} \| = 1$, $\| y^{**} - y_i^{**} \| < \frac{1}{4}$ for some i. Let $x_i^* \in X^*$, $\| x_i^* \| = 1$ be such that $y_i^{**}(x_i^*) > \frac{3}{4}$, $2 \leq i \leq n_2$. Then for every $y^{**} \in E_2$, we have

$$\max\{| y^{**}(x_i^*) | : 2 \leq i \leq n_2\} \geq \tfrac{1}{2} \| y^{**} \|.$$

Again there exists $a_2 \in A$ so that

$$\max\{|[x^{**} - J(a_2)](x_i^*) | : 1 \leq i \leq n_2\} < \tfrac{1}{2}.$$

Then we consider the space E_3 spanned by x^{**}, $x^{**} - J(a_1)$, and $x^{**} - J(a_2)$; we find $(x_i^*)_{i=n_2+1}^{n_3}$ and then choose $a_3 \in A$ as before so that for every $y^{**} \in E_3$, we have

$$\max\{| y^{**}(x_i^*) | : n_2 < i \leq n_3\} \geq \tfrac{1}{2} \| y^{**} \|$$

and

$$\max\{| [x^{**} - J(a_3)](x_i^*) | : 1 \leq i \leq n_3\} < \tfrac{1}{3}.$$

In this way, we continue to construct the sequence a_n.

By (c), the sequence (a_n) obtained above has a weak-limit point x. Clearly $x \in \overline{\text{sp}}(a_n)$ and so $x^{**} - J(x)$ is in the space $\overline{\text{sp}}(x^{**}, x^{**} - J(a_n)$ for $1 \leq n < \infty)$. Therefore by the construction of (a_n) above, we have

$$\sup_{1 \leq i < \infty} \{|(x^{**} - J(x))(x_i^*)|\} \geq \tfrac{1}{2} \| x^{**} - J(x) \|. \tag{5.7}$$

Also

$$\begin{aligned} | [x^{**} - J(x)](x_i^*) | &\leq | [x^{**} - J(a_p)](x_i^*) | + | x_i^*(a_p - x) | \\ &\leq 1/p + | x_i^*(a_p - x) | \text{ for } i \leq n_p. \end{aligned}$$

Since x is a weak-limit point of (a_n), it follows that $[x^{**} - J(x)](x_i^*) = 0$ for all i. By inequality (5.7), $x^{**} = J(x)$. The proof is complete. ∎

Remark 5.21. By Theorems 5.19 and 5.20, a Banach space X is reflexive if and only if its closed unit ball is weakly sequentially compact. This is true even for any normed linear space (Problem 5.5.4).

An application of the Eberlein–Šmulian Theorem characterizes weakly compact subsets of $C[0, 1]$. (See Problem 5.5.13.) The Banach–Alaoglu Theorem has already been utilized in obtaining Theorems 5.19 and 5.20. Another application of this theorem is outlined in Problem 5.5.14. In what follows, we present still another application of this theorem in the context of weak*-sequential compactness and then use this result to solve a problem in harmonic analysis.

First, we prove the following useful result.

• **Theorem 5.21.** Let X be a normed linear space and $S^* = \{f \in X^*: \|f\| \leq 1\}$. Then X is separable if and only if the weak* topology on X^* restricted to S^* is a metric topology. ∎

Proof. Suppose X is separable. Let $A \subset X$ be a countable dense subset of X. Then the weak* topology restricted to S^* is the topology on S^* that is induced by A; i.e., as in the proof of Theorem 5.18, the topology induced on S^* is a subset of the set $P_A = \prod_{x \in A} I_x$, where I_x is the set of all scalars c with $| c | \leq \| x \|$, with product topology. Since P_A, as a countable product of metric spaces, is metrizable, the "only if" part of the theorem follows.

To prove the "if" part, we assume that the weak* topology on S^* is metrizable and therefore has a countable local base at 0. Then there exist real numbers r_n and finite subsets $A_n \subset X$ such that

$$\{0\} = \bigcap_{n=1}^{\infty} W_n, \qquad W_n = \{x^* \in S^*: | x^*(x) | < r_n, \ x \in A_n\}.$$

Let $A = \bigcup_{n=1}^{\infty} A_n$. It is clear that $x^* = 0$ whenever $x^* \in S^*$ and $x^*(x) = 0$ for every $x \in A$. Hence, by an application of the Hahn–Banach Theorem, it follows that the closed linear subspace spanned by A is X. Since A is countable, X is separable. ∎

In a metric space, compactness and sequential compactness are equivalent and therefore an application of the Banach–Alaoglu Theorem leads to the following.

• **Corollary 5.12.** In a separable normed linear space X, the closed unit ball in X^* is weak*-sequentially compact. ∎

We will now close this section with an application to a problem of harmonic analysis. The problem is to characterize all those operators in $L(X, Y)$, $X = L_1(R)$ and $Y = L_p(R)$ where $1 < p \leq \infty$, which commute with convolution. Note that we have already introduced the notion of convolution in Problem 5.2.15. We recall that for $f \in L_1$ and $g \in L_p$ $(1 \leq p \leq \infty)$, $f * g$ is the convolution of f and g, and

$$f * g(x) = \int f(x - y)g(y)\, dy.$$

• **Theorem 5.22.** Let $T \in L(X, Y)$, where $X = L_1(R)$, $Y = L_p(R)$ and $1 < p \leq \infty$. Then the following are equivalent:

(i) For f and g in L_1, $T(f * g) = T(f) * g$.

(ii) There exists h in L_p such that for f in L_1, $T(f) = h * f$. ∎

Proof. Since (ii) ⇒ (i) by Problem 5.2.15, we prove only that (i) ⇒ (ii). We assume (i) and define the sequence

$$u_n(x) = (n/2)\chi_{[-1/n, 1/n]}(x), \qquad n = 1, 2, 3, \ldots.$$

Then $\| u_n \|_1 = 1$ and for any $f \in C_c(R)$ (and so for $f \in L_1$),

$$\lim_{n\to\infty} \| u_n * f - f \|_1 = 0.$$

Hence

$$\lim_{n\to\infty} \| T(f) - T(u_n) * f \|_p = \lim_{n\to\infty} \| T(f) - T(u_n * f) \|_p$$
$$\leq \| T \| \cdot \lim_{n\to\infty} \| f - u_n * f \|_1 = 0.$$

Since the sequence $T(u_n)$ is bounded in L_p-norm, it follows (after considering L_p as $L_q{}^*$, $pq = p + q$) by Corollary 5.12 that some subsequence $T(u_{n_i})$ converges to some h in L_p in the weak* topology of L_p. This means that for

$f \in L_1$ and $g \in L_q$, we have

$$\begin{aligned}\int T(f)(x)g(-x)\,dx &= \lim_{i\to\infty} \int T(u_{n_i}) * f(x)g(-x)\,dx \\ &= \lim_{i\to\infty} \{[T(u_{n_i}) * f] * g\}(0) \\ &= \lim_{i\to\infty} [T(u_{n_i}) * (f * g)](0) \\ &= \lim_{i\to\infty} \int T(u_{n_i})(x) f * g(-x)\,dx \\ &= \int h(x) f * g(-x)\,dx \\ &= [(h * f) * g](0) \\ &= \int (h * f)(x)g(-x)\,dx.\end{aligned}$$

It follows easily that $T(f) = h * f$ whenever $f \in L_1$. We leave the details to the reader. ∎

Problems

5.5.1. Let $X = \{x_1, x_2\}$ and μ be a measure on 2^X such that $\mu(\{x_1\}) = 1$ and $\mu(\{x_2\}) = \infty$. Show that dim $L_1(\mu) =$ dim $L_1{}^*(\mu) = 1$, whereas dim $L_\infty(\mu) =$ dim $L_\infty{}^*(\mu) = 2$.

5.5.2. Prove that BV$[a, b]$ is a Banach space under the norm $\| g \| = V(g) + g(a)$. Prove Lemma 5.4.

5.5.3. Show that in a finite-dimensional normed linear space a sequence is convergent if and only if it is weakly convergent.

5.5.4. Prove Remarks 5.15 and 5.21.

5.5.5. Let f be a real-valued measurable function in a σ-finite measure space such that for all g in L_p $(1 \leq p < \infty)$, $f \cdot g \in L_1$. Show that $f \in L_q$ where $1/p + 1/q = 1$. What happens when the measure is semifinite? (Hint: Write $X = \cup X_n$, $X_n \subset X_{n+1}$ and $\mu(X_n) < \infty$. Let $f_n(x) = \chi_{X_n}(x) \cdot \inf\{| f(x) |, n\}$. Define $T_n(g) = \int f_n g\, d\mu$. Use the uniform boundedness principle.)

5.5.6. Let S be a linear subspace of $C[0, 1]$ which is closed as a subspace of $L_2[0, 1]$. Show that S is finite-dimensional. (Hint: Show that S is closed as a subspace of $C[0, 1]$ and that therefore there exists $k > 0$ such that $\| f \|_\infty \leq k \| f \|_2$ for all f in S. Use this to show that the closed unit ball of S in L_2 is compact.)

5.5.7. Let $g, f_1, f_2, \ldots, f_n$ be linear functionals on a vector space X such that $\bigcap_{i=1}^{n}\{x: f_i(x) = 0\} \subset \{x: g(x) = 0\}$. Show that g is a linear combination of the f_i. [Hint: Consider the mapping $(g(x), f_1(x), \ldots, f_n(x)) \to (f_1(x), \ldots, f_n(x))$, which is injective.]

5.5.8. Show that if X is reflexive and separable, then so is X^*.

5.5.9. Show that if

$$\begin{aligned} f_n(t) &= nt, && 0 \leq t \leq 1/n \\ &= 2 - nt, && 1/n \leq t \leq 2/n \\ &= 0, && 2/n \leq t \leq 1, \end{aligned}$$

then $f_n \xrightarrow{w} 0$ in $C[0, 1]$; but $f_n(t) \not\to 0$ in $C[0, 1]$.

5.5.10. Let Y be a closed subspace of X. Show that $(X/Y)^*$ is linearly isometric onto $Y^\circ = \{x^* \in X^*: x^*(y) = 0 \text{ for all } y \text{ in } Y\}$. {Hint: Consider the mapping $T(V^*) = x^*$ where $x^*(x) = V^*([x])$.}

5.5.11. Let Y be a closed subspace of X. Show that X is reflexive if and only if Y and X/Y are reflexive. (Use Problem 5.5.10 for the "only if" part.)

5.5.12. (i) If f is a linear functional on X, then show that f is continuous relative to the weak topology if and only if $f \in X^*$.

(ii) If g is a linear functional on X^*, then show that g is continuous relative to the weak*-topology if and only if there is $x \in X$ such that $g(x^*) = x^*(x)$ for every $x^* \in X^*$.

5.5.13. *Weak Compactness in* $C[0, 1]$. Prove that a subset E of $C[0, 1]$ is weakly compact if and only if E is weakly closed, norm bounded, and every sequence (f_n) in E has a subsequence (f_{n_k}) such that $\lim_{k\to\infty} f_{n_k}(x) = f(x)$ for some f in E and all $x \in [0, 1]$. Also, is a pointwise convergent and uniformly bounded sequence in $C[0, 1]$ weakly convergent?

5.5.14. *Banach Spaces as Spaces of Continuous Functions.* Show that given a real Banach space X there exists a compact Hausdorff space S such that X is linearly isometric onto a closed linear subspace of $C(S)$ with uniform norm. (Hint: Take $S = \{x^* \in X^*: \| x^* \| \leq 1\}$, which is compact in the relative weak*-topology. Let $f_x(x^*) = x^*(x)$ for $x \in X$ and $x^* \in S$. Consider $\varrho(x) = f_x$.)

5.5.15. Let X be an infinite-dimensional Banach space. Prove that the weak closure of $\{x \in X: \| x \| = 1\}$ is the unit ball $\{x \in X: \| x \| \leq 1\}$. [Hint: Suppose that $\| y \| < 1$ and $| x_i^*(x - y) | \geq r > 0$ for $i = 1, 2, \ldots, n$ and all x with norm 1. Since for any nonzero x, there is a real t such that $\| y + tx \| = 1$, the mapping $x \to (x_1^*(x), \ldots, x_n^*(x))$ is injective.]

5.5.16. *Weak Convergence in L_p, $1 < p < \infty$.* Show that a norm bounded sequence (f_n) in L_p converges weakly to f in L_p if $f_n \to f$ in measure.

5.5.17. Prove the following result due to E. Hewitt: Let f be a real-valued measurable function in a semifinite measure space such that $f \notin L_p(\mu)$ for some $p > 1$. Then the set $\{g \in L_q: fg \in L_1 \text{ and } \int fg\, d\mu = 0\}$ is dense in L_q, where $1/p + 1/q = 1$. (Hint: The set $E = \{g \in L_q: fg \in L_1\}$ is dense in L_q since it contains all functions χ_A, where

$$\mu(A) < \infty \quad \text{and} \quad A \subset \{x: n \leq |f(x)| < n + 1\}.$$

Then T, where $T(g) = \int fg\, d\mu$, is not continuous, but linear on E. Now use Problem 5.3.3b.)

5.5.18. Let X be a compact Hausdorff space. Let $\varrho: X \to C(X)^*$ be defined by $\varrho(x)(f) = f(x)$. Show that ϱ is a homeomorphism from X onto $\varrho(X) \subset C(X)^*$ (with weak*-topology). (Since X need not be sequentially compact, this shows that the Eberlein–Smulian theorem is false for weak*-compact subsets.)

5.5.19. Show that a Banach space X is reflexive if and only if every total subspace of X^* is dense in X^*. (Hint: For the "if" part, suppose $x^{**} \in X^{**} - J(X)$. Then the subspace $\{x^* \in X^*: x^{**}(x^*) = 0\}$ is not dense in X^*. Show that it is total.)

5.5.20. *Banach–Saks Theorem.* Every weakly convergent sequence (f_n) in $L_2[0, 1]$ has a subsequence $(f_{n_k} = g_k)$ such that the sequence (h_n), $h_n = (1/n)\sum_{k=1}^{n} g_k$, converges in L_2 norm. (Hint: Consider a subsequence (f_{n_k}) such that for $j \geq n_{i+1}$ and $1 \leq k \leq i$, $|\int f_j(x) f_{n_k}(x)\, dx| < 1/2^{i+1}$.) This result remains true in L_p, $p > 1$. The reader can later observe that the proof in L_2 extends easily to any Hilbert space.

5.5.21. Prove that $L_1(\mu)$ is weakly sequentially complete if μ is σ-finite. (Hint: If (f_n) is a weak Cauchy sequence in L_1, then for each measurable set E, $\lim_{n\to\infty} \int_E f_n\, d\mu$ exists. Use Problem 4.3.16 to show that $\nu(E) = \lim_{n\to\infty} \int_E f_n\, d\mu$ defines a bounded signed measure absolutely continuous with respect to μ. Now apply the Radon–Nikodym theorem.)

5.5.22. *Weak Convergence in L_1.* Prove that in a σ-finite measure space, a sequence (f_n) in L_1 converges weakly to f in L_1 if and only if $\sup_n \| f_n \|_1 < \infty$ and for each measurable set E, the sequence $\int_E f_n\, d\mu$ converges.

5.5.23. Prove that X^* is weak*-sequentially complete if X is a Banach space. [It is relevant to mention here that though L_1 as well as any reflexive Banach space is weakly sequentially complete, a normed linear space X is weakly complete (that is, every weak Cauchy net converges weakly) if and only if X is finite-dimensional. A similar statement holds for the weak* topology of X^*.]

5.6. Compact Operators and Spectral Notions

To prove some of Fredholm's results on integral equations, F. Riesz devised vector space techniques which easily extend and can be applied to a special class of linear operators called compact operators. These operators are very useful and often find applications in classical integral equations as well as in nonsingular problems of mathematical physics.

In this section we will derive basic properties of compact operators and then consider the Riesz–Schauder theory of such operators. The connection between the classical approximation problem for compact operators (by finite-dimensional operators) and the Schauder-basis problem in Banach spaces will be briefly discussed. We will finally introduce the spectral notions for a bounded linear operator on a Banach space and then consider briefly the spectral theory of compact operators.

Let X and Y be normed linear spaces over the same scalars.

Definition 5.15. A linear operator A from X into Y is called *compact* (or *completely continuous*) if A maps bounded sets of X into relatively compact (that is, having compact closure) sets of Y. ∎

Remarks

5.22. If A is compact then A is continuous.

5.23. If $A \in L(X, Y)$ and $A(X)$ is finite-dimensional, then A is compact.

5.24. An operator $A \in L(X, Y)$ need not be compact. For example, the identity operator on an infinite-dimensional normed linear space is not compact. (See Theorem 5.2.)

Example 5.15. Let $A: C[0, 1] \to C[0, 1]$ be defined by

$$Af(x) = \int_0^1 k(x, y) f(y)\, dy,$$

where $k(x, y)$ is a continuous function of (x, y) on $[0, 1] \times [0, 1]$. The reader can easily check that if $S = \{f \in C[0, 1]: \|f\|_\infty \leq 1\}$, then $A(S)$ is uniformly bounded and equicontinuous. By the Arzela–Ascoli Theorem $A(S)$ is relatively compact and therefore A is compact.

The next few results give some basic properties of compact operators.

Proposition 5.22. Any finite linear combination of compact operators is compact. ∎

The proof is left to the reader.

Proposition 5.23. Let A and B be in $L(X, X)$ with A compact. Then AB and BA are both compact. ∎

Proof. If S is a bounded set in X, then $AB(S) = A\big(B(S)\big)$ is relatively compact since $B(S)$ is bounded and A is compact. Also $BA(S) \subset B\big(\overline{A(S)}\big)$, which is compact since B is continuous and $\overline{A(S)}$ is compact. ∎

Proposition 5.24. Let $A \in L(X, Y)$ and $(A_n)_N$ be a sequence of compact operators from X into Y, where Y is complete and $\lim_{n\to\infty} \| A_n - A \| = 0$. Then A is compact. ∎

Proof. We will show that $A(S)$, for S bounded in X, is totally bounded and hence relatively compact in Y, a complete metric space. Let $\varepsilon > 0$. Then there is a positive integer n such that for each $x \in S$, $\| A_n(x) - A(x) \| < \varepsilon$. Since A_n is compact, $\overline{A_n(S)}$ is compact and hence totally bounded. Therefore there exist $x_1, x_2, \ldots, x_m$ in S such that for each $x \in S$,

$$\inf_{1\le i\le m} \| A_n(x) - A_n(x_i) \| < \varepsilon.$$

This means that for each $x \in S$

$$\inf_{1\le i\le m} \| A(x) - A(x_i) \| < 3\varepsilon.$$

Hence $A(S)$ is totally bounded. ∎

The following example shows that completeness of Y is essential in Proposition 5.24.

Example 5.16. Let B be the operator from $c_0 (\subset l_\infty)$ into l_2 defined by

$$B\big((\alpha_k)\big) = (\alpha_k/k).$$

Let $X = c_0$, $Y =$ the range of B, and $A \in L(X, Y)$ be defined by $A(x)$

$= B(x)$. However, A is not compact. To see this, let

$$x_n = (\underbrace{1, 1, 1, \dots, 1}_{n}, 0, 0, \dots) \in c_0;$$

then the sequence $A(x_n) = (1, 1/2, \dots, 1/n, 0, 0, \dots)$ converges to $(1/k)_{k\in N}$ in l_2. However, $(1/k)_{k\in N} \notin Y$, since if $B((\alpha_k)) = (1/k)$ then $(1, 1, 1, \dots) \in c_0$, which is a contradiction. Hence the sequence $A(x_n)$ cannot have a convergent subsequence in Y or A is *not* compact. Nevertheless if we define $A_n \in L(X, Y)$ by

$$A_n((\alpha_k)) = (\beta_k),$$

where

$$\beta_k = \begin{cases} \alpha_k/k, & 1 \le k \le n \\ 0, & k > n, \end{cases}$$

then the A_n's are compact and $\lim_{n\to\infty} \| A_n - A \| = 0$.

Proposition 5.24 above shows that the compact operators form a closed linear subspace of $L(X, Y)$, when Y is complete. Also every operator in $L(X, Y)$ with finite-dimensional range is compact. It is therefore natural to ask the following:

(A) Is every compact operator T in $L(X, Y)$ a limit in the norm of operators with finite-dimensional range?

This is the famous *approximation problem* in Banach spaces and formerly was one of the most widely known unsolved problems in the theory of Banach spaces. The approximation problem was studied in detail by Grothendieck.† He conjectured that (A) is not true in general. Only recently has it been solved in the negative by Per Enflo.‡

This problem is really a problem of the structure of Banach spaces. To clarify this let us say that a Banach space Y is said to have the *approximation property* if for every compact set $K \subset Y$ and every $\varepsilon > 0$, there is $P \in L(Y, Y)$, depending on K and ε, with finite-dimensional range such that $\| P(x) - x \| \le \varepsilon$ for every $x \in K$. Now if Y has the approximation property, then the answer to (A) is affirmative. Indeed, if T is compact and K is the compact set $\overline{T(S_X)}$, where S_X is the closed unit ball of X, then $\dim PT(X) < \infty$ and $\| PT - T \| \le \varepsilon$.

Also it is clear that a Banach space has the approximation property if every separable subspace of it has this property. To see this, one has

† A. Grothendieck, *Canad. J. Math.* **7**, 552–561 (1955).

‡ Per Enflo, *Acta Math.* **130**, 3–4, 309–317 (1973).

to consider the closed subspace (separable) Y_K spanned by the compact set K in the definition above and then use the Hahn–Banach Theorem to extend P from an operator in $L(Y_K, Y_K)$ to one in $L(Y, Y)$. It can be easily verified that this is possible since the range of P is finite dimensional. This shows that (A) is related to the classical basis problem:

(B) Does every separable Banach space Y have a *Schauder basis* $(x_i)_{i=1}^{\infty}$ (that is, can every $y \in Y$ be written uniquely in the form $\sum_{i=1}^{\infty} \lambda_i x_i$)?

Actually, if Y has a Schauder basis, then Y has the approximation property. To see this, let $(x_i)_{i=1}^{\infty}$ be the Schauder basis of Y and $P_n \in L(Y, Y)$ be defined by $P_n(\sum_{i=1}^{\infty}\lambda_i x_i) = \sum_{i=1}^{n}\lambda_i x_i$. Then $\| P_n(x) - x \| \to 0$ uniformly as $n \to \infty$ on every compact subset of Y (see Problem 5.6.2), and, consequently, Y has the approximation property. Hence a positive answer to (B) must give a positive answer to (A). Per Enflo (in the paper mentioned above) solved (A) in the negative [and therefore also (B) in the negative] by giving an example of a separable reflexive Banach space that does not have the approximation property.

In the next chapter, we will see that every separable subspace of a Hilbert space has a Schauder basis, and as such Problem (A) has an affirmative solution in a Hilbert space. A different proof of this fact will also be considered there.

We now leave Problem (A) and consider other basic properties of a compact operator.

Proposition 5.25. If A is a compact operator, then the range of A is separable. ∎

Proof. The range of A is contained in $\bigcup_{n=1}^{\infty}\overline{A(S_n)}$, where $S_n = \{x: \| x \| \leq n\}$. Since this is a countable union of compact sets, and a compact metric space is separable, the result follows. ∎

Proposition 5.26. Let $A \in L(X, Y)$, where X is infinite dimensional and A is compact. Then A^{-1}, if it exists, is not bounded. ∎

The proof is left to the reader.

Let us now recall the definition of an adjoint operator (Problems 5.4.9, 5.4.12). If $A \in L(X, Y)$, then $A^* \in L(Y^*, X^*)$ is defined by

$$A^*(y^*) = y^* \circ A.$$

Then $\| A^* \| = \| A \|$. The concept of the adjoint of a bounded linear operator A is useful in obtaining information about the range and inverse of A, as was outlined in Problems 5.4.12–5.4.15. There is a further duality between A and A^* as the following theorem shows.

Theorem 5.23. Let $A \in L(X, Y)$. If A is compact, then A^* is compact. Conversely, if A^* is compact and Y complete, then A is compact. ∎

Proof. Let A be compact. It suffices to show that $A^*(W)$, for W bounded in Y^*, is totally bounded in X^*. Let $\varepsilon > 0$ and $S = \{x \in X: \| x \| \leq 1\}$. Since A is compact, there exist $x_1, x_2, \ldots, x_n \in X$ such that for each $x \in S$,

$$\inf_{1 \leq i \leq n} \| A(x) - A(x_i) \| < \varepsilon. \tag{5.8}$$

Let B be defined from Y^* into F^n by

$$B(y^*) = \big(y^* \circ A(x_1), \ldots, y^* \circ A(x_n)\big),$$

then $B \in L(Y^*, F^n)$. Hence $B(W)$ is totally bounded. Thus there exist $y_1^*, \ldots, y_m^* \in W$ such that for each $y^* \in W$ and each i, $1 \leq i \leq n$,

$$\inf_{1 \leq j \leq m} | y^* \circ A(x_i) - y_j^* \circ A(x_i) | < \varepsilon. \tag{5.9}$$

Now let $y^* \in W$ and let y_j^* for $1 \leq j \leq m$ be such that for each i, $1 \leq i \leq n$,

$$| y^* \circ A(x_i) - y_j^* \circ A(x_i) | < \varepsilon. \tag{5.10}$$

Then

$$\| A^*(y^*) - A^*(y_j^*) \| = \sup_{\|x\| \leq 1} | y^* \circ A(x) - y_j^* \circ A(x) |,$$

which is, by inequalities (5.8) and (5.9), less than or equal to $(2M + 1)\varepsilon$, where M is an upper bound for $\| y^* \|$, $y^* \in W$. This proves that $A^*(W)$ is totally bounded and hence A^* is compact.

Conversely, let A^* be compact. By what we have just proved, A^{**} is compact. Let (x_n) be a bounded sequence in X. Then the sequence $\big(J(x_n)\big)$ is also bounded in X^{**}, J being the natural mapping from X into X^{**}. Since $A^{**} \in L(X^{**}, Y^{**})$ and A^{**} is compact, there exists a subsequence $[A^{**}(J(x_{n_k}))]_{k \in N}$ which converges in Y^{**}. Now $J_1\big(A(x_{n_k})\big) = A^{**}\big(J(x_{n_k})\big)$, where J_1 is the natural mapping from Y into Y^{**}. Since J_1 is an isometry, $\big(A(x_{n_k})\big)_{k \in N}$ is a Cauchy sequence in Y and therefore must converge since Y is complete. This means that A is compact. ∎

Corollary 5.13. If the range of a compact operator A is complete, then the range of A is finite dimensional. ∎

Proof. Let $A \in L(X, Y)$ be compact and let Z be the range of A. Then if Z is complete, A as an operator in $L(X, Z)$ is compact and therefore by Theorem 5.23 A^* is compact. Also, by Problem 5.4.15, A^* has a bounded inverse. Hence by Proposition 5.26 Z^* (and therefore Z) is finite dimensional. ∎

Next we present the famous Fredholm Alternative Theorem, which was developed by F. Riesz as a tool for the study of linear integral equations.

The proof of this theorem uses the properties of S° (the annihilator of S) as given in Problems 5.4.12 and 5.4.13.

Theorem 5.24. *The Fredholm Alternative Theorem.* Let X be complete and $K \in L(X, X)$ be compact. Then $R(I - K)$ is closed and $\dim N(I - K) = \dim N(I - K^*) < \infty$, where N denotes the null space, R the range, and I the identity operator. In particular, *either* $R(I - K) = X$ and $N(I - K) = \{0\}$, *or* $R(I - K) \neq X$ and $N(I - K) \neq \{0\}$. ∎

Proof. We will prove the theorem in several steps. We will write A for $I - K$.

Step I. In this step we will show that $R(A)$ is closed [or equivalently, there exists $k > 0$ such that for every $x \in X$, $d(x, N(A)) \leq k \| A(x) \|$, where d denotes the usual distance]. (See Problem 5.4.11.)

Suppose $R(A)$ is not closed. Then there exists a sequence $(x_n)_{n\in N} \in X$ such that $d(x_n, N(A)) = 1$ and $\| A(x_n) \| \to 0$ as $n \to \infty$. Therefore, there also exists a sequence (z_n) such that $d(z_n, N(A)) = 1$, $\| A(z_n) \| \to 0$ as $n \to \infty$, and $1 \leq \| z_n \| \leq 2$. Since K is compact, there exists a subsequence $(z_{n_i})_{i\in N}$ such that $K(z_{n_i})$ converges to some $z \in X$. Then z_{n_i}, which is $A(z_{n_i}) + K(z_{n_i})$, also converges to z; or $A(z_{n_i})$ converges to $A(z)$. Hence $A(z) = 0$ or $z \in N(A)$. This is a contradiction, since for each i, $d(z_{n_i}, N(A)) = 1$ and $z_{n_i} \to z$ as $i \to \infty$.

Step II. In this step we will show that $\dim N(A) < \infty$ and $\dim N(A^*) < \infty$.

Consider a sequence $(x_n)_N$ in $N(A)$ with $\| x_n \| \leq 1$. Then for each n, $A(x_n) = 0$ or $K(x_n) = x_n$. Since K is compact, there exists a subsequence $(x_{n_i})_{i\in N}$ such that $K(x_{n_i}) = x_{n_i}$ is convergent. This means that the unit ball (closed) of $N(A)$ (considered as a normed linear space) is compact, and therefore $\dim N(A) < \infty$. Similarly, $\dim N(A^*) < \infty$.

Step III. In this step we will prove that $\dim N(A) = 0$ if and only if $\dim N(A^*) = 0$.

Suppose that $\dim N(A^*) = 0$. Since $R(A)$ is closed by Step I, $R(A) = {}^\circ N(A^*) = X$. (See Problem 5.4.12.) Suppose $\dim N(A) > 0$. Then there exists $x_1 \neq 0$ such that $A(x_1) = 0$. Since $R(A) = X$, there exists a sequence $(x_n)_N$ such that $A(x_{n+1}) = x_n$. Now $A^{n+1}(x_{n+1}) = A^n(x_n) = \cdots = A(x_1) = 0$, and $A^n(x_{n+1}) = A^{n-1}(x_n) = \cdots = A(x_2) = x_1 \neq 0$. This means that for each positive integer n, $N(A^n)$ is a *proper* closed subspace of $N(A^{n+1})$. By Riesz's lemma (Proposition 5.1), there exists $z_n \in N(A^n)$ such that $\| z_n \| = 1$ and $d(z_n, N(A^{n-1})) > 1/2$. Since $n > m$ implies

$$\begin{aligned} \| K(z_n) - K(z_m) \| &= \| z_n - A(z_n) - z_m + A(z_m) \| \\ &= \| z_n - [z_m - A(z_m) + A(z_n)] \| \\ &\geq d(z_n, N(A^{n-1})) > 1/2, \end{aligned}$$

we have a contradiction to the fact that K is compact. Hence $\dim N(A) = 0$.

Conversely, suppose that $\dim N(A) = 0$. Then $R(A^*) = N(A)^\circ = X^*$ (see Problem 5.4.13). Since $A^* = I - K^*$ and K^* is also compact, by a similar argument to that above, $\dim N(A^*) = 0$.

Step IV. In this step we will show that $\dim N(A) = \dim N(A^*)$, and this will complete the proof of the theorem.

Suppose $\dim N(A) = n > 0$ and $\dim N(A^*) = m > 0$. Let $\{x_1, x_2, \ldots, x_n\}$ be a basis for $N(A)$ and $\{x_1^*, x_2^*, \ldots, x_m^*\}$ be a basis for $N(A^*)$. Then we claim the following:

(i) There exists $x_0^* \in X^*$ such that $x_0^*(x_i) = 0$ for $1 \leq i < n$, and $x_0^*(x_n) \neq 0$.

(ii) There exists $x_0 \in X$ such that $x_i^*(x_0) = 0$ for $1 \leq i < m$, and $x_m^*(x_0) \neq 0$.

(iii) If $A_1 = A - K_0$ where $K_0(x) = x_0^*(x) \cdot x_0$, then $\dim N(A_1) = n - 1$ and $\dim N(A_1^*) = m - 1$.

Statement (i) follows easily. We prove (ii) by an inductive argument. Let $p(k)$ be the statement "There exist $\{z_j\colon j = 1, 2, \ldots, k\}$ such that for $1 \leq i, j \leq k$, $x_i^*(z_j)$ is equal to zero if $i \neq j$ and is equal to one if $i = j$." Clearly $p(k)$ holds for $k = 1$. Suppose that $p(k)$ holds for $k = m - 1$. Then for $1 \leq i \leq m - 1$, we have

$$x_i^*\left(x - \sum_{j=1}^{m-1} x_j^*(x) z_j\right) = 0$$

for each $x \in X$. If for each $x \in X$, $x_m^*(x - \sum_{j=1}^{m-1} x_j^*(x)z_j) = 0$, then $x_m^* = \sum_{j=1}^{m-1} x_m^*(z_j)x_j^*$, a contradiction. Hence there exists z_m' such that $x_m^*(z_m') = 1$ and for $1 \leq i \leq m-1$, $x_i^*(z_m') = 0$. Now letting $z_j' = z_j - x_m^*(z_j)z_m'$ for $1 \leq j \leq m-1$, we have for $1 \leq i, j \leq m$, $x_i^*(z_j')$ equals zero if $i \neq j$ and equals one if $i = j$. The argument is complete and (ii) is established.

To prove (iii), let $x \in N(A_1)$. Then $A(x) = K_0(x) = x_0^*(x)x_0$. Since $x_0 \notin {}^\circ[N(A^*)] = R(A)$ (see problem 5.4.12), $x_0^*(x) = 0$ and therefore $x \in N(A)$. So we can write $x = \sum_{j=1}^{n} \alpha_j x_j$ for some scalars α_j, and since

$$0 = x_0^*(x) = \sum_{j=1}^{n} \alpha_j x_0^*(x_j) = \alpha_n \cdot x_0^*(x_n),$$

we have $\alpha_n = 0$ or $x = \sum_{j=1}^{n-1} \alpha_j x_j$. This means that $\dim N(A_1) = n - 1$. To prove that $\dim N(A_1^*) = m - 1$, let $x^* \in N(A_1^*)$ or $A^*x^* = K_0^*x^* = x^*(x_0) \cdot x_0^*$. Since $x_0^* \notin N(A)^\circ = R(A^*)$ (see Problem 5.4.13), $x^*(x_0) = 0$ or $x^* \in N(A^*)$. Therefore, we can write $x^* = \sum_{j=1}^{m} \beta_j x_j^*$ for some scalars β_j. Then since $x^*(x_0) = 0$ and $x^*(x_0) = \beta_m x_m^*(x_0)$, $\beta_m = 0$. This means that $\dim N(A_1^*) = m - 1$.

Now the proof of the theorem will follow easily. If $n < m$, then $\dim N(A_1) \neq \dim N(A_1^*)$. Repeating the above process a finite number of times, we end up with an operator $A_n = I - (K + K_0 + \cdots + K_{n-1}) = I -$ (a compact operator) such that $\dim N(A_n) = 0$ and $\dim N(A_n^*) > 0$. But this contradicts the result in Step III. Hence $n \geq m$. Similarly, $n \leq m$. This proves the theorem. ∎

At the end of this section, we will outline some applications of the above theorem in linear integral equations.

Now we will consider briefly what is called the spectral theory of linear operators, which is the systematic study of various connections between $T - \lambda I$ and $(T - \lambda I)^{-1}$, where $T \in L(X, X)$, λ is a scalar, and I is the identity operator. A large part of the theory of bounded linear operators is centered around their spectral theory. The most highly developed spectral theory is that for a class of operators called self-adjoint operators on Hilbert spaces (this will be discussed in depth in the next chapter).

Definition 5.16. Let $T \in L(X, X)$ where X is a nonzero *complex* normed linear space. The set $\{\lambda \in C: (T - \lambda I)^{-1} \in L(X, X)\}$ is called the *resolvent* set of T. The complement in the complex plane of the resolvent set of T is called the *spectrum* of T and is denoted by $\sigma(T)$. If $T(x) = \lambda x$ for some $x \neq 0$, then λ is called an *eigenvalue* of T and x is called the corresponding *eigenvector*. ∎

We denote $(T - \lambda I)^{-1}$ by $R(\lambda, T)$ whenever $(T - \lambda I)^{-1} \in L(X, X)$.

Remarks

5.25. We note that when X is finite dimensional, $\lambda \in \sigma(T)$ if and only if $T - \lambda I$ is not one-to-one, which is true if and only if λ is an eigenvalue of T. Since the eigenvalues of $T - \lambda I$ are the solutions of the equation

$$\det (T_m - \lambda I) = 0,$$

where I is the identity matrix and T_m is the matrix representing T (with respect to some fixed basis of X), it follows that $\sigma(T)$ is nonempty. (Note that every nonconstant polynomial with complex coefficients has a complex root.)

5.26. $\sigma(T)$ can also be defined in *real* normed linear spaces X, but the difficulty is that $\sigma(T)$ can be empty even when X is finite dimensional. For example, let T be defined from R^2 into R^2 linearly by $T\big((1, 0)\big) = (0, -1)$ and $T\big((0, 1)\big) = (1, 0)$; then $\sigma(T)$ is empty.

5.27. When X is infinite dimensional, there can be elements in $\sigma(T)$ that are not eigenvalues. For *example*, let $X = l_2$ and T be defined on l_2 by

$$T\big((x_1, x_2, \ldots)\big) = (0, x_1, x_2, \ldots).$$

Since for $x \in l_2$ $\| T(x) \| = \| x \|$, T is one-to-one and bounded so that 0 is not an eigenvalue of T. But since $(1, 0, 0, \ldots) \notin$ the range of T, T^{-1} is is not defined on X and therefore $0 \in \sigma(T)$.

5.28. When X is infinite-dimensional, $\sigma(T)$ can be *uncountably infinite*. For example, let $X = l_2$ and let T be defined on l_2 by

$$T\big((x_1, x_2, \ldots)\big) = (x_2, x_3, \ldots).$$

Then $T \in L(X, X)$. If λ is a complex number with $|\lambda| < 1$, then $x = (1, \lambda, \lambda^2, \ldots) \in l_2$ and $T(x) = (\lambda, \lambda^2, \ldots) = \lambda x$. This means that $\sigma(T) \supset \{\lambda: |\lambda| < 1\}$. (The operator T is called the *shift* on l_2.)

Next we will present a basic result concerning $\sigma(T)$, namely, the fact that it is always nonempty. First we need the following important result.

Proposition 5.27. Let X be complete and $T \in L(X, X)$. Then the resolvent set $\varrho(T)$ is open and if $\lambda, \mu \in \varrho(T)$, then [writing $R_\mu = R(\mu, T)$]

$$R_\mu - R_\lambda = (\mu - \lambda) R_\lambda R_\mu.$$

Moreover, R_λ as a function from $\varrho(T)$ to $L(X, X)$ has derivatives of all orders. (See Remark 5.4 for definition.) ∎

Proof. Suppose $\lambda \in \varrho(T)$ and $|\mu - \lambda| < 1/\| R_\lambda \|$. Then $T - \mu I = (T - \lambda I)[I - (\mu - \lambda)R_\lambda]$. Now $\sum_{n=0}^{\infty}(\mu - \lambda)^n R_\lambda{}^n$ is convergent in $L(X, X)$, and it follows easily that $[I - (\mu - \lambda)R_\lambda]^{-1} = \sum_{n=0}^{\infty}(\mu - \lambda)^n R_\lambda{}^n$. Hence $(T - \mu I)^{-1} \in L(X, X)$ and $\mu \in \varrho(T)$; consequently, $\varrho(T)$ is open. Also for λ, $\mu \in \varrho(T)$ we have

$$\begin{aligned} R_\mu - R_\lambda &= R_\lambda[R_\lambda{}^{-1} - R_\mu{}^{-1}]R_\mu \\ &= R_\lambda[T - \lambda I - (T - \mu I)]R_\mu \\ &= (\mu - \lambda)R_\lambda R_\mu . \end{aligned}$$

Hence $(R_\mu - R_\lambda)/(\mu - \lambda) = R_\lambda R_\mu$ or as $\mu \to \lambda$, $(R_\lambda - R_\mu)/(\lambda - \mu) \to R_\lambda{}^2$, in the $L(X, X)$ norm. [Note that $\| R_\mu \| \le \| R_\lambda \| (1 - |\lambda - \mu| \, \| R_\lambda \|)^{-1}$, when $|\lambda - \mu|$ is sufficiently small.] By induction, it can be shown that R_λ has derivatives of all higher orders. ∎

Proposition 5.28. Let X be complete and $T \in L(X, X)$. Then $|\lambda| > \| T \|$ implies that $\lambda \in \varrho(T)$ and $R_\lambda = -\sum_{n=1}^{\infty}\lambda^{-n}T^{n-1}$. Hence $\sigma(T)$ is a compact subset of the complex plane. ∎

We leave the proof to the reader.

Now we use Propositions 5.27 and 5.28 to prove that $\sigma(T)$ is nonempty.

Theorem 5.25. If X is complete and $T \in L(X, X)$, then $\sigma(T)$ is not empty. ∎

Proof. We use Liouville's theorem from complex analysis and the Hahn–Banach Theorem. Let $x \in X$ and $x^* \in X^*$. Then the complex-valued function $x^*(R_\lambda(x))$ is, by Proposition 5.27, an analytic (or differentiable) function on $\varrho(T)$. By Proposition 5.28 for $|\lambda| > \| T \|$, we have

$$\| R_\lambda \| \le \sum_{n=1}^{\infty} \frac{\| T \|^{n-1}}{|\lambda|^n} = \frac{1}{|\lambda|(1 - \| T \|/|\lambda|)} \to 0$$

as $|\lambda| \to \infty$. This means that $x^*(R_\lambda(x))$ is a bounded function on the entire complex plane if $\sigma(T)$ is empty, since then $\varrho(T)$ is the entire complex plane. By Liouville's theorem, $x^*(R_\lambda(x))$, being a bounded entire function, must be a constant ($= 0$, in this case). An application of Corollary 5.4 then asserts that for all $x \in X$, $R_\lambda(x) = 0$, which is a contradiction. ∎

Now we show that Proposition 5.28 can be given a more precise form. We show that $\lim_{n\to\infty}\| T^n \|^{1/n}$ exists and $= \sup\{|\lambda| : \lambda \in \sigma(T)\}$. Let C be any circle with origin as center and radius greater than $\| T \|$. For $x^* \in X^*$ and $x \in X$, we consider the complex-valued analytic function $x^*(R_\lambda(x))$ on $\varrho(T)$. Using the line integral over C taken in a suitable direction, we have from Proposition 5.28

$$\int_C \lambda^n \cdot x^*(R_\lambda(x))\, d\lambda = - \sum_{k=1}^{\infty} x^*(T^{k-1}(x)) \cdot \int_C \lambda^{n-k}\, d\lambda = 2\pi i \cdot x^*(T^n(x)).$$

It follows that if C_0 is any circle containing $\sigma(T)$ in its interior, then

$$x^*(T^n(x)) = \frac{1}{2\pi i} \int_{C_0} \lambda^n \cdot x^*(R_\lambda(x))\, d\lambda,$$

where the line integral is taken in a suitable direction.

We can now prove the following.

Proposition 5.29. Let X be complete and $T \in L(X, X)$. Let $r(T) = \sup_{\lambda \in \sigma(T)} | \lambda |$. Then $\lim_{n\to\infty} \| T^n \|^{1/n}$ exists and is equal to $r(T)$. [Here $r(T)$ is called the *spectral radius* of T.] ∎

Proof. Let $\varepsilon > 0$, C be the circle $| \lambda | = r(T) + \varepsilon$. Then for $x^* \in X^*$ with $\| x^* \| = 1$ and $x \in X$ with $\| x \| = 1$, we have

$$| x^*(T^n(x)) | = \left| \frac{1}{2\pi i} \int_C \lambda^n \cdot x^*(R_\lambda(x))\, d\lambda \right|$$
$$\leq \frac{1}{2\pi} \cdot [r(T) + \varepsilon]^n \cdot 2\pi(r(T) + \varepsilon) \cdot M(C),$$

where $M(C) = \sup_{\lambda \in C} \| R_\lambda \| < \infty$, since R_λ is a continuous function of λ. This means that

$$\limsup_{n\to\infty} \| T^n \|^{1/n} \leq r(T) + \varepsilon.$$

Since $\varepsilon > 0$ is arbitrary, we have

$$\limsup_{n\to\infty} \| T^n \|^{1/n} \leq r(T).$$

We will now complete the proof by showing that

$$r(T) \leq \liminf_{n\to\infty} \| T^n \|^{1/n}.$$

To show this, we note that $\lambda \in \sigma(T) \Rightarrow \lambda^n \in \sigma(T^n)$. The reason is that

$$T^n - \lambda^n I = (T - \lambda I)A = A(T - \lambda I),$$

where

$$A = \sum_{k=0}^{n-1} \lambda^k T^{n-k-1},$$

so that $\lambda^n \in \varrho(T^n) \Rightarrow T - \lambda I$ is bijective $\Rightarrow \lambda \in \varrho(T)$, by the Open Mapping Theorem. Hence, if $\lambda \in \sigma(T)$, by Proposition 5.28

$$|\lambda^n| \leq \| T^n \|,$$

for each positive integer n. It follows that

$$r(T) \leq \liminf_{n \to \infty} \| T^n \|^{1/n}$$

This completes the proof. ∎

Our next result is what is called the Spectral Mapping Theorem. This theorem answers a natural question, namely, when the equation $p(T)(x) = y$ has a unique solution for each y in X, where p is a polynomial and $T \in L(X, X)$. It is clear that the equation is solvable if 0 is not in the spectrum of $p(T)$. The Spectral Mapping Theorem answers the question more precisely: The above equation can be solved uniquely for each y in X if and only if no λ in $\sigma(T)$ is a root of p. Here X is a Banach space.

Theorem 5.26. *The Spectral Mapping Theorem.* For $T \in L(X, X)$ and any polynomial p, $p(\sigma(T)) = \sigma(p(T))$. [Here $p(\sigma(T)) = \{p(\lambda): \lambda \in \sigma(T)\}$.] ∎

Proof. Let $\lambda \in \sigma(T)$. Since λ is a root of the polynomial $p(t) - p(\lambda)$, we can write

$$p(t) - p(\lambda) = q(t) \cdot (t - \lambda)$$

and

$$p(T) - p(\lambda)I = q(T)(T - \lambda I) = (T - \lambda I)q(T),$$

for some polynomial $q(t)$. Now if $p(\lambda) \in \varrho(p(T))$, then $T - \lambda I$ is bijective and therefore, by the Open Mapping Theorem, $\lambda \in \varrho(T)$. This proves that $p(\sigma(T)) \subset \sigma(p(T))$. For the opposite inclusion, let $\lambda \in \sigma(p(T))$. Suppose that $\lambda_1, \lambda_2, \ldots, \lambda_n$ are the complex roots of $p(t) - \lambda$. Then

$$p(T) - \lambda I = c \cdot (T - \lambda_1 I) \cdots (T - \lambda_n I), \qquad c \neq 0.$$

If each λ_i is in $\varrho(T)$, then for each i, $(T - \lambda_i I)^{-1} \in L(X, X)$ and therefore, by the above, $\lambda \in \varrho(p(T))$, which is a contradiction. Thus, one of the λ_i must be in $\sigma(T)$. Since this λ_i is a root of $p(t) - \lambda$, it follows that $\lambda \in p(\sigma(T))$. The proof is complete. ∎

Now we will study the spectrum of a compact operator T on a Banach space X. The spectrum of such operators is at most countable and contains 0 when X is infinite dimensional; this follows from Proposition 5.26 and the next result.

Theorem 5.27. Let T be a compact operator on a Banach space X. If $\lambda \neq 0$, then $\lambda \in \varrho(T)$ or λ is an eigenvalue of T. Moreover, $\sigma(T)$ is at most countable and 0 is its only possible limit point. ▮

Proof. Suppose $\lambda \neq 0$ is in the spectrum of T. If $T - \lambda I$ is not one-to-one, then clearly λ is an eigenvalue of T. On the other hand, if $(T - \lambda I)^{-1}$ exists as a function but is not bounded, then by Proposition 5.7, for each positive integer n, there exists an x_n in X with $\| x_n \| = 1$ and with $\| (T - \lambda I)x_n \| < 1/n$. This means $Tx_n - \lambda x_n \to 0$ in X. Since T is compact, Tx_n has a convergent subsequence $(Tx_{n_k})_{k \in N}$ converging to y in X. Since also $Tx_{n_k} - \lambda x_{n_k} \to 0$, it is clear that $\lambda x_{n_k} \to y$. Since T is continuous,

$$Ty = \lim_{k\to\infty} T(\lambda x_{n_k}) = \lambda \lim_{k\to\infty} T(x_{n_k}) = \lambda y.$$

Since $y \neq 0$ as $\| y \| = \lim_{k\to\infty} \| \lambda x_{n_k} \| = \lambda \lim_{k\to\infty} \| x_{n_k} \| = \lambda$, λ is an eigenvalue of T.

To prove the rest of the theorem, it is sufficient to prove that for any $\varepsilon > 0$, the set

$$P_\varepsilon = \{\lambda : | \lambda | \geq \varepsilon \text{ and } \lambda \in \sigma(T)\}$$

is finite. Suppose this is false. Then there exists $\varepsilon > 0$ such that P_ε is infinite. Let $(\lambda_i)_{i=1}^{\infty}$ be a sequence of distinct eigenvalues in P_ε, with $(x_i)_{i=1}^{\infty}$ the corresponding eigenvectors. Since eigenvectors corresponding to distinct eigenvalues are linearly independent, the subspace X_n spanned by $\{x_1, x_2, \ldots, x_n\}$ is properly contained in X_{n+1} spanned by $\{x_1, x_2, \ldots, x_{n+1}\}$. By Riesz's result (Proposition 5.1), there exist $y_n \in X_n$ with $\| y_n \| = 1$ and $\inf_{x \in X_{n-1}} \| y_n - x \| \geq 1/2$. We write $y_n = \sum_{i=1}^{n} \alpha_i x_i$; then

$$\begin{aligned} \lambda_n y_n - T(y_n) &= \sum_{i=1}^{n} \alpha_i \lambda_n x_i - \sum_{i=1}^{n} \alpha_i \lambda_i x_i \\ &= \sum_{i=1}^{n-1} \alpha_i (\lambda_n - \lambda_i) x_i \in X_{n-1}. \end{aligned}$$

Therefore, for $n > m$,

$$\| T(y_n) - T(y_m) \| = \| \lambda_n y_n - [\lambda_n y_n - T(y_n) + T(y_m)] \| \geq \tfrac{1}{2} | \lambda_n | \geq \varepsilon/2,$$

which is a contradiction to the compactness of T. The theorem follows. ▮

Finally, we give some applications of the preceding theory to the study of integral equations.

• **Remark 5.29.** *Applications to Integral Equations: The Dirichlet Problem.* A number of problems in applied mathematics and mathematical physics can be reduced to equations of the type

$$f(s) - \lambda \int_a^b K(s, t) f(t)\, dt = g(s), \tag{5.11}$$

where $K(s, t)$ is a complex-valued Lebesgue-measurable function on $[a, b] \times [a, b]$ such that

$$\| K \|_2^2 = \int_a^b \int_a^b | K(s, t) |^2 \, ds\, dt < \infty,$$

$f, g \in L_2[a, b]$. Here λ is a nonzero complex number and f is the unknown function. These equations are usually called Fredholm equations of the second kind and the function K is called the kernel of the equation.

In what follows, we will apply Theorems 5.24 and 5.27 to solve the problem of existence of solutions of equation (5.11); the results on integral equations will then be useful in studying a fundamental problem of mathematical physics–the Dirichlet problem.

First, we define the operator T by

$$Tf(s) = \int_a^b K(s, t) f(t)\, dt, \qquad f \in L_2[a, b]. \tag{5.12}$$

It follows from Fubini's Theorem (Theorem 3.7) and the Hölder Inequality (Proposition 5.3) that for almost all s,

$$| Tf(s) |^2 \leq \int_a^b | K(s, t) |^2 \, dt \cdot \int_a^b | f(t) |^2 \, dt$$

and therefore

$$\| Tf \|_2 \leq \| K \|_2 \cdot \| f \|_2. \tag{5.13}$$

Thus, T is a bounded linear operator on L_2. We claim that T is compact. To prove our claim, we assume with no loss of generality that K is a continuous function of (s, t). This is possible since by Lusin's Theorem (see Problem 3.1.13) we can approximate the kernel by a continuous kernel in L_2-norm and then Proposition 5.24 applies. Now by Theorem 1.26, a continuous kernel $K(s, t)$ can be approximated uniformly by kernels $K_n(s, t)$

of the form $\sum_{i=1}^{n} u_i(s)v_i(t)$. If we define

$$T_n f(s) = \int_a^b K_n(s, t) f(t)\, dt,$$

then

$$T_n f(s) = \sum_{i=1}^{n} u_i(s) \int_a^b v_i(t) f(t)\, dt.$$

This means that T_n is finite dimensional and therefore, by Remark 5.23 T_n is compact. By the same argument as used in obtaining equation (5.13), we have

$$\| Tf - T_n f \|_2 \leq \| K - K_n \|_2 \cdot \| f \|_2 .$$

It follows that $\lim_{n\to\infty} \| T - T_n \| = 0$. By Proposition 5.24, T is compact.

Now we write equation (5.11) as

$$(I - \lambda T) f = g \tag{5.14}$$

or, equivalently,

$$(T - \lambda^{-1} I) f = - \lambda^{-1} g. \tag{5.15}$$

Taking λT as the operator K in Theorem 5.24, we obtain easily the following.

Theorem A. Either the equation (5.14) has a unique solution

$$f = (I - \lambda T)^{-1} g$$

for each $g \in L_2$ or the homogeneous equation

$$f(s) - \lambda \int_a^b K(s, t) f(t)\, dt = 0 \tag{5.16}$$

has a nonzero solution f in L_2. In the latter case, the number of linearly independent solutions of equation (5.16) is finite. ∎

Now an application of Theorem 5.27 gives us immediately the following.

Theorem B. The equation (5.16) can have nonzero solutions for at most countably many values of λ. If there is an infinite sequence (λ_n) of such values, then $| \lambda_n | \to \infty$. ∎

To find some more information on equation (5.11), we need to find the adjoint of T. For $h \in L_2$, let h^* denote the linear functional on L_2 defined by

$$h^*(f) = \int_a^b f(t)\overline{h(t)}\, dt, \qquad f \in L_2.$$

Then it can be verified by a simple computation that $T^*h^* = g^*$, where g^* is the linear functional induced by g as above and g is given by

$$g(s) = \int_a^b \overline{K(t, s)}h(t)\, dt. \tag{5.17}$$

Another application of Theorem 5.24 leads to our next result.

Theorem C. The equation

$$f(s) - \bar{\lambda} \int_a^b \overline{K(t, s)}\, f(t)\, dt = 0 \tag{5.18}$$

and equation (5.16) have the same number of linearly independent solutions. Moreover, if λ^{-1} is an eigenvalue of T, then equation (5.11) has a solution in L_2 for a given g in L_2 if and only if

$$\int_a^b g(t)\,\overline{f(t)}\, dt = 0$$

whenever f is a solution of equation (5.18). ∎

[Note that this last result follows since $(I - \lambda T)(L_2) = {}^0[N(I - \lambda T^*)]$.]

We remark that the preceding Fredholm theory is also valid if the kernel $K(s, t)$ is a continuous function on $G \times G$, where G is a compact set in R^n, and the operator T acts on $C(G)$.

We now consider the Dirichlet Problem, a fundamental problem in mathematical physics and one of the oldest problems in potential theory.

The Dirichlet Problem. This is the first boundary-value problem of potential theory. The problem is to find a harmonic function on an open connected set E in R^n, which is continuous on $\bar{E}$ and coincides with a given continuous function g on the boundary of E. The problem originates in the study of various physical phenomena from electrostatics, fluid dynamics, heat conduction, and other areas of physics. During the last hundred years or so, this problem has been studied by many celebrated mathematicians including Dirichlet, Poincaré, Lebesgue, Hilbert, and Fredholm, and many different methods have been discovered for solving this problem. Though this problem is not solvable for all domains E, the existence of solutions has been proven in many important cases. In what follows, we shall consider only the two-dimensional case and show how the Fredholm theory can be applied in showing the existence of a solution under certain general assumptions. We shall assume (without proving) several facts from potential theory, our intent being to give the reader only an idea of the applicability of the Fredholm theory. For a detailed discussion of the Dirichlet Problem, the reader can consult *Partial Differential Equations.*†

Let E be an open connected set in R^2 bounded by and in the interior of a simple closed curve C with continuous curvature [i.e., points of C have rectangular coordinates $x(s)$, $y(s)$ (in terms of arc length s), possessing continuous second derivatives]. A function $u(x, y)$ on E with continuous second-order derivatives and satisfying the equation

$$\Delta u = \frac{\partial^2 u}{\partial x^2} + \frac{\partial^2 u}{\partial y^2} = 0$$

in E is called a *harmonic function* on E.

Let f be a continuous function on C. Then it is a fact from potential theory that the function

$$v(p) = \int_C f(t) \frac{\partial}{\partial n_t} \log\left(\frac{1}{|p-t|}\right) dt \tag{5.19}$$

is a harmonic function in E as well as in $(\bar{E})^c$. Here $\partial/\partial n_t$ represents the derivative in the direction of the interior normal n_t at t. For $s \in C$, let us write

$$v^-(s) = \lim_{\substack{t \to s \\ t \in E}} v(t)$$

and

$$v^+(s) = \lim_{\substack{t \to s \\ t \notin \bar{E}}} v(t).$$

It is known that these limits exist and the following equalities are valid. [Note that the integral in equation (5.19) defines $V(p)$ even when $p \in C$.]

$$\begin{aligned} &\text{(i)} \quad v^-(s) = v(s) + \pi f(s); \\ &\text{(ii)} \quad v^+(s) = v(s) - \pi f(s); \\ &\text{(iii)} \quad \text{the normal derivative of } v \text{ is continuous on } C. \end{aligned} \tag{5.20}$$

The reader can find detailed proofs of similar equalities in Garabedian's book.

† P. R. Garabedian, *Partial Differential Equations*, John Wiley (1967).

It is clear from equations (5.19) and (5.20) that the function $u(t)$ given by

$$u(t) = \begin{cases} v(t), & t \in E \\ v^-(t), & t \in C \end{cases}$$

will be a solution of the Dirichlet problem if we find a solution f of the equation

$$\frac{1}{\pi} g(s) = f(s) + \int_C K(s, t) f(t)\, dt, \tag{5.21}$$

where

$$K(s, t) = \frac{1}{\pi} \frac{\partial}{\partial n_t} \log\left(\frac{1}{|s - t|}\right)$$

and g is defined as the given continuous function on the boundary. Here one can show by straightforward computations and by using the continuous curvature of C that $K(s, t)$ is a continuous function of (s, t), even when $s = t$. To apply the Fredholm theory, we consider the homogeneous equation

$$f(s) + \int_C K(s, t) f(t)\, dt = 0 \tag{5.22}$$

and show that this equation does not have any nonzero continuous solution. To show this, we note that any continuous solution f of equation (5.22) will define, as in equation (5.19), a function F harmonic in E as well as in $(\bar{E})^c$ such that by equation (5.20), $F^-(s) = F(s) + f(s) = 0$, $s \in C$. Since a harmonic function is known to assume its maximum and minimum values on the boundary, the function $F(t) = 0$ for all $t \in E$. This means that $(\partial F/\partial n)^- = 0$ on C, and by equation (5.20) (iii), $(\partial F/\partial n)^+ = 0$ on C. Now using the harmonic property of F in $(\bar{E})^c$, it can be proven by using the classical divergence theorem (or Green's first identity) that

$$\iint_{(\bar{E})^c} \left[\left(\frac{\partial F}{\partial x}\right)^2 + \left(\frac{\partial F}{\partial y}\right)^2\right] dx\, dy = -\int_C F^+(t) \left(\frac{\partial F}{\partial n}\right)^+ dt$$
$$= 0.$$

This means that

$$\frac{\partial F}{\partial x} = 0 = \frac{\partial F}{\partial y} \quad \text{in } (\bar{E})^c$$

and therefore F is constant in $(\bar{E})^c$. Since F, because of its representation as in equation (5.19), is known to be zero at infinity, $F^+(s) = 0$, $s \in C$ and by equation (5.20)

$$f(s) = \tfrac{1}{2}[v^-(s) - v^+(s)] = 0, \qquad s \in C.$$

This proves that equation (5.22) cannot have any nonzero solutions. Since an analog of Theorem A holds also for continuous kernels and for operators T acting on $C(\bar{E})$, the following result is immediate.

Theorem D. For every continuous function g given on the boundary C, the Dirichlet problem has a solution. ∎

Problems

5.6.1. Prove Proposition 5.22.

5.6.2. Suppose $(x_i)_{i=1}^{\infty}$ is a Schauder basis for a Banach space X. Let

$$P_n\left(\sum_{i=1}^{\infty} \lambda_i x_i\right) = \sum_{i=1}^{n} \lambda_i x_i.$$

Show that $\| P_n(x) - x \| \to 0$ uniformly on every compact set K as $n \to \infty$.

5.6.3. Prove Proposition 5.26.

5.6.4. Let T be the operator

$$T(f)(x) = \int_0^1 K(x, y) f(y)\, dy$$

from $L_p[0, 1]$ into $L_q[0, 1]$, where $1/p + 1/q = 1$, $1 < p < \infty$, and $K(x, y) \in L_q([0, 1] \times [0, 1])$. Show that T is compact. (Hint: First prove the result when K is continuous; then approximate K by continuous functions and use Proposition 5.24.)

5.6.5. Show that a compact operator $T \in L(X, Y)$ maps weakly convergent sequences onto convergent sequences.

5.6.6. Let T be a bounded linear operator in a reflexive Banach space. If T maps weakly convergent sequences onto convergent sequences, then show that T is compact.

5.6.7. Consider the following Fredholm integral equation:

$$f(x) = g(x) + \lambda \int_0^1 K(x, y) f(y)\, dy,$$

where $g \in L_2[0, 1]$ and $K \in L_2([0, 1] \times [0, 1])$. Prove that if $g = 0$ implies $f = 0$, then there exists a unique solution of the equation for any $g \in L_2[0, 1]$.

5.6.8. Let X be a compact metric space and μ be a finite measure on it. Let $K(x, y)$ be continuous on $X \times X$, and suppose the only continuous solution of

$$f(x) = \lambda \int K(x, y) f(y)\, d\mu(y)$$

is $f = 0$. Prove that for every continuous function $g(x)$ on X, there exists a unique continuous solution $f(x)$ of the integral equation in Problem 5.6.7.

5.6.9. Consider the Volterra integral equation

$$f(x) = g(x) + \int_0^x K(x, t) f(t)\, dt, \qquad 0 \le x \le 1,$$

where $K(x, t)$ is continuous on $[0\ 1] \times [0, 1]$. Prove that for any continuous function g, there exists a unique continuous solution f of the Volterra equation.

5.6.10. Let $T \in L(X, X)$, X a complex Banach space. Show that $\varrho(T) = \varrho(T^*)$ and $R(\lambda, T^*) = [R(\lambda, T)]^*$.

5.6.11. Let X, Y, and Z be Banach spaces, $K \in L(X, Y)$ and $T \in L(Z, Y)$. If K is compact and $T(Z) \subset K(X)$, then prove that T is compact. (Hint: Let $N = K^{-1}(\{0\})$. Then K_0, defined by $K_0(x + N) = K(x)$, is a compact operator from X/N into Y.)

5.6.12. *Weakly Compact Operators.* A linear operator mapping bounded sequences onto sequences having a weakly convergent subsequence is called weakly compact. Prove that: (i) Weakly compact operators are continuous. (ii) If $T \in L(X, Y)$ and either X or Y is reflexive, then T is weakly compact. (iii) If T is the operator from $L_1[0, 1]$ into $L_p[0, 1]$, $1 \le p < \infty$, defined by $T(f)(x) = \int_0^1 K(x, y) f(y)\, dy$, where $K(x, y)$ is a bounded measurable function on $[0, 1] \times [0, 1]$, then T is weakly compact. (Hint: Use Problem 5.5.22 for $p = 1$.)

5.6.13. Let $T \in L(X, Y)$ be compact and $Z = T(X)$. Define T_0: $X \to Z$ by $T_0(x) = T(x)$. Is T_0 compact? What if X is reflexive? What if Z is closed in Y? What happens if "compact" is replaced by "weakly compact"?

6

Hilbert Spaces

In this chapter we will study aspects of the theory of Hilbert spaces. Roughly we may say that a Hilbert space is a Banach space whose norm is defined in a particular manner. We shall give a characterization in terms of the norm of those Banach spaces that are actually Hilbert spaces. This well-known result (Proposition 6.2) is due to Jordan and von Neumann.

Infinite-dimensional Hilbert spaces are natural generalizations of the finite-dimensional spaces R^n and C^n with the usual "Euclidean norms." Their study was initiated in the early 1900's by Hilbert, who studied the particular spaces l_2 and L_2. The abstract axiomatization of Hilbert space was later given by von Neumann in the separable case in the 1920's,[†] and in general by Löwig[‡] and Rellich.[§] Many others have made significant contributions.

Our aim in this chapter is to study Hilbert spaces starting with very basic properties of the structure of Hilbert spaces and ending with a brief exposition of some essential data concerning the spectral theory of self-adjoint operators. Primarily our aim is to prove the spectral theorem for bounded self-adjoint operators—an important tool in the further study of bounded linear operators in Hilbert space theory in that self-adjoint operators are represented as a sum (integral) of projection operators.

[†] J. von Neumann, Allgemeine Eigenwerttheorie Hermitescher Functionaloperen, *Math. Ann.* **102**, 49–131 (1929–1930); Mathematische Begründung der Quantenmechanik, *Nachr. Ges. Wiss. Göttingen Math.-Phys. Kl.*, 1–57 (1927).

[‡] H. Löwig, Komplexe euklidische Räume von beliebiger endlicher oder unendlicher Dimensionzahl, *Acta Sci. Math.* (*Szeged.*) **7**, 1–33 (1934).

[§] F. Rellich, Spectraltheorie in nichtseparabeln Räumen, *Math. Ann.* **110**, 342–356 (1935).

6.1. The Geometry of Hilbert Space

In this section V and W will denote vector spaces over the field F of real or complex numbers. $\bar{\alpha}$ will denote the complex conjugate of the complex number α.

Definition 6.1. A *sesquilinear form* B on $V \times W$ is a mapping B: $V \times W \to F$ such that for all α and β in F, x and y in V, and w and z in W,

(i) $B(\alpha x + \beta y, z) = \alpha B(x, z) + \beta B(y, z)$

and

(ii) $B(x, \alpha w + \beta z) = \bar{\alpha} B(x, w) + \bar{\beta} B(x, z)$. ∎

In case $V = W$, a sesquilinear form on $V \times W$ is referred to as a sesquilinear form *on* V. A sesquilinear form on V is called *Hermitian* if $B(x, y) = \overline{B(y, x)}$ for all x and y in V. Since $B(x, x)$ is necessarily a real number if B is Hermitian, we say that a Hermitian form on V is *positive* if $B(x, x) \geq 0$ for all x in V and strictly positive if $B(x, x) > 0$ when $x \neq 0$.

Sometimes when F is the field of real numbers so that $\bar{\alpha} = \alpha$ for all scalars, a sesquilinear form on V is called a *bilinear form* and a Hermitian form is called a *symmetric form* since $B(x, y) = B(y, x)$ for all x and y in V. Since the development to follow is true—unless specifically indicated—for real and complex vector spaces, we will continue to use the terms sesquilinear and Hermitian regardless of whether $F = R$.

The following proposition gives in a nutshell some facts we will find extremely useful regarding sesquilinear forms on V.

Proposition 6.1. Let B be a sesquilinear form on V.

(i) *Polarization Identity.* If V is a complex vector space, then for all x and y in V

$$B(x, y) = \tfrac{1}{4}[B(x + y, x + y) - B(x - y, x - y) + iB(x + iy, x + iy) - iB(x - iy, x - iy)]. \tag{6.1}$$

If V is a real vector space, then for all x and y

$$B(x, y) = \tfrac{1}{4}[B(x + y, x + y) - B(x - y, x - y)] \tag{6.2}$$

provided B is Hermitian.

(ii) *Parallelogram Law*. For all x and y in V

$$B(x + y, x + y) + B(x - y, x - y) = 2B(x, x) + 2B(y, y). \tag{6.3}$$

(iii) *Cauchy–Schwarz Inequality*. If B is a positive Hermitian sesquilinear form on V, then for all x and y in V

$$| B(x, y) |^2 \leq B(x, x)B(y, y). \tag{6.4}$$

(iv) If B is a positive Hermitian sesquilinear form on V, then for all x and y in V,

$$[B(x + y, x + y)]^{1/2} \leq [B(x, x)]^{1/2} + [B(y, y)]^{1/2}.$$ ∎

Proof. Statements (i) and (ii) are verified by direct computation and the verifications are left to the reader. Assuming momentarily that (iii) has been verified, we can easily prove (iv). Indeed using (iii)

$$\begin{aligned} B(x + y, x + y) &= B(x, x) + B(x, y) + B(y, x) + B(y, y) \\ &= B(x, x) + 2\text{Re}[B(x, y)] + B(y, y) \\ &\leq B(x, x) + 2 \,| B(x, y) | + B(y, y) \\ &\leq B(x, x) + 2[B(x, x)]^{1/2}[B(y, y)]^{1/2} + B(y, y) \\ &= \{[B(x, x)]^{1/2} + [B(y, y)]^{1/2}\}^2. \end{aligned}$$

It remains therefore to establish (iii). For all real numbers r and for $\alpha \in F$ with $| \alpha | = 1$,

$$\begin{aligned} 0 \leq B(r\alpha x + y, r\alpha x + y) &= r^2B(x, x) + r\alpha B(x, y) + r\bar{\alpha}B(y, x) + B(y, y) \\ &= r^2B(x, x) + 2r\, \text{Re}[\alpha B(x, y)] + B(y, y). \end{aligned} \tag{6.5}$$

Since equation (6.5) holds for all real numbers r, the quadratic function $f(r) = B(x, x)r^2 + 2\text{Re}[\alpha B(x, y)]r + B(y, y)$ has at most one distinct real root. Hence its discriminant must be nonpositive, that is,

$$\{\text{Re}[\alpha B(x, y)]\}^2 \leq B(x, x)B(y, y) \tag{6.6}$$

for all α with $| \alpha | = 1$. Choose α so that $\alpha B(x, y) = | B(x, y) |$. Then inequality (6.6) yields

$$| B(x, y) |^2 \leq B(x, x)B(y, y).$$ ∎

Remark 6.1. Clearly if B is a positive sesquilinear form on V and

$B(x, y) = B(x, z)$ for all x in V, then $y = z$. Indeed

$$B(y - z, y - z) = B(y - z, y) - B(y - z, z) = 0.$$

Other interesting and useful facts are given in the following corollary of Proposition 6.1 (i).

Corollary 6.1. Assume V is a complex vector space.

(i) If $B\colon V \times V \to C$ and $B'\colon V \times V \to C$ are sesquilinear forms such that $B(x, x) = B'(x, x)$ for all x, then $B = B'$.

(ii) A sesquilinear form $B\colon V \times V \to C$ is Hermitian if and only if $B(x, x)$ is real for all x. ▮

Proof. The proof of (i) is readily seen by examining equation (6.1). To prove (ii) note that $B(x, x)$ is real if B is Hermitian since $B(x, x) = \overline{B(x, x)}$. Conversely, if $B(x, x)$ is real for all x, the sesquilinear form $B'(x, y) = \overline{B(y, x)}$ is such that $B'(x, x) = B(x, x)$ for all x. By (i), $B = B'$ or B is Hermitian. ▮

With the information given in Proposition 6.1 we are in a good position to begin our study of Hilbert and pre-Hilbert spaces.

Definition 6.2. A *pre-Hilbert space P* over the field F is a vector space P over F together with a positive Hermitian sesquilinear form on P. ▮

The sesquilinear form in a pre-Hilbert space is often called an *inner product* and a pre-Hilbert space is accordingly called an *inner product space*. The image in F of the ordered pair (x, y) in $P \times P$ by the inner product B on P will be denoted by $(x \mid y)$ instead of $B(x, y)$.

Examples. Here are some simple yet important examples of inner product spaces:

6.1. For any positive integer n the space $C^n(R^n)$ of ordered n-tuples $x = (x_1, \ldots, x_n)$ of complex (real) numbers with inner product given by

$$(x \mid y) = \sum_{i=1}^{n} x_i \overline{y_i}.$$

6.2. The space l_2 of all complex (real) sequences $x = (x_i)_N$ such that $\sum_{i=1}^{\infty} | x_i |^2 < \infty$ with inner product given by

$$(x \mid y) = \sum_{i=1}^{\infty} x_i \overline{y_i}.$$

6.3. For any measure space $(X, \mathscr{A}, \mu)$, the space $L_2(\mu)$ of all measurable functions f for which $\int |f|^2\, d\mu < \infty$ with inner product given by

$$(f \mid g) = \int f\bar{g}\, d\mu.$$

[Note that Examples 6.1 and 6.2 are special cases of Example 6.3 if X is chosen to be $\{1, 2, \ldots, n\}$ and N, respectively—each with the counting measurc.]

6.4. The vector space of continuous functions f on an interval $[a, b]$ with inner product

$$(f \mid g) = \int_a^b f(t)\overline{g(t)}\, dt.$$

Any pre-Hilbert space P is a normed linear space by virtue of the following definition: If $x \in P$, define the norm of x by

$$\| x \| = (x \mid x)^{1/2}. \tag{6.7}$$

Since an inner product is a positive sesquilinear form, we have

$$\| x \| \geq 0 \text{ and } \| x \| = 0 \qquad \text{if and only if } x = 0.$$

Also,

$$\| \alpha x \| = | \alpha |\, \| x \| \qquad \text{since } \| \alpha x \|^2 = (\alpha x \mid \alpha x)$$
$$= \alpha\bar{\alpha}(x \mid x) = | \alpha |^2\, \| x \|^2.$$

Finally, Proposition 6.1 (iv) becomes the triangle inequality

$$\| x + y \| < \| x \| + \| y \|. \tag{6.8}$$

One should also note that in any inner product space the *Parallelogram Law* and the *Cauchy–Schwarz Inequality*, respectively, now have the following forms:

$$\| x + y \|^2 + \| x - y \|^2 = 2\, \| x \|^2 + 2\, \| y \|^2, \tag{6.9}$$

$$| (x \mid y) | \leq \| x \|\, \| y \|. \tag{6.10}$$

Examining the geometrical meaning of the Parallelogram Law in R^2 demonstrates the aptness of its title: The sum of the squares of the diagonals in a parallelogram is equal to the sum of the squares of the four sides.

Remark 6.2. If $x_n \to x$ and $y_n \to y$ in P, then $(x_n \mid y_n) \to (x \mid y)$ in F. This follows from the inequality

$$| (x \mid y) - (x_n \mid y_n) | = | (x \mid y) - (x \mid y_n) + (x \mid y_n) - (x_n \mid y_n) |$$
$$\leq \| x \|\, \| y - y_n \| + \| x - x_n \|\, \| y_n \|.$$

Definition 6.3. A *Hilbert space* is a complete pre-Hilbert space with norm $\| x \| = (x \mid x)^{1/2}$. ▮

Not all pre-Hilbert spaces are Hilbert spaces. For example the subspace of l_2 [Example 6.2] consisting of finitely nonzero sequences $x = (x_i)_N$ [a sequence $x = (x_i)_N$ is finitely nonzero if there exists some positive integer M such that $x_i = 0$ for all $i > M$] is a pre-Hilbert space that is not complete. Also Example 6.4 is not complete. The completion (see Problem 6.1.3) of this space is the space $L_2([a, b])$. The verifications of these statements are left as exercises (Problem 6.1.4).

Briefly we can say that a Hilbert space is a Banach space with the norm defined by an inner product as in equation (6.7). When is a Banach space a Hilbert space? The Parallelogram Law gives us one characterization.† Precisely, we have the following characterization, whose proof we have outlined in the Problems (Problem 6.1.5).

Proposition 6.2. A Banach space is a Hilbert space with its norm given by an inner product if and only if its norm satisfies the parallelogram identity (6.9). ▮

Definition 6.4. Two vectors x and y in a pre-Hilbert space P are said to be *orthogonal* (or perpendicular), written $x \perp y$, if $(x \mid y) = 0$. If E and F are subsets of P, then E and F are said to be *orthogonal* (to each other), written $E \perp F$, if $x \perp y$ for each x in E and y in F. A subset E of P is said to be an *orthogonal set* if $x \perp y$ for each nonequal pair of vectors x and y in E. If in addition $\| x \| = 1$ for each x in E, then E is said to be *orthonormal.* ▮

Remark 6.3. Any orthogonal set E in a pre-Hilbert space P that does not contain the zero vector is linearly independent. Indeed, if $\{x_1, x_2, \ldots, x_n\}$ is a finite subset of E and $\alpha_1 x_1 + \alpha_2 x_2 + \cdots + \alpha_n x_n = 0$, then $\alpha_j \| x_j \|^2 = (\sum_{i=1}^n \alpha_i x_i \mid x_j) = 0$ so that $\alpha_j = 0$ for each j.

In the space l_2 the countable set of vectors like $(0, 0, \ldots, 1, 0, 0, \ldots)$ where 1 is the ith coordinate for $i = 1, 2, \ldots$ is an orthonormal set. In R^3, the set $\{(1, 1, 0), (0, -1, 0)\}$ is independent but not orthogonal. In $L_2([0, 2\pi])$ the set $\{e^{int} \mid n = 0, \pm 1, \pm 2, \ldots\}$ is orthogonal, but not orthonormal.

Definition 6.5. A family of vectors $(x_i)_{i \in I}$ in a normed linear space is

† For other characterizations, see [37].

called *summable* to x, written $\sum_I x_i = x$, if for each $\varepsilon > 0$ there exists a finite subset $F(\varepsilon)$ of I such that if J is a finite subset of I containing $F(\varepsilon)$ then $\| \sum_{i \in J} x_i - x \| < \varepsilon$. ∎

It can be shown by the reader that in a Banach space a family $(x_i)_I$ is summable (to some x) if and only if for each number $\varepsilon > 0$ there exists a finite subset $F(\varepsilon)$ of I such that if J is a finite subset of I with $J \cap F(\varepsilon) = \varnothing$, then $\| \sum_{i \in J} x_i \| < \varepsilon$. From this criterion it follows that if $(x_i)_I$ is summable, the set of indices i for which $x_i \neq 0$ is at most countable. Indeed for each positive integer n, let $F(1/n)$ be the finite subset of I such that $\| \sum_J x_i \| < 1/n$ if $J \cap F(1/n) = \varnothing$ and J is finite. If $x \notin \bigcup_{n=1}^{\infty} F(1/n)$, a countable set, $\| x \| < 1/n$ for all n.

It is easy to verify the following rules in any pre-Hilbert space:

(i) If $\sum_I x_i = x$, then $\sum_I \alpha x_i = \alpha x$ for any scalar α.

(ii) If $\sum_I x_i = x$ and $\sum_I y_i = y$, then $\sum_I x_i + y_i = x + y$. (6.11)

(iii) If $\sum_I x_i = x$, then $\sum_I (x_i \mid y) = (x \mid y)$ and $\sum_I (y \mid x_i) = (y \mid x)$ for every vector y.

To verify (i) for instance, let $\varepsilon > 0$ be arbitrary. Then there is a finite subset $F(\varepsilon)$ such that if J is finite and $J \supset F(\varepsilon)$, then $\| \sum_J \alpha x_i - \alpha x \| = | \alpha | \, \| \sum_J x_i - x \| < | \alpha | \, \varepsilon$. Hence $\sum_I \alpha x_i = \alpha x$.

Proposition 6.3. *Pythagorean Theorem.*

(i) If $\{x_1, x_2, \ldots, x_n\}$ is any orthogonal family of vectors in Hilbert space H, then

$$\left\| \sum_{i=1}^{n} x_i \right\|^2 = \sum_{i=1}^{n} \| x_i \|^2.$$

(ii) Any orthogonal family $(x_i)_I$ of vectors in H is summable if and only if $(\| x_i \|^2)_I$ is summable. If $x = \sum_I x_i$, then $\| x \|^2 = \sum_I \| x_i \|^2$. ∎

Proof. An inductive argument proves (i). To prove (ii) note that $(x_i)_I$ is summable if and only if for each $\varepsilon > 0$ there exists a finite subset $F(\varepsilon)$ of I such that if J is a finite subset of I with $J \cap F(\varepsilon) = \varnothing$, then

$$\sum_J \| x_j \|^2 = \| \sum_J x_j \|^2 < \varepsilon^2. \tag{6.12}$$

By virtue of the equality in equation (6.12), this condition is also necessary and sufficient for $(\| x_i \|^2)_I$ to be summable.

If $x = \sum_I x_i$, then by equation (6.11) we have

$$\| x \|^2 = (x \mid x) = \left(\sum_I x_i \,\middle|\, x\right)$$
$$= \sum_I (x_i \mid x) = \sum_{i\in I} \left(x_i \,\middle|\, \sum_{j\in I} x_j\right) = \sum_{i\in I} \sum_{j\in I} (x_i \mid x_j) = \sum_I \| x_i \|^2. \quad \blacksquare$$

Proposition 6.4. *Bessel's Inequality*. Let P be a pre-Hilbert space.

(i) If $\{x_1, \ldots, x_n\}$ is any finite family of orthonormal vectors in P and x is any vector in P, then

$$\sum_{i=1}^{n} | (x \mid x_i) |^2 \leq \| x \|^2.$$

(ii) If $(x_i)_I$ is any orthonormal family of vectors in P and x is any vector in P, then

$$\sum_I | (x \mid x_i) |^2 \leq \| x \|^2. \quad \blacksquare$$

Proof. (i)

$$0 \leq \left\| x - \sum_{i=1}^{n} (x \mid x_i)x_i \right\|^2 = \| x \|^2 - \sum_{i=1}^{n} \overline{(x \mid x_i)}(x \mid x_i)$$
$$- \sum_{i=1}^{n} (x \mid x_i)(x_i \mid x) + \sum_{i=1}^{n} (x \mid x_i)\overline{(x \mid x_i)} = \| x \|^2 - \sum_{i=1}^{n} | (x \mid x_i) |^2,$$

from which (i) follows.

(ii) By (i), $\sum_{i\in F} | (x \mid x_i) |^2 \leq \| x \|^2$ for any finite subset F of I. Hence by the definition of summability the inequality must hold for I. $\blacksquare$

Using the concept of summability of an arbitrary family of vectors or scalars, we can give an example of a class of Hilbert spaces which we shall see later represents all Hilbert spaces.

Example 6.5. For any nonempty set I let $C^I(R^I)$ be the vector space of all complex- (real-) valued functions on I, that is, the set of all families of elements $(x_i)_{i\in I}$, where x_i is a scalar. Let $l_2(I)$ be the vector subspace of $C^I(R^I)$ of all families $(x_i)_{i\in I}$ such that $\sum_I | x_i |^2$ is summable (written $\sum_I | x_i |^2 < \infty$). $l_2(I)$ is a Hilbert space with inner product given by $(x \mid y) = \sum_I x_i \bar{y}_i$ for $x = (x_i)_I$ and $y = (y_i)_I$. Using the Hölder Inequality (Proposition 5.3 in Chapter 5), the reader can verify that this does in fact define an inner product, and in particular that $(x_i\bar{y}_i)_I$ is summable. We here establish the completeness of $l_2(I)$. [The completeness also follows from that of L_2 (see Theorem 5.4 in Chapter 5), where the measure is the counting measure; but we here give a different proof.]

To this end, let $x^k = (x_i^k)_{i\in I}$, $k = 1, 2, \ldots$ be a Cauchy sequence in $l_2(I)$. Sincc for each i

$$\| x_i^n - x_i^m \|^2 \leq \sum_{i\in I} | x_i^n - x_i^m |^2 = \| x^n - x^m \|^2,$$

$(x_i^k)_{k=1,2,\ldots}$ is a Cauchy sequence of scalars for each i in I. Hence there exists for each i in I an x_i such that $\lim_{k\to\infty} x_i^k = x_i$. Let $x = (x_i)_{i\in I}$. We wish to show $x \in l_2(I)$ and $x^k \to x$ in $l_2(I)$.

Let J be any finite subset of I and $\varepsilon > 0$ be arbitrary. Then there exists $N > 0$ such that if n, $m > N$, then

$$\sum_{i\in J} | x_i^n - x_i^m |^2 \leq \sum_{i\in I} | x_i^n - x_i^m |^2 \leq \varepsilon^2.$$

Letting $n \to \infty$, then for $m \geq N$

$$\sum_{i\in J} | x_i - x_i^m |^2 \leq \varepsilon^2.$$

Since J is an arbitrary finite subset of I, this means $\sum_I | x_i - x_i^m |^2 \leq \varepsilon^2$ and $\| (x_i - x_i^m)_I \| \to 0$ as $m \to \infty$. In particular $(x_i - x_i^m)_I$ is in $l_2(I)$, so that $(x_i)_I = (x_i - x_i^m)_I + (x_i^m)_I$ is in $l_2(I)$.

Problems

6.1.1. (i) If x $(\neq 0)$ and y are any vectors in a pre-Hilbert space, prove $| (x \mid y) = \| x \| \| y \|$ if and only if $y = \lambda x$ for some $\lambda \in F$. [Hint: Look at the proof of Proposition 6.1 (iii).]

(ii) If x and y are nonzero vectors, prove $\| x + y \| = \| x \| + \| y \|$ if and only if $y = \lambda x$ for some $\lambda > 0$.

(iii) Prove $\| x - z \| = \| x - y \| + \| y - z \|$ if and only if $y = \alpha x + (1 - \alpha)z$ for some α in $[0, 1]$.

6.1.2. (i) Prove that if (x_n) is an orthogonal sequence of vectors in a pre-Hilbert space such that $\sum_{i=1}^{\infty} \| x_i \|^2 < \infty$, then the sequence $(\sum_{i=1}^{n} x_i)_{n\in N}$ is a Cauchy sequence.

(ii) Give an example where the conclusion of (i) may fail if (x_n) is not orthogonal.

6.1.3. If P is a pre-Hilbert space with inner product $(x \mid y)$, prove that there is a Hilbert space H with inner product $B(x, y)$ and a linear map T: $P \to H$ such that $B(Tx, Ty) = (x \mid y)$ for all x and y in P and $T(P)$ is dense in H. Prove that if (H', B') is another Hilbert space satisfying these criteria, then H and H' are linearly isometric—that is, there is a linear map S from H onto H' such that $B'(Sx, Sy) = B(x, y)$ for all x and y in H. H is called the *completion* of P.

6.1.4. (i) Prove that the subspace of l_2 (see Example 6.2) consisting of finitely nonzero sequences is not complete, but its completion is l_2.

(ii) Prove that the space of Example 6.4 is not complete, but its completion is $L_2([a, b])$. {Hint: Look at the sequence $f_n(t) = 0$ if $a \leq t \leq (a + b)/2$; $= n[t - (a + b)/2]$ if $(a + b)/2 \leq t \leq (a + b)/2 + 1/n$; $= 1$ otherwise.}

6.1.5. (i) Prove Proposition 6.2 by showing that each Banach space whose norm satisfies the Parallelogram Law, equation (6.9), is a Hilbert space. [Hint: If B is a real Banach space, define $(x \mid y)$ as in equation (6.2) by $\frac{1}{4}\{\| x + y \|^2 - \| x - y \|^2\}$ while if B is a complex Banach space define $(x \mid y)$ as in equation (6.1) by $\frac{1}{4}\{\| x + y \|^2 - \| x - y \|^2 + i \| x + iy \|^2 - i \| x - iy \|^2\}$. In the real case show $(x \mid y) + (z \mid y) = (x + z \mid y)$, $(x_n \mid y) \to (x \mid y)$ if $x_n \to x$ in B, and conclude $(\alpha x \mid y) = \alpha(x \mid y)$ for all real α. Note that in the complex case $\text{Im}(x \mid y) = \text{Re}(x \mid iy)$.]

(ii) Prove $L_1[0, 1]$ is not a Hilbert space by showing that the Parallelogram Law is not satisfied.

6.1.6. Show that the result in Problem 6.1.5 (i) can be extended as follows: Let V be a real vector space and $\| \cdot \|$: $V \to R$ be a function satisfying the parallelogram law [equation (6.9)] and the following property: For every $x \in V$, the function $\alpha \to \| \alpha \cdot x \|$ on R is continuous at 0. Then $(x \mid y) = \frac{1}{4}\{\| x + y \|^2 - \| x - y \|^2\}$ defines a nonnegative Hermitian bilinear form on V. (This extension is due to D. Fearnley-Sander and J. Symons.)

6.1.7. State and prove a complex version of the result outlined in Problem 6.1.6.

6.2. Subspaces, Bases, and Characterizations of Hilbert Spaces

We will now turn our attention to subspaces of pre-Hilbert and Hilbert spaces. It is clear that any vector subspace of a pre-Hilbert space is a pre-Hilbert space with the restricted inner product.

Crucial to the study of the structure of Hilbert spaces and subspaces is the following result, not valid in every normed linear space.

Theorem 6.1. Let S be a complete and convex $\{x, y \in S$ implies $\alpha x + (1-\alpha)y \in S$ for all $\alpha \in [0, 1]\}$ subset of a pre-Hilbert space P. Given any vector x in P there exists one and only one vector $y_0 \in S$ such that $\| x - y_0 \| \leq \| x - y \|$ for all y in S. ▮

(In regard to this theorem see Problems 6.2.1 and 6.2.2.)

Proof. Let y_i be a sequence of vectors in S such that $\| x - y_i \|$ converges to δ, the inf of $\{\| x - y \|: y \text{ in } S\}$. We will show that y_i is a Cauchy sequence in S converging to the desired vector y_0. Using the Parallelogram Law, equation (6.9) of Section 6.1,

$$\| (y_i - x) + (x - y_j) \|^2 + \| (y_i - x) - (x - y_j) \|^2 = 2 \| y_i - x \|^2 + 2 \| x - y_j \|^2$$

or

$$\| y_i - y_j \|^2 = 2 \| y_i - x \|^2 + 2 \| x - y_j \|^2 - 4 \| \tfrac{1}{2}(y_i + y_j) - x \|^2.$$

Since

$$\tfrac{1}{2}(y_i + y_j) \in S, \quad \| \tfrac{1}{2}(y_i + y_j) - x \|^2 \geq \delta^2.$$

Hence

$$\| y_i - y_j \|^2 \leq 2 \| y_i - x \|^2 + 2 \| x - y_j \|^2 - 4\delta^2. \tag{6.13}$$

As $i, j \to \infty$, the right-hand side of equation (6.13) goes to zero so that $(y_i)_N$ is a Cauchy sequence in S. Since S is complete, y_i converges to some y_0 in S. Since $\| y_i - x \| \to \| y_0 - x \|$, $\| y_0 - x \| = \delta$.

If $y_0' \in S$ also satisfies $\| y_0' - x \| = \delta$, then using the Parallelogram Law again

$$\begin{aligned} \| y_0 - y_0' \|^2 &= 2 \| y_0 - x \|^2 + 2 \| x - y_0' \|^2 - 4 \| \tfrac{1}{2}(y_0 + y_0') - x \|^2 \\ &\leq 2 \| y_0 - x \|^2 + 2 \| x - y_0' \|^2 - 4\delta^2 = 4\delta^2 - 4\delta^2 = 0. \end{aligned}$$ ▮

Lemma 6.1. If S is a proper complete subspace of pre-Hilbert space P, then there exists x in $P - S$ such that $\{x\} \perp S$. ▮

Proof. By Theorem 6.1 for any vector z in $P - S$ there exists a unique vector $y_0(z)$ in S such that $\| z - y_0(z) \| \leq \|z - y \|$ for all y in S. Let $x = z - y_0(z)$. We will show $x \in P - S$ and $\{x\} \perp S$. Clearly $x \notin S$ since

$z \notin S$. Since for every scalar α, $y_0(z) + \alpha y \in S$ for every y in S,

$$\| \alpha y - x \|^2 = \| [y_0(z) + \alpha y] - z \|^2 \geq \| z - y_0(z) \|^2 = \| x \|^2.$$

Hence

$$0 \leq \| x - \alpha y \|^2 - \| x \|^2 = -\alpha(y \mid x) - \bar{\alpha}(x \mid y) + \alpha\bar{\alpha} \| y \|^2.$$

Letting $\alpha = -\beta(x \mid y)$ for any *real* β we get

$$\begin{aligned} 0 &\leq \beta(x \mid y)(y \mid x) + \beta\overline{(x \mid y)}(x \mid y) + \beta^2(x \mid y)\overline{(x \mid y)} \| y \|^2 \\ &= 2\beta \mid (x \mid y) \mid^2 + \beta^2 \mid (x \mid y)^2 \| y \|^2 \\ &= \beta \mid (x \mid y) \mid^2[2 + \beta \| y \|^2]. \end{aligned} \tag{6.14}$$

If β is chosen to be a negative number such that $\beta > -2/\| y \|^2$, equation (6.14) forces $\| (x \mid y) \| = 0$ or $x \perp y$. ∎

Definition 6.6. If S is any subset of a pre-Hilbert space P, the *annihilator* or *orthogonal complement* of S is the set

$$S^\perp = \{x \in P\colon x \perp y \text{ for all } y \text{ in } S\}. \qquad ∎$$

The orthogonal complement of any set S always contains the zero element of P. Clearly $S \cap S^\perp \subset \{0\}$ and $S \subset (S^\perp)^\perp$. More can be said.

Lemma 6.2. If S is any subset of P, then $S^\perp$ is a closed subspace of P. ∎

The proof is easy using Remark 6.2.

Note that in Theorem 6.1 and Lemma 6.1 if P is a Hilbert space, the word "complete" may be replaced by "closed" since in any complete metric space a subset is closed if and only if it is complete. In particular, Lemma 6.2 assures us that $S^\perp$ is complete in a Hilbert space.

If M and N are subspaces of a pre-Hilbert space P, then $M + N$ is the subspace defined as $\{m + n\colon m \in M, n \in N\}$. If $M \perp N$, then each element of $M + N$ has a unique representation as $m + n$ with $m \in M$ and $n \in N$. Indeed, if $m + n = m' + n'$ with $m, m' \in M$ and $n, n' \in N$, then $m - m' = n' - n$. Hence $m - m' \in M \cap N$ so that $(m - m' \mid m - m') = 0$. Hence $m = m'$. Similarly $n = n'$. In this case we write $M + N$ as $M \oplus N$.

Theorem 6.2. If M is a complete linear subspace of pre-Hilbert space P, then $P = M \oplus M^\perp$ and $M = (M^\perp)^\perp$. (See also Problem 6.2.6.) ∎

Proof. Let z be any vector in P. By Theorem 6.1, there exists a unique vector $y_0(z)$ in M such that

$$\| z - y_0(z) \| \leq \| z - y \| \text{ for all } y \text{ in } M.$$

As shown in the proof of Lemma 6.1, $x(z) = z - y_0(z)$ is in $M^\perp$. Hence $z = y_0(z) + x(z) \in M + M^\perp$. Since $M \perp M^\perp$, $P = M + M^\perp = M \oplus M^\perp$.

Clearly for any set S, $S \subset (S^\perp)^\perp$. If $z \in (M^\perp)^\perp$, then $z = y + x$ with $y \in M$ and $x \in M^\perp$. Since also $y \in (M^\perp)^\perp$, $x = z - y \in (M^\perp)^\perp$. But $(M^\perp) \cap (M^\perp)^\perp = 0$, so $x = 0$. Thus $z = y \in M$. Hence $(M^\perp)^\perp = M$. ∎

If S is a set in a pre-Hilbert space P, we denote by $[S]$ the smallest subspace of P containing S. It is easy to see that $[S]$ is the vector space of all finite linear combinations of elements of S.

Proposition 6.5. (*The Gram–Schmidt Orthonormalization Process*). If $\{x_i: i = 1, 2, \ldots, N\}$ for $1 \leq N \leq \infty$ is a linearly independent set in a pre-Hilbert space P, then there is an orthonormal set $\{z_i: i = 2, \ldots, N\}$ such that $[\{z_i: i = 1, 2, \ldots, n\}] = [\{x_i: i = 1, 2, \ldots, n\}]$ for each $n = 1, 2, \ldots, N$. ∎

Proof. For $N = \infty$ we proceed by induction. Let $z_1 = \| x_1 \|^{-1} x_1$. Clearly $[x_1] = [z_1]$. Assume an orthonormal set $\{z_i: i = 1, 2, \ldots, n - 1\}$ exists such that $[\{z_i: i = 1, \ldots, n - 1\}] = [\{x_i: i = 1, \ldots, n - 1\}]$. Define $z_n = \| y_n \|^{-1} y_n$, where $y_n = x_n - \sum_{i=1}^{n-1} (x_n \mid z_i) z_i$. This equation and the inductive hypothesis guarantee that each z_i for $i = 1, 2, \ldots, n$ is a linear combination of the set $\{x_i: i = 1, \ldots, n\}$ and each x_i for $i = 1, 2, \ldots, n$ is a linear combination of the set $\{z_i: i = 1, 2, \ldots, n\}$. Also for each $j = 1, 2, \ldots, n - 1$

$$\begin{aligned}(z_n \mid z_j) = \| y_n \|^{-1} (y_n \mid z_j) &= \| y_n \|^{-1} \left\{ (x_n \mid z_j) - \sum_{i=1}^{n-1} (x_n \mid z_i)(z_i \mid z_j) \right\} \\ &= \| y_n \|^{-1} \{ (x_n \mid z_j) - (x_n \mid z_j)(z_j \mid z_j) \} \\ &= 0.\end{aligned}$$

The modifications for the case when N is finite are obvious. ∎

If $\{x_i: i = 1, 2, \ldots, N\}$ for some N in $1 \leq N \leq \infty$ is a vector space basis for a pre-Hilbert space P, then by Proposition 6.5, P has a basis $\{z_i: i = 1, 2, \ldots, N\}$ which is an orthonormal set. This basis $\{z_i: i = 1, 2, \ldots, N\}$ has the following property: If x in P is orthonormal to $\{z_i:$

$i = 1, 2, \ldots, N\}$, then $x = 0$. Indeed if x is in P, $x = \sum_{i=1}^{N} \alpha_i z_i$ for some scalars α_i. However, for each $j = 1, 2, \ldots, N$

$$\alpha_j = \alpha_j \| z_j \| = (\alpha_j z_j \mid z_j) = \left(\sum_{i=1}^{N} \alpha_i z_i \mid z_j \right) = (x \mid z_j) = 0$$

so that $x = 0$. Because the set $\{z_i : i = 1, 2, \ldots, N\}$ has this property, it is an example of a "complete" orthonormal set. Precisely, we have the following definition.

Definition 6.7. An orthonormal set S in a pre-Hilbert space is *complete* if whenever $\{x\} \perp S$ then $x = 0$. A complete orthonormal set in a pre-Hilbert space is called a (pre-Hilbert space) *basis*. ∎

A word of caution and explanation is in order. The word "basis" is used with two different meanings. In one sense it means an algebraic or Hamel basis, that is, a linearly independent set in a vector space which spans the vector space. In the other sense it means a Hilbert space basis as defined in Definition 6.7.

Subsequently, whenever we speak of a basis we mean a pre-Hilbert space basis unless otherwise indicated.

Proposition 6.6. Let $(x_i)_I$ be an orthonormal family in a Hilbert space H. The following statements are equivalent.

(i) The family $(x_i)_I$ is a basis for H.

(ii) The family $(x_i)_I$ is a maximal orthonormal family in H.

(iii) If $x \in H$, then $x = \sum_I (x \mid x_i) x_i$ (the Fourier expansion of x).

(iv) If x and y are in H, then

$$(x \mid y) = \textstyle\sum_I (x \mid x_i)(x_i \mid y) \qquad \text{(Parseval's identity)}.$$

(v) If x is in H, then $\| x \|^2 = \sum_I | (x \mid x_i) |^2$. ∎

Proof. (i) ⇒ (ii). Suppose S is an orthonormal family containing $(x_i)_I$ and $s \in S - (x_i)_I$. Then $s \perp x_i$ for each $i \in I$ implies $s = 0$. However, this contradicts the fact that $\| s \| = 1$.

(ii) ⇒ (iii). By Bessel's Inequality (Proposition 6.4) the family

$$[| (x \mid x_i) |^2]_I$$

has a convergent sum since for any finite subset A of I,

$$\sum_{i\in A} |(x \mid x_i)|^2 \leq \|x\|^2.$$

Hence $\sum_I (x \mid x_i)x_i$ converges since for each $\varepsilon > 0$ there exists a finite subset $F(\varepsilon)$ of I so that if J is finite and $J \cap F(\varepsilon) = \varnothing$, then (using Proposition 6.3)

$$\Big\|\sum_J (x \mid x_i)x_i\Big\|^2 = \sum_J \|(x \mid x_i)x_i\|^2 = \sum_J |(x \mid x_i)|^2 < \varepsilon.$$

Let $y = \sum_I (x \mid x_i)x_i$. It remains to show $x = y$. For each $j \in I$,

$$(x - y \mid x_j) = \Big(x - \sum_I (x \mid x_i)x_i \,\Big|\, x_j\Big) = (x \mid x_j) - (x \mid x_j) = 0.$$

If $x - y \neq 0$, then $\{\|x - y\|^{-1}(x - y)\} \cup \{x_i\}_{i\in I}$ is an orthonormal family properly containing $\{x_i\}_{i\in I}$. Since this is impossible, $x = y$.

(iii) $\Rightarrow$ (iv).

$$\begin{aligned}(x \mid y) &= \Big(\sum_{i\in I} (x \mid x_i)x_i \,\Big|\, \sum_{j\in I} (y \mid x_j)x_j\Big) = \sum_{i\in I} \Big((x \mid x_i)x_i \,\Big|\, \sum_{j\in I} (y \mid x_j)x_j\Big) \\ &= \sum_{i\in I} (x \mid x_i)\Big(\sum_{j\in I} \overline{(y \mid x_j)}(x_i \mid x_j)\Big) = \sum_{i\in I} (x \mid x_i)(x_i \mid y).\end{aligned}$$

(iv) $\Rightarrow$ (v). Take $x = y$ in (iv).

(v) $\Rightarrow$ (i). If $(x \mid x_i) = 0$ for each i, then $\|x\|^2 = \sum (x \mid x_i)^2 = 0$. ∎

Does each pre-Hilbert space have a basis? It is sufficient to ask whether each pre-Hilbert space has a maximal orthonormal set. Consider the collection of all orthonormal sets in a given pre-Hilbert space. Order this collection by set inclusion. Since the union of an increasing family of orthonormal sets is orthonormal, Zorn's Lemma guarantees the existence of a maximal orthonormal set. We have proved the following theorem.

Theorem 6.3. Every pre-Hilbert space has a basis. ∎

Proposition 6.7. Any two bases of a pre-Hilbert space H have the same cardinality. ∎

Proof. If H is finitely generated, the result follows from the theory of finite-dimensional vector spaces. So assume that $\{x_i\}_I$ and $\{y_j\}_J$ are two bases of infinite cardinality. Since for each $k \in I$, $x_k = \sum_J (x_k \mid y_j)y_j$ by Proposition 6.6 (iii), the set J_k of those indices j for which $(x_k \mid y_j) \neq 0$

is at most countable. Since $\{x_i\}_I$ is a basis, no y_j can be orthogonal to each x_k. This means $J \subset \bigcup_{k\in I} J_k$. Hence

$$\text{card } J \leq \aleph_0 \cdot \text{card } I = \text{card } I.$$

A symmetrical argument gives card $I \leq$ card J. ∎

Proposition 6.7 gives meaning to the following definition.

Definition 6.8. The *dimension* of a pre-Hilbert space H is the cardinality of any basis for H. ∎

It is an interesting fact that in a pre-Hilbert space H the "distance" between any two distinct elements x_i and x_j of a basis $\{x_i\}_I$ is $2^{1/2}$. Indeed,

$$\| x_i - x_j \|^2 = (x_i - x_j \mid x_i - x_j) = (x_i \mid x_i) + (x_j \mid x_j) = 2.$$

It follows that an open neighborhood $N(x_i, 2^{1/2}/2) = \{x \in H: \| x - x_i \| < 2^{1/2}/2\}$ of x_i contains no other element of the basis $\{x_i\}_I$ except x_i. In fact the collection of such neighborhoods is pairwise disjoint. If S is a dense subset of H, each $N(x_i, 2^{1/2}/2)$ for $i \in I$ must contain a point of S. This means that the cardinality of I is no greater than that of S. In other words if H is a separable Hilbert (metric) space so that S is countable, card $I \leq \aleph_0$. We have partly proved the following proposition.

Proposition 6.8. The dimension of a Hilbert space H is less than or equal to $\aleph_0$ if and only if H is separable. ∎

The converse is left to the reader.

A *linear isometry* T from a Hilbert space H into a Hilbert space K is a linear mapping from H into K such that $\| Tx \| = \| x \|$ for all x in H. If $T: H \to K$ is a linear isometry, then it follows from the polarization identity (Proposition 6.1) and the equation $(Tx \mid Tx) = \| Tx \|^2 = \| x \|^2 = (x \mid x)$ that $(Tx \mid Ty) = (x \mid y)$ for all x and y in H. Hence $T: H \to K$ is a linear isometry if and only if T is linear and $(Tx \mid Ty) = (x \mid y)$ for all x and y in H. Two Hilbert spaces H and K are said to be *linearly isometric* if there is a linear isometry from H *onto* K.†

Proposition 6.9. Two Hilbert spaces H and K are linearly isometric if and only if they have the same dimension. ∎

† In this chapter we write Tx, rather than $T(x)$, as was done in Chapter 5, for ease of notation in the context of inner products.

Proof. If H and K are linearly isometric, let $T: H \to K$ be an isometry from H onto K. If $\{h_i: i \in I\}$ is a basis of H, then the set $\{Th_i: i \in I\}$ is a basis in K of the same cardinality.

Conversely, if the dimension of K equals the dimension of H, let $\{h_i: i \in I\}$ and $\{k_i: i \in I\}$ be bases of H and K, respectively, indexed by the same set I. If $h = \sum_I (h \mid h_i) h_i$ is in H, define Th as $\sum_I (h \mid h_i) k_i$. Clearly T is a linear mapping from H onto K and $\| Th \|^2 = \sum | (h \mid h_i) |^2 = \| h \|^2$. ▮

Theorem 6.4

(i) A Hilbert space of finite dimension n is linearly isometric to $l_2(I)$, where $I = \{1, 2, \ldots, n\}$.

(ii) A separable Hilbert space of infinite dimension is linearly isometric to l_2.

(iii) A Hilbert space of dimension α is linearly isometric to $l_2(I)$, where I is a set of cardinality α. ▮

Proof. For any set I, a basis for $l_2(I)$ is the set $B = \{f_i \in l_2(I): i \in I$ and $f_i(j) = \delta_{ij}$, the Kronecker delta. The dimension of $l_2(I)$ is then the cardinality of B, which is the cardinality of I. The result follows from Proposition 6.9. ▮

• **Remark 6.4.** *Applications of Proposition 6.6: Fourier Analysis in the Hilbert Space* L_2. Recall the definition of Fourier series from Chapter 5, Remark 5.5. The Fourier series of a function $f \in L_1 [-\pi, \pi]$ is sometimes written as the trigonometric series

$$\frac{1}{2} a_0 + \sum_{n=1}^{\infty} [a_n \cos nt + b_n \sin nt], \tag{6.15}$$

where

$$a_n = \frac{1}{\pi} \int_{-\pi}^{\pi} f(s) \cos ns \, ds, \; b_n = \frac{1}{\pi} \int_{-\pi}^{\pi} f(s) \sin ns \, ds.$$

Note that the series (6.15) can be easily derived from the series

$$\sum_{k=-\infty}^{\infty} \hat{f}(k) e^{ikt}, \qquad \hat{f}(k) = \frac{1}{2\pi} \int_{-\pi}^{\pi} f(s) e^{-iks} \, ds$$

by using the formula $e^{ins} = \cos ns + i \sin ns$. By the Riemann–Lebesgue theorem (see Problem 3.2.18), we have

$$\lim_{|n| \to \infty} \hat{f}(n) = 0, \qquad f \in L_1[-\pi, \pi]. \tag{6.16}$$

This fact also follows from Theorem A below. It can be proven (see Problem 6.2.8) that the family $\{(1/2\pi)^{1/2}e^{ikt}: k = 0, \pm 1, \pm 2, \dots\}$ is a basis for $L_2[-\pi, \pi]$. Using this fact, Theorem A below follows immediately from Proposition 6.6.

Theorem A. Let $f \in L_2[-\pi, \pi]$. Then

$$\lim_{n\to\infty} \int_{-\pi}^{\pi} \left| f(t) - \sum_{k=-n}^{n} \hat{f}(k)e^{ikt} \right|^2 dt = 0 \tag{6.17}$$

and

$$2\pi \sum_{k=-\infty}^{\infty} |\hat{f}(k)|^2 = \int_{-\pi}^{\pi} |f(t)|^2 \, dt. \tag{6.18}$$

For $g \in L_2[-\pi, \pi]$,

$$\int_{-\pi}^{\pi} f(t)\overline{g(t)}\, dt = 2\pi \sum_{k=-\infty}^{\infty} \hat{f}(k) \cdot \overline{\hat{g}(k)} \tag{6.19}$$

and

if $(a_k)_{k=-\infty}^{\infty}$ is a sequence of scalars such that $\sum_{k=-\infty}^{\infty} |a_k|^2 < \infty$, (6.20)

then there exists a unique $f \in L_2[-\pi, \pi]$ such that $f = \sum_{k=-\infty}^{\infty} a_k e^{ikt}$ and $a_k = \hat{f}(k)$. ▮

If we take $g = \chi_{[a,b]}$, $-\pi \le a < b < \pi$, in equation (6.19) above, we obtain the following striking result.

Theorem B. For $f \in L_2[-\pi, \pi]$,

$$\int_a^b f(t)\, dt = \sum_{k=-\infty}^{\infty} \int_a^b \hat{f}(k)e^{ikt}\, dt.$$ ▮

This theorem is surprising since we have obtained above the integral of f by integrating the terms of the Fourier series. Note that usually term by term integration of an infinite series is only possible by an assumption of uniform convergence, and in the case of the Fourier series above we have not assumed even pointwise convergence of the series.

Now we present a useful relationship between the convolution products of functions in L_2 and their Fourier coefficients. Let $f, g \in L_2[-\pi, \pi]$. We continue these functions on R with period 2π. For convenience, we define $f * g$ (a little differently from the definition given in Problem 5.2.15 in Chapter 5) as

$$f * g(t) = \frac{1}{2\pi} \int_{-\pi}^{\pi} f(t - s) g(s)\, ds. \tag{6.21}$$

Then $f * g$ is a continuous function on $[-\pi, \pi]$, since

$$2\pi \,|\, f * g(t_2) - f * g(t_1) \,| \leq \| f(t_2 - s) - f(t_1 - s) \|_2 \, \| g(s) \|_2 \to 0 \quad \text{as } | t_1 - t_2 | \to 0.$$

(Note that this convergence to zero is trivially justified when f is continuous, and therefore it follows for $f \in L_2$ since the continuous functions are dense in L_2.)

Theorem C. The Fourier series for $f * g$ [as defined in equation (6.21) above] is absolutely convergent (uniformly in t) and given by

$$f * g(t) = \sum_{k=-\infty}^{\infty} \hat{f}(k) \hat{g}(k) e^{ikt}. \quad \blacksquare \tag{6.22}$$

Proof. By using the periodicity of the functions, substitution, and Fubini's Theorem one can easily verify that

$$(\widehat{f * g})(k) = \hat{f}(k) \hat{g}(k);$$

and thus the series in equation (6.22) is the Fourier series of $f * g$. Since $(\hat{f}(n))$ and $(\hat{g}(n))$ are both in l_2 by equation (6.18), it follows that the series in equation (6.22) is absolutely convergent, uniformly in t, and thus has a continuous sum function, say $F(t)$. Since the series is uniformly convergent, term-by-term integration is possible and it follows that the series is the Fourier series of F. Now it is clear from equation (6.17) that $F = f * g$ almost everywhere. Since these functions are continuous, $F = f * g$. ∎

Problems

6.2.1. Show that Theorem 6.1 may be false if S is not complete. [Hint: Consider the set S of all real sequences (x_i) such that $\sum_{i=1}^{\infty} x_i = 1$ in the pre-Hilbert space of finitely nonzero sequences of real numbers. S contains no vector of minimum norm.]

6.2.2. (i) Show that the uniqueness conclusion of Theorem 6.1 is not valid in $R \times R$ with norm $\| (x, y) \| = \max\{\| x \|, \| y \|\}$.

(ii) Show that the existence conclusion of Theorem 6.1 is not valid in some normed linear spaces.

6.2.3. In the space $L_2[0, 1]$, let $\{f_1, f_2, \ldots\}$ be the orthonormal set obtained from the set $\{1, x, x^2, \ldots\}$ by the Gram–Schmidt orthonormalization process. Using the fact that the set of polynomials is dense in $L_2[0, 1]$, prove that $\{f_1, f_2, \ldots\}$ is a basis of $L_2[0, 1]$.

6.2.4. (i) Show that on any real or complex vector space an inner product can be defined.

(ii) Give an example of a vector space (necessarily infinite dimensional) that is a pre-Hilbert space under two different inner products but the completions of each of the pre-Hilbert spaces are not linearly isometric Hilbert spaces. [Hint: Consider l_2 and form a new inner product by considering a Hamel basis $(x_i)_{i \in I}$ such that $(x_i \mid x_j) = \delta_{ij}$.]

6.2.5. As in Example 6.4, let P be the pre-Hilbert space of continuous real-valued functions on $[-1, 1]$. Show that the set of odd functions O is orthogonal to the set of even functions E. Show $P = E \oplus O$. Is $E = O^\perp$?

6.2.6. Let P be the subspace of Example 6.2 of finitely nonzero real sequences. Let M be the subspace of P of all sequences $x = (x_k)$ such that $\sum_{k=1} x_k/k = 0$. Prove that M is closed but $M \neq (M^\perp)^\perp$. Compare Theorem 6.2. (Hint: Show $M^\perp = \{0\}$ since $\{1, 0, 0, \ldots, -n, 0, \ldots\}$ is in M where $-n$ is the nth coordinate.)

6.2.7. Prove Theorem 6.3 for separable Hilbert spaces without using Zorn's Lemma (or an equivalent).

6.2.8. Prove that $S = \{e^{ikt}/(2\pi)^{1/2}: k = 0, \pm 1, \pm 2, \ldots\}$ is a basis for $L_2([0, 2\pi])$. {Hint: Show S is complete by showing $f = 0$ a.e. if $\int_0^{2\pi} f(t) e^{-ikt} = 0$ and $f \in L_2([0, 2\pi])$. To accomplish this let $F(t)$ be the periodic absolutely continuous function $F(t) = \int_0^t f(u)\, du$. By partial integration show that $\int_0^{2\pi} [F(t) - c] e^{-ikt}\, dt = 0$ for $k = \pm 1, \pm 2, \ldots$. Pick c so that this is true also if $k = 0$. Since $F - c$ has period 2π, approximate $F - c$ uniformly on $[0, 2\pi]$ by a trigonometric polynomial $T(t) = \sum_{-n}^{n} c_k e^{ikt}$ (see Corollary 1.2 in Chapter 1). Show that $\int_0^{2\pi} | F(t) - c |^2\, dt$ is arbitrarily small. Hence $F(t) = c$ a.e.}

6.2.9. Show that (i)–(v) of Proposition 6.6 are equivalent in any pre-Hilbert space P and so are (i) and (ii).

6.2.10. Show that a maximal orthonormal set $\{x_\alpha\}$ in a *pre-Hilbert space P* need *not* be a basis for P, in the sense that it may not be possible to write each $x \in P$ uniquely as $\sum c_\alpha x_\alpha$. (Hint: Let $\{e_i\}$ be an orthonormal basis for a Hilbert space H and P be the linear subspace spanned by $\{\sum_{n=1}^{\infty} n^{-1} e_n, e_2, e_3, \ldots\}$. Then $\{e_2, e_3, \ldots\}$ is a maximal orthonormal set in P, though not a basis for P.)

★ **6.2.11.** Let H_1 and H_2 be Hilbert spaces over the same scalars, and let $T: H_1 \to H_2$ be a linear operator such that $T(B) = 0$, where B is an orthonormal basis for H_1. Suppose that $\overline{T(H_1)} = H_2$, G is the graph of T, and $H = H_1 \oplus H_2$ [the direct sum $\{(x, y): x \in H_1, y \in H_2\}$ with inner product $((x_1, y_1) \mid (x_2, y_2)) = (x_1 \mid x_2) + (y_1 \mid y_2)$]. Then show that $\bar{G} = H$ and $\dim G = \dim H_1$. (This result is taken from S. Gudder.†)

★ **6.2.12.** Is the dimension of a pre-Hilbert space the same as that of its completion? [Hint: Use Problem 6.2.11 to find the answer in the negative. Take $H_1 = l_2$, $\dim H_2 = c$, B an orthonormal basis for H_1, and $D(\supset B)$ a Hamel basis for H_1. Then $\dim D = c$. Let $T: H_1 \to H_2$ so that $T(B) = 0$ and $T(D - B)$ is an orthonormal basis for H_2. Then $\dim G$ (the graph of T) $= \aleph_0$, but $\dim \bar{G} = c$.]

6.2.13. Use Problem 6.2.12 to show that the "necessity" part of Proposition 6.8 is not necessarily true in a pre-Hilbert space.

6.2.14. *Measure in a Hilbert Space.* Let $(X, \mathscr{B}, \mu)$ be a measure space where $\mathscr{B}$ is the smallest σ-algebra containing the open sets of a Hilbert space X. Suppose $\mu(G) > 0$ for every open set G, $\mu(B + x) = \mu(B)$ for every $B \in \mathscr{B}$, $x \in X$, and X is infinite dimensional. Then prove that $\mu(G) = \infty$ for every nonempty open set G. [Hint: Let (x_i) be an infinite orthonormal set in X. Let $G = \{x: \| x \| < r\}$ and $G_n = \{x: \| x - (r/2) \cdot x_n \| < r/4\}$. Then each $G_n \subset G$, and G_n's are pairwise disjoint with the same positive measure.]

6.2.15. Prove that the linear subspace spanned by the set $\{x^n e^{-x^2/2}: n = 0, 1, 2, \ldots\}$ is dense in $L_2(-\infty, \infty)$. [Hint: Assume for some $f \in L_2(-\infty, \infty)$,

$$\int_{-\infty}^{\infty} f(t) e^{-t^2/2} t^n \, dt = 0 \text{ for } n = 0, 1, 2, \ldots.$$

Let

$$F(z) = \int_{-\infty}^{\infty} f(t) e^{-t^2/2} e^{itz} \, dt$$

† S. Gudder, *Amer. Math. Monthly* **81**(1), 29–36 (1974).

for complex numbers z. Show $F^{(k)}(z) = 0$ for $k = 1, 2, \ldots$ implying that $F(z)$ is identically zero. Therefore

$$\int_{-\infty}^{\infty} f(t)e^{-t^2/2}e^{itx} = 0$$

if $-\infty < x < \infty$. Multiply this equality by e^{-ixy}, where y is real and integrate with respect to x from $-w$ to w to get

$$\int_{-\infty}^{\infty} f(t)e^{-t^2/2} \frac{\sin w(t-y)}{t-y}\, dt = 0$$

for every w and y. Conclude $f(t) = 0$ a.e.]

6.3. The Dual Space and Adjoint Operators

In this section we will consider the dual of a Hilbert space and characterize each element of the dual space. Leading up to our study of operators in the next section, we will also define the adjoint of an operator and give its basic properties.

The reader will recall from Chapter 5 that the dual space E^* of a normed linear space E is the Banach space of all bounded linear functionals on E. In a pre-Hilbert space P each element y in P gives rise to a special element y^* of P^* defined by

$$y^*(x) = (x \mid y) \qquad \text{for all } x \text{ in } P. \tag{6.23}$$

It is easy to verify using the Cauchy–Schwarz Inequality [Proposition 6.1 (iii)] that y^* is a bounded linear functional with $\| y^* \| = \| y \|$. Thus we can define a mapping $y \to y^*$ of P into P^* that is an isometry and conjugate linear, that is,

$$(y + z)^* = y^* + z^* \qquad \text{for all } y, z \in P$$

$$(\alpha y)^* = \bar{\alpha} y^* \qquad \text{for all scalars } \alpha \text{ and all } y \text{ in } P.$$

Since P^* is always complete, P will be complete—a Hilbert space—if this mapping is surjective. The converse is also true, as shown in the next theorem.

Theorem 6.5. Let H be a Hilbert space. For each continuous linear functional f in H^*, there exists a unique y in H such that $f(x) = (x \mid y)$ for all x in H. Thus the conjugate linear isometry $y \to y^*$ [given by equation (6.23)] of a pre-Hilbert space P into its dual P^* is surjective if and only if P is actually a Hilbert space. ▮

Proof. Let N be the null space $\{x \in H: f(x) = 0\}$ of f, a closed linear subspace of H. If $N = H$, $f = 0$ and $y = 0$. If $N \neq H$, since $(N^\perp)^\perp = N$ by Theorem 6.2, $N^\perp$ does not equal the zero space. Let $z \in N^\perp$ with $z \neq 0$. Since $N \cap N^\perp = \{0\}$, $f(z) \neq 0$. Replacing z by $[f(z)]^{-1}z$ we may assume $f(z) = 1$.

Now for $x \in H, f(x - f(x)z) = f(x) - f(x) = 0$ so that $x - f(x)z \in N$. Since $z \in N^\perp$,

$$0 = (x - f(x)z \mid z) = (x \mid z) - f(x)(z \mid z)$$

or

$$f(x) = \frac{(x \mid z)}{(z \mid z)} = \left(x \,\middle|\, \frac{z}{\| z \|^2}\right).$$

Letting $y = z/\| z \|^2$, $f(x) = (x \mid y)$, where y is independent of x.

If also w is in H so that $f(x) = (x \mid w)$ for all x in H, then in particular $f(w - y) = (w - y \mid w) = (w - y \mid y)$ or $w = y$. ▮

Remark 6.5. The dual H^* of a Hilbert space H is a Hilbert space with inner product given by

$$(x^* \mid y^*) = (y \mid x)_H \tag{6.24}$$

for each x^* and y^* in H^* where $x \to x^*$ and $y \to y^*$ as in equation (6.23) under mapping $H \to H^*$. [Here $(\ \mid\ )_H$ denotes the inner product in H.] The inner product given by equation (6.24) is compatible with the already existent norm on H^* as $\| y^* \| = \| y \| = (y \mid y)_H = (y^* \mid y^*)$. We may say therefore that H and H^* are "conjugate" isomorphic since the isometric surjection $y \to y^*$ from H to H^* is conjugate linear and $(x^* \mid y^*) = \overline{(x \mid y)_H}$.

Corollary 6.2. Each Hilbert space is reflexive. ▮

Proof. The composition of the conjugate isometric surjections $H \to H^*$ and $H^* \to (H^*)^*$ is easily seen to be the natural mapping $J: H \to H^{**}$ given by $J(h) = \hat{h}$, where $\hat{h}(h^*) = h^*(h)$ for all h^* in H^*. ▮

Definition 6.9. If X and Y are normed linear spaces, a sesquilinear form B on $X \times Y$ is said to be bounded if there exists a constant M such that $| B(x, y) | \leq M \| x \| \| y \|$ for all x in X and y in Y. The norm $\| B \|$ of B is given by

$$\| B \| = \inf\{M: | B(x, y) | \leq M \| x \| \| y \| \text{ for all } x \text{ in } X, y \text{ in } Y\}. \blacksquare$$

Proposition 6.10

(i) If B is a bounded sesquilinear form on $X \times Y$, where X and Y are normed linear spaces, then

$$\begin{aligned} \| B \| &= \sup\{| B(x, y) |: \| x \| < 1 \text{ and } \| y \| < 1\} \\ &= \sup\{| B(x, y) |: \| x \| \leq 1 \text{ and } \| y \| \leq 1\} \\ &= \sup\{| B(x, y) |: \| x \| = 1 \text{ and } \| y \| = 1\}, \end{aligned}$$

and $| B(x, y) | \leq \| B \| \| x \| \| y \|$ for all x in X and y in Y.

(ii) If P is a pre-Hilbert space and B is a bounded Hermitian sesquilinear form on P then

$$\| B \| = \sup\{| B(x, x) |: \| x \| \leq 1\}.$$

[Problem 6.3.6 shows that a converse of (ii) is not always true.] $\blacksquare$

Proof. The proof of part (i) is analogous to the proof of the corresponding equalities for norms of linear operators on normed linear spaces and is left to the reader.

To prove (ii) we first note that by assumption $B(x, x)$ is real since B is Hermitian. If $\| x \| \leq 1$, then by part (i), $| B(x, x) | \leq \| B \|$ so that $S \leq \| B \|$ if S equals the $\sup\{| B(x, x) |: \| x \| \leq 1\}$. It suffices to show that for any x and y in P with $\| x \| \leq 1$ and $\| y \| \leq 1$ we have $| B(x, y) | \leq S$. We look at two cases.

Case 1. Suppose $B(x, y)$ is real. By Proposition 6.1 (i) we have

$$B(x, y) = \tfrac{1}{4}\{B(x + y, x + y) - B(x - y, x - y)\}.$$

Using the Parallelogram law

$$\begin{aligned} | B(x, y) | &\leq \tfrac{1}{4}\{| B(x + y, x + y) | + | B(x - y, x - y) |\} \\ &\leq \tfrac{1}{4}\{S \| x + y \|^2 + S \| x - y \|^2\} \\ &= \tfrac{1}{4}S\{2\| x \|^2 + 2\| y \|^2\} \leq S. \end{aligned}$$

Case 2. In general, write $|B(x, y)| = \alpha B(x, y)$, where α is a complex number of norm 1. Then $B(\alpha x, y)$, equal to $\alpha B(x, y)$, is a real number. Hence by case 1,

$$|B(x, y)| = |B(\alpha x, y)| \le S. \qquad \blacksquare$$

If H and K are Hilbert spaces each bounded linear operator $T: H \to K$ generates a bounded sesquilinear form B_T on $H \times K$ by the formula

$$B_T(x, y) = (Tx \mid y)_K, \tag{6.25}$$

where $(\ \mid\ )_K$ is the inner product of K. It is easy to verify that B_T is sesquilinear and that $\| B_T \| \le \| T \|$ using the Cauchy–Schwarz Inequality, Proposition 6.1. If $\| x \| \le 1$ and $\| y \| \le 1$, then by definition of $\| B_T \|$,

$$|B_T(x, y)| = |(Tx \mid y)_K| = |(y \mid Tx)_K| \le \| B_T \|.$$

Fixing x and taking the supremum over $\| y \| \le 1$, we get $\| Tx \| \le \| B_T \|$ since $(y \mid Tx)_K$ for fixed x is a continuous linear functional on K of norm $\| Tx \|$. Now taking the supremum over $\| x \| \le 1$, we get $\| T \| \le \| B_T \|$. Hence $\| T \| = \| B_T \|$.

It is interesting that every bounded sesquilinear form B on $H \times K$ is equal to B_T for some bounded linear operator T from H to K. This is the content of the next theorem.

Theorem 6.6. If H and K are Hilbert spaces, then for each bounded sesquilinear form B on $H \times K$ there exists a unique bounded linear operator $T: H \to K$ such that $B(x, y) = (Tx \mid y)$ for all x and y. Moreover $\| B \| = \| T \|$. $\blacksquare$

Proof. For each x in H, define $f_x: K \to F$ (the scalar field) by the formula

$$f_x(y) = \overline{B(x, y)}. \tag{6.26}$$

Then f_x is linear (easily checked) and

$$|f_x(y)| = |B(x, y)| \le \| B \| \, \| x \| \, \| y \|.$$

So f_x is bounded with $\| f_x \| \le \| B \| \, \| x \|$. By Theorem 6.5, there exists for each x in H a unique element z_x in K such that $\| f_x \| = \| z_x \|$ and $f_x(y) = (y \mid z_x)$ for all y in K. Define $T: H \to K$ by $Tx = z_x$. Then

$$\overline{B(x, y)} = f_x(y) = (y \mid z_x) = (y \mid Tx),$$

so that $(Tx \mid y) = B(x, y)$. We now assert that T is linear, bounded with norm $\| B \|$, and is the unique operator satisfying $B(x, y) = (Tx \mid y)$:

T is linear: for all y, $\big(T(\alpha_1 x_1 + \alpha_2 x_2) \mid y\big) = (\alpha_1 Tx_1 + \alpha_2 Tx_2 \mid y)$.

T is bounded: $\| Tx \| = \| f_x \| \leq \| B \| \; \| x \|$. (Clearly now $\| T \| = \| B \|$ by our remarks before the theorem.)

T is unique: If S is also a bounded linear operator satisfying $(Sx \mid y) = B(x, y)$, then the equality $(Sx - Tx \mid y) = 0$ for all y implies $Sx = Tx$ for all x. ∎

Given a bounded linear operator $T\colon H \to K$, we have shown there is a unique bounded sesquilinear form B_T on $H \times K$ satisfying the formula $B_T(x, y) = (Tx \mid y)$. Moreover, $\| B_T \| = \| T \|$. In the same manner

$$B^*(y, x) = \overline{B_T(x, y)} = \overline{(Tx \mid y)} = (y \mid Tx)$$

defines a bounded sesquilinear form B^* on $K \times H$ with norm $\| B^* \|$ satisfying

$$\begin{aligned} \| B^* \| &= \inf\{M\colon |(y \mid Tx)| \leq M \| x \| \; \| y \|\} \\ &= \inf\{M\colon |(Tx \mid y)| \leq M \| x \| \; \| y \|\} = \| B_T \| = \| T \|. \end{aligned}$$

However, by Theorem 6.6 we know that to B^* there corresponds a unique operator $S\colon K \to H$ satisfying $B^*(y, x) = (Sy \mid x)$ and with $\| S \| = \| B^* \|$. Hence corresponding to $T\colon H \to K$ there corresponds a (necessarily unique) bounded linear operator $S\colon K \to H$ satisfying

$$(Tx \mid y) = B_T(x, y) = \overline{B^*(y, x)} = \overline{(Sy \mid x)} = (x \mid Sy)$$

and

$$\| T \| = \| B_T \| = \| B^* \| = \| S \|.$$

The operator S is called the *adjoint* of T and is generally denoted by T^*. Formally we have proved the following result.

Theorem 6.7. To each bounded linear operator $T\colon H \to K$ there corresponds one and only one linear operator $T^*\colon K \to H$, called the adjoint of T satisfying $(Tx \mid y)_K = (x \mid T^* y)_H$ for all x in H and y in K. Moreover $\| T \| = \| T^* \|$. ∎

Problem 6.3.7 gives some examples of adjoints of operators.

Remark 6.6. In Chapter 5, the adjoint $T^*\colon Y^* \to X^*$ of a continuous linear operator $T\colon X \to Y$ was defined for normed space X and Y by the equation $T^*(f^*)(x) = f^*\big(T(x)\big)$ for $f^* \in Y^*$ and $x \in X$. Denoting this adjoint momentarily by T' instead of T^*, we remark that T' is *not* the same operator defined in Theorem 6.7 when X and Y are Hilbert spaces. Indeed the domains are different—that of T' being the dual space Y^*. To emphasize this distinction the T^* of Theorem 6.7 is sometimes called the Hilbert space adjoint. The relationship between T^* and T' is examined in Problem 6.3.8.

We conclude this section with several easily verified results giving some properties of bounded linear operators and their adjoints on Hilbert spaces.

Proposition 6.11. For any bounded linear operators $T: H \to K$ and $S: H \to K$ and scalar α we have the following:

(i) $(y \mid Tx)_K = (T^*y \mid x)_H$ for all $x \in H$ and $y \in K$;

(ii) $(S + T)^* = (T^* + S^*)$;

(iii) $(\alpha T)^* = \bar{\alpha} T^*$;

(iv) $(T^*)^* = T$;

(v) $\| T^*T \| = \| TT^* \| = \| T \|^2$;

(vi) $T^*T = 0$ if and only if $T = 0$. ∎

Proof

(i) Left to the reader.

(ii)
$$\begin{aligned}(x \mid (S + T)^*y)_H &= ((S + T)x \mid y)_K = (Sx \mid y)_K + (Tx \mid y)_K \\ &= (x \mid S^*y)_H + (x \mid T^*y)_H = (x \mid S^*y + T^*y)_H\end{aligned}$$

for all x in H and y in K. Hence

$$(S + T)^*y = (S^* + T^*)y \qquad \text{for all } y.$$

(iv) $$((T^*)^*x \mid y)_K = \overline{(y \mid (T^*)^*x)_K} = \overline{(T^*y \mid x)_K} = (Tx \mid y)_K$$

for all x in H and y in K. Hence $(T^*)^* = T$.

(v) For $\| x \| \leq 1$,

$$\| Tx \|^2 = (Tx \mid Tx)_K = (T^*Tx \mid x)_H \leq \| T^*Tx \| \; \| x \| \leq \| T^*T \|.$$

Hence $\| T \|^2 \leq \| T^*T \|$. However, $\| T^*T \| \leq \| T^* \| \; \| T \| = \| T \|^2$ so $\| T^*T \| = \| T \|^2$. Replacing T by T^* and T^* by $(T^*)^*$ or T gives $\| TT^* \| = \| T^* \|^2 = \| T \|^2$.

(vi) follows from (v) and the proof of (iii) is similar to that of (ii). ∎

Proposition 6.12. If $T\colon H \to K$ and $S\colon K \to L$ are continuous linear operators (H, K, and L are Hilbert spaces), then $(ST)^* = T^*S^*$. ▮

Proof. For all x in H and y in L,

$$((ST)^*y \mid x)_H = (y \mid STx)_L = (S^*y \mid Tx)_K = (T^*S^*y \mid x)_H. \quad ▮$$

Proposition 6.13. If $T\colon H \to K$ is a continuous linear mapping and $M \subset H$ and $N \subset K$ with $T(M) \subset N$, then $T^*(N^\perp) \subset M^\perp$. ▮

Proof. If $y \in N^\perp$ and $m \in M$,

$$(T^*y \mid m)_H = (y \mid Tm)_K = 0 \text{ so that } T^*y \in M^\perp. \quad ▮$$

A stronger conclusion in Proposition 6.13 can be obtained if N is a closed subspace, as Proposition 6.14 shows.

Proposition 6.14. If $T\colon H \to K$ is a continuous linear mapping, M is a linear subspace of H, and N is a closed linear subspace of K, then $TM \subset N$ if and only if $T^*(N^\perp) \subset M^\perp$. ▮

Proof. The necessity is proved by Proposition 6.13. If $T^*(N^\perp) \subset M^\perp$, then also $T^{**}(M^{\perp\perp}) \subset N^{\perp\perp}$ by Proposition 6.13. Since $M^{\perp\perp} \supset M$, $N^{\perp\perp} = N$ by Theorem 6.2 and $T^{**} = T$, $T(M) \subset N$. ▮

Proposition 6.15. If $T\colon H \to K$ is a continuous linear mapping, then

(i) $\{x\colon Tx = 0\} = [T^*(K)]^\perp$;

(ii) $\{x\colon Tx = 0\}^\perp = \overline{T^*(K)}$;

(iii) $\{y\colon T^*y = 0\} = [T(H)]^\perp$;

(iv) $\{y\colon T^*y = 0\}^\perp = \overline{T(H)}$. ▮

Proof. Let M be the closed linear subspace $\{x\colon Tx = 0\}$. Since $T(M) \subset \{0\}$, Proposition 6.14 gives $T^*(\{0\}^\perp) \subset M^\perp$ or

$$T^*(K) \subset \{x\colon Tx = 0\}^\perp. \tag{6.27}$$

In Proposition 6.13, letting $M = H$ and $N = T(H)$, we have $T^*([T(H)]^\perp) \subset H^\perp$ or $T^*([T(H)]^\perp) \subset \{0\}$. This means

$$[T(H)]^\perp \subset \{y\colon T^*y = 0\}. \tag{6.28}$$

Replacing T by T^* in inclusions (6.27) and (6.28) and noting that $T^{**} = T$, we get

$$T(H) \subset \{y\colon T^*y = 0\}^\perp \tag{6.29}$$

and

$$[T^*(K)]^\perp \subset \{x\colon Tx = 0\}. \tag{6.30}$$

Now from (6.27) and (6.30),

$$\{x\colon Tx = 0\} = \{x\colon Tx = 0\}^{\perp\perp} \subset [T^*(K)]^\perp \subset \{x\colon Tx = 0\}$$

so that (i) is proved. From (i), $\{x\colon Tx = 0\}^\perp = [T^*(K)]^{\perp\perp}$. Since $[T^*(K)]^\perp = [\overline{T^*(K)}]^\perp$ and $[\overline{T^*(K)}]^{\perp\perp} = \overline{T^*(K)}$, we get (ii). Similarly (iii) follows from (6.28) and (6.29) and (iv) follows from (iii). ∎

Having introduced the idea of the adjoint in this section and having given some of its essential properties, we will use it in the next section to define special classes of operators. In particular we will show in Section 6.5 that each bounded linear operator $T\colon H \to H$ which equals its adjoint T^* has a neat representation.

Problems

6.3.1. If P is a pre-Hilbert space such that $N^{\perp\perp} = N$ for every closed linear subspace N of P, show P is a Hilbert space by showing that every continuous linear functional y^* in P^* is of the form $y^*(x) = (x \mid y)$ for some y in P.

6.3.2. Prove that if f is a linear functional on a Hilbert space H, then the null space $N = \{x\colon f(x) = 0\}$ is dense in H if f is not continuous.

6.3.3. (i) Prove that in a Hilbert space H, a sequence (x_n) converges weakly to x in H if and only if the sequence $((x_n \mid y))$ converges to $(x \mid y)$ for each y in H.

(ii) Give an example of a sequence (x_n) in the infinite-dimensional Hilbert space H that converges weakly but not strongly in H. [Hint: Use (i) and Bessel's Inequality.]

6.3.4. (i) Let $P \times Q$ be the direct product of pre-Hilbert spaces P and Q. For $x = (p_1, q_1)$ and $y = (p_2, q_2)$ in $P \times Q$, define $(x \mid y)$ as $(p_1 \mid p_2)_P + (q_1 \mid q_2)_Q$. Show that $P \times Q$ becomes a pre-Hilbert space with this inner product and that $\| (x \mid y) \|^2 = \| x \|^2 + \| y \|^2$. Show that $P \times Q$ equals $P_0 \oplus Q_0$, where P_0 is the range of the mapping from P into $P \times Q$ given by $p \to (p, 0)$ and similarly Q_0 is the range of the mapping from Q

into $P \times Q$. Show that $P \times Q$ is a Hilbert space if and only if P and Q are Hilbert spaces.

(ii) Generalize (i) to an arbitrary collection (P_α) of pre-Hilbert spaces.

6.3.5. If X and Y are normed linear spaces, prove that the set $B(X, Y)$ of bounded sesquilinear forms on $X \times Y$ is a normed linear space linearly isometric to $L(X, (\bar{Y})^*)$, where $\bar{Y}$ is the complex conjugate of Y—that is, the same space as Y except that addition $\oplus$ and scalar multiplication $\odot$ in $\bar{Y}$ is given by $\alpha \odot x \oplus \beta \odot y = \bar{\alpha} \cdot x + \bar{\beta} \cdot y$ for all scalars α, β and vectors x and y, where $+$ and $\cdot$ represent the operations in Y. Conclude $B(X, Y)$ is a Banach space. Compare Theorem 6.6.

6.3.6. (i) Let P be a complex pre-Hilbert space and define $B(x, y) = i(x \mid y)$. Prove B is a bounded sesquilinear form on $P \times P$ with $\| B \| = \sup\{| B(x, x) |: \| x \| \leq 1\}$ but B is not Hermitian. Compare Proposition 6.10(ii).

(ii) Let P be any pre-Hilbert space. Suppose $T: P \to P$ is a continuous linear mapping with $\| T \| = 1$ and $Tp = p$ for some nonzero p in P. Define $B_T(x, y) = (Tx \mid y)$. Show that B_T is sesquilinear with $\| B_T \| = \sup\{| B_T(x, x) |: \| x \| \leq 1\}$, but B_T is not necessarily Hermitian.

6.3.7. (i) Let T be a mapping from a separable Hilbert space H into itself and let (e_i) be a basis of H. Then $Te_j = \sum_{i=1} \alpha_{ij} e_i$ as in Proposition 6.6. The collection (α_{ij}) for $i = 1, 2, \ldots$ and $j = 1, 2, \ldots$ is called the matrix of T with respect to the basis (e_i). Show $(Te_j \mid e_k) = \alpha_{kj}$ and that the matrix of T^* with respect to (e_i) is the collection of scalars (β_{ij}) for $i = 1, 2, \ldots,$ and $j = 1, 2, \ldots$ where $\beta_{ij} = \bar{\alpha}_{ji}$.

(ii) What is the adjoint of the shift operator $T: l_2 \to l_2$ given by $T(a_1, a_2, \ldots) = (0, a_1, a_2, \ldots)$?

6.3.8. If H and K are Hilbert spaces, and $U: H \to H^*$ and $V: K \to K^*$ are the conjugate isometries discussed in Theorem 6.5, prove that $T^* = U^{-1}T'V$, where T^* is the Hilbert space adjoint of T, and T' is the Banach space adjoint from K^* to H^* given by $T'(k^*)h = k^*(Th)$.

6.3.9. Let H and K be Hilbert spaces and $\{h_i\}_{i \in I}$ and $\{k_j\}_{j \in J}$ be bases for H and K, respectively. If $T: H \to K$ is a bounded linear mapping, $\sum_I \| Th_i \|^2$ converges if and only if $\sum_J \| T^*k_j \|^2$ converges and in this case both sums equal $\sum_{i,j} | (Th_i \mid k_j) |^2$.

6.3.10. A bounded linear operator $T: H \to K$ is said to be a *Hilbert–Schmidt operator* if there exists a basis $\{h_i\}$ in H such that $\sum_I \| Th_i \|^2 < \infty$. Prove the following using Problem 6.3.9:

(i) If T is a Hilbert–Schmidt operator, show $\sum_I \| Tg_i \|^2 < \infty$ for every basis $\{g_i\}_I$ of H.

(ii) T is a Hilbert–Schmidt operator if and only if T^* is a Hilbert–Schmidt operator.

(iii) The class of Hilbert–Schmidt operators is a pre-Hilbert space with the inner product $(S \mid T) = \sum_I (Sh_i \mid Th_i)$ where $\{h_i\}_I$ is a basis of H.

(iv) The pre-Hilbert space of (iii) is complete. Prove this by showing that it is isomorphic to $l_2(I \times J)$ via $T \to (x_{ij})_{I\times J}$, where $x_{ij} = (Th_i \mid k_j)$ and where $\{h_i\}_I$ and $\{k_j\}_J$ are bases of H and K, respectively.

★ **6.3.11.** Prove that the following statements are equivalent in a *pre-Hilbert* space P.

(i) P is complete.
(ii) If M is a closed subspace of P, then $M \oplus M^\perp = P$.
(iii) If M is a closed subspace of P, then $M = M^{\perp\perp}$.
(iv) If M is a proper closed subspace of P, then $M^\perp \neq \{0\}$.
(v) If f is a continuous linear functional on P, then there exists $x \in P$ such that $f(y) = (y \mid x)$ for all $y \in P$.

6.3.12. Suppose that $S \subset R$ and that for each s in S there is a bounded linear operator $T(s)$ on H. An operator A on H is the *weak limit* of the function T: $S \to H$ as $s \to s_0$, written $A = w\text{-}\lim_{s\to s_0} T(s)$, if for any pair h and k in H

$$(Ah \mid k) = \lim_{s\to s_0} \big(T(s)h \mid k\big).$$

An operator A on H is the *strong limit* of $T(s)$ as $s \to s_0$, written $A = s\text{-}\lim_{s\to s_0} T(s)$ if for all h in H

$$\lim_{s\to s_0} \| \big(T(s) - A\big)h \| \to 0.$$

An operator A on H is the *uniform limit* of $T(s)$ as $s \to s_0$ if

$$\lim_{s\to s_0} \| T(s) - A \| \to 0.$$

Show that if A is the uniform limit of $T(s)$ as $s \to s_0$, then A is the strong limit of $T(s)$ as $s \to s_0$. Show also that if A is the strong limit of $T(s)$ as $s \to s_0$, then A is the weak limit of $T(s)$ as $s \to s_0$. (Note here that s_0 may be $\pm\infty$ and S may be the natural numbers N.)

6.3.13. Let A be a linear operator from a Hilbert space H into itself with $(x \mid Ay) = (Ax \mid y)$ for all x and $y \in H$. Prove that A is bounded. (Hint: Use the Closed Graph Theorem.)

6.4. The Algebra of Operators. The Spectral Theorem and the Approximation Theorem for Compact Operators

In this section we will first examine some special classes of operators on a Hilbert space H—that is, special classes of operators from H into itself. These special classes of operators can all be defined by use of the adjoint. Secondly, we will prove a spectral theorem for compact operators which the spectral theorem of the next section will generalize. Finally, we will show how each completely continuous operator can be approximated by operators with finite-dimensional ranges.

Let us first give in capsule form the definition of all types of operators we will consider in this section.

Definition 6.10. Let $T: H \to H$ be a continuous linear operator with adjoint T^*.

(i) T is *isometric* if and only if $T^*T = I$, the identity of H.
(ii) T is *unitary* if and only if $T^*T = TT^* = I$.
(iii) T is *self-adjoint* (or Hermitian) if and only if $T = T^*$.
(iv) T is a *projection* if and only if $T^2 = T$ and $T^* = T$.
(v) T is *normal* if and only if $T^*T = TT^*$. ∎

Note that when we speak of T as being any one of the types of operators of our Definition 6.10, T is understood to be continuous and linear. The reader should consult Problem 6.4.1 for various examples of these types of operators.

Clearly, every unitary, every self-adjoint, and every projection operator is normal. The following four propositions give equivalences of isometric, unitary, self-adjoint, and normal operators, respectively. We first prove the following lemma.

Lemma 6.3. If H is a complex Hilbert space and $S, T: H \to H$ are bounded linear operators such that $(Sx \mid x) = (Tx \mid x)$ for all x, then $S = T$. ∎

Proof. The sesquilinear forms $B_S(x, y) = (Sx \mid y)$ and $B_T(x, y) = (Tx \mid y)$ are such that $B_S(x, x) = B_T(x, x)$ for all x in H. By the polarization identity [Proposition 6.1(i)], $B_S(x, y) = B_T(x, y)$ for all x and y in H. By Remark 6.1, $S = T$. ∎

Proposition 6.16. The following conditions on $T: H \to H$ are equivalent.

(i) T is isometric.

(ii) $(Tx \mid Ty) = (x \mid y)$ for all x and y in H.

(iii) $\| Tx \| = \| x \|$ for all x in H. ▮

Proof. (i) ⇒ (ii) since $(Tx \mid Ty) = (x \mid T^*Ty) = (x \mid y)$. Trivially (ii) implies (iii). If we assume (ii), then $(T^*Tx \mid y) = (x \mid y)$ for all x and y so that $T^*Tx = x$ for all x, or $T^*T = I$. Hence (ii) ⇒ (i). If H is a real Hilbert space, then by Proposition 6.1 (i)

$$(Tx \mid Ty) = \tfrac{1}{4}\{\| Tx + Ty \|^2 - \| Tx - Ty \|^2\}.$$

Clearly (iii) implies (ii) if H is real. Similarly if H is complex Proposition 6.1 (i) gives that (iii) ⇒ (ii). ▮

Proposition 6.17. The following conditions on $T: H \to H$ are equivalent.

(i) T is unitary.

(ii) T^* is unitary.

(iii) T and T^* are isometric.

(iv) T is isometric and T^* is injective.

(v) T is isometric and surjective.

(vi) T is bijective and $T^{-1} = T^*$. ▮

Proof. Since $T^{**} = T$, the equivalences of (i), (ii), and (iii) are trivial, as are the implications (iii) ⇒ (iv) and (vi) ⇒ (i). The proof is completed by showing (iv) ⇒ (v) and (v) ⇒ (vi).

(iv) ⇒ (v): Since T is isometric, $T(H)$ is closed. Hence by Proposition 6.15,

$$H = \{0\}^\perp = \{x: T^*x = 0\}^\perp = \overline{T(H)} = T(H).$$

(v) ⇒ (vi): Let $S = T^{-1}$. To show $T^* = S$. Since T is isometric,

$$T^* = T^*I = T^*(TS) = (T^*T)S = IS = S.$$ ▮

Proposition 6.18. Let $T: H \to H$ be a continuous linear operator. Let (i), (ii), (iii), and (iv) represent the following statements:

(i) T is self-adjoint.

(ii) $(Tx \mid y) = (x \mid Ty)$ for all x and y in H.

(iii) $(Tx \mid x) = (x \mid Tx)$ for all x in H.

(iv) $(Tx \mid x)$ is real for all x in H.

Then (i) is equivalent to (ii), (iii) is equivalent to (iv), and if H is a complex Hilbert space, all the statements are equivalent. ∎

Proof. All the implications are trivial except perhaps (iv) $\Rightarrow$ (i) in the complex case. Suppose (iv) holds. Then

$$(T^*x \mid x) = (x \mid Tx) = (Tx \mid x)$$

for each x in H. By Lemma 6.3, $T = T^*$. ∎

Proposition 6.19. If H is a Hilbert space, then $T: H \to H$ is normal if and only if $\| T^*x \| = \| Tx \|$ for each x in H. ∎

Proof. If T is normal,

$$\| T^*x \|^2 = (T^*x \mid T^*x) = (TT^*x \mid x) = (T^*Tx \mid x) = (Tx \mid Tx) = \| Tx \|^2. \tag{6.31}$$

Conversely, if $\| T^*x \| = \| Tx \|$, equation (6.31) shows that T is normal using Lemma 6.3 in the case H is complex. However, in either case the Hermitian sesquilinear forms B and B', given by $B(x, y) = (Tx \mid Ty)$ and $B'(x, y) = (T^*x \mid T^*y)$, respectively, are such that

$$B(x, x) = (Tx \mid Tx) = \| Tx \|^2 = \| T^*x \|^2 = (T^*x \mid T^*x) = B'(x, x)$$

for all x in H. By the polarization identity, Proposition 6.1 (i), $B(x, y) = B'(x, y)$ for all x and y in H. Hence

$$(T^*Tx \mid y) = (Tx \mid Ty) = (T^*x \mid T^*y) = (TT^*x \mid y)$$

for all x and y in H. Hence $T^*T = TT^*$ by Remark 6.1. ∎

We will take a closer look at self-adjoint operators in what follows. Before doing so, let us examine a few pertinent theorems regarding some special self-adjoint operators—the projection operators.

Recall that if N is a closed linear subspace of H, then $H = N \oplus N^\perp$. We can define $P: H \to H$ by the following rule: If $x = y + z$ with $y \in N$ and $z \in N^\perp$, let $Px = y$. P is easily checked to be a bounded linear operator, but also satisfies the equations $P^2 = P$ and $P^* = P$. Indeed, $Px \in N$ implies $P(Px) = Px$ and

$$(Px_1 \mid x_2) = (y_1 \mid y_2 + z_2) = (y_1 + z_1 \mid y_2) = (x_1 \mid Px_2),$$

where $x_1 = y_1 + z_1$ and $x_2 = y_2 + z_2$ with $y_1, y_2 \in N$ and $z_1, z_2 \in N^\perp$. P is called the projection† operator associated with N and is denoted by P_N.

Proposition 6.20. If $T: H \to H$ is a projection operator, then there is one and only one closed linear subspace N such that $T = P_N$. In fact $N = T(H)$ and $N^\perp = \text{Ker } T$. ∎

Proof. Let $N = \{x: (T - I)x = 0\} = \{x: Tx = x\}$, a closed subspace. If y is in the range $R(T)$ of T, then for some x, $y = Tx = T(Tx) = Ty$ so that y is in N. Conversely, if $y \in N$, $y = Ty \in R(T)$. Therefore $N = R(T)$. Now by Proposition 6.15

$$\{x: Tx = 0\} = [T^*(H)]^\perp = [T(H)]^\perp$$

or $N^\perp = \{x: Tx = 0\}$. Let $x \in H$. Then $x = y + z$ with $y \in N$ and $z \in N^\perp$. Hence $Tx = T(y + z) = Ty + Tz = y$. Hence $T = P_N$.

If $T = P_N = P_M$ for some subspaces N and M then $N = P_N(H) = P_M(H) = M$. ∎

• **An Application of Proposition 6.20: The Mean Ergodic Theorem.** In Chapter 5, we discussed measure-preserving mappings and the Individual Ergodic Theorem showing the existence of an a.e. pointwise limit for L_1 functions. Here we show how the subject of unitary operators comes up naturally in the context of ergodic theory. We present below the Mean Ergodic Theorem, proven by von Neumann in 1933, which will consider L_2 functions and convergence in the L_2 norm. We will use Proposition 6.20.

First, we notice that if Φ is a measure-preserving mapping on a measure space $(X, \mathscr{A}, \mu)$ and if

$$T(f)(x) = f(\Phi(x)), \qquad f \in L_2(\mu),$$

then T is a unitary operator on L_2. Thus the Mean Ergodic Theorem in L_2 can be presented in the following Hilbert space setup.

† P is also called the orthogonal projection on N.

The Mean Ergodic Theorem. Let T be a unitary operator on a Hilbert space H. Let P be the orthogonal projection onto the closed linear subspace $N = \{x: Tx = x\}$. Then, for any $x \in H$,

$$T_n x = \frac{1}{n} \sum_{k=0}^{n-1} T^k x \to Px \qquad \text{as } n \to \infty. \qquad \blacksquare$$

Proof. First, we notice that by Proposition 6.15,

$$[(I - T)(H)]^\perp = \operatorname{Ker}(I - T^*) = \{x: T^*x = x\} = \{x: T^{-1}x = x\} = N(= P(H)).$$

Now if $x = y - Ty$, then

$$\| T_n x \| = \| (1/n)[y - T^n y] \| \leq (2/n) \| y \| \to 0 \qquad \text{as } n \to \infty.$$

By Theorem 6.2 and Proposition 6.20,

$$\operatorname{Ker} P = N^\perp = \overline{(I - T)(H)}.$$

Therefore, for $x \in N^\perp$, $\lim_{n\to\infty} T_n x = 0 = Px$. Also, for $x \in N = P(H)$, $\lim_{n\to\infty} T_n x = Tx = x = Px$. Since $H = N \oplus N^\perp$, the theorem now follows easily. ∎

Having established a one-to-one correspondence between projections on H and closed linear subspaces of H, it is interesting to investigate the preservation properties of this correspondence. Before giving these properties we define a partial ordering on the class of self-adjoint operators.

Definition 6.11. A self-adjoint operator T is said to be positive in case $(Tx \mid x) \geq 0$ for all $x \in H$. If S and T are self-adjoint operators, $S \leq T$ if $T - S$ is positive, written $T - S \geq 0$. ∎

Proposition 6.21. Suppose M_1 and M_2 are closed linear subspaces corresponding, respectively, to projections P_1 and P_2 on H. Then the following hold:

(i) The assertions $M_1 \perp M_2$, $P_1 P_2 = 0$, $P_2 P_1 = 0$, $P_1(M_2) = \{0\}$, and $P_2(M_1) = \{0\}$ are equivalent.

(ii) $P_1 P_2$ is a projection if and only if $P_1 P_2 = P_2 P_1$. If this condition is satisfied the range of $P_1 P_2$ is $M_1 \cap M_2$.

(iii) The assertions $P_1 \leq P_2$, $\| P_1 x \| \leq \| P_2 x \|$ for all x, $M_1 \subset M_2$, $P_2 P_1 = P_1$, and $P_1 P_2 = P_1$ are equivalent. ∎

Proof. (i) $M_1 \perp M_2$ implies $M_2 \subset M_1{}^\perp$. As shown in the proof of Proposition 6.20, $M_2 = \{x: P_2x = x\} = R(P_2)$ and $M_1{}^\perp = \{x: P_1x = 0\}$. Since $M_2 \subset M_1{}^\perp$ and $P_2x \in M_2$ for any x in H, $P_1P_2 = 0$. Now suppose $P_1P_2 = 0$. Then if $x \in M_2$, $P_2x = x$ and therefore $P_1x = P_1P_2x = 0$. Hence $P_1(M_2) = \{0\}$. Now suppose $P_1(M_2) = \{0\}$. Then $M_2 \subset M_1{}^\perp$ and therefore $M_2 \perp M_1$. The other equivalences of (i) follow by symmetry.

(ii) P_1P_2 a projection implies

$$P_1P_2 = (P_1P_2)^* = P_2{}^*P_1{}^* = P_2P_1.$$

Conversely, $P_1P_2 = P_2P_1$ implies

$$(P_1P_2)^2 = P_1(P_2P_1)P_2 = P_1{}^2P_2{}^2 = P_1P_2$$

and

$$(P_1P_2)^* = P_2{}^*P_1{}^* = P_2P_1 = P_1P_2.$$

Finally if P_1P_2 is a projection, then the range M of $P = P_1P_2$ is contained in M_1 and M_2 since P_1 and P_2 commute. Also if $x \in M_1 \cap M_2$, then $P_1x = x = P_2x$ or $Px = x$ so that $x \in M$. Hence $M = M_1 \cap M_2$.

(iii) If $P_1 \le P_2$, then $\| P_1x \|^2 = (P_1x \mid x) \le (P_2x \mid x) = \| P_2x \|^2$ for all x. If $\| P_1x \| \le \| P_2x \|$ for all x, then for $x \in M_1$,

$$\| x \| = \| P_1x \| \le \| P_2x \| \le \| x \|$$

so that $\| P_2x \| = \| x \|$. Since $\| x \|^2 = \| P_2x \|^2 + \| x - P_2x \|^2$, $P_2x = x$ and $x \in M_2$. If $M_1 \subset M_2$, then for all x, $P_1x \in M_2$. Hence $P_2P_1x = P_1x$. If $P_2P_1 = P_1$, then $P_1P_2 = (P_2P_1)^* = P_1{}^* = P_1$. If $P_1P_2 = P_1$, then

$$(P_1x \mid x) = \| P_1x \|^2 = \| P_1P_2x \|^2 \le \| P_2x \|^2 = (P_2x \mid x). \qquad ▮$$

We have therefore seen that the one-to-one correspondence between subspaces of H and projections on H preserves order: $M \subset N$ if and only if $P_M \le P_N$. Thus the set of projections is a partially ordered set such that for every family of projections $\{P_i\}_I$ there exists a "greatest" projection "smaller than" each P_i and a "smallest" projection "greater than" each P_i.

Self-adjoint operators are also called *Hermitian* operators. This terminology is first of all consistent with that used for square complex matrices (a_{ij}) where $a_{ij} = \bar{a}_{ji}$. Indeed if T is a self-adjoint operator on the Hilbert space C^n, let (a_{ij}) $i = 1, \ldots, n$ and $j = 1, \ldots, n$ be the matrix that represents T relative to some orthonormal basis $\{c_1, \ldots, c_n\}$ in C^n—that is, $Tc_j = \sum_{i=1}^n a_{ij}c_i$. Problem 6.3.7 (i) shows that the matrix of T^* is (b_{ij}), where $b_{ij} = \bar{a}_{ji}$. Thus T is Hermitian if and only if (a_{ij}) is a Hermitian matrix. In particular we know from the theory of diagonalization of Hermitian matrices that T can be represented with respect to some orthonormal basis, say $\{x_1, \ldots, x_n\}$, by a diagonal matrix so that $Tx_i = \lambda_i x_i$, $i = 1, \ldots, n$, for some scalars $\lambda_1, \ldots, \lambda_n$. Hence if $x \in C^n$, $x = \sum_{i=1}^n (x \mid x_i)x_i$ by Proposition 6.6 so that

$$Tx = \sum_{i=1}^{n} \lambda_i (x \mid x_i) x_i. \tag{6.32}$$

It is formula (6.32) we wish to generalize to compact self-adjoint operators in this section. We return to it in Theorem 6.8.

The term "Hermitian" for self-adjoint operators is also consistent with that used for sesquilinear forms. Indeed, if B_T is defined on $H \times H$ by $B_T(x, y) = (Tx \mid y)$, then $\overline{B_T(x, y)} = \overline{(Tx \mid y)} = (y \mid Tx) = (T^*y \mid x)$ so that $\overline{B_T(x, y)} = B_T(y, x)$ if and only if $T = T^*$, by Remark 6.1. Hence if T is self-adjoint, then by Proposition 6.10 (ii),

$$\| T \| = \| B_T \| = \sup\{| (Tx \mid x) |: \| x \| \leq 1\}.$$

These remarks enable us to prove the next results.

Lemma 6.4. Let T be a self-adjoint operator on H. Let $m_T = \inf\{(Tx \mid x): \| x \| = 1\}$ and $M_T = \sup\{(Tx \mid x): \| x \| = 1\}$. Then $m_T I \leq T \leq M_T I$ and $\| T \| = \max\{M_T, - m_T\}$. ∎

Proof. For all x in H,

$$m_T(x \mid x) = m_T \| x \|^2 \leq (Tx \mid x) \leq M_T \| x \|^2 = M_T(x \mid x)$$

which shows $m_T I \leq T \leq M_T I$. The rest is clear. ∎

Lemma 6.5. Let T be a positive self-adjoint operator on H. Then

$$| (Tx \mid y) |^2 \leq (Tx \mid x)(Ty \mid y)$$

for all $x, y \in H$. ∎

Proof. Apply the Cauchy–Schwarz inequality, Proposition 6.1 (iii), to the positive Hermitian sesquilinear form $B(x, y) = (Tx \mid y)$. ∎

Lemma 6.6. Suppose T is a self-adjoint operator.

(i) If μ is an eigenvalue of T, then μ is real and $m_T \leq \mu \leq M_T$,

$$m_T = \inf\{(Tx \mid x): \; \| x \| = 1\}, \quad M_T = \sup\{(Tx \mid x): \; \| x \| = 1\}.$$

Eigenvectors corresponding to distinct eigenvalues are orthogonal.

(ii) $\sigma(T) \subset [m_T, M_T]$ and the endpoints m_T and M_T are both in $\sigma(T)$. ∎

Proof. (i) Since $(Tx \mid x)$ is real and equals $\mu(x \mid x)$ with $(x \mid x) \geq 0$, μ must be real. Also if x is an eigenvector with $\| x \| = 1$,

$$\mu = \mu \| x \|^2 = (\mu x \mid x) = (Tx \mid x)$$

so that $m_T \leq \mu \leq M_T$.

Now suppose μ_0 is an eigenvalue with $\mu_0 \neq \mu$ and let x_0 and x be eigenvectors corresponding to μ_0 and μ, respectively. Since μ_0 is real,

$$\mu(x \mid x_0) = (Tx \mid x_0) = (x \mid Tx_0) = (x \mid \mu_0 x_0) = \mu_0(x \mid x_0).$$

Thus $(x \mid x_0) = 0$ since $\mu \neq \mu_0$.

(ii) Suppose $\lambda \in F - [m_T, M_T]$. Then $\delta = d(\lambda, [m_T, M_T]) > 0$ and for every x in H with $\| x \| = 1$,

$$\| (T - \lambda I)x \| \geq |\big((T - \lambda I)x \mid x\big)| = |(Tx \mid x) - \lambda| \geq \delta. \qquad (6.33)$$

Hence by Proposition 5.7 in Chapter 5, $T - \lambda I$ is a bijection onto the range $R(T - \lambda I)$ of $T - \lambda I$ with a bounded inverse on $R(T - \lambda I)$. In particular $R(T - \lambda I)$ is complete and closed. Suppose $R(T - \lambda I)^\perp \neq \{0\}$. Then for $y \in R(T - \lambda I)^\perp$ with $\| y \| = 1$, we have $\big((T - \lambda I)(y) \mid y\big) = 0$, which contradicts (6.33). Hence $R(T - \lambda I) = H$ and $\lambda \notin \sigma(T)$.

Finally we show that m_T and M_T are in $\sigma(T)$. Since by Lemma 6.4 $M_T I - T \geq 0$, we have, using Lemma 6.5 that for $\| x \| = 1$

$$\begin{aligned} \| (M_T I - T)x \|^4 &= |\big((M_T I - T)x \mid (M_T I - T)x\big)|^2 \\ &\leq \big((M_T I - T)x \mid x\big)\big((M_T I - T)^2 x \mid (M_T I - T)x\big) \\ &\leq \| M_T I - T \|^3 \big(M_T - (Tx \mid x)\big). \end{aligned}$$

By virtue of the definition of M_T,

$$\sup\{\| (M_T I - T)x \|: \| x \| = 1\} \leq \sup\{M_T - (Tx \mid x): \| x \| = 1\} = 0.$$

Thus $(M_T I - T)^{-1}$ does not exist, so $M_T \in \sigma(T)$. Similarly $m_T \in \sigma(T)$. ∎

In particular if T is a nonzero compact, self-adjoint operator on H, then Lemmas 6.4, 6.6 and Theorem 5.27 show us that either $\| T \|$ or $-\| T \|$ is an eigenvalue of T. Hence there is a real number λ such that $|\lambda| = \| T \|$ and a vector x in H with $\| x \| = 1$ and $Tx = \lambda x$. We have proved the following result.

Lemma 6.7. Suppose T is a nonzero compact, self-adjoint operator on H. Then either $\| T \|$ or $-\| T \|$ is an eigenvalue of T and there is a corresponding eigenvector x such that $\| x \| = 1$ and $|(Tx \mid x)| = \| T \|$. ▮

Before proving the spectral theorem for compact self-adjoint operators, let us establish two results concerning self-adjoint operators which will be needed in the next section.

Lemma 6.8. If $T: H \to H$ is a self-adjoint operator, then T is positive if and only if $\sigma(T) \subset [0, \infty)$. ▮

Proof. Let $m_T = \inf\{(Tx \mid x): \| x \| = 1\}$. If T is positive, $m_T \geq 0$ and so by Lemma 6.6, $\sigma(T) \subset [0, \infty)$. On the other hand, if $\sigma(T) \subset [0, \infty)$ then $m_T \geq 0$ by Lemma 6.6 and $(Tx \mid x) \geq 0$ for all x. ▮

Lemma 6.9. Suppose (T_n) is a bounded sequence of self-adjoint operators with $T_n \leq T_{n+1}$ (as in Definition 6.11) for $n = 1, 2, \ldots$. Then there is a self-adjoint operator T such that T_n converges *strongly* to T (that is, $T_n x \to Tx$ for each x in H). ▮

Proof. By assumption there exists a positive number M such that $\| T_n \| \leq M$ or $-MI \leq T_n \leq MI$. If $\frac{1}{2}M^{-1}(T_n + MI)$ converges strongly to a self-adjoint operator S, then T_n converges strongly to the self-adjoint operator $2MS - MI$. Thus, replacing T_n by $\frac{1}{2}M^{-1}(T_n + MI)$, we may suppose $0 \leq T_n \leq I$ for $n = 1, 2, \ldots$.

If m and n are positive integers with $m > n$, then $0 \leq T_{mn} \leq I$ if $T_{mn} = T_m - T_n$, and $\| T_{mn} \| \leq 1$ by Lemma 6.4. Using Lemma 6.5 and Proposition 6.1 (iii)

$$\begin{aligned} \| T_m x - T_n x \|^4 &= \| T_{mn} x \|^4 = \| (T_{mn} x \mid T_{mn} x) \|^2 \\ &\leq (T_{mn} x \mid x)(T_{mn}^2 x \mid T_{mn} x) \\ &\leq (T_{mn} x \mid x) \| T_{mn}^2 x \| \, \| T_{mn} x \| \\ &\leq \| x \|^2 [(T_m x \mid x) - (T_n x \mid x)]. \end{aligned} \tag{6.34}$$

Since $0 \leq T_n \leq T_{n+1} \leq I$, $(T_n x \mid x)$ is a bounded increasing sequence of real numbers which converges so that $(T_n x)$ is a Cauchy sequence by equation (6.34). Let $Tx = \lim_{n\to\infty} T_n x$. It is easy to check that the operator T thus defined is a bounded self-adjoint operator. ∎

We have now come to a main objective of this section.

Theorem 6.8. Suppose $T: H \to H$ is a nonzero, compact, self-adjoint operator.

(i) If T is of finite rank, then the nonzero eigenvalues of T form a real finite sequence $\lambda_1, \lambda_2, \ldots, \lambda_n$ and $Tx = \sum_{i=1}^{n} \lambda_i (x \mid x_i) x_i$ for all x in H, where $x_1, x_2, \ldots, x_n$ is a corresponding orthonormal set of eigenvectors.

(ii) If T is not finite dimensional, then the nonzero eigenvalues of T form a real infinite sequence $\lambda_1, \lambda_2, \ldots$ with $|\lambda_i| \to 0$ such that $Tx = \sum_{i=1}^{\infty} \lambda_i (x \mid x_i) x_i$ for each x in H where $x_1, x_2, \ldots$ is an orthonormal set of eigenvectors with x_i corresponding to λ_i. ∎

Proof. Suppose first that T is not of finite rank. We show first that for each n, there exists a nonzero closed subspace X_n of H, an eigenvalue λ_n of T with $|\lambda_n| = \| T|_{X_n} \|$, and a corresponding eigenvector x_n for λ_n with $\| x_n \| = 1$ such that

$$|\lambda_n| \geq |\lambda_{n+1}|,$$
$$X_{n+1} = \{x\colon (x \mid x_i) = 0 \quad \text{for } i = 1, 2, \ldots, n\}, \tag{6.35}$$

and

$$x_n \in X_n \qquad \text{for } n = 1, 2, \ldots.$$

Let $X_1 = X$ and $T_1 = T$. By Lemma 6.7, there exists a vector x_1 and a scalar λ_1 such that $\| x_1 \| = 1$, $|\lambda_1| = \| T_1 \|$, and $Tx_1 = \lambda_1 x_1$.

Proceeding inductively, suppose that for each $i = 1, 2, \ldots, n-1$, X_i, λ_i, and x_i have been chosen satisfying equation (6.35). Define

$$X_n = \{x\colon (x \mid x_i) = 0 \text{ for } i = 1, 2, \ldots, n-1\}.$$

$X_n \neq 0$, for if $X_n = 0$ then the set $\{x_1, x_2, \ldots, x_{n-1}\}$ forms a basis for H, which contradicts the fact that T is of infinite rank. Clearly, $X_n \subset X_{n-1}$ and $T\colon X_n \to X_n$ since for $x \in X_n$

$$(Tx \mid x_i) = (x \mid Tx_i) = (x \mid \lambda_i x_i) = 0$$

for $i = 1, 2, \ldots, n-1$. Letting $T_n = T|_{X_n}$, T_n is compact and self-adjoint. Now if $T_n = 0$, then letting $y_{n-1} = x - \sum_{i=1}^{n-1}(x \mid x_i)x_i$ we see that $y_{n-1} \in X_n$ so that $Ty_{n-1} = 0$ or $Tx = \sum_{i=1}^{n-1}\lambda_i(x \mid x_i)x_i$ for each x in H. This is impossible since T has infinite rank. Hence T_n is nonzero. By Lemma 6.7 again, there is a vector $x_n \in X_n$ and a scalar λ_n such that $\| x_n \| = 1$, $| \lambda_n | = \| T_n \|$, and $Tx_n = \lambda_n x_n$. Since $X_n \subseteq X_{n-1}$,

$$| \lambda_n | = \| T_n \| \leq \| T_{n-1} \| = | \lambda_{n-1} |.$$

The sequence (λ_i) of eigenvalues thus formed converges to zero. If not $x_n = T(x_n/\lambda_n)$ would have a convergent subsequence x_{n_k} since (x_n/λ_n) is bounded and T is compact. However, since the x_{n_k} are orthonormal, $\| x_{n_k} - x_{n_{k'}} \|^2 = 2$ if $k \neq k'$ and the sequence (x_{n_k}) can never converge.

Setting again $y_n = x - \sum_{k=1}^{n}(x \mid x_k)x_k$, we have

$$\| y_n \|^2 = \| x \|^2 - \sum_{k=1}^{n} | (x \mid x_k) |^2 \leq \| x \|^2.$$

Since $y_n \in X_{n+1}$ and $| \lambda_{n+1} | = \| T|_{X_{n+1}} \|$,

$$\| Ty_n \| \leq | \lambda_{n+1} | \; \| y_n \| \leq | \lambda_{n+1} | \; \| x \|,$$

which means $Ty_n \to 0$. Hence

$$Tx = \sum_{k=1}^{\infty} \lambda_k (x \mid x_k) x_k.$$

Finally, if λ is a nonzero eigenvalue excluded from (λ_k), then $Tx = \lambda x$ for some vector x with $\| x \| = 1$ and $(x \mid x_k) = 0$ for all k. Hence $\lambda x = Tx = 0$, which is a contradiction. We have proved (ii).

Suppose now T has finite rank. Clearly T has only a finite set of distinct eigenvalues, for if $\lambda_1, \lambda_2, \ldots$ is an infinite set of distinct nonzero eigenvalues with $x_1, x_2, \ldots$ a corresponding orthogonal set of eigenvectors, then $T(H)$ is infinite dimensional as each x_k is in $T(H)$ since $x_k = T(\lambda_k^{-1}x_k)$.

Just as in expression (6.35), using the same procedure, let (λ_k) be the finite set of all eigenvalues with corresponding eigenvectors (x_k) such that $| \lambda_{k+1} | \leq | \lambda_k | = \| T|_{X_k} \|$. If $m+1$ is the smallest integer such that $\lambda_{m+1} = 0$, then $T|_{X_{m+1}} = 0$, and, as before, $Tx = \sum_{i=1}^{m}\lambda_i(x \mid x_i)x_i$. In the other case, there is a smallest integer m such that $X_{m+1} = \{0\}$; then $\{x_1, x_2, \ldots, x_m\}$ is a basis for H. ∎

At this juncture we can prove another significant result using the theorem just proved. In short, this result says that each Hilbert space satisfies the approximation property (see Chapter 5). More precisely, we have the following theorem.

Theorem 6.9. Let $T: H \to H$ be a compact operator. Then there is a sequence T_n of finite-rank linear operators converging to T in the norm (uniformly). ∎

Proof. First assume H is separable. Then H contains a complete orthonormal sequence (x_k). Define T_n by

$$T_n x = \sum_{k=1}^{n} (x \mid x_k) x_k .$$

Clearly $\| T_n \| \leq 1$, $\| I - T_n \| \leq 1$, and $T_n x \to x$ as $n \to \infty$ for each x in H. If the conclusion of the theorem is false, then there exists a $\delta_0 > 0$ such that $\| T - A \| > \delta_0$ for all finite-rank linear operators A. Since $T_n T$ has finite rank, there exists for each n a y_n in H with $\| y_n \| = 1$ and $\| (T - T_n T) y_n \| \geq \delta_0/2$. Since the sequence (y_n) is bounded, there exists a subsequence (y_{n_i}) of (y_n) such that Ty_{n_i} converges to some z in H. However,

$$\| (T - T_{n_i} T) y_{n_i} \| = \| (I - T_{n_i}) T y_{n_i} \| \leq \| (I - T_{n_i})(Ty_{n_i} - z) \| + \| (I - T_{n_i}) z \| \qquad (6.36)$$

and the left side of equation (6.36) goes to zero as $i \to \infty$. Since this is a contradiction, the theorem is proved when H is separable.

Now let H be any Hilbert space. Let $S = T^*T$. Then $S^* = S$, so S is self-adjoint and compact. By Theorem 6.8, $Sx = \sum_{k=1}^{\infty} \lambda_k (x \mid x_k) x_k$ for each x in H, where (x_k) is some orthonormal sequence of eigenvectors with corresponding eigenvalues (λ_k). Let H_0 be the subspace of H spanned by $\{T^n x_k: n = 0, 1, \ldots, \text{ and } k = 1, 2, \ldots\}$. Then $\bar{H}_0$ is a separable Hilbert space and $T: \bar{H}_0 \to \bar{H}_0$. However, we know by what we have just proved that there is a sequence $\{F_n\}$ of finite-rank operators converging to $T_{\bar{H}_0}$ in norm. Now $H = \bar{H}_0 \oplus \bar{H}_0{}^{\perp}$. Therefore for each x in H, $x = y + z$ with $y \in \bar{H}_0$ and $z \in \bar{H}_0{}^{\perp}$. Since $z \perp x_k$ for all $k = 1, 2, \ldots$, $Sz = 0$. Therefore $T^*Tz = 0$ so that $Tz = 0$ as $(Tz \mid Tz) = (T^*Tz, z) = 0$.

Define G_n on H by $G_n x = F_n y$ for $n = 1, 2, \ldots$. Then G_n are finite-rank operators, and since $\| x^2 \| = \| y \|^2 + \| z \|^2$,

$$\| T - G_n \| = \sup_{\|x\|=1} \| Tx - G_n x \| = \sup_{\|x\|=1} \| Ty - F_n y \| \leq \| T_{\bar{H}_0} - F_n \|.$$ ∎

• Application of Theorem 6.8: Fredholm Integral Equations and the Stürm-Liouville Problem

Let us first consider here solving what is known as a Fredholm integral equation with a Hilbert–Schmidt-type kernel. It is an equation of the form

$$f(s) - \lambda \int_a^b K(s, t) f(t)\, dt = g(s), \tag{6.37}$$

where $K(s, t)$ is a complex-valued measurable function on $[a, b] \times [a, b]$ such that

$$\| K \|_2 = \int_a^b \int_a^b | K(s, t) |^2\, ds\, dt < \infty,$$

$f \in L_2[a, b]$ and $g \in L_2[a, b]$. The function K is called the *kernel* of equation (6.37). λ is a nonzero complex parameter. Solving equation (6.37) means finding a function f in $L_2[a, b]$ that satisfies equation (6.37).

As shown in Remark 5.29 of Chapter 5, the operator T given by

$$Tf(s) = \int_a^b K(s, t) f(t)\, dt$$

is a compact bounded linear operator from $L_2[a, b]$ to $L_2[a, b]$ with norm $\| T \|$ satisfying $\| T \| \le \| K \|_2$. Moreover T is a self-adjoint operator if and only if $K(s, t) = \overline{K(t, s)}$ a.e. with respect to (s, t). In this case the kernel K is called symmetric.

If we assume that the kernel K is symmetric, the equation (6.37) takes the form

$$f = \lambda T f + g$$

or

$$Tx - \mu x = y, \tag{6.38}$$

where $\mu = 1/\lambda$, $x = f$, $y = -(1/\lambda) g$, and T is a compact self-adjoint operator. Clearly $y = 0$ if and only if $g = 0$. For $g = 0$, the values of λ for which equation (6.37) has a nontrivial solution are called *characteristic values* of equation (6.37). For $y = 0$, the values of μ for which equation (6.38) has a nonzero solution are the eigenvalues of T. Clearly characteristic values and nonzero eigenvalues are reciprocally related.

We can consider the equation (6.38) as an equation in any Hilbert space H where T is a self-adjoint compact operator on H and x and y are elements of H. Since T is compact, the nonzero values of μ are either *regular* values of T or eigenvalues of T; that is, $T - \mu I$ either has a bounded inverse on H or μ is an eigenvalue of T. If $(T - \mu I)^{-1}$ exists, then equation (6.38) has a unique solution for each y in H, and in particular 0 is the only solution of $Tx - \mu x = 0$.

If μ is an eigenvalue of T, let H_μ be the eigenspace of T, the subspace of H consisting of all eigenvectors corresponding to μ. Let $H_\mu^\perp$ be the orthogonal complement of H_μ. Note that $T(H_\mu) \subset H_\mu$ and $T(H_\mu^\perp) \subset H_\mu^\perp$. If we consider T restricted to $H_\mu^\perp$, then μ is not an eigenvalue for this restricted operator. Consequently as an operator from $H_\mu^\perp$ to $H_\mu^\perp$, $T - \mu I$ has a bounded inverse on $H_\mu^\perp$ and equation (6.38) has a unique solution for any y in $H_\mu^\perp$. Since any x in H can be written uniquely as $x_1 + x_2$ with $x_1 \in H_\mu$ and $x_2 \in H_\mu^\perp$ and $(T - \mu I)x_1 = 0$ we have

$$(T - \mu I)x = (T - \mu I)x_2 \in H_\mu^\perp.$$

Hence equation (6.38) has a solution only for y in $H_\mu^\perp$. If x_2 is any solution of (6.38) and x_1 is any element of H_μ, then $x_1 + x_2$ is also a solution. Consequently, as in the case when μ was not an eigenvalue, we can say that (6.38) has a unique solution for y in H if and only if the homogeneous equation $(T - \mu I)x = 0$ has a unique solution. Moreover if $\mu \neq 0$ is an eigenvalue, then (6.38) has a solution if and only if y is in the subspace $H_\mu^\perp$ of vectors orthogonal to the space of eigenvectors.

Theorem 6.8 furnishes a technique for solving equation (6.38) and of course equation (6.37). First of all, suppose μ is not an eigenvalue. By Theorem 6.8, for each x in H

$$Tx = \sum_i \mu_i(x \mid x_i)x_i.$$

From (6.38), $Tx = y + \mu x$ or

$$x = \frac{1}{\mu}(Tx - y) = \frac{1}{\mu}\sum_i \mu_i(x \mid x_i)x_i - y. \tag{6.39}$$

For any n, by taking scalar products of both sides of this equation, we have

$$\begin{aligned}\mu_n(x \mid x_n) &= \frac{\mu_n}{\mu}\left(\left[\sum_i \mu_i(x \mid x_i)x_i - y\right] \mid x_n\right)\\ &= \frac{\mu_n}{\mu}[\mu_n(x \mid x_n) - (y \mid x_n)]\end{aligned}$$

from which

$$\mu_n(x \mid x_n) = \frac{(\mu_n/\mu)(y \mid x_n)}{\mu_n/\mu - 1} = \frac{\mu_n(y \mid x_n)}{\mu_n - \mu}. \tag{6.40}$$

Hence from equation (6.39) the solution x is given by

$$x = \frac{1}{\mu}\sum_i \frac{\mu_i(y \mid x_i)}{\mu_i - \mu}x_i - y.$$

Secondly if μ is an eigenvalue, then equation (6.39) is still valid and as before we can solve for $\mu_n(x \mid x_n)$ as in equation (6.40) if $\mu \neq \mu_n$. If $\mu = \mu_n$, the coefficient of eigenvector x_n can be taken as arbitrary since this term represents a vector in H_μ. Consequently, a solution of equation (6.38) is given by

$$x = (1/\mu)[\sum c_i x_i - y]$$

where $c_n = \mu_n(y \mid x_n)/(\mu_n - \mu)$ if $\mu_n \neq \mu$ and c_n is arbitrary if $\mu_n = \mu$.

These results are summarized in the following theorem.

Theorem A. Suppose T is a compact self-adjoint operator on H. If $\mu \neq 0$ is not an eigenvalue of T and y is an element of H, then the solution of $(T - \mu I)x = y$ exists and is the unique vector

$$x = \frac{1}{\mu}\left[\sum_i \frac{\mu_i(y \mid x_i)x_i}{\mu_i - \mu} - y\right],$$

where the μ_i are the eigenvalues of T and the x_i are corresponding eigenvectors. If $\mu \neq 0$ is an eigenvalue of T and $y \in H_\mu^\perp$, then a solution of $(T - \mu I)x = y$ exists and is given by

$$x = \frac{1}{\mu}\left(\sum_i c_i x_i - y\right)$$

where $c_i = \mu_i(y \mid x_i)/(\mu_i - \mu)$ if $\mu_i \neq \mu$ and c_i is arbitrary if $\mu_i = \mu$. ∎

An equation which is solved by means of the theory for solving Fredholm integral equations is the so-called Stürm–Liouville equation. It is an equation of the form

$$-[\psi(x)f'(x)]' + \varphi(x)f(x) - \mu f(x) + g(x) = 0, \tag{6.41}$$

where $\psi(x)$, $\psi'(x)$, and $\varphi(x)$ are real continuous functions on $[a, b]$, $g(x)$ is a complex-valued continuous function on $[a, b]$, and $\psi(x) > 0$ for x in $[a, b]$. Equations of this sort arise in the study of vibrating strings and membranes, transmission lines, and resonance in a cavity. Solving the Stürm–Liouville problem means finding a function f in $L_2[a, b]$ satisfying equation (6.41) such that

$$a_1 f(a) + a_2 f'(a) = 0, \tag{6.42}$$

$$b_1 f(b) + b_2 f'(b) = 0, \tag{6.43}$$

and

$$f'' \text{ exists and is continuous on } [a, b], \tag{6.44}$$

where a_1, a_2, b_1 and b_2 are real constants such that $|a_1| + |a_2| > 0$ and $|b_1| + |b_2| > 0$.

Together with equation (6.41) the associated homogeneous equation (6.45) in which $g(x) = 0$ can also be considered:

$$-[\psi(x)f'(x)]' + \varphi(x)f(x) - \mu f(x) = 0. \tag{6.45}$$

In consideration of this equation, two possibilities arise. Either $\mu = 0$ admits the existence of a nontrivial solution to equation (6.45) that satisfies equations (6.42)–(6.44) or $\mu = 0$ admits only the trivial solution $f = 0$. Alternatively, we say, respectively, that either 0 is an eigenvalue of equation (6.45) or 0 is not an eigenvalue. We deal with each of these cases below.

Let us observe first that if μ_1 and μ_2 are two distinct eigenvalues of equation (6.45), that is, two scalars that admit nontrivial solutions f_1 and f_2, respectively, of equation (6.45), then f_1 and f_2 are orthogonal elements of $L_2[a, b]$. Indeed, since

$$[\psi(x)f_1'(x)]' - \varphi(x)f_1(x) + \mu_1 f_1(x) = 0,$$
$$[\psi(x)\overline{f_2'(x)}]' - \varphi(x)\overline{f_2(x)} + \bar{\mu}_2\overline{f_2(x)} = 0,$$

it follows that upon multiplying the first equation by $\bar{f}_2$ and the second by f_1 and then subtracting,

$$\{\psi(x)[\bar{f}_2(x)f_1'(x) - f_1(x)\overline{f_2'(x)}]\}' = (\bar{\mu}_2 - \mu_1)f_1\bar{f}_2. \tag{6.46}$$

Since $a_1 f_1(a) + a_2 f_1'(a) = 0$, $a_1\overline{f_2(a)} + a_2\overline{f_2'(a)} = 0$, and $|a_1| + |a_2| > 0$, the determinant $f_1(a)\overline{f_2'(a)} - f_1'(a)\overline{f_2(a)}$ is zero. Similarly $\overline{f_2(b)}f_1'(b) - f_1(b)\overline{f_2'(b)} = 0$. Upon integrating both sides of equation (6.46), we find therefore that $\int_a^b f_1(x)\overline{f_2(x)}\,dx = 0$.

Since $L_2[a, b]$ is separable, it contains at most a countable set of orthogonal elements. Consequently there are at most countably many distinct eigenvalues of equation (6.45) and uncountably many real numbers that are not eigenvalues of equation (6.45).

Let us first consider now the case where 0 is an eigenvalue of equation (6.45) and show that this case can be reduced to the case where 0 is not an eigenvalue. Suppose λ_0 is a real number that is not an eigenvalue of equation (6.45), that is,

$$-[\psi(x)f'(x)]' + \varphi(x)f(x) - \lambda_0 f(x) = 0$$

has only the trivial solution that satisfies equations (6.42) and (6.43). Letting $\gamma(x) = \varphi(x) - \lambda_0$, this equation has the form

$$[-\psi(x)f'(x)]' + \gamma(x)f(x) = 0,$$

and equation (6.41) has the form

$$-[\psi(x)f'(x)]' + \gamma(x)f(x) - (\mu - \lambda_0)f(x) + g(x) = 0. \qquad (6.47)$$

By assumption the homogeneous counterpart of equation (6.47), i.e., with $g(x) = 0$, has only the trivial solution when $\mu - \lambda_0 = 0$, which is precisely the assumption made when the case in which 0 is not an eigenvalue of equation (6.45) is considered.

Let us then assume the latter case: $\mu = 0$ admits only the trivial solution $f = 0$ to equation (6.45) that satisfies equations (6.42)–(6.44). Let $\mu = 0$. From the theory of elementary differential equations, a nontrivial real solution u_1 of equation (6.45) exists that satisfies equation (6.42) and a nontrivial solution u_2 of equation (6.45) exists that satisfies equation (6.43). Moreover, the Wronskian of u_1 and u_2 is given by $c/\psi(x)$ for each x in $[a, b]$, where c is a constant. Solutions u_1 and u_2 are linearly independent, for if one were a constant multiple of the other, each would satisfy equation (6.45) and conditions (6.42)–(6.44). This means that $c \neq 0$. By the proper choice of u_1 and u_2 we can assume that $c = -1$, whereby u_1 and u_2 satisfy the equation

$$u_1(x)u_2'(x) - u_1'(x)u_2(x) = -1/\psi(x), \qquad x \in [a, b]. \qquad (6.48)$$

Using the method of variation of parameters from elementary differential equations, a solution of equation (6.41) with $\mu = 0$ is given by

$$u(x) = c_1(x)u_1(x) + c_2(x)u_2(x),$$

where c_1 and c_2 are functions on $[a, b]$ satisfying

$$c_1'(x)u_1(x) + c_2'(x)u_2(x) = 0$$

and

$$c_1'(x)u_1'(x) + c_2'(x)u_2'(x) = -g(x)/\psi(x).$$

From equation (6.48), the solution of these equations is given by

$$c_1'(x) = -u_2(x)g(x) \qquad \text{and} \qquad c_2'(x) = u_1(x)g(x)$$

or

$$c_1(x) = \int_x^b u_2(y)g(y)\,dy \qquad \text{and} \qquad c_2(x) = \int_a^x u_1(y)g(y)\,dy.$$

Hence

$$u(x) = u_1(x) \int_x^b u_2(y)g(y)\,dy + u_2(x) \int_a^x u_1(y)g(y)\,dy$$
$$= \int_a^b G(x, y)g(y)\,dy,$$

where

$$G(x, y) = \begin{cases} u_1(x)u_2(y) & \text{if } x \leq y \\ u_1(y)u_2(x) & \text{if } y \leq x. \end{cases} \tag{6.49}$$

Observe that the function u satisfies the boundary conditions (6.42) and (6.43). Indeed,

$$u(a) = u_1(a) \int_a^b u_2(y)g(y)\,dy$$

and

$$u'(a) = u_1'(a) \int_a^b u_2(y)g(y)\,dy$$

so that

$$a_1u(a) + a_2u'(a) = [a_1u_1(a) + a_2u_1'(a)] \int_a^b u_2(y)g(y)\,dy = 0.$$

Similarly $b_1u(b) + b_2u'(b) = 0$. Observe also that u is the only solution of equation (6.41) satisfying equations (6.42) and (6.43), for if v is any other solution, then $u - v = 0$ as $u - v$ satisfies equation (6.45) with $\mu = 0$.

The function $G(x, y)$ is called *Green's function*. It is easily seen to be continuous on $[a, b] \times [a, b]$ and to satisfy the symmetry property $G(x, y) = G(y, x)$. The solution of equation (6.41) is seen therefore to be the operator value of the compact self-adjoint linear operator $G: L_2[a, b] \to L_2[a, b]$ given by

$$(Gg)(x) = \int_a^b G(x, y)g(y)\,dy \tag{6.50}$$

corresponding to the continuous function g in $L_2[a, b]$. Specifically we have the following theorem relating the Stürm–Liouville equation to an integral equation.

Theorem B. (i) Assume $\mu = 0$ is not an eigenvalue of equation (6.45). Then there is a real continuous function G defined on $[a, b] \times [a, b]$ and satisfying $G(x, y) = G(y, x)$ such that $f(x)$ is a solution of

$$-[\psi(x)f'(x)]' + \varphi(x)f(x) + g(x) = 0$$

satisfying equations (6.42) and (6.43) if and only if

$$f(x) = \int_a^b G(x, y)g(y)\, dy.$$

In particular, there is a real symmetric function G on $[a, b]\times[a, b]$ such that f is a solution of

$$[\psi(x)f'(x)]' - \varphi(x)f(x) = \lambda f(x) \tag{6.51}$$

satisfying equations (6.42) and (6.43) if and only if f is a solution of the integral equation

$$f(x) = \lambda \int_a^b G(x, y)f(y)\, dy. \tag{6.52}$$

(ii) Assume 0 is an eigenvalue of equation (6.45). If λ_0 is not an eigenvalue, there is a real symmetric function $G(x, y, \lambda_0)$ such that $f(x)$ is a solution of

$$-[\psi(x)f'(x)]' + \varphi(x)f(x) - \lambda_0 f(x) + g(x) = 0$$

if and only if

$$f(x) = \int_a^b G(x, y, \lambda_0)g(y)\, dy.$$

In particular, $G(x, y, \lambda_0)$ exists such that f is a solution of

$$-[\psi(x)f'(x)]' + \varphi(x)f(x) = \lambda f(x)$$

if and only if f is a solution of

$$f(x) = (\lambda_0 - \lambda)\int_a^b G(x, y, \lambda_0)f(y)\, dy. \qquad \blacksquare$$

In the foregoing discussion of the Stürm–Liouville problem, the first two terms of equation (6.41) can be considered as the values of an operator S in $L_2[a, b]$ whose values are given by

$$Sf(x) = -[\psi(x)f'(x)]' + \varphi(x)f(x).$$

S is actually an unbounded self-adjoint operator on the domain of functions in $L_2[a, b]$ satisfying equations (6.42), (6.43), and (6.44). (This type of operator and its eigenvalues are studied in the Appendix.) Equation (6.45) thus has the form

$$(S - \mu)f = 0. \tag{6.53}$$

We have seen in Theorem B that if 0 is not an eigenvalue then f is a solution of equation (6.53) if and only if f is a solution of equation

$$(G - 1/\mu)f = 0, \tag{6.54}$$

where G is a compact operator on $L_2[a, b]$. Hence we see that the values of μ for which equation (6.53) has a nontrivial solution (the eigenvalues of S) are the reciprocals of the nonzero eigenvalues of the integral operator G. In other words, the eigenvalues of S are the characteristic values of the Fredholm integral equation (6.52). When zero is not an eigenvalue of G, the set of eigenvalues of G form a bounded real sequence converging to zero by Theorem 6.8, whereby the eigenvalues of S then form a real sequence $(\mu_i)_N$ such that $|\mu_i| \to \infty$.

Problems

6.4.1. In this problem A, S, T, U, and Q are the continuous linear operators on l_2 given by the following rules:

$$A(x_1, x_2, \ldots) = (\alpha_1 x_1, \alpha_2 x_2, \ldots),$$

where (α_i) is a bounded sequence of scalars;

$$S(x_1, x_2, \ldots) = (x_2, x_3, \ldots);$$
$$T(x_1, x_2, \ldots) = (0, x_1, x_2, x_3, \ldots);$$
$$U(x_1, x_2, \ldots) = (x_{i_1}, x_{i_2}, x_{i_3}, \ldots),$$

where the set $\{i_1, i_2, \ldots\}$ is a permutation of $\{1, 2, \ldots\}$; and

$$Q(x_1, x_2, \ldots) = (0, x_2, x_3, \ldots).$$

(i) Show that A is always normal but A is self-adjoint if and only if $\alpha_i = \overline{\alpha_i}$ for each i, and A is unitary if and only if $|\alpha_i| = 1$ for all i.

(ii) Show that $S = T^*$ and $S^* = T$. Show that T is isometric but not unitary.

(iii) Show that U is unitary and a projection.

(iv) Show that Q is a projection but not isometric. (Hint: See Problem 6.3.7.)

6.4.2. Prove Lemma 6.3. is false if H is a real Hilbert space. (Hint: Consider a rotation in R^2.) Show that the conclusion of Lemma 6.3 is true for all Hilbert spaces if T and S are self-adjoint.

6.4.3. (i) Prove that the product (that is, composition) of two normal operators is normal if and only if one commutes with the adjoint of the other.

(ii) Prove that the product of two self-adjoint operators is self-adjoint if and only if the given operators commute.

6.4.4. Prove that every continuous linear operator $T: H \to H$, where H is a complex Hilbert space, can be expressed uniquely as $A + iB$, where A and B are self-adjoint, and that A is normal if and only if A and B commute.

6.4.5. Let H be any Hilbert space and $(x_i)_{i\in I}$ be a basis for H. Let $T: H \to H$ be a bounded linear operator.

(i) Prove that T is isometric if and only if $\{Tx_i: i \in I\}$ is an orthonormal set.

(ii) Prove that T is unitary if and only if $\{Tx_i: i \in I\}$ is a basis for H.

6.4.6. (i) Prove that if $T: H \to H$ is any bounded linear operator, then $\| T^*T = \| TT^* \| = \| T \|^2$.

(ii) Prove that if T is self-adjoint then for any positive integer n, $\| T^n \| = \| T \|^n$.

6.4.7. If A and B are linear operators from H into H such that $(Ax \mid y) = (x \mid By)$ for all x and y in H, show that A is continuous and that $A^* = B$. [Hint: Show that A has a closed graph and use the Closed Graph Theorem (Theorem 5.11 in Chapter 5).]

6.4.8. If T_1 and T_2 are self-adjoint operators on H such that $T_1 \leq T_2$ and $T_2 \leq T_1$, show that $T_1 = T_2$ if H is a complex Hilbert space. Is this true if H is a real Hilbert space? (See Problem 6.4.2.)

6.4.9. If H is any Hilbert space, let $(x_n)_N$ be an orthonormal set and $(\lambda_n)_N$ be any sequence of real numbers such that $\lambda_n \to 0$. Define A: $H \to H$ by

$$Ax = \sum_{k=1}^{\infty} \lambda_k (x \mid x_k) x_k .$$

Prove that A is self-adjoint and compact.

6.4.10. If M is a closed subspace of an infinite-dimensional Hilbert space, prove that P_M, the projection on M, is compact if and only if M is finite dimensional.

6.4.11. If $T: H \to K$ is a bounded linear operator, show that T is compact if and only if T^* is compact.

6.4.12. Let T be the shift operator of Problem 6.4.1 and let $T_n = (T)^n$. Let V_i be the operator on the subspace N of l_2 of finitely nonzero sequences given by $V_i((x_1, x_2, \ldots)) = (1x_1, 2x_2, \ldots, ix_i, x_{i+1}, x_{i+2}, \ldots)$. Finally let U_i be the operator given by $U_i((x_1, x_2, \ldots)) = (0, 0, 0, \ldots, x_i, x_{i+1}, \ldots)$. Prove the following:

(i) The sequence (U_i) converges *strongly* to the zero operator (that is, $U_i x \to 0$ for each x in H), but (U_i) does not converge uniformly to 0 (that is, $\| U_i \| \not\to 0$).

(ii) (V_i) converges strongly to the operator V on N given by $V(x_1, x_2, \ldots) = (x_1, 2x_2, 3x_3, \ldots)$, but V is not bounded.

(iii) (T_n) converges neither strongly nor uniformly to the zero operator but (T_n) converges weakly to 0 [that is, $h^*(T_n f) \to 0$ for all $h^* \in l_2{}^*$ and f in l_2].

6.4.13. Let T be the shift operator of Problem 6.4.1. Prove the following:

(i) T has no eigenvalue, but $0 \in \sigma(T)$. (Hint: If $Tx = \mu x$, show that $x = 0$.]

(ii) 0 is an eigenvalue of T^*.

(iii) μ is an eigenvalue for $T^* \Leftrightarrow | \mu | < 1$.

6.4.14. (i) If T is a normal operator, prove that μ is an eigenvalue of T if and only if $\bar{\mu}$ is an eigenvalue of T^*.

(ii) If A is the operator of Problem 6.4.1, show $\{\alpha_i: i \in N\}$ are the only eigenvalues of A. Conclude that $\{\bar{\alpha}_i: i \in N\}$ are the only eigenvalues of A^*.

(iii) Prove $\sigma(A)$ is the closure of $\{\alpha_i: i \in N\}$.

6.4.15. Let T be the operator on the complex Hilbert space $L_2[0, 1]$ given by $Tf(x) = xf(x)$. Prove the following:

(i) T is self-adjoint.

(ii) $\| T \| = 1$.

(iii) T has no eigenvalue.

(iv) $\sigma(T) = [0, 1]$. [Hint: If $(T - \lambda)^{-1}$ exists as a bounded linear operator, show $\| (T - \lambda)^{-1} \| \geq 1/| x - \lambda |$ a.e.]

6.4.16. Let T be a linear operator on H. A complex number λ is called an *approximate* eigenvalue of T if there exists a sequence (h_n) of unit vectors in H (i.e., $\| h_n \| = 1$) such that $\lim_{n\to\infty}(T - \lambda)h_n = 0$.

(i) Prove that every approximate eigenvalue belongs to the spectrum $\sigma(T)$ of T.

(ii) Prove that if A is a bounded normal operator, then every point of the spectrum is an approximate eigenvalue.

6.4.17. If A: $H \to H$ is a compact operator, prove that a set of orthonormal eigenvectors (x_i) corresponding to the set of eigenvalues of A is a complete orthonormal set if 0 is not an eigenvalue of A. (Hint: Use Theorem 6.8.)

6.5. Spectral Decomposition of Self-Adjoint Operators

Our goal in this section is simply stated: to prove what is known as the Spectral Theorem for bounded self-adjoint operators. It can be said that all our preceding work in Hilbert space theory—although important in itself—has been preparatory to proving this outstanding theorem. As seen earlier, self-adjoint operators are generalizations of Hermitian matrices. The Spectral Theorem is a generalization of the diagonalization theorem for such matrices.

This theorem was originally proved by Hilbert between 1904 and 1910.[†] Proofs of this result were also given by F. Riesz in 1910 and 1913.[‡] Since that time other proofs of spectral theorems have been given by many others for self-adjoint, unitary, and normal operators both in the bounded and unbounded cases. The interested reader may also consult for further study the works of Dunford and Schwartz [17], Halmos [23], Riesz and Sz-Nagy [48], Stone [59], Prugovečki [46], among others, as well as the Appendix.

The spectral theory of certain classes of operators has been studied extensively since it was initiated by Hilbert in the early 1900's. The theory has profound applications in the study of operators on Hilbert space and in areas of classical analysis such as differential equations. See for example the work of Dunford and Schwartz [17].

Although the Spectral Theorem can be generalized to broader classes of operators, as mentioned above, we consider here only self-adjoint bounded operators. In the Appendix other formulations of the Spectral Theorem for bounded and unbounded operators will be given—the proofs of which involve measure theoretic techniques. We will here prove the Spectral Theorem using an approach that makes use of what is known as the "functional" or "operational" calculus and involves no measure theory. After proving this theorem, some applications of this key result will be outlined in the Problems.

Recall from Theorem 6.8 that if $T\colon H \to H$ is a nonzero compact, self-adjoint operator, then for each x in H,

$$Tx = \sum_{k=1}^{\infty} \lambda_k(x \mid x_k)x_k, \tag{6.55}$$

† D. Hilbert, Grundzuge einer allgemeinen Theorie der linearen Integralgleichungen, *Nachr. Akad. Wiss. Göttingen Math.-Phys. IV, Kl.* 157–227 (1906).

‡ F. Riesz, Über quadratische Formen von unendlich vielen Veränderlichen, *Nachr. Akad. Wiss. Göttingen Math.-Phys. Kl.* 190–195 (1910); *Les systèmes d'equations linéaires à une infinité d'inconnues*, Gauthier-Villars, Paris (1913).

where the λ_k are eigenvalues of T and the x_k form a corresponding set of eigenvectors. After Definition 6.13 we will show how the Spectral Theorem generalizes the formula (6.55) to arbitrary self-adjoint operators.

There is one stipulation we must make, however, before proceeding. *All Hilbert spaces considered in this section are complex Hilbert spaces.* The need for this stipulation arises from the fact that the complex numbers are algebraically closed while the real numbers are not—meaning every real polynomial has complex roots but not necessarily real roots. The necessity for using complex Hilbert spaces becomes already evident in the proof of Lemma 6.10.

In this section T will denote an arbitrary self-adjoint (bounded linear) operator on H, and m and M will denote real numbers such that $mI \leq T \leq MI$ (see Lemma 6.4).

To prove a form of the Spectral Theorem for bounded self-adjoint operators, we first prove a theorem that describes the so-called "Continuous Functional Calculus." It gives us much more information than actually needed to prove the Spectral Theorem for a bounded self-adjoint operator, but we prove it here since we will have need of it in our more general considerations of spectral theory in the Appendix. We first must present two lemmas.

Lemma 6.10. Suppose $p(x) = \sum_{k=0}^{n} a_k x^k$ is a polynomial with real or complex coefficients. If T is a bounded self-adjoint operator, let $p(T) = \sum_{k=0}^{N} a_k T^n$. Then

$$\sigma(p(T)) = \{p(\lambda);\ \lambda \in \sigma(T)\} \equiv p(\sigma(T)). \qquad \blacksquare$$

For proof, see Theorem 5.26 in Chapter 5.

Lemma 6.11. Let T be a bounded self-adjoint operator. Then if p is as in Lemma 6.10,

$$\| p(T) \| = \sup_{\lambda \epsilon \sigma(T)} | p(\lambda) |. \qquad \blacksquare$$

Proof.
$$\begin{aligned} \| p(T) \|^2 &= \| p(T)^* p(T) \| = \| (\bar{p}p)T \| \\ &= \sup_{\lambda \epsilon \sigma(\bar{p}p(T))} | \lambda | \qquad \text{by Lemma 6.6 (ii)} \\ &= \sup_{\lambda \epsilon \sigma(T)} | \bar{p}p(\lambda) | \qquad \text{by Lemma 6.10} \\ &= (\sup_{\lambda \epsilon \sigma(T)} | p(\lambda) |)^2. \end{aligned}$$

[Here $\bar{p}(x) = \overline{p(x)}$, where $\overline{p(x)}$ is the complex conjugate of $p(x)$.] $\blacksquare$

If p is a polynomial and T is a bounded (self-adjoint) operator, $p(T)$ is defined as in Lemma 6.10.

Theorem 6.10. (*Continuous Functional Calculus*) Let T be a self-adjoint operator on a Hilbert space H. There exists a unique map φ: $C_1[\sigma(T)] \to L(H, H)$ with the following properties:

(i) $\varphi(\alpha f + \beta g) = \alpha\varphi(f) + \beta\varphi(g)$,
$\varphi(fg) = \varphi(f)\varphi(g)$,
$\varphi(1) = I_H$,
$\varphi(\bar{f}) = \varphi(f)^*$ for all f, g in $C_1[\sigma(T)]$ and scalars α and β.

(ii) $\| \varphi(f) \| \leq C \| f \|_\infty$ for all f, for some constant C.

(iii) If $f(x) = x$, $\varphi(f) = T$.

Moreover, φ satisfies the following additional properties.

(iv) If $Th = \lambda h$ for all h, $\varphi(f)h = f(\lambda)h$.

(v) $\sigma(\varphi(f)) = f(\sigma(T))$.

(vi) If $f \geq 0$, $\varphi(f) \geq 0$.

(vii) $\| \varphi(f) \| = \| f \|_\infty \equiv \sup\{| f(\lambda) |: \lambda \in \sigma(T)\}$.

[Part (v) is called the *Spectral Mapping Theorem*.] ∎

Proof. Define $\varphi(p)$ to be $p(T)$ for each polynomial p in $C_1(\sigma(T))$. By Lemma 6.11, $\| \varphi(p) \| = \| p \|_\infty$ so that φ has a unique continuous linear extension to the closure of the set of polynomials in $C_1(\sigma(T))$. By the Stone–Weierstrass Theorem (Theorem 1.26), this closure is all of $C_1(\sigma(T))$. Clearly, the extension of φ to $C_1(\sigma(T))$ satisfies (i), (ii), and (iii) and is the unique function satisfying these properties.

Since (iv) and (vii) are valid for polynomials, they are valid by the continuity of φ for all continuous functions. To prove (vi), observe that if $f \geq 0$, then $f = g^2$, where g is real-valued and $g \in C_1(\sigma(T))$. Hence, $\varphi(f) = \varphi(g^2) = [\varphi(g)]^2$, where $\varphi(g)$ is self-adjoint so that $\varphi(f) \geq 0$.

It remains to show (v). To this end suppose λ is a scalar and $\lambda \neq f(u)$ for any u in $\sigma(T)$. Let $g = (f - \lambda)^{-1}$. Then

$$\varphi(f - \lambda)\varphi(g) = \varphi((f - \lambda)g) = \varphi(1) = I_H = \varphi(g(f - \lambda)) = \varphi(g)\varphi(f - \lambda)$$

so that $\varphi(g) = [\varphi(f) - \lambda]^{-1}$. This means $\lambda \notin \sigma(\varphi(f))$. Hence $\sigma(\varphi(f))$

$\subset f(\sigma(T))$. Conversely, suppose $\lambda \in f(\sigma(T))$. Let $\varepsilon > 0$ be arbitrary. There exists a polynomial $p(x)$ such that $|p(x) - f(x)| < \varepsilon/3$ for all $x \in \sigma(T)$. From Lemma 6.10, $p(\sigma(T)) = \sigma(p(T))$. If $\lambda = f(u)$ for $u \in \sigma(T)$, let $\lambda_1 = p(u)$ so that $\lambda_1 \in \sigma(p(T))$ and $|\lambda - \lambda_1| < \varepsilon/3$. Hence, $[p(T) - \lambda_1]^{-1}$ does not exist, or by Proposition 5.7 there exists $h \in H$ with $\| h \| = 1$ such that $\| [p(T) - \lambda_1]h \| < \varepsilon/3$. Therefore,

$$\| [f(T) - \lambda]h \| \leq \| [p(T) - \lambda_1]h \| + \| [f(T) - p(T)]h \| + \| (\lambda - \lambda_1)h \| < \varepsilon/3 + \varepsilon/3 + \varepsilon/3 = \varepsilon$$

so that $[f(T) - \lambda]^{-1}$ does not exist by Proposition 5.7. Hence, $\lambda \in \sigma(f(T))$. [Here the image of f under φ is denoted by $f(T)$.] ▮

It should be emphasized that any real-valued continuous function is mapped by φ to a self-adjoint operator, as (i) shows; and, in particular, a nonnegative continuous function is mapped to a positive self-adjoint operator, as shown by (vi).

An immediate consequence of Theorem 6.10 is the following corollary.

Corollary 6.3. If $T \geq 0$, then there exists a positive operator S such that $S^2 = T$. (S is called the *square root* of T.) ▮

Proof. If $T \geq 0$, then $\sigma(T) \subset [0, \infty)$ by Lemma 6.8. Let $f(x) = x^{1/2}$ on $[0, \infty)$. Then if $S = f(T)$, then $S^2 = T$. ▮

We wish now to "extend" (see Problem 6.5.2) the mapping φ to certain nonnegative discontinuous functions defined below. To do so we will need the following lemma.

Let P^+ denote the class of all real-valued nonnegative polynomials defined on $\sigma(T)$.

Lemma 6.12

(i) If $p \in P^+$, $p(T)$ is a positive self-adjoint operator.

(ii) If (p_n) is a sequence in P^+ with $p_{n+1} \leq p_n$ for $n = 1, 2, \ldots$, then $p_n(T)$ converges strongly to a positive self-adjoint operator.

(iii) Let (p_n) and (q_n) be sequences in P^+ with $p_{n+1} \leq p_n$ and $q_{n+1} \leq q_n$ for $n = 1, 2, \ldots$, and let S_p and S_q be the strong limits of the sequences $(p_n(T))$ and $(q_n(T))$, respectively. If $\lim_{n\to\infty} p_n(t) \leq \lim_{n\to\infty} q_n(t)$ for all t in $\sigma(T)$, then $S_p \leq S_q$. ▮

Proof

(i) By (i) and (vi) of Theorem 6.10.

(ii) As in Lemma 6.9., the sequence $(p_n(T))$ converges strongly to a self-adjoint operator S_p. Since $p_n(T) \geq 0$,

$$(S_p x \mid x) = \lim_{n\to\infty} (p_n(T)x \mid x) \geq 0 \quad \text{or} \quad S_p \geq 0.$$

(iii) For each t in $\sigma(T)$, the sequences $(p_n(t))$ and $(q_n(t))$—as nonnegative, nonincreasing sequences—must converge. Let k be a fixed positive integer. For each t in $\sigma(T)$

$$\lim_{n\to\infty} [p_n(t) - q_k(t)] \leq \lim_{n\to\infty} p_n(t) - \lim_{n\to\infty} q_n(t) \leq 0. \tag{6.56}$$

Let $m_n(t) = \max\{(p_n - q_k)(t), 0\}$ for each t and $n = 1, 2, \ldots$. From equation (6.56) it follows that $\lim_{n\to\infty} m_n(t) = 0$ for each t in $\sigma(T)$. Clearly, $m_{n+1} \leq m_n$. Hence, by Dini's Theorem, Problem 1.5.18, the sequence m_n converges uniformly to 0 in $\sigma(T)$. This means that for any $\varepsilon > 0$ there is a positive integer N so that whenever $n \geq N$,

$$p_n(t) - q_k(t) < \varepsilon$$

for any t in $\sigma(T)$. Hence, by (vi) of Theorem 6.10,

$$p_n(T) < \varepsilon I + q_k(T)$$

for $n \geq N$. Hence $S_p \leq \varepsilon I + q_k(T)$ and since k is arbitrary $S_p \leq \varepsilon I + S_q$. The fact that ε is arbitrary means $S_p \leq S_q$. ∎

Now let L^+ represent the class of all real-valued functions f on $\sigma(T)$ for which there exists a sequence (p_n) of nonnegative polynomials defined on $\sigma(T)$ such that

$$\text{(i)}\quad 0 \leq p_{n+1}(t) \leq p_n(t) \qquad \text{for } n = 1, 2, \ldots$$

and (6.57)

$$\text{(ii)}\quad \lim p_n(t) = f(t) \qquad \text{for each } t \text{ in } \sigma(T).$$

The preceding lemma enables us to define $f(T)$ for each f in L^+ as the strong limit of $(p_n(T))$, where (p_n) is a sequence as in condition (6.57). Indeed, if (q_n) is another sequence of nonnegative polynomials satisfying condition (6.57), then Lemma 6.12 assures us that the strong limit of

$(p_n(T))$ and $(q_n(T))$ are the same positive self-adjoint operator. Hence, $f(T)$ is well-defined as this positive self-adjoint operator.

Lemma 6.13. The mapping $f \to f(T)$ of L^+ into the class of positive self-adjoint operators satisfies the following properties:

(i) $(f+g)T = f(T) + g(T)$ for all f, g in L^+.

(ii) $(\alpha f)T = \alpha f(T)$ for all f in L^+ and $\alpha \geq 0$.

(iii) $(fg)T = f(T)g(T)$ for all f, g in L^+.

(iv) If $f \leq g$ then $f(T) \leq g(T)$ for all $f, g \in L^+$. ∎

Proof. The proofs of (i) and (ii) follow from the corresponding statements for p and q in P^+. Statement (iv) is an easy consequence of Lemma 6.12 (iii). We here prove (iii). Choose any nonincreasing sequences (p_n) and (q_n) in P^+ with $f(t) = \lim_n p_n(t)$ and $g(t) = \lim_n q_n(t)$. Then, $p_{n+1}q_{n+1} \leq p_n q_n$ and $\lim_n (p_n q_n)(t) = fg(t)$. For all x and y in H

$$\begin{aligned}(f(T)g(T)x \mid y) &= (g(T)x \mid f(T)y) = \lim_{n\to\infty} (q_n(T)x \mid p_n(T)y) \\ &= \lim_{n\to\infty} (p_n(T)q_n(T)x \mid y) = \lim_{n\to\infty} (p_n q_n(T)x \mid y) = ((fg)Tx \mid y).\end{aligned}$$

Hence $f(T)g(T) = (fg)T$ by Remark 6.1. ∎

Let L be the set of all bounded real-valued functions on $\sigma(T)$ of the form $f - g$ with $f, g \in L^+$. Clearly L is a subspace of the real linear space of all bounded real-valued functions on $\sigma(T)$. If p is any real-valued polynomial on $\sigma(T)$, then $p \in L$ since $p + \alpha 1$ is in P^+ for some positive α and $p = (p + \alpha 1) - \alpha 1$.

Our development has led us to the point where we can extend the mapping $p \to p(T)$ of real-valued polynomials to L. Let $f \in L$ and choose g and h in L^+ so that $f = g - h$. As expected define $f(T)$ as $g(T) - h(T)$. Notice that if also $f = g' - h'$, then $g + h' = h + g'$ so that by Lemma 6.13 (i), $g(T) + h'(T) = h(T) + g'(T)$. Hence $g(T) - h(T) = g'(T) - h'(T)$ and $f(T)$ is well defined.

Proposition 6.22. The mapping $f \to f(T)$ on L is a linear mapping into the class of self-adjoint operators, and for all f and g in L,

(i) $f \leq g$ implies $f(T) \leq g(T)$,

(ii) $(fg)T = f(T)g(T)$. ∎

Proof. The proof follows readily from Lemma 6.13. ∎

There are special functions in L^+ that will be used to prove the Spectral Theorem. For each real number s define e_s on $\sigma(T)$ as follows: If $m \le s \le M$ put

$$e_s(t) = \begin{cases} 1 & \text{for } t \in [m, s] \cap \sigma(T) \\ 0 & \text{for } t \in (s, M] \cap \sigma(T). \end{cases} \tag{6.58}$$

If $s < m$, set $e_s = 0$; if $s \ge M$, set $e_s = 1$.

It is easy to see that e_s is in L^+ if $s < m$ or $s \ge M$. For $s \in [m, M]$, we appeal to the Weierstrass approximation theorem (p. 73). For $s \in [m, M)$ let N be the least positive integer such that $s + 1/N \le M$. For $n \ge N$ let

$$f_n(t) = \begin{cases} 1 & \text{for } t \in [m, s] \cap \sigma(T) \\ -nt + ns + 1 & \text{for } t \in (s, s + 1/n) \cap \sigma(T) \\ 0 & \text{for } t \in [s + 1/n, M] \cap \sigma(T). \end{cases} \tag{6.59}$$

For each such n, f_n is a real-valued continuous function on $\sigma(T)$ with $0 \le f_{n+1} \le f_n$ for $n = N, N + 1. \ldots$ $e_s(t)$ is the limit of $f_n(t)$ for each t in $[m, M]$. Utilizing the Weierstrass approximation theorem, we can find a sequence (p_n) in P with

$$e_s(t) \le f_n(t) + 2^{-n-1} < p_n(t) < f_n(t) + 2^{-n} \tag{6.60}$$

for each t in $\sigma(T)$ and $n = N, N + 1, \ldots.$ From inequalities (6.60) it follows that $p_n \in P^+$ and $p_{n+1} \le p_n$ for $n = N, N + 1, \ldots.$ Also $e_s(t) = \lim_n p_n(t)$. Thus, $e_s \in L^+$.

Definition 6.12. For every real number s, define $E(s)$ to be the positive self-adjoint operator given by $E(s) = e_s(T)$, where e_s is the element of L^+ given by equation (6.58). ∎

Proposition 6.23. Let T be a self-adjoint operator on H and suppose $mI \le T \le MI$. For each real number s there is a projection $E(s)$ on H such that

(i) $E(s)T = TE(s)$,

(ii) $E(s) \le E(s')$ for $s \le s'$,

(iii) $E(s) = 0$ for $s < m$,

(iv) $E(s) = I$ for $s > M$,

(v) $\lim_{h\to 0+} E(s + h) = E(s)$ (in the strong sense). ∎

Proof. For each s, $E(s)$ is the positive self-adjoint operator $e_s(T)$. Since $T = f(T)$, where $f(x) = x$ on $\sigma(T)$, (i) follows from Proposition 6.22 (ii). Since $e_s^2 = e_s$, $e_s \leq e_{s'}$, for $s \leq s'$, $e_s = 0$ for $s < m$, and $e_s = 1$ for $s \geq M$, it is clear that $E(s)$ is a projection and that statements (ii), (iii), and (iv) are true. To prove (v) we note that analogous to (6.60) we can construct a sequence (p_n) in P^+ with $p_{n+1} \leq p_n$ and $e_{s+1/n} \leq p_n$ such that $\lim_{n\to\infty} p_n(t) = e_s(t)$. This means $p_n(T) \geq E(s + 1/n) \geq E(s)$. Since $\lim_n p_n(T) = E(s)$, we have $E(s + 1/n) \to E(s)$ as $n \to \infty$. Using (ii) we see that $\lim_{h\to 0+} E(s + h) = E(s)$. ∎

Any projection-valued function E satisfying the conditions of Proposition 6.23 is called a *resolution of the identity* associated with T.

Definition 6.13. Suppose E is a function on R that assigns to each real number s a self-adjoint operator $E(s)$ on H. Let a and b be real numbers with $a < b$ and let f be a real-valued function on $[a, b]$. f *is E-integrable* if and only if there is a self-adjoint operator S with the following property: For each $\varepsilon > 0$ there exists a $\delta > 0$ such that for all partitions $\{a = s_0 < s_1 < \cdots < s_n = b\}$ of $[a, b]$ with $s_k - s_{k-1} < \delta$ for $k = 1, \ldots, n$ and for all numbers $t_1, t_2, \ldots, t_n$ with $s_{k-1} \leq t_k \leq s_k$ for $k = 1, \ldots, n$

$$\left\| S - \sum_{k=1}^{n} f(t_k)[E(s_k) - E(s_{k-1})] \right\| < \varepsilon.$$

The operator S is called the *integral of f* with respect to E and is denoted by $\int_a^b f(t)\, dE(t)$. ∎

Example 6.6. *A Concrete Example of the Resolution of the Identity for a Compact Self-Adjoint Operator.* Let us pause a moment in our development and see that the function $f(t) = t$ is E-integrable with respect to some operator-valued function E on R and its integral $\int_a^b t\, dE(t)$ is equal to a given compact self-adjoint operator T. To this end let T be a fixed nonzero compact, self-adjoint operator on H. Recall from Theorem 6.8 that, for each x in H,

$$Tx = \sum_i \lambda_i (x \mid x_i) x_i, \tag{6.61}$$

where the λ_i are the eigenvalues of T and the x_i form an orthonormal set of eigenvectors with x_i corresponding to λ_i. For each real number s define $E(s)$ by

$$E(s)(x) = \begin{cases} \sum\limits_{\lambda_k \le s} (x \mid x_k)x_k & \text{for } s < 0 \\ x - \sum\limits_{\lambda_k > s} (x \mid x_k)x_k & \text{for } s \ge 0. \end{cases} \tag{6.62}$$

It is understood that if there are no $\lambda_k \le s$ for $s < 0$ or no $\lambda_k > s$ for $s \ge 0$, then the respective sums in equation (6.62) are zero. If B is the orthonormal set $\{x_i : i = 1, 2, \ldots\}$ and G is the closed linear span of B, then B is an orthonormal basis for G and each element x in H can be written in the form

$$x = y_x + \sum_k (x \mid x_k)x_k, \tag{6.63}$$

where $y_x \in G^\perp$ by Theorem 6.2. Clearly for $s < 0$, $E(s)$ is the projection onto the closed linear span G_s of $\{x_k \mid \lambda_k \le s\}$, and for $s \ge 0$, $E(s)$ is the projection on $G^\perp \oplus G_s$. It is left as an exercise (Problem 6.5.1) for the reader to show that E is a projection-valued function on R that satisfies the criteria of Proposition 6.23. In short, we may say that the values of E "increase" from the zero projection when $s < m$ to the identity projection when $s > M$ where $mI \le T \le MI$.

Now let a and b be real numbers with $a < m$ and $b \ge M$. If $P = \{a = s_0 < s_1 < \cdots < s_n = b\}$ is any partition of $[a, b]$, using equation (6.63) we easily see that

$$[E(s_i) - E(s_{i-1})](x) = \sum_{s_{i-1} \le \lambda_k \le s_i} (x \mid x_k)x_k \quad \text{if } s_k < 0, \tag{6.64}$$

$$[E(s_i) - E(s_{i-1})](x) = y_x + \sum_{s_{i-1} \le \lambda_k \le s_i} (x \mid x_k)x_k \quad \text{if } s_{i-1} < 0 \le s_i, \tag{6.65}$$

$$[E(s_i) - E(s_{i-1})](x) = \sum_{s_{i-1} \le \lambda_k \le s_i} (x \mid x_k)x_k \quad \text{if } 0 \le s_{i-1} < s_i. \tag{6.66}$$

We show now that the compact operator in equation (6.61) is the integral $\int_a^b t\, dE(t)$ as in Definition 6.13, where E is as in equation (6.62). Let $\varepsilon > 0$ be arbitrary. With $\delta = \varepsilon$, let $P = \{a = s_0 < s_1 < s_2 \cdots < s_n = b\}$ be any partition of $[a, b]$ with $s_i - s_{i-1} < \delta$ for $i = 1, 2, \ldots, n$. Let i_0 be the index with $s_{i_0-1} < 0 \le s_{i_0}$. If then t_i is arbitrary in $[s_{i-1}, s_i]$, we have, using equation (6.61) and the orthonormality of the set $\{x_1, x_2, \ldots\}$, that

$$\left\| Tx - \sum_{i=1}^{n} t_i[E(s_i) - E(s_{i-1})](x) \right\|^2$$

$$= \left\| \sum_{\substack{i=1 \\ i \neq i_0}}^{n} \left[\sum_{s_{i-1} < \lambda_k \leq s_i} (\lambda_k - t_i)(x \mid x_k)x_k \right] + \sum_{s_{i_0-1} < \lambda_k \leq s_{i_0}} (\lambda_k \quad t_{i_0})(x \mid x_k)x_k + t_{i_0} \cdot y_x \right\|^2$$

$$= \sum_{\substack{i=1 \\ i \neq i_0}}^{n} \left(\sum_{s_{i-1} \leq \lambda_k \leq s_i} (\lambda_k - t_i)^2 \mid (x \mid x_k) \mid^2 \right) + \sum_{s_{i_0-1} < \lambda_k \leq s_{i_0}} (\lambda_k - t_{i_0})^2 \mid (x \mid x_k) \mid^2$$

$$+ \mid t_{i_0} \mid^2 \| y_x \|^2$$

$$\leq \varepsilon^2 \sum_{\substack{i=1 \\ i \neq i_0}}^{n} \sum_{s_{i-1} \leq \lambda_k \leq s_i} \mid (x \mid x_k) \mid^2 + \varepsilon^2 \sum_{s_{i_0-1} \leq \lambda_k \leq s_{i_0}} \mid (x \mid x_k) \mid^2 + \varepsilon^2 \| y_x \|^2$$

$$= \varepsilon^2 \left[\sum_k \mid (x \mid x_k) \mid^2 + \| y_x \|^2 \right] = \varepsilon^2 \| x \|^2.$$

Hence

$$\left\| T - \sum_{i=1}^{n} t_i[E(s_i) - E(s_{i-1})] \right\| < \varepsilon$$

provided $s_i - s_{i-1} < \delta$ and $t_i \in [s_{i-1}, s_i]$. Hence $T = \int_a^b t \, dE(t)$.

Using this example as a motivation, we now come to the goal—and the most outstanding theorem—of this section.

Theorem 6.11. *The Spectral Theorem.* (Resolution of the identity formulation.) Let T be a self-adjoint operator, and m and M be real numbers with $mI \leq T \leq MI$. Then there exists a unique resolution of the identity E on R associated with T such that if a and b are real numbers with $a < m$ and $b \geq M$, then the mapping $t \to t$ of $[a, b]$ into R is E-integrable and

$$T = \int_a^b t \, dE(t). \qquad \blacksquare \quad (6.67)$$

Proof. If e_s represents the function defined in equation (6.58), it is easy to verify that if $s < u$ then

$$e_u(t) - e_s(t) = \begin{cases} 1 & \text{for} \quad t \in (s, u] \cap \sigma(T) \\ 0 & \text{for} \quad t \in \sigma(T) - (s, u]. \end{cases}$$

Hence for all $t \in \sigma(T)$,

$$s[e_u(t) - e_s(t)] \leq t[e_u(t) - e_s(t)] \leq u[e_u(t) - e_s(t)].$$

Defining E by Definition 6.12, and using Proposition 6.22, we obtain

$$s[E(u) - E(s)] \leq T[E(u) - E(s)] \leq u[E(u) - E(s)]. \tag{6.68}$$

We know from the proof of Proposition 6.23 that E satisfies conditions (i)–(v) of Proposition 6.23. We must show that the mapping $t \to t$ of $[a, b]$ into R is E-integrable. To this end, let $\varepsilon > 0$ be arbitrary and let $\{a = s_0 < s_1 < \cdots < s_n = b\}$ be a partition of $[a, b]$ with $s_k - s_{k-1} < \varepsilon$ for $k = 1, 2, \ldots, n$. Then from (6.68) we get the inequality

$$\sum_{k=1}^{n} s_{k-1}[E(s_k) - E(s_{k-1})] \leq T \sum_{k=1}^{n} [E(s_k) - E(s_{k-1})] \leq \sum_{k=1}^{n} s_k[E(s_k) - E(s_{k-1})].$$

However, since $a < m$ and $b \geq M$, Proposition 6.23 tells us that

$$\sum_{k=1}^{n} [E(s_k) - E(s_{k-1})] = E(b) - E(a) = I. \tag{6.69}$$

Hence,

$$\sum_{k=1}^{n} s_{k-1}[E(s_k) - E(s_{k-1})] \leq T \leq \sum_{k=1}^{n} s_k[E(s_k) - E(s_{k-1})].$$

Now let t_k be a real number with $s_{k-1} \leq t_k \leq s_k$ for $k = 1, 2, \ldots, n$. Then

$$\begin{aligned} \sum_{k=1}^{n} (s_{k-1} - t_k)[E(s_k) - E(s_{k-1})] &\leq T - \sum_{k=1}^{n} t_k[E(s_k) - E(s_{k-1})] \\ &\leq \sum_{k=1}^{n} (s_k - t_k)[E(s_k) - E(s_{k-1})]. \end{aligned} \tag{6.70}$$

However, $s_{k-1} - t_k > -\varepsilon$, $s_k - t_k < \varepsilon$, and $E(s_k) - E(s_{k-1}) \geq 0$ so that using equation (6.69), the inequalities (6.70) reduce to

$$-\varepsilon I \leq T - \sum_{k=1}^{n} t_k[E(s_k) - E(s_{k-1})] \leq \varepsilon I$$

so that from Lemma 6.4

$$\left\| T - \sum_{k=1}^{n} t_k[E(s_k) - E(s_{k-1})] \right\| \leq \varepsilon.$$

Hence $T = \int_a^b t \, dE(t)$.

It remains to show the uniqueness conclusion of the theorem. The proof of this is outlined in Problems 6.5.4 and 6.5.5. ∎

Two applications of the preceding results are outlined in Problems

6.5.6 and 6.5.7. For a further study of the spectral theory of self-adjoint operators in Hilbert space along with applications of this theory, the reader may wish to consult the exhaustive work of Dunford and Schwartz [17].

To conclude this section, we present an example to illustrate much of the preceding theory.

Example 6.7. *The Spectrum, the Eigenvalues, and the Resolution of Identity of a Bounded Self-Adjoint Operator.* Let $H = L_2([\alpha, \beta], \mu)$ where $-\infty < \alpha < \beta < \infty$ and μ is the Lebesgue measure of $[\alpha, \beta]$. Define T: $H \to H$ by $Tf = \lambda f(\lambda)$. T is called the multiplication operator. The equation

$$\int_\alpha^\beta |Tf|^2\, d\mu = \int_\alpha^\beta |\lambda f(\lambda)|^2\, d\mu(\lambda) \leq m^2 \int_\alpha^\beta |f(\lambda)|^2\, d\mu(\lambda) = m^2 \|f\|^2 < \infty, \tag{6.71}$$

where $m = \max\{|\alpha|, |\beta|\}$, shows that $Tf \in L_2[\alpha, \beta]$ for every f and that T is a bounded operator. T is also self-adjoint, since

$$(Tf \mid g) = \int_\alpha^\beta \lambda f(\lambda)\overline{g(\lambda)}\, d\mu(\lambda) = \int_\alpha^\beta f(\lambda)\overline{\lambda g(\lambda)}\, d\mu(\lambda) = (f \mid Tg). \tag{6.72}$$

Let λ_0 be any scalar and let f be an element of $L_2[\alpha, \beta]$ such that $(T - \lambda_0 I)f = 0$. Then

$$\|(T - \lambda_0 I)f\|^2 = \int_\alpha^\beta |\lambda - \lambda_0|^2 |f(\lambda)|^2\, d\mu(\lambda) = 0,$$

and since $|\lambda - \lambda_0| > 0$ almost everywhere, $f(\lambda) = 0$ almost everywhere or $f = 0$. Hence T has no eigenvalues.

Next suppose that $\lambda_0 \in [\alpha, \beta]$. For each positive integer n, let f_n be the characteristic function of $[\lambda_0 - 1/n, \lambda_0 + 1/n] \cap [\alpha, \beta]$. Noting that $\|f_n\| > 0$, let g_n be the function $f_n/\|f_n\|$. If $\lambda_0 \notin \sigma(T)$, then

$$\begin{aligned} 1 = \|g_n\|^2 &= \|(T - \lambda_0 I)^{-1}(T - \lambda_0 I)g_n\|^2 \leq \|(T - \lambda_0 I)^{-1}\|^2 \|(T - \lambda_0 I)g_n\|^2 \\ &= \|(T - \lambda_0 I)^{-1}\|^2 \int_\alpha^\beta |\lambda - \lambda_0|^2 |g_n(\lambda)|^2\, d\mu(\lambda) \\ &\leq \|T - \lambda_0 I\|^2 \frac{1}{n^2} \int_\alpha^\beta |g_n(\lambda)|^2\, d\mu(\lambda) \\ &= \|(T - \lambda_0 I)^{-1}\|^2 \frac{1}{n^2}. \end{aligned}$$

This contradiction shows that $\lambda_0 \in \sigma(T)$ and that $[\alpha, \beta] \subset \sigma(T)$.

Finally, let ξ_0 be any scalar outside of $[\alpha, \beta]$. We want to show $\xi_0 \notin \sigma(T)$ and conclude $[\alpha, \beta] = \sigma(T)$. Let $\delta = \inf\{|\xi - \xi_0| : \xi \in [\alpha, \beta]\} > 0$. For any $g \in L_2[\alpha, \beta]$, define Sg to be $(\xi - \xi_0)^{-1}g(\xi)$. Since

$$\int_\alpha^\beta \left| \frac{g(\xi)}{(\xi - \xi_0)} \right|^2 d\mu(\xi) \leq \int_\alpha^\beta \frac{|g(\xi)|^2}{\delta^2} d\mu(\xi) = \frac{1}{\delta^2} \| g \|^2, \tag{6.73}$$

$Sg \in L_2[\alpha, \beta]$ for each g. Moreover,

$$\xi \frac{g(\xi)}{\xi - \xi_0} = g(\xi) + \xi_0 \frac{g(\xi)}{\xi - \xi_0},$$

which means that $(T - \xi_0 I)S = I$. In addition $S(T - \xi_0 I) = I$, as is easily verified. Moreover, the inequality (6.73) shows that $\| S \| \leq 1/\delta$. We conclude that $\xi_0 \notin \sigma(T)$.

To complete this example, let us calculate the unique resolution of the identity associated with T satisfying equation (6.67). Since $[\alpha, \beta] = \sigma(T) \subset [m_T, M_T]$ and m_T and M_T are in $\sigma(T)$, $\alpha = m_T$ and $\beta = M_T$. Define E: $R \to L(H, H)$ by

$$E(\lambda) = \begin{cases} 0 & \text{if } \lambda < \alpha \\ I & \text{if } \lambda \geq \beta \end{cases}$$

and by

$$E(\lambda) f(\xi) = \begin{cases} f(\xi) & \text{if } \alpha \leq \xi \leq \lambda \\ 0 & \text{if } \lambda < \xi < \beta. \end{cases}$$

It is easy to verify that E is a resolution of the identity associated with T. It remains to show that $T = \int_a^b t \, dE(t)$ if $a < \alpha$ and $b \geq \beta$.

Let $\varepsilon > 0$ be arbitrary and let $\{a = s_0 < s_1 < \cdots < s_n = b\}$ be a partition of $[a, b]$ with $s_i - s_{i-1} < \varepsilon$. For each i, let $s_{i-1} \leq t_i \leq s_i$. Now

$$\left\| Tf - \sum_{i=1}^n t_i [E(s_i) - E(s_{i-1})] f \right\|^2 = \left\| \lambda f(\lambda) - \sum_{i=1}^n t_i f(\lambda) \chi_{[\alpha,\beta] \cap [s_{i-1}, s_i]}(\lambda) \right\|^2$$

$$= \sum_{i=1}^n \int_{[\alpha,\beta] \cap [s_{i-1}, s_i]} |\lambda - t_i|^2 |f(\lambda)|^2 d\mu(\lambda) \leq \varepsilon^2 \int |f(\lambda)|^2 d\mu(\lambda) = \varepsilon^2 \| f \|^2.$$

Hence $T = \int_a^b \lambda \, dE(\lambda)$.

Problems

6.5.1. Prove that the function of R defined by equation (6.62) is a resolution of the identity associated with the compact self-adjoint operator T. [Hint: To prove (v) of Proposition 6.23 recall from Theorem 6.8 that $|\lambda_1| \geq |\lambda_2| > \cdots \to 0$.]

In the following problems T is a self-adjoint operator on H and $mI \leq T \leq MI$.

6.5.2. (i) Prove that each real-valued continuous function f on $\sigma(T)$ is an element of L, that is, $f = g - h$, where g and h are in L^+. [Hint: If $f \geq 0$, then by the Weierstrass approximation theorem, there exist polynomials $p_n(t)$ such that $f(t) + 1/(n+1) \leq p_n(t) \leq f(t) + 1/n$ for all t in $\sigma(T)$.]

(ii) Prove that the mapping in Proposition 6.22 of L into the class of self-adjoint operators is an extension of the mapping φ in Theorem 6.10 on the class of continuous real-valued functions.

6.5.3. Let E' be any projection-valued function on R satisfying the conditions (i), (ii), (iii), and (iv) of Proposition 6.23. If $a < m$ and $b \geq M$, suppose $T = \int_a^b t\, dE'(t)$. (This can be proved as in Theorem 6.11.) Prove the following:

(i) $T^m = \int_a^b t^m\, dE'(t)$ for every nonnegative integer m. {Hint: If $i < j$, $[E'(s_i) - E'(s_{i-1})][E'(s_j) - E'(s_{j-1})] = 0$. Hence

$$\left(\sum_{i=1}^{n} t_i[E'(s_i) - E'(s_{i-1})] \right)^m = \sum_{i=1}^{n} t_i^m[E'(s_i) - E'(s_{i-1})].$$

Show that

$$\left\| T^m - \sum_{i=1}^{n} t_i^m[E'(s_i) - E'(s_{i-1})] \right\| \leq K(m) \left\| T - \sum_{i=1}^{n} t_i[E'(s_i) - E'(s_{i-1})] \right\|,$$

where $K(m)$ is a constant dependent on m.}

(ii) $p_0(T) = \int_a^b p(t)\, dE'(t)$ for every real polynomial p on $[a, b]$, where $p_0 = p\,|_{\sigma(T)}$.

(iii) $f_0(T) = \int_a^b f(t)\, dE'(t)$ for every continuous real-valued function f on $[a, b]$, where $f_0 = f\,|_{\sigma(T)}$. [Note that $f_0(T)$ is defined by Proposition 6.22 since $f_0 \in L$ by Problem 6.5.2.] {Hint: There exists a real polynomial p so that $-\varepsilon/3 \leq p(t) - f(t) \leq \varepsilon/3$ for all t in $[a, b]$. By Lemma 6.4, $\| p_0(T) - f_0(T) \| < \varepsilon/3$. Also

$$-(\varepsilon/3)I \leq \sum[p(t_i) - f(t_i)][E'(s_i) - E'(s_{i-1})] \leq (\varepsilon/3)I$$

so that

$$\| \sum[p(t_i) - f(t_i)][E'(s_i) - E'(s_{i-1})] \| < \varepsilon/3.$$

Use part (ii).}

6.5.4. This problem outlines a proof of the uniqueness conclusion of the Spectral Theorem.

(i) Suppose E' is a projection-valued function on R satisfying conditions (i)–(iv) of Proposition (6.23), and $T = \int_a^b t\, dE'(t)$ $(a < m < M \leq b)$. Suppose $m \leq s < M$ and f_n are the continuous functions on $[a, b]$ given by

$$f_n(x) = \begin{cases} 1 & \text{if } a \leq x \leq s \\ n(s - x) + 1 & \text{if } s < x < s + 1/n \\ 0 & \text{if } s + 1/n \leq x \leq b \end{cases}$$

for $n = N + 1, N + 2, \ldots$, where N is the least positive integer so that $s + 1/N < M$. Note that $f_n|_{\sigma(T)}(T)$ is equal to $\int_a^b f_n(t)\, dE'(t)$ by (iii) of Problem 6.5.3. Prove that $\int_a^b f_n(t)\, dE'(t)$ converges strongly to $\lim_{h\to 0+} E'(s + h)$.

(ii) Establish the uniqueness conclusion of the Spectral Theorem.

6.5.5. (i) If E is any projection-valued function satisfying conditions (ii), (iii), and (iv) of Proposition 6.23, show that the function $t \to (E(t)x \mid y)$ is a bounded variation function on $[a, b]$, where x and y are fixed in H. [Hint: $(E(t)x \mid x)$ is a monotonic function; for any t, $(E(t)x \mid y)$ is a sesquilinear form on H. Use the polarization identity.]

(ii) If the assumptions are as in Problem 6.5.3 (using E instead of E'), prove that for all x and y in H

$$(f_0(T)x \mid y) = \int_a^b f(t)\, d(E(t)x \mid y)$$

in the Riemann–Stieltjes sense.

(iii) Give another proof of the uniqueness conclusion of Theorem 6.11 using part (ii).

6.5.6. Let E be the resolution of the identity satisfying Theorem 6.11.

(i) Prove that a real number λ is not in the spectrum $\sigma(T)$ of T if and only if there exists a positive number ε such that $E(s) = E(t)$ whenever $\lambda - \varepsilon \leq s < t \leq \lambda + \varepsilon$. {Hint: To prove the sufficiency, define a continuous function f on $[a, b]$ so that $f(t) = (t - \lambda)^{-1}$ for $t \notin [a, b] \cap [\lambda - \varepsilon, \lambda + \varepsilon]$. Let $g(t) = t - \lambda$ for all t. Show $(T - \lambda) f(t) = g(T) f(T) = \int_a^b f(t)g(t)\, dE(t) = I$ whenever $\lambda \in [m, M]$. To prove the necessity, assume that for each $\varepsilon > 0$, s and t exist in $[\lambda - \varepsilon, \lambda + \varepsilon]$ such that $E(s) \neq E(t)$. Exhibit a y such that $E(s)y = 0$ and $E(t)y = y$ if $s < t$. Using Problem 6.5.5 (ii), show $\| (T - \lambda I)(y) \|^2 \leq \varepsilon^2 \| y \|^2$.}

(ii) Deduce from (i) that if $\lambda \notin \sigma(T)$, $(T - \lambda)^{-1} = \int_a^b f(t)\, dE(t)$, where $f(t)$ is a continuous function on $[a, b]$ equal to $(t - \lambda)^{-1}$ for $t \notin [a, b] \cap [\lambda - \varepsilon, \lambda + \varepsilon]$.

6.5.7. Prove that a real number λ is an eigenvalue of T if and only if $E(\lambda) \neq E(\lambda-)$, where E is the resolution of the identity satisfying Theorem 6.11. {Hint: To prove the sufficiency, by Problem 6.5.5 (ii) $\| (T - \lambda)x \|^2 = \int_a^b (T - \lambda)^2 \, d\| E(t)x \|^2$ for any x in H. Apply this to a particular x, where $E(t)x = x$ if $t \geq \lambda$ and $E(t)x = 0$ if $t < \lambda$. To prove the necessity, if λ is an eigenvalue, suppose $Tx_0 - \lambda x_0 = 0$, where $x_0 \neq 0$. By Problem 6.5.5 (iii),

$$\big((T - \lambda)^2 x_0 \mid x_0\big) = \int_a^b (t - \lambda)^2 d\big(E(t)x_0 \mid x_0\big) = 0.$$

Since $\lambda \in (a, b]$, show $E(\lambda + \varepsilon)x_0 = x_0$ and $E(\lambda - \varepsilon)x_0 = 0$ by considering the integrals

$$\int_{\lambda+\varepsilon}^{M+\varepsilon} (t - \lambda)^2 d\big(E(t)x_0 \mid x_0\big) \quad \text{and} \quad \int_m^{\lambda-\varepsilon} (t - \lambda)^2 d\big(E(t)x_0 \mid x_0\big).\}$$

6.5.8. *A spectral function* on R is a function $E\colon R \to L(H, H)$ whose values $E(\lambda)$ are projections and that satisfies

(1) $E(\lambda) \leq E(\lambda')$ if $\lambda \leq \lambda'$,

(2) $E(\lambda) = (s) \lim_{\substack{\lambda' \to \lambda \\ \lambda' \geq \lambda}} E(\lambda')$ (in the strong sense),

and

(3) $E_{-\infty} \equiv (s) \lim_{\lambda \to -\infty} E(\lambda) = 0$ and $E_{+\infty} \equiv (s) \lim_{\lambda \to \infty} E(\lambda) = I$.

A *normalized spectral measure* on R is a function P from the Borel sets $B(R)$ of R into $L(H, H)$ whose values are projections such that

(a) $P(R) = I$,

(b) $P(\varnothing) = 0$,

(c) $P(M \cup N) = P(M) + P(N)$ whenever $M \cap N = \varnothing$,

and

(d) $P\left(\bigcup_{i=1}^{\infty} M_i\right) = (s) \lim_n \sum_{i=1}^{n} P(M_i)$ whenever $M_i \cap M_j = \varnothing$ for $i \neq j$.

Prove that to every spectral measure P on $B(R)$, there corresponds a unique spectral function E such that $E(\lambda) = P\big((-\infty, \lambda]\big)$, and conversely. (Spectral measures are studied in the Appendix.)

7

Measure and Topology

As the chapter title indicates, our purpose here is to study measures on classes of subsets of certain topological spaces. Given a topological space X, measures with pertinent properties will be studied on σ-rings and σ-algebras generated by the compact subsets of X, the closed subsets of X, the compact G_δ subsets of X, or others. For the most part, X will be taken as a locally compact topological space.

The primary results of this chapter are the various formulations in Section 7.4 of what are called Riesz representation theorems. These theorems show the relationship between linear functionals on certain vector lattices and measures on topological spaces. The name of F. Riesz (1880–1956) is generally attached to these theorems since he first represented a continuous linear functional on $C[0, 1]$ by an integral.[†] Even though earlier representations had been given of such functionals, the representation given by Riesz avoided earlier defects. Later Riesz gave other proofs of his result; and later extensions to more general spaces than [0, 1] were given by such notables as Radon, who in 1913 considered continuous functions on compact sets in R^n; Banach, who in 1937 considered the space $C(S)$, where S is a compact metric space; Kakutani, who in 1941 extended the theorem to compact Hausdorff spaces; and Markov, who in 1938 extended the result to some spaces other than compact spaces. More recent contributors are Halmos [24, Chap. 10], Hewitt,[‡] and Edwards.[§]

[†] F. Riesz, Sur les opérations fonctionnelles linéaires, *C. R. Acad. Sci. Paris* **149**, 974–977 (1909).

[‡] E. Hewitt, Linear functionals on spaces of continuous functions, *Fund. Math.* **37**, 161–189 (1950).

[§] R. Edwards, A theory of Radon measures on locally compact spaces, *Acta Math.* **89**, 133–164 (1953).

Although there are various proofs of the results of Section 7.4, we give proofs based on the Daniell (1889–1946) approach to integration theory and specifically the Daniell–Stone Representation Theorem of Section 7.1. The Daniell approach, expounded first in (1918)† defines the integral as a linear functional on a certain class of functions and then derives the notions of measure and measurability of functions in terms of this linear functional. Various modifications of this point of view have taken place throughout the years, notable of which is the contribution of M. H. Stone.‡ The name of Theorem 7.1 acknowledges the contributions of Daniell and Stone to this point of view.

Section 7.1 is preparatory to Sections 7.2, 7.3, and 7.4. Sections 7.2 and 7.3 give a comprehensive survey of various classes of subsets of a topological space and measures on these classes. The fifth section is largely due to R. A. Johnson in [31]. The Fubini–Tonelli Theorem is the principal result here and gives a beautiful extension of the Fubini–Tonelli Theorem of Chapter 3. The final section contains the Kakutani fixed point theorem and its application to show the existence of an invariant measure on a compact topological group.

7.1. The Daniell Integral

This section is basically preliminary for the sections to follow—in particular Section 7.4. Although it is possible to prove many of the results of Section 7.4 without making use of the Daniell approach to integration theory, the study of the Daniell integral is worth considering in its own right inasmuch as it gives an alternate approach to integration theory.

In Chapters 2 and 3 the concepts of measure, measurability of functions, and integrability of functions were studied extensively. In particular we know that if $(X, \mathscr{B}, \mu)$ is a measure space, then the set $\mathscr{L}^1(\mu)$ of *real-valued* integrable functions§ on X satisfies the following:

(i) If f, g are in $\mathscr{L}^1(\mu)$ and $\alpha, \beta \in R$, then $\alpha f + \beta g \in \mathscr{L}^1(\mu)$ and

$$\int_X (\alpha f + \beta g)\, d\mu = \alpha \int_X f\, d\mu + \beta \int_X g\, d\mu.$$

(ii) If f is in $\mathscr{L}^1(\mu)$, then $|f|$ is in $\mathscr{L}^1(\mu)$.

† P. J. Daniell, A general form of integral, *Ann. Math.* **19**, 279–294 (1918).

‡ M. H. Stone, Notes on integration I–IV, *Proc. Nat. Acad. Sci.* **34, 35** (1948–1949).

§ In Chapter 5, $L_1(\mu)$ denoted the space of *all* extended-real measurable functions f such that $|f|$ is μ-integrable.

The consequence of (i) is that $\mathscr{L}^1(\mu)$ is a vector space and the integral is a linear functional on $\mathscr{L}^1(\mu)$. The second implication guarantees that $\mathscr{L}^1(\mu)$ is also a "lattice"; that is, if f and g are any two functions in $\mathscr{L}^1(\mu)$, then the "meet" $f \wedge g = \inf(f, g)$ and the "join" $f \vee g = \sup(f, g)$ are in $\mathscr{L}^1(\mu)$. This follows since

$$\sup(f, g) = \tfrac{1}{2}(f + g + |f - g|) \tag{7.1}$$

and

$$\inf(f, g) = -\sup(-f, -g) = \tfrac{1}{2}(f + g - |f - g|).$$

In short we can say that $\mathscr{L}^1(\mu)$ is an example of what is called a "vector lattice."

More precisely, we have the following definition.

Definition 7.1. Let X be any set. A *vector lattice* VL is a vector space of real-valued functions on X such that

(i) f in VL implies $|f|$ is in VL and
(ii) $\inf(f, 1)$ is in VL for all f in VL. (1 is the function χ_X.) ∎

Remarks

7.1. Condition (ii) of Definition 7.1 is generally not part of the definition and is known as "Stone's" condition. We include it in the definition since all the vector lattices to be considered will have this property.

7.2. If f is in VL, so are $f^+ = \sup(f, 0)$ and $f^- = \sup(-f, 0)$, so f can be written as the difference of nonnegative functions in VL.

As noted above, the integral is a linear functional on $\mathscr{L}^1(\mu)$. Moreover it is a "positive" linear functional in the sense that $\int f\, d\mu \geq 0$ whenever $f \geq 0$. In addition, whenever (f_n) is a nonincreasing sequence of functions in $\mathscr{L}^1(\mu)$ converging to the zero function, then $\lim_n \int f_n\, d\mu = 0$ by the Lebesgue Convergence Theorem. The integral on $\mathscr{L}^1(\mu)$ is an example of a Daniell integral, defined as follows.

Definition 7.2. If VL is a vector lattice, a *Daniell integral I* on VL is a positive linear functional on VL such that condition (D) is satisfied.

(D) If (f_n) is a nonincreasing sequence in VL converging to zero, then $I(f_n)$ converges to zero. ∎

Remark 7.3. If I is a positive linear functional on VL, then I satisfies (D) if and only if I satisfies (D′).

(D′) If (f_n) is a nondecreasing sequence in VL with $f = \sup f_n$ in VL, then $I(f) = \sup_n I(f_n)$.

Examples

7.1. If $\mathscr{A}$ is an algebra of sets, μ is a measure on $\mathscr{A}$, VL is the class of simple functions that vanish outside a set of finite measure, and I: VL $\to R$ is given by

$$I(f) = \sum \alpha_i \mu(A_i)$$

for $f \in$ VL, then I is a Daniell integral on vector lattice VL.

7.2. The set of continuous functions on R, each vanishing outside some finite interval, is a vector lattice. The Riemann integral $\int f(x)\, dx$ gives a Daniell integral on this vector lattice.

7.3. Let $C_+'[0, 1]$ be the set of continuous functions f on X which are (right) differentiable at 0 and have $f(0) = 0$. $C_+'[0, 1]$ is a vector lattice. $I(f) = f'(0)$ is a positive linear functional on $C_+'[0, 1]$ but not a Daniell integral since $f_n \downarrow 0$, where $f_n(x) = \inf(1/n, x)$, but $I(f_n) = 1$.

7.4. These examples will be useful in Section 7.2 and following sections. If X is a topological space, the space $C(X)$ of continuous real functions on X, the space $C_b(X)$ of bounded continuous functions on X, and the space $C_c(X)$ of continuous functions with compact support are examples of vector lattices. We will characterize Daniell integrals on these spaces in Section 7.4.

If VL is a vector lattice of real-valued functions on a set X, let VLU be the collection of all extended real-valued functions on X of the form $\sup f_n$, where (f_n) is a nondecreasing sequence of nonnegative functions in VL. Let I be a Daniell integral on VL. If (f_n) and $(\bar{f}_n)$ are two nondecreasing sequences of nonnegative functions in VL with $\sup_n f_n \le \sup_n \bar{f}_n$, then $\sup_n I(f_n) \le \sup_n I(\bar{f}_n)$. Indeed, using condition (D'), for any integer m we have

$$\sup_n I(\bar{f}_n) \ge \sup_n I(\bar{f}_n \wedge f_m) = I(f_m) \tag{7.2}$$

since $\bar{f}_n \wedge f_m \uparrow [(\sup \bar{f}_n) \wedge f_m] = f_m$. Inequality (7.2) means

$$\sup_n I(\bar{f}_n) \ge \sup_n I(f_n).$$

We can therefore safely define $I^*(f)$ for any f in VLU to be $\sup I(f_n)$, where (f_n) is any nondecreasing sequence in VL of nonnegative functions whose $\sup f_n$ is f. This extension of I on VL to I^* on VLU has some important properties.

Proposition 7.1. The extension of I on vector lattice VL to I^* defined on VLU satisfies the following properties:

(i) $0 \leq I^*(f) \leq \infty$ for all $f \in$ VLU.

(ii) If f and g are in VLU and $f \leq g$, then $I^*(f) \leq I^*(g)$.

(iii) If $f \in$ VLU and $0 \leq c < \infty$, then $I^*(cf) = cI^*(f)$.

(iv) If $f, g \in$ VLU, then $f + g$, $f \wedge g$, and $f \vee g$ are in VLU and $I^*(f+g) = I^*(f \wedge g) + I^*(f \vee g) = I^*(f) + I^*(g)$.

(v) If (f_n) is a sequence in VLU and $f_n \uparrow f$, then $f \in$ VLU and $I^*(f_n) \uparrow I^*(f)$. ∎

The proofs are left to the reader.

Now let

$$\mathscr{G} = \{G \subset X: \chi_G \in \text{VLU}\}.$$

$\mathscr{G}$ will be called the collection of VL-open sets corresponding to vector lattice VL. For $G \in \mathscr{G}$ define $\mu(G)$ as $I^*(\chi_G)$. Then we have the following result.

Proposition 7.2. The VL-open sets and the extended nonnegative function μ on these sets satisfy the following properties:

(i) If G_1 and G_2 are VL-open sets, so are $G_1 \cup G_2$ and $G_1 \cap G_2$; and

$$\mu(G_1 \cup G_2) + \mu(G_1 \cap G_2) = \mu(G_1) + \mu(G_2).$$

(ii) If G_1 and G_2 are VL-open sets with $G_1 \subset G_2$, then $\mu(G_1) \leq \mu(G_2)$.

(iii) If G_n for $n = 1, 2, \ldots$ are VL-open sets so is $\bigcup_{n=1}^{\infty} G_n$, and

$$\mu\left(\bigcup_{K=1}^{n} G_K\right) \uparrow \mu\left(\bigcup_{K=1}^{\infty} G_K\right).$$ ∎

The proofs of these statements follow from (iv), (ii), and (v), respectively of Proposition 7.1.

In general the collection of VL-open sets has the following "measurability" relation in regard to VL and VLU.

Proposition 7.3. If VL is a vector lattice, then

(i) for every real number $\alpha \geq 0$ and for every $f \in$ VLU, $A_\alpha \equiv \{x \in X: f(x) > \alpha\}$ is VL-open;

(ii) the smallest σ-algebra $\sigma(\mathscr{G})$ containing all VL-open sets coincides with the smallest σ-algebra $\sigma(\text{VL})$ of subsets of X for which all functions in VL are measurable. ∎

Proof. (i) If $f \in$ VLU, then there exists a nondecreasing sequence (f_n) of nonnegative functions in VL such that $f = \sup f_n$. Since inf $(f_n, \alpha) \in$ VL for each $\alpha \geq 0$,

$$g_n \equiv n[f_n - \inf(f_n, \alpha)]$$

is in VL and $g_n \geq 0$. If $x \in X - A_\alpha$, then $\lim_n g_n(x) = 0$; while if $x \in A_\alpha$, $\lim_n g_n(x) = \infty$. Since the sequence (g_n) is also nondecreasing, so is the sequence $(\inf(g_n, 1))_{n \in N}$—a nonnegative sequence in VL. Inasmuch as χ_{A_α} equals $\sup_n[\inf(g_n, 1)]$, χ_{A_α} is in VLU or A_α is VL-open.

(ii) If f is in VLU and α is any real number

$$\{x: f(x) > \alpha\} = \{x: \lim_n f_n(x) > \alpha\} = \bigcup_n \{x: f_n(x) > \alpha\},$$

where (f_n) is a nondecreasing sequence of nonnegative functions in VL with $f = \sup f_n$. Since $\{x: f_n(x) > \alpha\} \in \sigma(\text{VL})$ for each n, $\{x: f(x) > \alpha\} \in \sigma(\text{VL})$. This means $\mathscr{G} \subset \sigma(\text{VL})$ or $\sigma(\mathscr{G}) \subset \sigma(\text{VL})$. By part (i), every function in VLU is $\sigma(\mathscr{G})$-measurable. Since every function f in VL is the difference of two nonnegative functions in VL, each f in VL is $\sigma(\mathscr{G})$-measurable. This means we also have $\sigma(\text{VL}) \subset \sigma(\mathscr{G})$. ∎

Our objective now is to define a measure μ on the σ-algebra $\sigma(\text{VL})$ such that $\text{VL} \subset \mathscr{L}^1(\mu)$ and $I(f) = \int_X f\, d\mu$ for all f in VL. To this end we construct an outer measure on 2^X and show that its restriction to $\sigma(\text{VL})$ is a measure. If $A \subset X$, define

$$\mu^*(A) = \inf\{\mu(G): G \in \mathscr{G} \text{ and } A \subset G\}. \tag{7.3}$$

(We understand the infimum of the empty set is $+\infty$.) Clearly if $A \in \mathscr{G}$, then $\mu^*(A)$ equals $\mu(A)$, which was defined above to be $I^*(\chi_A)$. We may say that μ^* extends μ on $\mathscr{G}$ to all of 2^X.

Lemma 7.1. μ^* is an outer measure on 2^X. ∎

Proof. The only significant item to prove here is that

$$\mu^*\left(\bigcup_{n=1}^{\infty} A_n\right) \leq \sum_{n=1}^{\infty} \mu^*(A_n)$$

for an arbitrary sequence (A_n) from 2^X. If $\sum_{n=1}^{\infty}\mu^*(A_n) = \infty$, there is nothing to prove. So we may assume this sum is finite. For each $\varepsilon > 0$ and each n there is a G_n in $\mathscr{G}$ such that

$$A_n \subset G_n \qquad \text{and} \qquad \mu(G_n) < \mu^*(A_n) + \varepsilon/2^n.$$

By Proposition 7.2, $\cup\, G_n \in \mathscr{G}$ and we may write by Proposition 7.1

$$\begin{aligned}\mu\left(\bigcup_{i=1}^{\infty} G_i\right) &= I^*\left(\chi_{\bigcup_{i=1}^{\infty} G_i}\right) \leq I^*\left(\sum_{i=1}^{\infty} \chi_{G_i}\right)\\ &= I^*\left(\sup_{n\in N} \sum_{i=1}^{n} \chi_{G_i}\right) = \sup_{n\in N} \sum_{i=1}^{n} I^*(\chi_{G_i})\\ &= \sum_{i=1}^{\infty} I^*(\chi_{G_i}) = \sum_{i=1}^{\infty} \mu(G_i).\end{aligned}$$

This means

$$\mu^*\left(\bigcup_{n=1}^{\infty} A_n\right) \leq \mu\left(\bigcup_{n=1}^{\infty} G_n\right) \leq \sum_{n=1}^{\infty} \mu(G_n) \leq \sum_{n=1}^{\infty} \mu^*(A_n) + \varepsilon. \qquad \blacksquare$$

We know from Chapter 2 (Theorem 2.2) that the collection $\mathscr{B}$ of μ^* measurable sets—that is, the collection of all sets E in 2^X satisfying

$$\mu^*(A) \geq \mu^*(A \cap E) + \mu^*(A \cap E^c)$$

for all A in 2^X—is a σ-algebra. In addition we know that μ^* restricted to this σ-algebra $\mathscr{B}$ is a measure. We now wish to show that $\mathscr{G}$—and therefore $\sigma(\mathscr{G})$—is contained in the σ-algebra of μ^*-measurable sets.

Lemma 7.2. Each set G in $\mathscr{G}$ is μ^*-measurable. $\blacksquare$

Proof. We must show that for each A in 2^X,

$$\mu^*(A) \geq \mu^*(A \cap G) + \mu^*(A \cap G^c).$$

By virtue of the definition [equation (7.3)] of μ^*, it suffices to show

$$\mu(E) \geq \mu(E \cap G) + \mu^*(E \cap G^c) \tag{7.4}$$

for each E in $\mathscr{G}$ with $\mu(E) < \infty$. Indeed, equation (7.4) means that for $\mu^*(A) < \infty$,

$$\begin{aligned}\mu^*(A) &= \inf\{\mu(E): E \in \mathscr{G}, E \supset A\}\\ &\geq \inf\{\mu(E \cap G): E \in \mathscr{G}, E \supset A\}\\ &\quad + \inf\{\mu^*(E \cap G^c): E \in \mathscr{G}, E \supset A\}\\ &\geq \mu^*(A \cap G) + \mu^*(A \cap G^c).\end{aligned}$$

(Note that $E \cap G^c$ is not necessarily in $\mathscr{G}$.) Since E and $E \cap G$ are in $\mathscr{G}$, there exist nondecreasing sequences (f_n) and (g_n) of nonnegative functions in VL such that

$$\chi_E = \sup f_n \quad \text{and} \quad \chi_{E \cap G} = \sup g_n.$$

Inasmuch as $\chi_{E\cap G} \leq \chi_E$, the function h_n given by $h_n \equiv \chi_E - g_n$ is nonnegative and in VLU since h_n also satisfies $h_n = \sup_k\{f_k - g_n, 0\}$. Since also

$$h_n \geq \chi_E - \chi_{E\cap G} = \chi_{E\cap G^c},$$

the sets $A_\alpha{}^n = \{x \in X: h_n(X) > \alpha\}$ for $0 < \alpha < 1$ (sets in $\mathscr{G}$ by Proposition 7.3) contain $E \cap G^c$. Realizing that $\alpha\chi_{A_\alpha{}^n} \leq h_n$ or equivalently that $\chi_{A_\alpha{}^n} \leq (1/\alpha)h_n$ we have

$$\mu^*(E \cap G^c) \leq \mu(A_\alpha{}^n) = I^*(\chi_{A_\alpha{}^n}) \leq 1/\alpha\, I^*(h_n)$$

which means $\mu^*(E \cap G^c) \leq I^*(h_n)$ for each n since α is arbitrary in (0, 1). This implies

$$\begin{aligned}\mu^*(E \cap G^c) &\leq \lim_n I^*(h_n) = \lim_n I^*(\chi_E - g_n)\\ &= I^*(\chi_E) - I^*(\chi_{E\cap G})\\ &= \mu(E) - \mu(E \cap G).\end{aligned}$$

Inequality (7.4) follows. ∎

We are now ready to prove the principal result of this section.

Theorem 7.1. *Daniell–Stone Representation Theorem.* Let VL be a vector lattice of functions on a set X and let I be a Daniell integral on VL. Then there exists a unique measure ν on the σ-algebra σ(VL) such that

$$\text{VL} \subset L_1(\nu), \tag{7.5}$$

$$I(f) = \int f\,d\nu \quad \text{for all } f \text{ in VL}, \tag{7.6}$$

and

$$\nu(A) = \inf \{\mu(G) \colon G \supset A \text{ and } G \text{ is VL-open}\}. \qquad \blacksquare \quad (7.7)$$

Proof. Let ν be the restriction of μ^* defined by equation (7.3) to the σ-algebra $\sigma(\mathscr{G})$ or to what is the same, $\sigma(\text{VL})$. Obviously equation (7.7) is satisfied. Suppose G is VL-open. Then

$$I^*(\chi_G) = \mu(G) = \mu^*(G) = \nu(G) = \int_X \chi_G \, d\nu. \qquad (7.8)$$

If $f \in \text{VL}$ and $f \geq 0$ then $f = \sup h_n$, where

$$h_n = \frac{1}{2^n} \sum_{K=1}^{n2^n} \chi_{\{K/2^n < f\}}.$$

Each $\{K/2^n < f\}$ is VL-open by Proposition 7.3 so that by Proposition 7.1, h_n is in VLU. Using equation 7.8 and Proposition 7.1, we have

$$I^*(h_n) = \frac{1}{2^n} \sum_{K=1}^{n2^n} I^*(\chi_{\{K/2^n < f\}}) = \frac{1}{2^n} \sum_{K=1}^{n2^n} \int_X \chi_{\{K/2^n < f\}} \, d\nu$$

$$= \frac{1}{2^n} \int_X \sum_{K=1}^{n2^n} \chi_{\{K/2^n < f\}} \, d\nu = \int_X h_n \, d\nu.$$

Since $I^*(h_n) \uparrow I^*(f)$ by Proposition 7.1, we have by the Monotone Convergence Theorem

$$I(f) = I^*(f) = \lim \int_X h_n \, d\nu = \int_X f \, d\nu.$$

Since $0 \leq I(f) < \infty$, f is ν-integrable or $f \in L_1(\nu)$. If f is an arbitrary function in VL (not necessarily nonnegative) the integrability of f and the equation $I(f) = \int_X f \, d\nu$ follows by writing f as the difference of the nonnegative functions f^+ and f^- in VL.

The uniqueness conclusion must yet be established. For any VL-open set G, there exists a sequence (f_n) from VL with $f_n \geq 0$ and $f_n \uparrow \chi_G$. Therefore if ν' is any measure such that $I(f) = \int_X f \, d\nu'$ for all f in VL then

$$\nu'(G) = \int \chi_G \, d\nu' = \lim_n \int f_n \, d\nu' = \lim_n I(f_n) = \mu^*(G).$$

Since $\nu'(G) = \mu^*(G)$ for each VL-open set G, equation (7.7) shows that $\nu = \nu'$. $\blacksquare$

The reader should have observed that the uniqueness conclusion of Theorem 7.1 is dependent on condition (7.7). In fact Problem 7.1.3 gives an example of a vector lattice VL and a Daniell integral I such that two measures exist on $\sigma(\text{VL})$ with equations (7.5) and (7.6) satisfied. We have the following sufficiency, which will guarantee uniqueness in lieu of condition (7.7).

• **Proposition 7.4.** Suppose VL is a vector lattice and I is a Daniell integral on VL. Suppose ν is a measure on $\sigma(\text{VL})$ such that

$$\text{VL} \subset L_1(\nu),$$

$$I(f) = \int_X f \, d\nu \qquad \text{for each } f \text{ in VL},$$

and

$$X = \bigcup_{i=1}^{\infty} X_i, \qquad \text{where } X_i \text{ are VL-open and } \nu(X_i) < \infty.$$

Then ν is the only measure on $\sigma(VL)$ satisfying these properties. ∎

Proof. If $\bar{\nu}$ is any such measure, notice that $\bar{\nu}(G) = \nu(G)$ for any VL-open set G. Indeed, if (f_n) is a nonnegative sequence in VL with $f_n \uparrow \chi_G$, then by the Monotone Convergence Theorem

$$\bar{\nu}(G) = \int \chi_G \, d\bar{\nu} = \lim_n \int f_n \, d\bar{\nu} = \lim_n I(f_n) = \nu(G).$$

Now, for each n, let

$$D_n = \{A \in \sigma(\text{VL}) \colon \bar{\nu}(A \cap X_n) = \nu(A \cap X_n)\}.$$

For each n, D_n is a Dynkin system containing the VL-open sets so that it contains $\sigma(\text{VL})$ (see Chapter 1). Hence for each $A \in \sigma(\text{VL})$

$$\bar{\nu}(A) = \lim_n \bar{\nu}(A \cap X_n) = \lim_n \nu(A \cap X_n) = \nu(A). \qquad ∎$$

Problems

7.1.1. Verify Remark 7.3.

7.1.2. Let μ be a measure on an algebra $\mathscr{A}$ and let VL be the vector lattice of simple functions as $f = \sum_{i=1}^{n} a_i \chi_{E_i}$ where $E_i \in \mathscr{A}$ and $\mu(E_i) < \infty$. Let I be given by $I(f) = \sum_{i=1}^{n} a_i \mu(E_i)$. Prove the following:

(i) I is a Daniell integral on VL.

(ii) $\sigma(\mathscr{A}) = \sigma(\text{VL})$.

(iii) Compare the extension of μ to $\sigma(\text{VL})$ as given by equation (7.3) and the Carathéodory extension of μ to $\sigma(\mathscr{A})$.

7.1.3. Let $X = (-\infty, \infty) \cup \{\omega\}$ and let VL consist of all real-valued functions on X that are Lebesgue integrable on $(-\infty, \infty)$ and zero at ω. Let I be defined on VL by $I(f) = \int_X f(x)\,dx$. Prove the following:

(i) VL is a vector lattice.
(ii) $\sigma(\text{VL}) = \{B\colon B \cap (-\infty, \infty) \text{ is Lebesgue measurable}\}$.
(iii) I is a Daniell integral on VL.
(iv) There are two measures ν_1 and ν_2 on $\sigma(\text{VL})$ satisfying conditions (7.5) and (7.6) of Theorem 7.1.

7.1.4. Let $\text{VL} = C[0, 1]$, the continuous functions on $[0, 1]$, and let $I(f)$ be the Riemann integral of f on $[0, 1]$. Prove the following:

(i) I is a Daniell integral.
(ii) $\sigma(\text{VL})$ is the σ-algebra containing all open subsets of $[0, 1]$.
(iii) The measure of Theorem 7.1 is the Lebesgue measure on $[0, 1]$.

7.2. Topological Preliminaries. Borel and Baire Sets

Preparatory to our study of measures on certain topological spaces, we will examine in this section special classes of subsets of a topological space. These sets are called Borel and Baire sets. These sets are defined so that the concepts of measurability and continuity of functions coalesce.

First let us recall some topological concepts and derive some auxiliary results that will be used extensively in following sections. Recall that a topological space X is said to be T_4 if X is a Hausdorff space such that whenever A and B are disjoint closed subsets of X, then there are disjoint open sets $U \supset A$ and $V \supset B$. *Urysohn's Lemma* states that if A and B are disjoint closed subsets of a T_4 space then there is a function f in $C(X)$ such that $0 \leq f \leq 1$ on X and $f \equiv 0$ on A while $f \equiv 1$ on B. The reader may readily verify that each compact Hausdorff space and each metric space is T_4.

Let us also recall that a *locally compact* topological space X is a topological space such that each point in X is contained in an open set whose closure is compact. Every compact space is locally compact while the Euclidean spaces R^n are examples of locally compact spaces that are not compact.

If X is a locally compact Hausdorff space we can form a compact Hausdorff space by adding to X a single point ω not in X. The result is a new compact space $X^* = X \cup \{\omega\}$ called the *one-point compactification* of X provided we put the following topology on X^*: $A \subset X^*$ is open if A is either an open subset of X or A is the complement of a compact subset of X. If X is originally compact, then ω is an isolated point of X^*; otherwise X is dense in X^*. ω is called the *point at infinity* in X^*.

A subset of a topological space X is an *$F\sigma$ set* (respectively, G_δ set) if it is the union (intersection) of a sequence of closed (open) sets in X. A set A is a *σ-compact* set (respectively, *σ-bounded* set) if A is (is contained in) the union of a sequence of compact sets. Set A is bounded if A is contained in a compact set.

The *support* S_f of a real-valued function f on a topological space X is the closure of the set $\{x \in X: f(x) \neq 0\}$. The *zero set* of a real function is $\{x: f(x) = 0\}$. $C_c(X)$ denotes the set of all continuous real-valued functions on X whose support is compact. Clearly $C_c(X) \subset C_b(X)$, the space of bounded continuous functions on X.

The following lemmas are crucial to our study of measure and topology.

Lemma 7.3. Let K be a compact subset of a locally compact Hausdorff space X. Then there is an open set O containing K such that $\bar{O}$ is compact. In addition, for any such O there is an $f \in C_c(X)$ with $0 \leq f \leq 1$, $f \equiv 1$ on K, and $S_f \subset 0$. If K is also a G_δ set, then we can take $f < 1$ on K^c. ▮

Proof. For each x in K there is an open set O_x such that $\bar{O}_x$ is compact. Let O be the union of a finite number of such O_x which will cover K. If O is any open set containing K with $\bar{O}$ compact, the sets K and $X^* - O$ are closed and disjoint in X^*, the one-point compactification of X. Since as a compact Hausdorff space, X^* is normal, Urysohn's Lemma gives the existence of a f^* in $C(X^*)$ such that $0 \leq f^* \leq 1$, $f^* \equiv 0$ on $X^* - O$ and $f^* \equiv 1$ on K. If f is the restriction of f^* to X then clearly f is the desired function of the lemma. The last statement is left for the reader to verify. ▮

Lemma 7.4. If K is compact, O is open, and $K \subset O$ in a locally compact Hausdorff space X, then there exist sets K_0 and O_0 in X such that K_0 is a compact G_δ, O_0 is a σ-compact open set, and

$$K \subset O_0 \subset K_0 \subset O.$$

In fact, O_0 is the countable union of compact G_δ sets. ▮

Proof. Since by Lemma 7.3 there exists a bounded open set U with $K \subset U \subset O$, we may assume O is bounded. Let f be a function of Lemma

7.3 such that $f \equiv 1$ on K, $0 \leq f \leq 1$, and $S_f \subset 0$. Let

$$K_0 = \{x : f(x) \geq 1/2\} = \bigcap_{n=1}^{\infty} \{x : f(x) > 1/2 - 1/n\}$$

and

$$O_0 = \{x : f(x) > 1/2\} = \bigcup_{n=1}^{\infty} \{x : f(x) \geq 1/2 + 1/n\}.$$

Then K_0 is a closed G_δ contained in the bounded set O so that K_0 is a compact G_δ. Clearly O_0 is σ-compact. ∎

Remark 7.4. It follows from Lemma 7.3 that the open Baire sets (see Definition 7.3) are a basis for the locally compact topology on X.

With these topological preliminaries we can now begin our study of special classes of sets in a topological space X. There are many classes of sets—both σ-rings and σ-algebras—that can be considered in a topological space. We have listed all these classes in Figure 4 together with re-

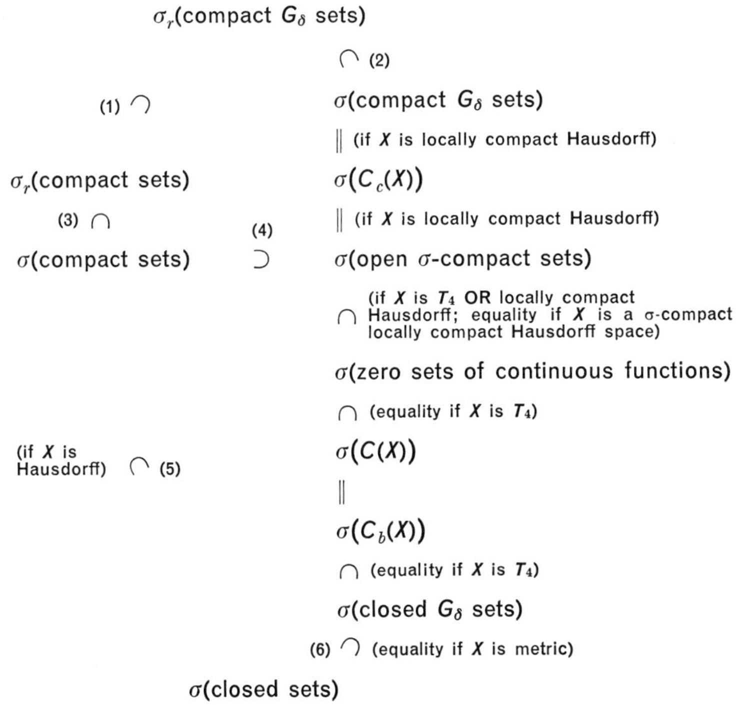

Fig. 4

lationships among these classes and under what conditions on topological space X these relationships are valid. In Figure 4 and the discussion to follow, if $\mathscr{C}$ is a class of sets, $\sigma_r(\mathscr{C})$ and $\sigma(\mathscr{C})$, respectively, represent the σ-ring generated by $\mathscr{C}$ and the σ-algebra generated by $\mathscr{C}$. Similarly if $\mathscr{L}$ is a class of real functions on X, $\sigma(\mathscr{L})$ is the smallest σ-algebra such that all functions in $\mathscr{L}$ are measurable. Our objective in this section is primarily to verify the relationships of Figure 4.

If X is a compact Hausdorff space, then since X is a compact G_δ set and every closed set is compact, we get equality in Figure 4 at all places except possibly at (1), (6), and (4). If X is a σ-compact locally compact Hausdorff space, we get equality at (2) and (3). In fact in this case we will show in Corollary 7.2 that we get equality along the right side of Figure 4 from σ_r (compact G_δ sets) to σ (closed G_δ sets.)

Definition 7.3. If X is a topological space, then

(i) the σ-algebra of *weakly Borel sets* $B_w(X)$ is the σ-algebra σ (closed sets),

(ii) the σ-algebra of *Borel sets* $B(X)$ is the σ-algebra σ (compact sets),

(iii) the σ-algebra of *weakly Baire sets* $\mathrm{Ba}_w(X)$ is the σ-algebra σ (closed G_δ),

(iv) the σ-algebra of *Baire sets* $\mathrm{Ba}(X)$ is the σ-algebra σ (compact G_δ).

The σ-rings σ_r (compact G_δ) and σ_r (compact) are called, respectively, the *σ-ring of Baire sets* and the *σ-ring of Borel sets.* ▮

Accordingly, we have the following scheme if X is a Hausdorff space:

$$\begin{array}{ccccc} \sigma_r\,(\text{compact } G_\delta) & \subset & \mathrm{Ba}(X) & \subset & \mathrm{Ba}_w(X) \\ \cap & & \cap & & \cap \\ \sigma_r\,(\text{compact}) & \subset & \mathrm{B}(X) & \subset & \mathrm{B}_w(X) \end{array}$$

It should be noted that not all authors use the same terminology. For some the Borel sets are what we have termed weakly Borel sets or the σ-ring σ_r (compact). For some the Baire sets are what we have termed weakly Baire sets or the σ-ring σ_r (compact G_δ).

It is understandable why we wish to consider such classes of sets. The objective is to study the interplay of measure and topology. It is natural therefore to consider σ-algebras generated by closed and compact sets. The σ-algebras $\mathrm{Ba}_w(X)$ and $\mathrm{Ba}(X)$ are so defined to ensure the measurability of continuous functions on X and continuous functions of compact support on X, respectively.

Examples

7.5. Let X be an infinite set with the discrete topology (every subset is open). Then

$$\mathrm{Ba}(X) = \mathrm{B}(X) = \{A \in 2^X: A \text{ or } A^c \text{ is countable}\} \subsetneqq 2^X = \mathrm{Ba}_w(X) = \mathrm{B}_w(X).$$

7.6. Let $X = R$ be the reals with the usual topology. Since every closed set is a G_δ (see proof of Corollary 7.1 below) we have

$$\mathrm{Ba}(X) = \mathrm{B}(X) \subset \mathrm{Ba}_w(X) = \mathrm{B}_w(X).$$

Moreover $\mathrm{Ba}(X) = \mathrm{Ba}_w(X)$ by Corollary 7.2 below.

7.7. Let $D = \{j/2^n: j \text{ an integer and } n \text{ a nonnegative integer}\}$. Let $E = R - D$ and for each integer j and nonnegative integer n let D_{nj} be the singleton set $\{[j/2^n, (j+1)/2^n]\}$. Let

$$X = E \cup \{D_{nj}\}.$$

Topologize X as follows: $N \subset X$ is a neighborhood of x in E if $x \in N$ and N contains all but finitely many of the D_{nj} with $j/2^n \leq x \leq (j+1)/2^n$; $N \subset X$ is a neighborhood of y from $\{D_{nj}\}$ if N is any set containing y. It is readily seen that X is then a locally compact Hausdorff space. Also

$$B(X) = \mathrm{Ba}(X) = \{A \subset X: A \text{ or } A^c \text{ is countable}\},$$

and

$$B_w(X) = 2^X.$$

The following propositions serve to verify the relationships of Figure 4. Some of the inclusions of course are trivial.

Proposition 7.5. If X is any topological space, $\sigma(C_b(X)) = \sigma(C(X))$. ∎

Proof. Obviously $\sigma(C_b(X)) \subset \sigma(C(X))$. On the other hand, each f in $C(X)$ is measurable with respect to $\sigma(C_b(X))$ since it is the pointwise limit of the sequence (f_n) from $C_b(X)$, where

$$f_n \equiv \inf(\sup(f, -n), n).$$ ∎

Proposition 7.6. If X is a T_4 topological space, $\mathrm{Ba}_w(X) = \sigma(C(X))$. ∎

Proof. Obviously $\sigma(C(X)) \subset \mathrm{Ba}_w(X)$, the σ-algebra generated by closed G_δ sets, since for any $\alpha \in R$

$$\{x \in X: f(x) \geq \alpha\} = \bigcap_{n=1}^{\infty} \{x: f(x) > \alpha - 1/n\}$$

so that $\{x \in X: f(x) \geq \alpha\}$ is a closed G_δ set. (Normality is not needed here.) The converse is accomplished by showing that if F is an open F_σ set (that is, the complement of a closed G_δ set), then $F = \{x \in X: f(x) > 0\}$ for some nonnegative function f in $C_b(X)$. As an open F_σ set, F equals $\bigcup_{n=1}^{\infty} F_n$, where each F_n is closed. Since F_n and F^c are disjoint closed sets in X, by Urysohn's Lemma there is for each n an f_n in $C_b(X)$ with $0 \leq f_n \leq 1$, $f_n \equiv 1$ on F_n, and $f_n \equiv 0$ on F^c. Let f be defined as

$$f(x) = \sum_{n=1}^{\infty} \frac{f_n(x)}{n^2}$$

for each x in X. Since the convergence is uniform, $f \in C_b(X)$ and $F = \{x \in X: f(x) > 0\}$. ∎

Corollary 7.1. If (X, d) is a metric space, then

$$B_w(X) = \mathrm{Ba}_w(X) = \sigma(C(X)).$$ ∎

Proof. Since a metric space is T_4, the right equality follows from Proposition 7.6. Each closed subset F of X is a G_δ set as

$$F = \bigcap_{n=1}^{\infty} \{x \in X: \mathrm{dist}(x, F) < 1/n\},$$

whence follows the left equality. ∎

Remark 7.5. Problem 7.2.5 gives an example of a compact *nonmetrizable* Hausdorff space such that each closed set is a G_δ set so that in this case the conclusion of Corollary 7.1 is again true.

In case X is a locally compact Hausdorff space, the following lemma is useful.

Lemma 7.5. Suppose X is a locally compact Hausdorff space.

(i) A subset G of X is $C_c(X)$-open (see Section 7.1) if and only if G is an open σ-compact set.

(ii) If X is σ-compact, then X can be written as the union of compact G_δ sets K_n with $K_n \subset K_{n+1}^\circ$, the interior of K_{n+1}, for each n. In particular, each compact subset K of X is contained in some K_n. ▮

Proof. (i) If G is $C_c(X)$-open, then $\chi_G = \lim f_n$, where (f_n) is a nondecreasing sequence of nonnegative functions in $C_c(X)$. Hence

$$G = \bigcup_{n=1}^{\infty} \{x \in X: f_n(x) > 0\} = \bigcup_{n,k=1}^{\infty} \{x \in X: f_n(x) \geq 1/k\}$$

so that G is open and a σ-compact set, since the set $\{x \in X: f_n(x) \geq 1/k\}$ is compact as a closed subset of the compact support of f_n. Conversely, if $G = \bigcup_{n=1}^{\infty} K_n$, where each K_n is compact, then by Lemma 7.3 for each n there exists an f_n in $C_c(X)$ with $0 \leq f_n \leq 1$, $f_n \equiv 1$ on K_n, and $S_{f_n} \subset G$. Letting $g_n = \sup\{f_1, f_2, \ldots, f_n\}$, $\chi_G = \lim g_n$ and (g_n) is a nondecreasing sequence in $C_c(X)$, whence G is $C_c(X)$-open.

(ii) By (i) X is $C_c(X)$-open. Hence there exists a nondecreasing sequence (f_n) in $C_c(X)$ such that $\chi_X = \lim f_n$. Let $K_n = \{x \in X: f_n(x) \geq 1/n\}$; K_n is a G_δ set and compact as a closed subset of the support of f_n. Also

$$K_n \subset \{x \in X: f_{n+1}(x) \geq 1/n\} \subset \{x \in X: f_{n+1}(x) > 1/n + 1\} \subset K_{n+1}$$

so that $K_n \subset K_{n+1}^\circ$. If K is compact in X, then $K \subset K_n$ for large n, as otherwise $\bigcup_{n=1}^{\infty} K_n^\circ$ is an open covering of K with no finite subcovering. ▮

Proposition 7.7. Let X be a locally compact Hausdorff space.

(i) The σ-algebra of Baire sets is identical to $\sigma(C_c(X))$.

(ii) The σ-algebra of Baire sets is identical to the σ-algebra generated by open σ-compact sets.

(iii) If X is σ-compact, the σ-algebra of Baire sets coincides with $\sigma(C(X))$. ▮

Proof. (i) The σ-algebra $\sigma(C_c(X))$ is the σ-algebra generated by sets of the form $\{x: f(x) \geq \alpha\}$ for $\alpha \in R$ and $f \in C_c(X)$. If $\alpha > 0$ these are compact G_δ sets and it follows by a simple argument that $\text{Ba}(X)$ contains all sets of this form. Conversely, it follows from Lemma 7.3 that each compact G_δ set K equals $\{x \in X: f(x) \geq 1\}$ for some $f \in C_c(X)$.

(ii) By Proposition 7.3 the smallest σ-algebra containing $C_c(X)$-open sets equals $\sigma(C_c(X))$. The result now follows from (i) and Lemma 7.5 (i).

(iii) By Proposition 7.3, $\sigma(C_c(X))$ coincides with the σ-algebra generated by $C_c(X)$-open sets and $\sigma(C(X))$ coincides with the σ-algebra generated by the $C(X)$-open sets. It suffices to show that each $C(X)$-open set is $C_c(X)$-open. To this end let G be a $C(X)$-open set and let (f_n) be a nondecreasing sequence of nonnegative functions in $C(X)$ such that $\chi_G = \lim f_n$. Since X is an open σ-compact set, X is $C_c(X)$-open by Lemma 7.5 (i). This means there is a nondecreasing sequence (g_n) of nonnegative functions in $C_c(X)$ whose limit is χ_X. The result is that $\chi_G = \lim f_n h_n$ and $(f_n h_n)$ is a nonnegative sequence in $C_c(X)$ so that G is a $C_c(X)$-open set. ∎

Corollary 7.2. If X is a σ-compact locally compact Hausdorff space, then

$$\begin{aligned}\sigma_r\text{ (compact } G_\delta \text{ sets)} &= \sigma\text{ (open } \sigma\text{-compact sets)} = \sigma(C_c(X)) \\ &= \sigma\text{ (compact } G_\delta) \\ &= \sigma\text{ (zero sets of continuous functions)} \\ &= \sigma(C(X)) = \sigma\text{ (closed } G_\delta \text{ sets)}.\end{aligned}$$

In particular, $\mathrm{Ba}(X) = \mathrm{Ba}_w(X)$. ∎

Proof. Proposition 7.7 establishes the second and third equalities. Since by Lemma 7.4, X is the countable union of compact G_δ sets, whenever X is σ-compact the first equality is true. From Lemma 7.3, if G is a compact G_δ set, then there exists $f \in C_c(X)$ such that

$$G = \{x: (f - 1)(x) = 0\}$$

so that σ (compact G_δ) $\subset \sigma$ (zero sets). Trivially σ (zero sets) $\subset \sigma(C(X))$ for any topological space X. By Proposition 7.7 (iii), the fourth and fifth equalities hold. Since each σ-compact locally compact Hausdorff space is normal, the last equality holds by Proposition 7.6. ∎

It is natural to ask whether there are compact Baire sets that are not G_δ sets in the case that X is a locally compact Hausdorff space. If X is also second countable it is easy to show that each compact set is a G_δ so that in this case $\mathrm{Ba}(X) = B(X)$. Regardless of whether X is second countable, we always have the following result.

Proposition 7.8. If X is a locally compact Hausdorff space, each compact Baire set C is a G_δ set. ∎

Proof. By Problem 7.2.6, there are compact G_δ sets C_n such that $C \in \sigma((C_n))$ and $C_n = \{x \in X: f_n(x) = 0\}$ for some f_n in $C(X)$ with $0 \leq f_n \leq 1$. Define the pseudometric [$d(x, y) = 0$ does not always imply $x = y$] on X by

$$d(x, y) = \sum_{n=1}^{\infty} \frac{1}{2^n} | f_n(x) - f_n(y) |.$$

For each x in X, let $[x] = \{y \in X: d(x, y) = 0\}$; that is, $[x]$ is the equivalence class of x with respect to the equivalence relation: $x \sim y$ if and only if $d(x, y) = 0$. Let $\hat{X}$ denote the set of all such equivalence classes and define metric δ on $\hat{X}$ by

$$\delta([x], [y]) = d(x, y).$$

If $n: X \to \hat{X}$ is the natural mapping given by $n(x) = [x]$, then n is continuous since if $E = \{[y]: \delta([y], [x]) < r\}$ for some $[x]$ in $\hat{X}$ and positive number r, then $n^{-1}(E) = \{y: d(y, x) < r\}$.

A subset S of X equals $n^{-1}(\hat{S})$ for some $\hat{S} \subset \hat{X}$ if and only if S contains the set $[x]$ for any x in S. Since each C_n has this property, each C_n is the inverse image of some set in $\hat{X}$. Since the class of all inverse image sets is a σ-ring, $C = n^{-1}(\hat{C})$ for some subset $\hat{C}$ of $\hat{X}$. Since $n: X \to \hat{X}$ is continuous, $n(C) = n(n^{-1}\hat{C}) = \hat{C}$ is compact in $\hat{X}$. Since every compact subset of a metric space is a G_δ set, $\hat{C} = \bigcap_{n=1}^{\infty} \hat{O}_n$ for some open sets $\hat{O}_n$ in $\hat{X}$. This means

$$C = \bigcap_{n=1}^{\infty} O_n,$$

where O_n are the open sets $n^{-1}(\hat{O}_n)$ in X. ∎

The next proposition and the discourse following further illustrates the contrast between Borel and Baire sets. We will show that Baire sets "multiply" as described in Proposition 7.9, whereas Borel sets fail to "multiply." This information will motivate our study of product measures and Fubini's Theorem in Section 7.5.

Proposition 7.9. If X and Y are locally compact Hausdorff spaces, then

$$\sigma_r\,(\text{compact } G_\delta \text{ of } X) \times \sigma_r\,(\text{compact } G_\delta \text{ of } Y) = \sigma_r\,(\text{compact } G_\delta \text{ of } X \times Y). \quad (7.9)$$

[The left side of this equation is by definition the σ-ring generated by sets $A \times B$, where $A \in \sigma_r$ (compact G_δ of X) and $B \in \sigma_r$ (compact G_δ of Y).] ∎

Proof. From Problem 1.3.8 of Chapter 1 we know that the left side of equation (7.9) is identical to the σ-ring generated by the sets of the form $A\times B$, where A is a compact G_δ set in X, and B is a compact G_δ set in Y. Since $A\times B$ is a compact G_δ set in $X\times Y$ whenever A and B are compact G_δ sets in X and Y, respectively, clearly the left side of equation (7.9) is contained in the right side.

Conversely if C is a compact G_δ set in $X\times Y$, let O be an open set in $X\times Y$ containing C. Then from Remark 7.4, $C\subset E\subset O$, where E is a finite union of sets of the form $U\times V$, where U and V are open countable unions of compact G_δ sets in X and Y, respectively. Since $C=\bigcap_{i=1}^{\infty}O_i$ where each O_i is open, $C=\bigcap_{i=1}^{\infty}E_i$, where each E_i is a finite union of sets as $U\times V$. This means C is an element of the left side of equation (7.9). ∎

Corollary 7.3. If X and Y are σ-compact locally compact Hausdorff spaces, then

$$\mathrm{Ba}(X)\times\mathrm{Ba}(Y)=\mathrm{Ba}(X\times Y)$$

and

$$\mathrm{Ba}_w(X)\times\mathrm{Ba}_w(Y)=\mathrm{Ba}_w(X\times Y) \qquad ∎$$

Proof. See Corollary 7.2. ∎

Example 7.8. This example shows that the condition that X and Y both be σ-compact is essential in Corollary 7.3. Let X be an uncountable set with the discrete topology and let Y be an infinite countable set with the discrete topology. Note that both X and Y are locally compact and Hausdorff; but whereas Y is σ-compact, X is not σ-compact. We have, since compact sets coalesce with finite sets,

$$\mathrm{Ba}(X)=B(X)=\{A\subset X\colon A \text{ or } A^c \text{ is countable}\},$$

$$\mathrm{Ba}(Y)=B(Y)=\{B\subset Y\colon B \text{ or } B^c \text{ is countable}\},$$

$$\mathrm{Ba}(X\times Y)=B(X\times Y)=\{C\subset X\times Y\colon C \text{ or } C^c \text{ is countable}\}.$$

However, if A is an uncountable set in X with countable complement and $c\in Y$, then $A\times\{c\}$ is an element of $\mathrm{Ba}(X)\times\mathrm{Ba}(Y)$ but not an element of $\mathrm{Ba}(X\times Y)$.

In contrast, the result of Proposition 7.9 is not true for Borel sets even though X and Y are σ-compact. If X and Y are σ-compact locally compact Hausdorff spaces, we know from Problem 1.3.8 in Chapter 1 that

$B(X)\times B(Y)$ is generated by the sets of the form $A\times B$, where A is compact in X and B is compact in Y. Hence

$$B(X)\times B(Y) \subset B(X\times Y).$$

However, this inclusion is sometimes proper. Let us consider the following interesting examples.

Examples

7.9. Let X be the one-point compactification of a discrete space having cardinality greater than c. As every subset of X is either open or closed, $B(X) = 2^X$, the power set of X. As every subset of $X\times X$ is the difference of two compact subsets, $B(X\times X) = 2^{X\times X}$. Nevertheless, $B(X\times X) \neq B(X)\times B(X)$ as the diagonal $D = \{(x, y)\in X\times X\colon x = y\}$ is not in $B(X)\times B(X)$ (see Problem 7.2.8).

7.10. The classes of Borel and Baire sets can correspond in a σ-compact space X and yet be distinct in $X\times X$. If this is the case, we clearly have

$$B(X)\times B(Y) = \mathrm{Ba}(X)\times \mathrm{Ba}(Y) = \mathrm{Ba}(X\times Y) \subsetneqq B(X\times Y).$$

As an example, let X be a compact nonmetrizable space such that every closed subset is a G_δ set (see Problem 7.2.5, for instance). The diagonal D is compact and is thus a Borel set. If D were a Baire set, it would be a G_δ set by Proposition 7.8. Now the neighborhoods of D define a uniform structure which yields the given topology of X. (See Chapter 6 of [34].) If D were a G_δ set, it would have a fundamental sequence of neighborhoods so that X would be metrizable.

Do Borel sets ever multiply? A positive answer is given in the following proposition. It says in effect that if bounded subspaces of one of the factor spaces are second countable, then the Borel sets multiply.

Proposition 7.10. Suppose X and Y are locally compact Hausdorff spaces. Suppose that each bounded subspace of Y is second countable. (Since any locally compact Hausdorff space with a countable base is metrizable, it is equivalent to assume that each bounded subspace of Y is metrizable.) Then

$$\sigma_r\,(\text{compact sets in } X)\times \sigma_r\,(\text{compact sets in } Y) = \sigma_r\,(\text{compact sets in } X\times Y).$$

If X and Y are also σ-bounded, then

$$B(X)\times B(Y) = B(X\times Y).$$ ∎

Proof. For simplicity of notation we proved the result for the σ-bounded case. The other case is almost identical. Since the Borel sets are precisely the σ-algebra generated by the bounded open sets, it is sufficient to show that if W is a bounded open set in $X\times Y$, then $W \in B(X)\times B(Y)$. Let $F = \mathrm{Pr}_Y(W)$, where Pr_Y is the projection of $X\times Y$ onto Y, a continuous open mapping. F is then a bounded open set in Y. By hypothesis, the subspace F has a countable base for the open sets, say $\mathscr{V}$. As F is a bounded open set, observe that each member of $\mathscr{V}$ is a bounded open set in Y. Since W is also an open subset of $X\times F$,

$$W = \bigcup\{U\times V\colon\ U \text{ is open in } X,\ V\in\mathscr{V},\ U\times V\subset W\}.$$

For each $V\in\mathscr{V}$, define

$$U_V = \bigcup\{U\colon\ U \text{ is open in } X,\ U\times V\subset W\}$$

an open and bounded subset of X. Obviously, $U_V\times V\in B(X)\times B(Y)$. Now

$$\begin{aligned} W &= \bigcup\{U\times V\colon\ U \text{ open in } X,\ V\in\mathscr{V},\ U\times V\subset W\} \\ &= \bigcup_{V\in\mathscr{V}}\bigcup\{U\times V\colon\ U \text{ open in } X,\ U\times V\subset W\} \\ &= \bigcup_{V\in\mathscr{V}}\left[\bigcup\{U\colon\ U \text{ is open in } X,\ U\times V\subset W\}\right]\times V \\ &= \bigcup_{V\in\mathscr{V}} U_V\times V. \end{aligned}$$

Since the last union is countable, $W\in B(X)\times B(Y)$. ∎

Corollary 7.4. If X and Y are locally compact σ-compact Hausdorff spaces and Y is metrizable, then

$$B(X\times Y) = B(X)\times B(Y).$$ ∎

Problems

7.2.1. Prove that if K is a compact subset of a locally compact Hausdorff space and $\{O_\alpha\}$ is an open covering of K, then there are a finite number of nonnegative continuous functions $f_1, f_2, \ldots, f_n$ on X, each f_i vanishing outside a compact set and outside some O_{α_i} such that $f_1+f_2+\cdots+f_n\equiv 1$ on K. (Hint: For each $x\in K$, there is a nonnegative continuous function g that is positive at x and zero outside some O_α. Choose a finite number $g_1, g_2, \ldots, g_n$ of such g's so that $g_1+g_2+\cdots+g_n$ is positive on K. Let $h = 1/g$ and set $f_i = hg_i$.)

7.2.2. Prove that each element of σ_r (compact G_δ) and each element of σ_r (compact) is σ-bounded. Conclude that these are σ-algebras if and only if X is σ-bounded.

7.2.3. If $\mathscr{A}$ is a σ-ring or σ-algebra of subsets of a set X, then a subset M of X is said to be *locally measurable* with respect to $\mathscr{A}$ if $M \cap A \in \mathscr{A}$ for each $A \in \mathscr{A}$. Prove the following:

(i) The class of locally measurable sets with respect to $\mathscr{A}$ is a σ-algebra containing $\mathscr{A}$.

(ii) Each weakly Borel set is locally measurable with respect to the σ-ring of Borel sets if X is a Hausdorff space.

(iii) Each weakly Baire set is locally measurable with respect to the σ-ring of Baire sets if X is Hausdorff.

7.2.4. Prove that if X is Hausdorff, the elements of σ_r (compact) are precisely the weakly Borel sets of X that are σ-bounded; the elements of σ_r (compact G_δ) are precisely the weakly Baire sets that are contained in a countable union of compact G_δ sets.

7.2.5. Let $X = [-1, 1]$ with topology given by the subbase $\mathscr{S}$ given by

$$\mathscr{S} = \{A \subset [-1, 1]: A \text{ or } A^c \text{ has the form } [-b, b) \text{ for } 0 \leq b \leq 1\}.$$

Then:

(i) Prove that X is Hausdorff with this topology.

(ii) Show that X is compact by showing that every covering of X by subbasic members can be reduced to a covering of not more than two such members (see [34], page 139).

(iii) Show that X is not metrizable. (Hint: If X were compact and metrizable, it would have a countable base.)

(iv) Show that each closed subset of X is a G_δ set by showing that each open set is a countable union of basic sets.

7.2.6. Let X be a locally compact Hausdorff space.

(i) Show that if C is a compact Baire set in X, then there is a sequence (C_n) of compact G_δ sets such that C is in the smallest σ-algebra $\sigma((C_n))$ containing each C_n.

(ii) Show that each compact G_δ set C equals $\{x: f(x) = 0\}$ for some continuous function f on X with $0 \leq f \leq 1$.

7.2.7. Show that if X is a second countable locally compact Hausdorff space, each compact set is a G_δ set.

7.2.8. Let X be the one-point compactification of a discrete space having cardinality greater than c. Prove that $D = \{(x, y) \in X \times X: x = y\}$

is not an element of $B(X) \times B(X)$. [Hint: If $D \in B(X) \times B(X)$, then there is a countable collection (R_n) of rectangles $R_n = A_n \times B_n$ such that $D \in \sigma((R_n))$. Let $\mathscr{E}$ be the collection of the sets A_n and B_n for $n = 1, 2, \ldots$. Then $D \in \sigma(\mathscr{E}) \times \sigma(\mathscr{E})$ so that $D_y = \{x \in X: (x, y) \in D\}$ is an element of $\sigma(\mathscr{E})$ for each $y \in X$. However, $\sigma(\mathscr{E})$ has cardinality no greater than c. (See Proposition 1.10 in Chapter 1.)]

7.3. Measures on Topological Spaces; Regularity

On any topological space X we have defined six primary classes of sets in Definition 7.3—the σ-ring and σ-algebra of Borel sets, the σ-ring and σ-algebra of Baire sets, the σ-algebra of weakly Borel sets, and the σ-algebra of weakly Baire sets. In this section we wish to consider measures on these classes of sets.

Throughout this section X is assumed to be a locally compact Hausdorff space.

Definition 7.4. A (weakly) *Baire measure on* X is any measure on the σ-algebra of (weakly) Baire sets that is finite valued on compact Baire sets. A (weakly) *Borel measure on* X is any measure defined on the σ-algebra of (weakly) Borel sets that is finite valued on compact Borel sets. ▌

We will also concern ourselves with measures on σ_r (compact G_δ) and σ_r (compact) that are finite valued on compact sets belonging to these σ-rings. Such measures will be called *Baire measures on the σ-ring of Baire sets* and *Borel measures on the σ-ring of Borel sets*, respectively, in contrast to Baire and Borel measures on X.

Of primary interest are Borel and Baire measures whose values on Borel and Baire sets, respectively, can be approximated by the measures of the generating sets—the compact and compact G_δ sets, respectively. For this reason we introduce the concepts of inner and outer regularity.

Remark 7.6. In the following definition and in Examples 7.11–7.15 and Lemmas 7.6 and 7.7 below, the use of the term "measurable set" refers exclusively throughout to sets in whatever class is under consideration —the σ-ring of Baire sets, the σ-algebra of Baire sets, the σ-ring of Borel sets, or the σ-algebra of Borel sets.

Definition 7.5. Let μ be a Baire measure on X or the σ-ring of Baire sets, or let μ be a Borel measure on X or the σ-ring of Borel sets. A measurable set F is *inner regular* with respect to μ if

$$\mu(F) = \sup \{\mu(C)\colon C \subset F,\ C \text{ a compact measurable set}\}. \tag{7.10}$$

F is *outer regular* with respect to μ if

$$\mu(F) = \inf \{\mu(O)\colon F \subset O,\ O \text{ an open measurable set}\}. \tag{7 11}$$

μ is *outer* (inner) *regular* if every measurable set is *outer* (inner) *regular* and μ is *regular* if it is both inner and outer regular. ∎

Our first goal in this section is to give conditions for which Borel or Baire measures are inner or outer regular. First some examples.

Examples

7.11. Not every Borel measure is regular. Let $\bar{X}$ be the set of all ordinals less than or equal to Ω, the first uncountable ordinal. Let $X = \bar{X} - \{\Omega\}$. On $\bar{X}$ put the order topology—the subbase consists of sets of the form $\{x \in \bar{X}\colon x < a\}$ or $\{x \in \bar{X}\colon a < x\}$ for some $a \in \bar{X}$. Since every cover of $\bar{X}$ by subbasic sets has a finite subcover, $\bar{X}$ is compact (see [34], Theorem 6, Chapter 5). Clearly $\bar{X}$ is Hausdorff. For each Borel set E in $\bar{X}$ define

$$\mu(E) = \begin{cases} 0 & \text{if } E \text{ contains no unbounded closed subset of } X \\ 1 & \text{otherwise.} \end{cases}$$

μ is then a Borel measure that is not regular. (See Problems 7.3.9 and 7.3.13.)

7.12. Let X be as in Example 7.11 with the order topology. X is normal and locally compact, but not paracompact (see [34], page 172). Define μ on Baire sets $\mathrm{Ba}(X)$ as follows:

$$\mu(E) = \begin{cases} 1 & \text{if } E \text{ contains an uncountable closed set} \\ 0 & \text{otherwise.} \end{cases}$$

μ is a finite nonregular Baire measure.

7.13. Let X be any space that is not σ-compact. Then the measure μ on $\mathrm{Ba}(X)$ given by

$$\mu(E) = \begin{cases} 0 & \text{if } E \text{ is a subset of a } \sigma\text{-compact set} \\ \infty & \text{otherwise} \end{cases}$$

is an infinite Baire measure that is not regular.

7.14. Let X be the locally compact space of Example 7.7. Then

$$\mathrm{Ba}(X) = \{A \subset X: A \text{ or } A^c \text{ is countable}\}.$$

Define μ on $\mathrm{Ba}(X)$ by

$$\mu(\{D_{nj}\}) = 1/2^n,$$

and

$$\mu(B) = \sum_{D_{nj} \in B} \mu(\{D_{nj}\}) \qquad \text{for any } B \in \mathrm{Ba}(X).$$

μ is a Baire measure that is inner regular but not outer regular. Since $B(X) = \mathrm{Ba}(X)$, μ is also a Borel measure on X that is inner regular but not outer regular.

7.15. Let X be an uncountable set with the discrete topology. Define μ on $\mathrm{Ba}(X)$ or $B(X)$

$$\mu(A) = \begin{cases} 0 & \text{if } A \text{ is countable} \\ 1 & \text{if } A^c \text{ is countable.} \end{cases}$$

μ is a finite Baire and Borel measure that is outer regular but not inner regular.

It turns out that for Borel measures defined on the σ-ring of Borel sets or for Baire measures defined on the σ-ring of Baire sets outer and inner regularity are equivalent. In fact we show below that every such Baire measure is regular. Clearly these statements are not true for Baire measures on $\mathrm{Ba}(X)$ or Borel measures on $\mathrm{B}(X)$, as Examples 7.11–7.15 above show.

First we need two lemmas whose proofs are straightforward and are left for the reader. The terminology is that of Remark 7.6 and Definition 7.5.

Lemma 7.6. If (E_n) is a sequence of outer (inner) regular sets, then $\bigcup_{n=1}^{\infty} E_n$ is outer (inner) regular. In addition if $\mu(E_n) < \infty$ for all n, then $\bigcap_{n=1}^{\infty} E_n$ is also outer (inner) regular. ∎

Lemma 7.7

(i) If C and D are compact measurable sets and C is outer regular, then $C - D$ is outer regular.

(ii) If every bounded open measurable set is inner regular, then $C - D$ is inner regular for all compact measurable sets C and D. ∎

Proposition 7.11. The following statements are equivalent if μ is either a Borel measure on the σ-ring of Borel sets or a Baire measure on the σ-ring of Baire sets.

(i) μ is regular.

(ii) Every bounded open measurable set U is inner regular.

(iii) Every compact measurable set C is outer regular.

(iv) For each compact measurable set C there is a compact G_δ set G such that $C \subset G$ and $\mu(C) = \mu(G)$. ∎

Proof. Trivially (i) $\Rightarrow$ (ii) and (i) $\Rightarrow$ (iii). We prove first that (iii) implies (ii). Let U be a bounded open measurable set and let $\varepsilon > 0$ be arbitrary. Let C be a compact measurable set such that $U \subset C$. Since $C - U$ is measurable and compact, there exists by (iii) an open measurable set V such that $C - U \subset V$ and

$$\mu(V) \leq \mu(C - U) + \varepsilon.$$

Now $U = C - (C - U) \supset C - V$ and $C - V$ is a compact measurable set. Hence we can write

$$\begin{aligned}\mu(U) - \mu(C - V) &= \mu\big(U - (C - V)\big) = \mu(U \cap V)\\ &\leq \mu\big(V - (C - U)\big) = \mu(V) - \mu(C - U) \leq \varepsilon.\end{aligned}$$

This means U is inner regular.

Next we show that (ii) implies (iii). Let C be a compact measurable set and let $\varepsilon > 0$ be arbitrary. Using Lemma 7.4 there is an open bounded measurable set U such that $C \subset U$. Since $U - C$ is an open bounded measurable set, there exists a compact measurable set D such that

$$D \subset U - C \quad \text{and} \quad \mu(U - C) \leq \mu(D) + \varepsilon.$$

Now $C = U - (U - C) \subset U - D$ so that

$$\begin{aligned}\mu(U - D) - \mu(C) &= \mu\big((U - D) - C\big) = \mu\big((U - C) - D\big)\\ &= \mu(U - C) - \mu(D) < \varepsilon.\end{aligned}$$

Hence C is outer regular since $U - D$ is open.

Next we show that (ii) implies (i). By the above, for compact measurable sets C and D, the set $C - D$ is regular by Lemma 7.7. If $\mathscr{R}$ is the ring of all finite unions of the form $\bigcup_{i=1}^{n} C_i - D_i$, where C_i and D_i are compact measurable sets, then by Lemma 7.6 each element of $\mathscr{R}$ is regular. Now if for each compact measurable set C

$$\mathscr{M}_C = \{E\colon\ E \text{ is measurable and } E \cap C \text{ is regular}\},$$

then $\mathscr{M}_C$ is a monotone class by Lemma 7.6 since $\mu(E \cap C) < \infty$ for all measurable sets E. Since $\mathscr{R} \subset \mathscr{M}_C$, the smallest σ-ring containing $\mathscr{R}$ is contained in $\mathscr{M}_C$. This smallest σ-ring is the σ-ring of measurable sets. Now if E is any measurable set, there is a sequence of compact measurable sets (C_n) such that $E \subset \bigcup_{n=1}^{\infty} C_n$ (by Problem 1.3.2 in Chapter 1). Since $E = \bigcup_{n=1}^{\infty}(C_n \cap E)$ and since each $C_n \cap E$ is regular, E is regular by Lemma 7.6.

We now show that (iv) implies (iii). Let G be a compact G_δ set. Then $G = \cap\, O_i$ where the O_i are open sets. By Problem 7.3.4 we may assume the O_i are open measurable sets with $\mu(O_i) < \infty$. Since open measurable sets are outer regular, G is outer regular by Lemma 7.6. Assuming (iv), each compact measurable set is outer regular.

Finally we show (iii) implies (iv). Let C be a compact measurable set. For each n, there exists a sequence of open measurable sets U_n such that $C \subset U_n$ and $\mu(U_n - C) < 1/n$. For each n choose a compact G_δ set G_n such that $C \subset G_n \subset U_n$. Then $G = \cap\, G_n$ is the required G_δ set. ▮

Theorem 7.2. Every Baire measure on the σ-ring of Baire sets is regular. ▮

Proof. Every compact Baire set is a compact G_δ set by Proposition 7.8. The theorem now follows from Proposition 7.11.

When is a Baire measure on the σ-algebra $\mathrm{Ba}(X)$ of Baire sets regular? A sufficient condition for outer regularity is given in the next result. Recall that a space X is *paracompact* if and only if each open cover $\mathscr{C}$ of X has an open locally finite refinement $\mathscr{C}'$—an open cover $\mathscr{C}'$ of X each member of which is a subset of a member of $\mathscr{C}$ and such that each point of X has a neighborhood intersecting only finitely many members of $\mathscr{C}'$. Each paracompact space is normal, while compact Hausdorff spaces and metric spaces are examples of paracompact spaces.

Theorem 7.3. Any Baire measure μ on a locally compact paracompact Hausdorff space X is outer regular.† ▮

Note that even metrizability is not enough to give us a corresponding result for inner regularity, as Example 7.15 shows.

Proof. As every member of $\mathrm{Ba}(X)$ is itself σ-bounded or its complement is σ-bounded (see Problem 7.3.3), the proof is divided into these

† For this and other related results, the reader may also consult [35].

two cases. Note that if $\mu(E) = \infty$ for $E \in \mathrm{Ba}(X)$ it is trivially true that E is outer regular. Accordingly, it may be assumed that $\mu(E) < \infty$. Suppose first, E in $\mathrm{Ba}(X)$ is σ-bounded. Then $E \subset \bigcup G_i$, where G_i are disjoint sets in σ_r(compact G_δ) (see Lemma 7.4). Hence

$$E = E \cap \left(\bigcup_i G_i\right) = \bigcup_i (E \cap G_i),$$

where each $E \cap G_i$ is in the σ-ring of Baire sets (see Problem 7.2.3). Since every Baire measure on the σ-ring of Baire sets is outer regular, for each i there is an open Baire set such that $E \cap G_i \subset O_i$, and

$$\mu(O_i) \leq \mu(E \cap G_i) + \varepsilon/2^i.$$

Hence if $O = \cup O_i$, $E \subset O$, and

$$\mu(O) \leq \sum_{i=1}^{\infty} \mu(O_i) \leq \sum_{i=1}^{\infty} \mu(E \cap G_i) + \varepsilon/2^i = \mu(E) + \varepsilon.$$

Now suppose E^c is σ-bounded. Let $A = E^c$. Since X is paracompact and locally compact, $\bar{A}$ is σ-bounded. By Lemma 7.4, $\bar{A}$ is contained in a countable union U of open sets in σ_r(compact G_δ). Since a paracompact space is normal, Urysohn's Lemma can be used to obtain a continuous function f such that $f \equiv 1$ on $\bar{A}$ and $f \equiv 0$ on U^c. Let $B = \{x: f(x) = 1\}$. B is a closed G_δ set which contains $\bar{A}$ and since $B \subset U$, B is also σ-bounded. This means $B \in \sigma_r$(compact G_δ) and by the first part of this proof there is an open set O in $\mathrm{Ba}(X)$ such that $B \cap E \subset O$ and

$$\mu(O) \leq \mu(B \cap E) + \varepsilon.$$

Now $O \cup B^c$ is an open set in $\mathrm{Ba}(X)$ containing E and since $B^c = E - B$,

$$\mu(O \cup B^c) \leq \mu(O) + \mu(B^c) < \mu(B \cap E) + \varepsilon + \mu(E - B) \leq \mu(E) + \varepsilon. \blacksquare$$

We now turn our attention toward the extension of Baire measures to Borel measures. The hope is that the extension of a regular measure on a class of Baire sets to a larger class will also be regular. We will prove three extension theorems in the following discussion, namely, the extension of a regular Baire measure on the σ-ring of Baire sets to a unique regular Borel measure on the σ-ring of Borel sets, the extension of a regular Borel measure on the σ-ring of Borel sets to a unique inner regular weakly Borel measure, and the extension of an inner regular Baire measure to an outer regular weakly Borel measure.

Given a Borel measure μ on the σ-ring of Borel sets, it is a trivial matter to extend μ to a weakly Borel measure μ_w on $B_w(X)$ by simply defining $\mu_w(A) = \infty$ for each A in $B_w(X)$ but not in the σ-ring of Borel sets. However, the following proposition shows how to extend μ to $B_w(X)$ to ensure preservation of regularity.

• **Proposition 7.12.** If μ is any Borel measure on the σ-ring of Borel sets, then the formula for A in $B_w(X)$ given by

$$\mu_w(A) = \sup\{\mu(E)\colon E \subset A,\ E \in \sigma_r(\text{compact})\}$$

defines an extension of μ to a measure μ_w on $B_w(X)$. If μ is inner regular, then μ_w is the only extension of μ satisfying

$$\mu_w(A) = \sup\{\mu(E)\colon E \subset A,\ E \text{ compact}\}. \qquad \blacksquare \quad (7.12)$$

Proof. The countable additivity of μ_w is essentially all there is to show in order that μ_w be a measure. It is important to observe that if $E \in \sigma_r$(compact sets) and $A \varepsilon B_w(X)$, then $E \cap A \in \sigma_r$(compact sets) (see Problem 7.2.3). Let (A_i) be a disjoint countable collection of sets in $B_w(X)$. Then

$$\begin{aligned}\mu_w\left(\bigcup A_i\right) &= \sup\left\{\mu(E)\colon E \subset \bigcup A_i;\ E \in \sigma_r(\text{compact})\right\}\\ &= \sup\left\{\sum_{i=1}^{\infty} \mu(E \cap A_i)\colon E \subset \bigcup_i A_i;\ E \in \sigma_r(\text{compact})\right\}\\ &\le \sum_{i=1}^{\infty} \mu_w(A_i).\end{aligned}$$

On the other hand

$$\begin{aligned}\sum_{i=1}^{\infty} \mu_w(A_i) &= \sum_{i=1}^{\infty} \sup\{\mu(E_i)\colon E_i \subset A_i;\ E_i \in \sigma_r(\text{compact})\}\\ &= \sup\left\{\sum_{i=1}^{\infty} \mu(E_i \cap A_i)\colon E_i \subset A_i;\ E_i \in \sigma_r(\text{compact})\right\}\\ &= \sup\left\{\mu\left[\left(\bigcup_{i=1}^{\infty} E_i\right) \cap \left(\bigcup_{j=1}^{\infty} A_j\right)\right]\colon E_i \subset A_i\right\}\\ &\le \mu_w\left(\bigcup_{j=1}^{\infty} A_j\right).\end{aligned}$$

It is clear that μ_w satisfies equation (7.12) whenever μ is inner regular. The uniqueness of μ_w satisfying equation (7.12) follows from the fact that its values are determined by its values on compact sets. ∎

Now let us extend a Baire measure (necessarily regular by Theorem 7.2) on the σ-ring of Baire sets to a regular Borel measure on the σ-ring of Borel sets and hence to an inner regular weakly Borel measure by Proposition 7.12.

• **Theorem 7.4.** Let μ be a Baire measure on σ_r(compact G_δ). Then there exists a unique regular Borel measure $\bar{\mu}$ on σ_r(compact) which extends μ. ∎

Proof. Define the set function μ^* on the class of open sets in σ_r (compact) by

$$\mu^*(O) = \sup\{\mu(K)\colon\ K \subset O,\ K \text{ a compact } G_\delta \text{ set}\}.$$

If $O \in \sigma_r$(compact G_δ), obviously $\mu^*(O) = \mu(O)$ since μ is regular. Using Problem 7.3.5, it is easy to verify that μ^* is countably subadditive. Let

$$\mathscr{M} = \{M \subset X\colon \text{for each } \varepsilon > 0, \text{ there is an open set } O_\varepsilon \text{ in } \sigma_r(\text{compact}) \text{ containing } M \text{ with } \mu^*(O_\varepsilon) < \varepsilon\}.$$

Observe that each subset of a set in $\mathscr{M}$ is also in $\mathscr{M}$, and $\mathscr{M}$ is closed under countable unions. Notice also that if $E \in \mathscr{M} \cap \sigma_r$(compact G_δ) and $\varepsilon > 0$ is arbitrary, then inasmuch as each compact G_δ set K contained in E is also in $\mathscr{M}$, for each such K there is an open set O_K in σ_r(compact) containing K with $\mu^*(O_K) < \varepsilon$. This means $\mu(E) = O$ as by the regularity of μ we have

$$\mu(E) = \sup\{\mu(K)\colon\ K \subset E,\ K \text{ compact } G_\delta\} \leq \varepsilon.$$

Let $\mathscr{R}$ be the class of sets

$$\mathscr{R} = \{E \triangle M\colon\ E \,\varepsilon\, \sigma_r(\text{compact } G_\delta) \text{ and } M \in \mathscr{M}\}.$$

Define on $\mathscr{R}$ the set function $\bar{\mu}$ given by

$$\bar{\mu}(E \triangle M) = \mu(E).$$

That $\bar{\mu}$ is well-defined follows from the fact that if $E \triangle M$ and $F \triangle N$ are equal elements of $\mathscr{R}$, then $E \triangle F = M \triangle N$ so that $E \triangle F \,\varepsilon\, \sigma_r$(compact G_δ) $\cap\, \mathscr{M}$. This means $\mu(E \triangle F) = 0$ and hence $\mu(E) = \mu(F)$. Observe that

(a) $\mathscr{R}$ is the smallest σ-ring containing σ_r(compact G_δ) and $\mathscr{M}$;

(b) $\mathscr{R}$ contains σ_r(compact); and

(c) the restriction of $\bar{\mu}$ to σ_r(compact) is a regular Borel measure on σ_r(compact) extending μ.

The verification of (a) is left for the reader, while the proof of (b) is accomplished by showing that each compact set is contained in $\mathscr{R}$. To this end let C be compact in X. By Lemma 7.4 we know there exists a compact G_δ set K_0 and an open σ-compact set O_0 in σ_r(compact G_δ) such that

$$C \subset O_0 \subset K_0.$$

Since $O_0 - C$ is an open set in σ_r(compact), by the definition of μ^* we can find a sequence of compact G_δ sets K_i such that $\mu^*(O_0 - C) = \lim_i \mu(K_i)$ with $K_i \subset O_0 - C$. Let $K = \cup K_i$. Then since

$$\mu^*\big((O_0 - C) - K\big) \leq \mu^*\big((O_0 - C) - K_i\big) \leq \mu^*(O_0 - C) - \mu(K_i) \to 0,$$

we have $\mu^*\big((O_0 - C) - K\big) = 0$ and $(O_0 - C) - K \in \mathscr{M}$. As $C = (O_0 - K) \triangle [(O_0 - C) - K]$ and $O_0 - K \in \sigma_r$(compact G_δ), $C \in \mathscr{R}$.

We complete the proof by showing that the restriction of $\bar{\mu}$ to σ_r(compact) is regular. (See Problem 7.3.7.) It suffices to show in light of Proposition 7.11 that each compact set is outer regular. If C is compact, we know from the above that

$$C = (O_0 - K) \triangle [(O_0 - C) - K],$$

where $O_0 - K \in \sigma_r$(compact G_δ). Hence for any $\varepsilon > 0$ there exists an open Baire set O with

$$O \supset O_0 - K \supset (O_0 - C) - K$$

and

$$\mu(O) \leq \mu(O_0 - K) + \varepsilon.$$

This means $O \supset C$ and $\mu(O) \leq \mu(O_0 - K) + \varepsilon = \mu(C) + \varepsilon$. ∎

Although a Baire measure on the σ-ring of Baire sets can be extended uniquely to an inner regular weakly Borel measure by virtue of Proposition 7.12 and Theorem 7.4, such an extension is not always possible from Ba(X) to $B_w(X)$ without additional assumptions. The remaining portion of this section discusses this situation and gives various additional assumptions making this extension possible.

Remember that X is always assumed to be a locally compact Hausdorff space in this section.

• **Theorem 7.5.** If μ is a regular Baire measure on $\text{Ba}(X)$, then there exists an extension of μ to a unique weakly Borel measure $\bar{\mu}$ on $B_w(X)$ satisfying

(i) $\bar{\mu}(O) = \sup\{\mu(K): K \subset O, K \text{ compact } G_\delta\}$ for each open set O,

and (7.13)

(ii) $\bar{\mu}(E) = \inf\{\bar{\mu}(O): E \subset O, O \text{ an open set}\}$ for each weakly Borel set E.

If μ is finite, $\bar{\mu}$ is finite and regular. (See also Problem 7.3.8.) ▮

Proof. The proof is almost a replay of that of Theorem 7.4. The reader can supply omitted details. Define μ^* on the class of open sets by

$$\mu^*(O) = \sup\{\mu(K): K \subset O, K \text{ a compact } G_\delta \text{ set}\}, \tag{7.14}$$

μ^* is countably subadditive and agrees with μ on open sets in $\text{Ba}(X)$. Let

$\mathscr{M} = \{M \subset X$: for each $\varepsilon > 0$, there is an open set O_ε containing M with $\mu^*(O_\varepsilon) < \varepsilon\}$.

Let $\mathscr{R}$ equal $\{E \triangle M: E \in \text{Ba}(X) \text{ and } M \in \mathscr{M}\}$. $\mathscr{R}$ is a σ-algebra containing $\text{Ba}(X)$ and $\mathscr{M}$. Defining the set function μ_1 on $\mathscr{R}$ by

$$\mu_1(E \triangle M) = \mu(E),$$

it can be verified that μ_1 is a measure (Problem 7.3.7). Observe that if $\mu_1(E \triangle M) < \infty$, then by the outer regularity of μ and the definition of $\mathscr{M}$, there is an open set O containing $E \triangle M$ with $\mu^*(O) < \infty$. Observe in addition that each open set O with $\mu^*(O) < \infty$ is in $\mathscr{R}$. Indeed if $\mu^*(O)$ is finite, then for each positive integer n there is a compact G_δ sct K_n with $K_n \subset O$ and

$$\mu^*(O - K_n) \leq \mu^*(O) - \mu^*(K_n) < 1/n.$$

Letting $K = \bigcup_{n=1}^\infty K_n$, $K \in \text{Ba}(X)$ and $O - K$ is in $\mathscr{M}$. Hence O is in $\mathscr{R}$ since

$$O = K \triangle (O - K).$$

Let $\mathscr{A}$ be the σ-algebra defined as

$$\mathscr{A} = \{A \subset X: A \cap B \in \mathscr{R} \text{ for each } B \in \mathscr{R} \text{ with } \mu_1(B) < \infty\}.$$

Then $\mathscr{A}$ contains each open set O' in X since if B is in $\mathscr{R}$ with $\mu_1(B) < \infty$, let O be the open set in $\mathscr{R}$ with $B \subset O$ and $\mu^*(O) < \infty$. Then

$$O' \cap B = (O' \cap O) \cap B$$

with $O' \cap O$ and B in $\mathscr{R}$. This means $\mathscr{A}$ contains $B_w(X)$.

Finally define $\bar{\mu}$ on $\mathscr{A}$ and hence on $B_w(X)$ by

$$\bar{\mu}(E) = \begin{cases} \mu_1(E) & \text{if } E \in \mathscr{R} \\ \infty & \text{otherwise.} \end{cases}$$

$\bar{\mu}$ satisfies the conclusion of the theorem. (i) is easily verified. To show (ii) suppose first that E is a weakly Borel set not in $\mathscr{R}$. Then $\bar{\mu}(E) = \infty$ and each open set O containing E has $\bar{\mu}(O) = \infty$. If not, then $\mu^*(O) < \infty$ and $O \in \mathscr{R}$, implying that E, equal to $E \cap O$, is in $\mathscr{R}$. Secondly, suppose E is a weakly Borel set in $\mathscr{R}$. Then $E = F \triangle M$, where $F \in \text{Ba}(X)$ and $M \in \mathscr{M}$. From the outer regularity of μ and the definition of $\mathscr{M}$, respectively, we get open sets O_1 and O_2 with $F \subset O_1$ and $M \subset O_2$, and

$$\bar{\mu}(O_1 \cup O_2) \leq \bar{\mu}(O_1) + \bar{\mu}(O_2) \leq \mu(F) + \varepsilon. \qquad ∎$$

• **Example 7.16.** Consider the space X of Examples 7.7 and 7.14. Let $E = R - D$ where $D = \{j/2^n \mid j \text{ an integer and } n \text{ a nonnegative integer}\}$. It is easy to verify that $\mu(E) = 0$, with μ being the measure on $\text{Ba}(X)$ defined in Example 7.14. Recall that μ is inner regular but not outer regular. There is no outer regular extension $\bar{\mu}$ of μ to $B_w(X)$. For let $\bar{\mu}$ be any extension of μ to the class of weakly Borel sets. Let O be any open set containing E. For each n let

$$E_n = \{x \in E \colon x \in [j/2^m, (j+1)/2^m] \text{ and } D_{mj} \in O \text{ for all } m \geq n \text{ and for some } j \text{ depending on } m\}.$$

Since $E = \bigcup_{n=1}^{\infty} E_n$ (recall the topology defined on X), the outer Lebesgue measure λ^* of some E_{n_0} is not zero, that is, $\lambda^*(E_{n_0}) > 0$. This means that for each $m \geq n_0$

$$\lambda^*\Big(\bigcup_j [j/2^m,\ (j+1)/2^m] \colon D_{mj} \in O\Big) \geq \lambda^*(E_{n_0}) > 0,$$

whereby for each $m \geq n_0$, since $\lambda^*[j/2^m, (j+1)/2^m] = \mu(\{D_{mj}\})$,

$$\mu(O \cap X_m) \geq \lambda^*(E_{n_0}),$$

where $X_m = \{D_{mj} \colon j \text{ is an integer}\}$. (Notice that $O \cap X_m$ is a Baire set since it is a countable subset of X.) This means that

$$\bar{\mu}(O) \geq \sup\{\mu(K): K \subset O, K \text{ compact } G_\delta \text{ sets}\} = \infty.$$

The preceding example shows that without regularity—even with inner regularity—the conclusion of Theorem 7.5 is invalid. However, we have the following corollaries of Theorem 7.5.

• **Corollary 7.5.** If X is a paracompact space and μ is an inner regular Baire measure on $\text{Ba}(X)$, then there exists a unique extension $\bar{\mu}$ of μ to $B_w(X)$ satisfying equation (7.13). ∎

Proof. See Theorem 7.3 and use Theorem 7.5. ∎

• **Corollary 7.6.** If μ is a semifinite measure on $\text{Ba}(X)$ such that for each E in $\text{Ba}(X)$

$$\mu(E) = \inf\{\mu(G): G \text{ open } \sigma\text{-compact set}, E \subset G\}, \tag{7.15}$$

then μ is regular and there exists a unique extension $\bar{\mu}$ of μ to $B_w(X)$ satisfying equations (7.13). If μ is finite, so is $\bar{\mu}$. ∎

Proof. The outer regularity of μ follows from the following inequality:

$$\begin{aligned}\mu(E) &\leq \inf\{\mu(O): E \subset O, O \text{ an open set in } \text{Ba}(X)\}\\ &\leq \inf\{\mu(G): E \subset G, G \text{ open } \sigma\text{-compact}\}\\ &= \mu(E).\end{aligned} \tag{7.16}$$

We now show that μ is inner regular. Assume first $E \in \text{Ba}(X)$ with $\mu(E) < \infty$. By equation (7.15), there exists for each $\varepsilon > 0$ an open σ-compact set $G = \bigcup_{i=1}^{\infty} K_i$ (where each K_i is a compact G_δ set) such that $E \subset G$ and $\mu(G - E) < \varepsilon$. Applying equation (7.15) again to $G - E$, there exists an open (σ-compact) set O containing $G - E$ such that $\mu(O) < 2\varepsilon$. Since $\mu(G) = \lim_n \mu(\bigcup_{i=1}^{n} K_i)$, there exists N such that $\mu(G - \bigcup_{i=1}^{N} K_i) < \varepsilon$. Let C be the compact set $(\bigcup_{i=1}^{N} K_i) \cap O^c$ contained in E. Now

$$\begin{aligned}\mu(E) &= \mu(C) + \mu(E - C)\\ &\leq \mu(C) + \mu\left(O \cup \left[G - \bigcup_{i=1}^{N} K_i\right]\right)\\ &\leq \mu(G) + \mu(O) + \mu\left(G - \bigcup_{i=1}^{N} K_i\right)\\ &\leq \mu(G) + 2\varepsilon + \varepsilon = \mu(G) + 3\varepsilon.\end{aligned}$$

Secondly, if $\mu(E) = \infty$, then as μ is semifinite there exists for each positive integer n a set F contained in E with finite measure greater than n. According to the preceding argument there exists a compact Baire set contained in F and hence in E with measure greater than n.

The conclusion of the corollary now follows from Theorem 7.5. ∎

The hypothesis of Corollary 7.5 includes the cases where X is σ-bounded or μ is finite or σ-finite. The only case left is that given in the next theorem.

• **Theorem 7.6.** Suppose μ is a nonfinite Baire measure on a non-σ-bounded space X. Assume also the hypothesis (7.15). Then μ has an extension to a unique weakly Borel measure $\bar{\mu}$ on $B_w(X)$ satisfying

$$\bar{\mu}(O) = \sup\{\mu(K): K \subset O,\ K \text{ a compact } G_\delta \text{ set}\} \tag{7.17}$$

for each open σ-bounded set O and

$$\bar{\mu}(E) = \inf\{\mu(G): E \subset G,\ G \text{ an open } \sigma\text{-bounded set}\} \tag{7.18}$$

for each E in $B_w(X)$. ∎

Proof. The proof is almost identical to that of Theorem 7.5 with the following exceptions. Replace equation (7.14) by

$$\mu^*(O) = \begin{cases} \sup\{\mu(K): K \subset O,\ K \text{ compact } G_\delta \text{ set}\} & \text{if } O \text{ is } \sigma\text{-bounded} \\ +\infty & \text{otherwise} \end{cases} \tag{7.19}$$

for each open set O. Using the fact that for each open Baire set O either O or O^c is σ-bounded (Problem 7.3.3), it is easy to show that μ^* agrees with μ on open Baire sets (Problem 7.3.10). Define $\mathscr{M}$, $\mathscr{R}$, μ_1, $\mathscr{A}$, and $\bar{\mu}$ exactly as in the proof of Theorem 7.5 and obtain the desired conclusions (Problem 7.3.10.). ∎

Problems

7.3.1. Prove Lemma 7.6.

7.3.2. Prove Lemma 7.7.

7.3.3. (i) Prove that every element of Ba(X) and $B(X)$ is either σ-bounded or its complement is σ-bounded.

(ii) Prove that a Borel measure on the σ-ring of Borel sets is σ-finite.

7.3.4. Prove that any compact G_δ set G can be written as $\bigcap_{i=1}^{\infty} O_i$, where each O_i is an open set in σ_r(compact G_δ) and $\mu(O_i) < \infty$ for any Baire measure μ on σ_r(compact G_δ).

7.3.5. Prove that if K is a compact G_δ set and $K \subset \bigcup_{i=1}^{\infty} O_i$ with each O_i open, then $K = \bigcup_{i=1}^{n} K_i$, where each K_i is a compact G_δ and $K_i \subset O_i$. (Hint: Use Problem 7.2.1.)

7.3.6. Prove that every finite inner regular Baire measure is outer regular.

• **7.3.7.** Prove that $\bar{\mu}$ and μ_1 of Theorems 7.4 and 7.5, respectively, are measures on $\mathscr{R}$.

• **7.3.8.** Prove that the measure $\bar{\mu}$ of Theorem 7.5 also satisfies the following properties:

(i) For each weakly Borel set E of finite measure, $\bar{\mu}(E) = \sup\{\bar{\mu}(K)$: $K \subset E$ and K is compact$\}$.

(ii) For each weakly Borel set A with $\bar{\mu}(A) < \infty$, there is a Baire set E and a weakly Borel set N with $\bar{\mu}(N) = 0$ and $A = E \triangle N$.

(iii) If μ is finite, $\bar{\mu}$ is finite. In this case $\bar{\mu}$ is regular.

7.3.9. (i) Prove that the set function μ defined in Example 7.11 is a measure.

(ii) Prove that μ is not regular. (Hint: Consider any interval containing Ω.) Compare Problem 7.3.13.

• **7.3.10.** Let μ be a Baire measure on Ba(X) satisfying equation (7.15).

(i) Prove that each σ-bounded set in Ba(X) is inner regular.

(ii) Prove that μ^* of equation (7.19) agrees with μ on open Baire sets if μ is nonsemifinite.

(iii) Complete the proof of Theorem 7.6.

★ **7.3.11.** *On the Measurability of Functions in Two Variables.* Let X and Y be two compact topological spaces and μ, ν be finite regular measures defined on the Borel sets of X and Y, respectively. By a *measurable modification* $\tilde{f}(x, y)$ of a function $f(x, y)$, we mean a $(\mu \times \nu)$-measurable function such that $\tilde{f}(x, \cdot) = f(x, \cdot)$ a.e. (ν) for every x. Prove the following result due to Mark Mahowald: If Y is metric, and if $f(x, y)$ has a measurable modification, and if $f(x, \cdot)$ is continuous for almost all x, then $f(x, y)$ is $(\mu \times \nu)$-measurable. [Hint: We may and do assume that $f(x, \cdot)$ is continuous for all x. Use Lusin's Theorem (Problem 3.1.13 in Chapter 3) to find a "large" compact set C in $X \times Y$ such that $\tilde{f}(x, y)$ is continuous on C and $\mu \times \nu(C \cap G) > 0$ whenever G is open and $C \cap G \neq \varnothing$. Let (U_n) be a countable basis for Y, and let $A_n = \{x$: $\nu((X \times U_n) \cap C)_x = 0\}$ and $B_n = (A_n \times Y) \cap (X \times U_n) \cap C$. Then $\cup B_n \equiv D$ has $(\mu \times \nu)$-measure zero. Let $E = C - D$. If $N_x = \{n\colon x \in A_n\}$, then

$E_x = \bigcap_{n \in N_x}(C_x - U_n)$, which is compact. Also $\nu(E_x \cap H) > 0$ whenever H is open and $H \cap E_x \neq \emptyset$. If $\nu(E_x) > 0$, $f(x, \cdot) = \tilde{f}(x, \cdot)$ *everywhere* (by continuity). It follows that $\tilde{f}(x, y) = f(x, y)$ a.e. $(\mu \times \nu)$ on C.]

7.3.12. Let X be a locally compact Hausdorff space and μ a weakly Borel measure such that for any weakly Borel B with $\mu(B) < \infty$,

$$\begin{aligned}\mu(B) &= \inf\{\mu(O)\colon O \text{ open} \supset B\}\\ &= \sup\{\mu(K)\colon K \text{ compact} \subset B\}.\end{aligned}$$

Then for $1 \leq p < \infty$, $C_c(X)$ is dense in $L_p(\mu)$. [Hint: Let $\mu(B) < \infty$ and $K \subset B \subset O$ such that there exists $f \in C_c(X)$ with $f(x) = 1$ for $x \in K$, $= 0$ for $x \notin O$ and $0 \leq f \leq 1$. If $\mu(O - K) < \varepsilon$, then $\| \chi_B - f \|_p \leq 2 \cdot \varepsilon^{1/p}$.]

★ **7.3.13.** *The Structure of Measures and Measurable Functions on the Borel Sets of* $[0, \Omega]$. Let $\mathscr{B}$ be the Borel sets of $[0, \Omega]$, Ω the first uncountable ordinal, with order topology. Let μ be a set function as in Example 7.11 on $\mathscr{B}$ such that $\mu(E) = 1$ or 0 according as E does or does not contain an unbounded, closed subset of $[0, \Omega)$. Then μ is a measure that is *not* regular. Prove the following assertions due to M. B. Rao and K. P. S. B. Rao:

(i) $\mathscr{B} = \{A \subset [0, \Omega]\colon A$ or A^c contains an unbounded closed subset of $[0, \Omega)\}$. [Hint: Let the set on the right be $\mathscr{F}$. Show that $\mathscr{F}$ is a σ-algebra and $\mathscr{F} \supset \mathscr{B}$. For the converse, let A be an unbounded closed subset of $[0, \Omega)$. Let $B = A^c - \Omega$. Assume $0 \in A$. Let α' stand for the first succeeding ordinal of α in A. Define a set-valued function g on A such that $g(\alpha) = \{\beta \in B\colon \alpha < \beta < \alpha'\}$. Then $g(A) = B$. Enumerate the elements of $g(\alpha)$ [whenever $\alpha \in A$ and $g(\alpha) \neq \emptyset$] and write $A_n = \bigcup_{\alpha \in A}\{$the nth element in the enumeration of $g(\alpha)\}$. Then $g(A) = \bigcup_{n=1}^{\infty} A_n$; also A_n is Borel since

$$A_n = \bigcup_{\alpha \in A} \bigcap_{i=1}^{\infty} A_{i\alpha}^{(n)} = \bigcap_{i=1}^{\infty} \bigcup_{\alpha \in A} A_{i\alpha}^{(n)},$$

where, $A_{i\alpha}^{(n)}$ are open sets $\subset (\alpha, \alpha')$ such that $\bigcap_{i=1}^{\infty} A_{i\alpha}^{(n)} = \{$the nth element in the enumeration of $g(\alpha)\}$].

(ii) If λ is any Borel measure on $[0, \Omega]$ such that $\lambda(\{x\}) = 0$ for every singleton $\{x\}$, then $\lambda = c \cdot \mu$ for some $c \geq 0$. {Hint: By the Lebesgue Decomposition Theorem, $\lambda = \lambda_1 + \lambda_2$, $\lambda_1 << \mu$, and $\lambda_2 \perp \mu$. Then $\lambda_1 = c\mu$ for some $c \geq 0$; also there is $Y \subset [0, \Omega]$ such that $\lambda_2(Y) = \lambda_2([0, \Omega])$ and $\mu(Y) = 0$ and $Y \cap \mathscr{B} =$ the power set of Y by (i). By the Ulam Theorem, $\lambda_2 = 0$.}

(iii) There is *no* monotonic Borel measure on $[0, \Omega]$.

(iv) Any Borel measure λ on $[0, \Omega]$ can be expressed as $c \cdot \mu + \nu$, where $c \geq 0$ and $\nu(E^c) = 0$ for some countable set $E \subset [0, \Omega]$.

(v) Every regular Borel measure on $[0, \Omega]$ is of the form ν in (iv).

(vi) A real-valued function on $[0, \Omega]$ is Borel measurable if and only if it is constant on an unbounded closed subset of $[0, \Omega)$.

7.4. Riesz Representation Theorems

In this section we will use the Daniell–Stone Representation Theorem to obtain results that establish relationships between linear functionals on certain vector lattices and measures on topological spaces. We will list several results of this nature which are generally called Riesz representation theorems. Proposition 7.13 and Theorems 7.7, 7.8, and 7.10 constitute various formulations of a result generally called the Riesz Representation Theorem.

Our first result is an almost immediate consequence of the Daniell–Stone Theorem.

Proposition 7.13. Let X be any topological space and let I be a Daniell integral on the vector lattice $C_b(X)$. There exists a unique measure μ on $\sigma(C_b(X))$ such that

$$C_b(X) \subset L_1(\mu) \tag{7.20}$$

and

$$I(f) = \int_X f\, d\mu \qquad \text{for all } f \text{ in } C_b(X). \tag{7.21}$$

μ is necessarily finite. If X is T_4, μ is a weakly Baire measure. ∎

Proof. The Daniell–Stone Theorem gives the existence of a measure μ on $\sigma(C_b(X))$ satisfying conditions (7.20) and (7.21). Since $\chi_X \in C_b(X)$,

$$\mu(X) = \int \chi_X\, d\mu = I(\chi_X) < \infty.$$

If ν is any other measure on $\mathrm{Ba}_w(X)$ satisfying conditions (7.20) and (7.21), then $\nu = \mu$ by Proposition 7.4. The last assertion follows from Propositions 7.5 and 7.6. ∎

Proposition 7.13 actually gives a one-to-one correspondence between Daniell integrals on $C_b(X)$ and finite measures on $\sigma(C_b(X))$. Indeed, if μ

is any finite measure on $\sigma(C_b(X))$, $I(f) = \int_X f\,d\mu$ defines a Daniell integral on $C_b(X)$.

Let us now consider the vector lattice $C_c(X)$, where X is a locally compact Hausdorff space. If X is, moreover, σ-compact, then we know

$$\mathrm{Ba}(X) = \sigma(C_c(X)) = \sigma(C_b(X)) = \mathrm{Ba}_w(X).$$

Whether X is σ-compact or not, it is interesting that every positive linear functional on $C_c(X)$ is already a Daniell integral.

Proposition 7.14. If X is a locally compact Hausdorff space, then every positive linear functional on $C_c(X)$ is a Daniell integral. ∎

Proof. We must verify condition (D) of Definition 7.2: If (f_n) is a nonincreasing sequence in $C_c(X)$ converging to zero, then $I(f_n)$ converges to zero. To this end, let S_n be the support of the function f_n. Note that each S_n is compact and $S_{n+1} \subset S_n \subset S_1$ for each n. By Dini's Theorem (Problem 1.5.18 in Chapter 1) the sequence (f_n) converges uniformly to zero on S_1. Hence for any $\varepsilon > 0$ there exists an N such that for all $n \geq N$, $|f_n(x)| < \varepsilon$ for all x in S_1 and hence for all x in X, since $S_n \subset S_1$. Corresponding to S_1 there exists a function g in $C_c(X)$ with $0 \leq g \leq 1$ and $g \equiv 1$ on S_1 by Lemma 7.3. Therefore for $n \geq N$, $|f_n| < \varepsilon g$ and $|I(f_n)| \leq \varepsilon I(g)$. Since ε is arbitrary, $I(f_n) \to 0$. ∎

The next result is one of the key theorems of this section. It gives a one-to-one correspondence between the class of Baire measures on a σ-compact Hausdorff space X and positive linear functionals on $C_c(X)$.

Theorem 7.7. Let X be a locally compact Hausdorff space.

(i) For every Baire measure μ, the function

$$g \to \int_X g\,d\mu$$

is a Daniell integral on $C_c(X)$.

(ii) Corresponding to every positive linear functional I on $C_c(X)$ there is a Baire measure on $\mathrm{Ba}(X)$ such that

$$I(f) = \int_X f\,d\mu$$

for all $f \in C_c(X)$. If X is σ-compact, μ is unique.

(iii) Corresponding to every positive linear functional I on $C_c(X)$ there is a unique Baire measure μ on $\mathrm{Ba}(X)$ such that

$$I(f) = \int_X f\,d\mu \qquad \text{for all } f \text{ in } C_c(X) \tag{7.22}$$

and

$$\mu(A) = \inf\{\mu(G)\colon G \supset A,\ G \text{ open } \sigma\text{-compact}\} \tag{7.23}$$

for each $A \in \mathrm{Ba}(X)$. μ is also outer regular. ▮

Proof. (i) If $g \in C_c(X)$, then g is measurable with respect to Ba(X) as $\mathrm{Ba}(X) = \sigma(C_c(X))$, the smallest σ-algebra such that all such g are measurable (see Figure 4 of Section 7.2). Also g is integrable with respect to μ. This follows since by Lemma 7.4 there is a compact G_δ set and $S_g \subset A$. Letting $\| g \|_\infty = \sup\{| g(x) |: x \in X\}$ we have $| g | \leq \| g \|_\infty \chi_A$. This means that g is integrable, since

$$\int | g |\, d\mu \leq \| g \|_\infty \int \chi_A\, d\mu = \| g \|_\infty\, \mu(A) < \infty.$$

Clearly, $g \to \int_X g\, d\mu$ is a positive linear functional on $C_c(X)$ and hence by Proposition 7.14 is a Daniell integral.

(ii) Since a positive linear functional on $C_c(X)$ is a Daniell integral, by the Daniell–Stone Theorem 7.1 there exists a measure μ on $\mathrm{Ba}(X)$ such that

$$C_c(X) \subset L_1(\mu)$$

and

$$I(f) = \int_X f\,d\mu \qquad \text{for all } f \in C_c(X).$$

If K is any compact Baire set, then $\chi_K \leq f$ where f in $C_c(X)$ is a function given by Lemma 7.3 such that $f \equiv 1$ on K. Hence

$$\mu(K) = \int \chi_K\, d\mu \leq \int_X f\, d\mu = I(f) < \infty.$$

This means that μ is a Baire measure.

The uniqueness of μ remains to be proved in the case where X is σ-compact. By Lemma 7.5, X can be written as $\bigcup K_n$ with each K_n a compact Baire set and $K_n \subset K_{n+1}^\circ$. By Lemma 7.3, for each n there exists $f_n \in C_c(X)$ with $0 \leq f_n \leq 1$, $K_n \subset S_{f_n} \subset K_{n+1}^\circ$, and $\chi_{K_n} \leq f_n$. Clearly,

$\chi_X = \sup f_n$. Letting G_n be the $C_c(X)$-open set $\{x \in X: f_n(x) > 1/n\}$ (see Lemma 7.5), we have $X = \bigcup_{n=1}^{\infty} G_n$. Since $G_n \subset S_{f_n} \subset K_{n+1}$, the measure of G_n is finite for any Baire measure on Ba(X). By Proposition 7.4, μ is unique.

(iii) According to Proposition 7.14 and Theorem 7.1, there exists a unique measure μ on $\sigma(C_c(X))$ or on what is the same as Ba(X), such that

$$I(f) = \int_X f\,d\mu \qquad \text{for all } f \in C_c(X)$$

and

$$\mu(A) = \inf\{\mu(G)\colon G \supset A \text{ and } G,\ C_c(X)\text{-open}\}$$

for each A in Ba(X). Equations (7.22) and (7.23) follow from these relations and Lemma 7.5. That μ is outer regular follows from Lemma 7.5 and the following inequalities:

$$\begin{aligned}\mu(A) &\le \inf\{\mu(O)\colon A \subset O \text{ and } O \text{ an open Baire set}\}\\ &\le \inf\{\mu(G)\colon A \subset G \text{ and } G \text{ an open } \sigma\text{-compact set}\}\\ &= \mu(A).\end{aligned}$$ ∎

Using the extension theorems of Section 7.3, we can also correspond to any positive linear functional I on $C_c(X)$ a weakly Borel measure. More precisely, we have the following result.

Theorem 7.8. Corresponding to each positive linear functional I on $C_c(X)$ is a unique weakly Borel measure $\bar{\mu}$ on $B_w(X)$ satisfying

$$I(f) = \int f\,d\bar{\mu} \qquad \text{for all } f \text{ in } C_c(X), \tag{7.24}$$

$$\bar{\mu}(O) = \sup\{\bar{\mu}(K)\colon K \text{ compact } G_\delta \text{ set}, K \subset O\} \text{ for each open } \sigma\text{-bounded set } O, \tag{7.25}$$

and

$$\bar{\mu}(E) = \inf\{\bar{\mu}(O)\colon E \subset O, O \text{ open } \sigma\text{-bounded set}\} \text{ for each weakly Borel set } E. \tag{7.26}$$ ∎

Proof. According to Theorem 7.7 (iii), there exists a unique Baire measure μ such that equations (7.22) and (7.23) hold. If X is σ-compact, Corollary 7.6 guarantees a weakly Borel measure $\bar{\mu}$ satisfying equations (7.25) and (7.26) since in this case each open set is σ-bounded. If X is not σ-compact, then equation (7.23) implies that $\mu(X) = \infty$ or μ is not finite.

In this case Theorem 7.6 gives a measure $\bar{\mu}$ satisfying equations (7.25) and (7.26). In any case the unique measure $\bar{\mu}$ extending μ satisfies equation (7.24). That this is so follows from the fact that if f is any nonnegative function in $C_c(X)$, then

$$\int f\,d\bar{\mu} = \lim_n \int \Phi_n\,d\bar{\mu} = \lim_n \int \Phi_n\,d\mu = \int f\,d\mu$$

by the Monotone Convergence Theorem, where Φ_n is the simple function

$$\Phi_n = 2^{-n} \sum_{k=0}^{2^{2n}} k\chi_{E_{n,k}}$$

where $E_{n,k} = \{x\colon\ k2^{-n} \leq f(x) < (k+1)2^{-n}\}$ for nonnegative integers n and k.

The uniqueness part can also be established directly. Indeed, by equations (7.25) and (7.26) it is sufficient to show that any two measures μ_1 and μ_2 satisfying equations (7.24), (7.25), and (7.26) agree on compact G_δ sets. Let K be a compact G_δ set. By equation (7.26) for each $\varepsilon > 0$ there exists an open σ-bounded set O such that $K \subset O$ and

$$\mu_1(O) < \mu_1(K) + \varepsilon.$$

By Lemma 7.3 there exists f in $C_c(X)$ such that $f \equiv 1$ on K and $S_f \subset O$. Hence

$$\begin{aligned}\mu_2(K) &= \int \chi_K\,d\mu_2 \leq \int f\,d\mu_2 = I(f) = \int f\,d\mu_1 \\ &\leq \int \chi_O\,d\mu_1 = \mu_1(O) < \mu_1(K) + \varepsilon.\end{aligned}$$

Hence $\mu_2(K) \leq \mu_1(K)$. Similarly $\mu_1(K) \leq \mu_2(K)$. ∎

Examples

7.17. Let X be an uncountable set with the discrete topology. Define ν on $\mathrm{Ba}(X)$ as the zero measure and I on $C_c(X)$ by $I(f) = \int f\,d\nu$. The unique measures guaranteed by Theorems 7.7 (iii) and 7.8 are the measures μ and $\bar{\mu}$ on $\mathrm{Ba}(X)$ and $B_w(X)$, respectively, given by

$$\mu(A) = \begin{cases} 0 & \text{if } A \text{ is countable} \\ \infty & \text{if } A^c \text{ is countable} \end{cases}$$

and

$$\bar{\mu}(A) = \begin{cases} 0 & \text{if } A \text{ is countable} \\ \infty & \text{if } A \text{ is not countable.} \end{cases}$$

7.18. Let X be any set with the discrete topology and let $x_0 \in X$. Define I on $C_c(X)$ by $I(f) = f(x_0)$. I is a Daniell integral by Proposition 7.14. The measures μ and $\bar{\mu}$ on $\text{Ba}(X)$ and $B_w(X)$, respectively, given by Theorems 7.7 (iii) and (7.8) are the measures defined by

$$\mu(A) = \bar{\mu}(A) = \begin{cases} 0 & \text{if } A \text{ is countable not containing } x_0 \\ 1 & \text{if } A \text{ is countable containing } x_0 \\ \infty & \text{if } A \text{ is not countable.} \end{cases}$$

Example 7.17 shows that in Theorem 7.8 the measure $\bar{\mu}$ corresponding to a *positive* (*even bounded*) linear functional need not be finite unless the space X is σ-compact. Despite the beauty of Theorems 7.7 and 7.8, this seems to be a slight drawback for these theorems. Fortunately, our next two theorems will remedy this. The proofs do not depend upon the Daniell theory.

Theorem 7.9. Let I be a positive linear functional on $C_c(X)$, where X is a locally compact Hausdorff space. There exists a *unique* weakly Borel measure μ satisfying the following conditions:

(i) $\mu(V) = \sup\{I(g): 0 \leq g \leq 1, g \in C_c(X) \text{ and } S_g \subset V\}$ for each open V.

(ii) $\mu(B) = \inf\{\mu(V): B \subset V \text{ open}\}$ for each weakly Borel set B.

(iii) If (A is open) *or* (A is weakly Borel and σ-finite), then $\mu(A) = \sup\{\mu(K): K \text{ compact} \subset A\}$. ∎

Proof. We give the proof in three steps.

Step I. We define, for any $B \subset X$,

$$\mu^*(B) = \inf\{\mu(V): B \subset V \text{ open}\}, \tag{7.27}$$

where $\mu(V)$ is defined as in the statement (i) of the theorem. We claim that μ^* is an outer measure on 2^X. Obviously, $\mu^*(\varnothing) = 0$; and whenever $A \subset B$, then $\mu^*(A) \leq \mu^*(B)$. For open V_1 and V_2, we notice that

$$\mu(V_1 \cup V_2) \leq \mu(V_1) + \mu(V_2). \tag{7.28}$$

Indeed, to verify inequality (7.28), let $h \in C_c(X)$ with $S_h \subset V_1 \cup V_2$ and $0 \leq h \leq 1$. Clearly we can find $g_1, g_2 \in C_c(X)$ with $0 \leq g_1, g_2 \leq 1$ and $h \leq g_1 + g_2$ such that $S_{g_1} \subset V_1$ and $S_{g_2} \subset V_2$. Hence $I(h) \leq I(g_1 + g_2)$

$\leq I(g_1) + I(g_2) \leq \mu(V_1) + \mu(V_2)$, which implies inequality (7.28). Now let $B_n \subset X$ and $B = \bigcup_{n=1}^{\infty} B_n$. Given $\varepsilon > 0$, there exists open $V_n \supset B_n$ such that

$$\mu(V_n) \leq \mu^*(B_n) + \varepsilon/2^n.$$

Then $B \subset V = \bigcup_{n=1}^{\infty} V_n$ and $\mu^*(B) \leq \mu(V)$. If $0 \leq f \leq 1$, $f \in C_c(X)$ and $S_f \subset V$, then $S_f \subset \bigcup_{n=1}^{N} V_n$ (by compactness of the support of f); therefore,

$$\begin{aligned} I(f) &\leq \mu\left(\bigcup_{n=1}^{N} V_n\right) \leq \sum_{n=1}^{N} \mu(V_n) \quad \text{[by inequality (7.28)]} \\ &\leq \sum_{n=1}^{\infty} \mu^*(B_n) + \varepsilon, \end{aligned}$$

implying $\mu^*(B) \leq \mu(V) \leq \sum_{n=1}^{\infty} \mu^*(B_n) + \varepsilon$.

Step II. Let $\mathscr{A} = \{A \subset X\colon \mu^*(A) < \infty \text{ and } \mu^*(A) = \sup\{\mu^*(K)\colon K \text{ compact} \subset A\}\}$. Then $\mathscr{A}$ is a ring of subsets of X containing open sets (with finite measure μ^*) and all compact sets. Furthermore, if (A_n) is a disjoint sequence in $\mathscr{A}$, then $\mu^*(\bigcup_{n=1}^{\infty} A_n) = \sum_{n=1}^{\infty} \mu^*(A_n)$; and if $\mu^*(\bigcup_{n=1}^{\infty} A_n) < \infty$, then $\bigcup_{n=1}^{\infty} A_n \in \mathscr{A}$.

The reader can verify the above assertions easily. Let us show only

$$\mu(V) = \sup\{\mu^*(K)\colon K \text{ compact} \subset V\} \tag{7.29}$$

for *all* open V, and

$$\mu^*\left(\bigcup_{n=1}^{\infty} A_n\right) = \sum_{n=1}^{\infty} \mu^*(A_n) \tag{7.30}$$

if (A_n) are in $\mathscr{A}$ and are pairwise disjoint.

To prove equation (7.29), let $0 < a < \mu(V)$. By the definition of $\mu(V)$, there exists $f \in C_c(X)$ with $0 \leq f \leq 1$ and $S_f = K \subset V$ such that $a < I(f) < \mu(V)$. By equation (7.27), $\mu^*(K) \geq I(f)$, and thus equation (7.29) is proved.

To prove equation (7.30), first let K_1, K_2 be two disjoint compact sets with W open such that $K_1 \cup K_2 \subset W$, and

$$\mu(W) \leq \mu^*(K_1 \cup K_2) + \varepsilon.$$

Let V_1, V_2 be open, $K_1 \subset V_1$, $K_2 \subset V_2$, and $V_1 \cap V_2 = \varnothing$. Let $g_1, g_2 \in C_c(X)$ with $0 \leq g_1, g_2 \leq 1$, $S_{g_1} \subset V_1 \cap W$, and $S_{g_2} \subset V_2 \cap W$ such that

$$\mu(W \cap V_i) \leq I(g_i) + \varepsilon, \qquad i = 1, 2.$$

Then

$$\begin{aligned}\mu^*(K_1) + \mu^*(K_2) &\le \mu(W \cap V_1) + \mu(W \cap V_2)\\ &\le I(g_1) + I(g_2) + 2\varepsilon\\ &= I(g_1 + g_2) + 2\varepsilon\\ &\le \mu(W) + 2\varepsilon \le \mu^*(K_1 \cup K_2) + 3\varepsilon\end{aligned}$$

showing μ^* is finitely additive for disjoint compact sets. Now let $(A_n) \in \mathscr{A}$ with the A_n's pairwise disjoint. Then there are compact $K_n \subset A_n$ such that

$$\mu^*(A_n) \le \mu^*(K_n) + \varepsilon/2^n$$

and

$$\begin{aligned}\sum_{n=1}^{k} \mu^*(A_n) &\le \sum_{n=1}^{k} \mu^*(K_n) + \varepsilon\\ &= \mu^*\left(\bigcup_{i=1}^{n} K_i\right) + \varepsilon\\ &\le \mu^*\left(\bigcup_{i=1}^{\infty} A_i\right) + \varepsilon, \qquad \text{for all } k.\end{aligned}$$

This proves equation (7.30).

Step III. The outer measure μ^* is a measure on the σ-algebra $\mathscr{F} = \{A \subset X\colon A \cap K \in \mathscr{A} \text{ for all compact } K\}$ containing all the weakly Borel sets; and $A \in \mathscr{F}$ with $\mu^*(A) < \infty$ implies $A \in \mathscr{A}$.

To prove these assertions, let $(A_n) \in \mathscr{F}$. Then $A_n \cap K \in \mathscr{A}$ for compact K. Since $\mathscr{A}$ is a ring, $\bigcup_{n=1}^{\infty}(A_n \cap K)$ can be written as $\bigcup_{n=1}^{\infty} B_n$, where (B_n) is a disjoint sequence of sets in $\mathscr{A}$. By step (II), $\bigcup_{n=1}^{\infty} B_n (\subset K) \in \mathscr{A}$. This proves that $\mathscr{F}$ is closed under countable unions. Since $(X - A) \cap K = K - K \cap A \in \mathscr{A}$ if K is compact and $A \in \mathscr{F}$, $\mathscr{F}$ is also closed under complementation. Thus $\mathscr{F}$ is a σ-algebra. Since every closed set is in $\mathscr{F}$, $\mathscr{F}$ contains all the weakly Borel sets. To prove that μ^* is a measure on $\mathscr{F}$, it is sufficient to show [because of equation (7.30)] that

$$A \in \mathscr{F},\ \mu^*(A) < \infty \Leftrightarrow A \in \mathscr{A}. \tag{7.31}$$

Since $\mathscr{A}$ is a ring containing all the compact sets, $\mathscr{A} \subset \mathscr{F}$. To prove identity (7.31), let $\mu^*(A) < \infty$. Let V be open $\supset A$ such that $\mu^*(V) < \infty$. Then by equation (7.29), let $K \subset V$ with K compact such that

$$\mu(V) < \mu^*(K) + \varepsilon.$$

Since $A \cap K \in \mathscr{A}$, there is compact $K_1 \subset A \cap K$ such that

$$\mu^*(A \cap K) < \mu^*(K_1) + \varepsilon$$

or

$$\mu^*(A) \leq \mu^*(A \cap K) + \mu^*(V - K) < \mu^*(K_1) + 2\varepsilon$$

proving that $A \in \mathscr{A}$, and therefore (7.31) is true. The proof of the theorem is now complete. ∎

Theorem 7.10. *The Riesz Representation Theorem.* Let I be a positive linear functional on $C_c(X)$, where X is a locally compact Hausdorff space. Then there is a *unique* weakly Borel measure μ satisfying (ii) and (iii) of Theorem 7.9 such that $I(f) = \int f\, d\mu$ for all $f \in C_c(X)$. ∎

Proof. Let μ be the weakly Borel measure of Theorem 7.9. First we show that $I(f) \leq \int f\, d\mu$ for all $f \in C_c(X)$. Let $\varepsilon > 0$, $f \in C_c(X)$, and $m = \sup_{x \in X} |f(x)| > 0$. Then if $K = S_f$, we can find a partition of K by weakly Borel sets $B_1, B_2, \ldots, B_n$ and constants $c_1, c_2, \ldots, c_n$ such that $f < \sum_{i=1}^n c_i \chi_{B_i} \leq f + \varepsilon$. Let V_i be an open set $\supset B_i$ such that $x \in V_i \Rightarrow f(x) < c_i$ and $\mu(V_i) < \mu(B_i) + \varepsilon/nm$. Let $\{g_1, g_2, \ldots, g_n\}$ be functions in $C_c(X)$ such that $h(x) = \sum_{i=1}^n g_i(x) = 1$ for all $x \in K$ with $0 \leq g_i \leq 1$, and $S_{g_i} \subset V_i$ with $1 \leq i \leq n$. (See Problem 7.2.1.) Now let $W = \{x: h(x) > 1 - \varepsilon\}$ and $p(x) \in C_c(X)$ with $0 \leq p \leq 1$ and $S_p \subset W$ such that $\mu(W) \leq I(p) + \varepsilon$. Then $\mu(K) \leq [1/(1 - \varepsilon)]I(h) + \varepsilon$, implying $\mu(K) \leq \sum_{i=1}^n I(g_i)$. Now we have

$$\begin{aligned}
I(f) &= \sum_{i=1}^n I(fg_i) \leq \sum_{i=1}^n I(c_i g_i) \\
&= \sum_{i=1}^n c_i I(g_i) \\
&\leq \sum_{i=1}^n (c_i + m) I(g_i) - m\mu(K) \\
&\leq \sum_{i=1}^n (c_i + m)\left[\mu(B_i) + \frac{\varepsilon}{nm}\right] - m\mu(K) \\
&= \sum_{i=1}^n c_i \mu(B_i) + \varepsilon \\
&\leq \int f\, d\mu + \varepsilon \cdot \mu(K) + \varepsilon.
\end{aligned}$$

This proves that $I(f) \leq \int f\, d\mu$ for all $f \in C_c(X)$. Changing f to $-f$, we have $I(f) = \int f\, d\mu$.

To prove uniqueness, let μ_1 and μ_2 be two weakly Borel measures satisfying (ii) and (iii) of Theorem 7.9, and let $I(f) = \int f\,d\mu_1 = \int f\,d\mu_2$ for all $f \in C_c(X)$. If K is a compact set and V open such that $V \supset K$ and

$$\mu_2(V) < \mu_2(K) + \varepsilon,$$

then taking $f \in C_c(X)$ such that $\chi_K(x) \leq f(x) \leq \chi_V(x)$ we have

$$\mu_1(K) \leq \int f\,d\mu_1 = \int f\,d\mu_2 \leq \mu_2(V) < \mu_2(K) + \varepsilon.$$

Hence $\mu_1(K) \leq \mu_2(K)$. By symmetry, $\mu_1 = \mu_2$ on compact sets, and they are equal on all weakly Borel sets by (ii) and (iii) of Theorem 7.9. ∎

Our final consideration in this section is a representation theorem for bounded linear functionals on $C_c(X)$ which is *not necessarily* positive. First we need a lemma.

Lemma 7.8. For each bounded linear functional I on $C_c(X)$, where X is a locally compact Hausdorff space, there exist two positive bounded linear functionals I_+ and I_- such that $I = I_+ - I_-$ and $\| I \| = \| I_+ \| + \| I_- \|$. ∎

Proof. For $f \geq 0$ and $f \in C_c(X)$ let

$$I_+(f) = \sup\{I(g)\colon 0 \leq g \leq f,\ g \in C_c(X)\}.$$

For $f_1, f_2 \in C_c(X)$ with $f_1 \geq 0$, $f_2 \geq 0$ and for any $c \geq 0$, we have

(a) $I_+(f_1) \geq 0$,

(b) $I_+(f_1) \geq I(f_1)$,

(c) $I_+(cf_1) = cI_+(f_1)$,

(d) $I_+(f_1 + f_2) = I_+(f_1) + I_+(f_2)$.

We establish only (d). If $0 \leq g_1 \leq f_1$ and $0 < g_2 < f_2$, then $0 \leq g_1 + g_2 \leq f_1 + f_2$; and so

$$I_+(f_1 + f_2) \geq I(g_1 + g_2) = I(g_1) + I(g_2).$$

Taking the supremum over all such g_1 and g_2,

$$I_+(f_1 + f_2) \geq I_+(f_1) + I_+(f_2).$$

Conversely, if $0 \leq g \leq f_1 + f_2$, $g_1 = g \wedge f_1$, and $g_2 = g - g_1$, then $0 \leq g_1$

$\leq f_1$ and $0 \leq g_2 \leq f_2$ and so

$$I(g) = I(g_1) + I(g_2) \leq I_+(f_1) + I_+(f_2),$$

which implies (d).

To extend I_+ to all $f \in C_c(X)$, let $f = f_1 - f_2$, where $f_1 \geq 0$ and $f_2 \geq 0$ are in $C_c(X)$. By taking $I_+(f) = I_+(f_1) - I_+(f_2)$, it is obvious that I_+ is *well defined* because of (d). I_+ is also linear. We define $I_- = I_+ - I$. Then I_+ and I_- are both positive linear and $I = I_+ - I_-$.

Now we will prove $\| I \| = \| I_+ \| + \| I_- \|$. Since $| I_+(f) | \leq \| I \| \, \| f \|$ for $f \geq 0$, I_+ is obviously bounded and so is I_-. Also, since $| I(f) | = | I_+(f) - I_-(f) | \leq \| I_+ \| \, \| f \| + \| I_- \| \, \| f \|$, $\| I \| \leq \| I_+ \| + \| I_- \|$. For the converse inequality, let $\varepsilon > 0$; then we can find $f_1 \in C_c(X)$ with $0 \leq f_1 \leq 1$ such that

$$I(f_1) \geq \| I_+ \| - \varepsilon.$$

Also we can find $f_2 \in C_c(X)$ with $0 \leq f_2 \leq 1$ and $f_2 \geq f_1$ such that

$$I_-(f_2) \geq \| I_- \| - \varepsilon.$$

Let $f = 2f_1 - f_2$. Then $|f| \leq 1$ and

$$\begin{aligned} I(f) &= 2I(f_1) - I(f_2) \\ &\geq \| I_+ \| + I(f_1) - I(f_2) - \varepsilon \\ &\geq \| I_+ \| + I_+(f_2) - I(f_2) - 2\varepsilon \\ &= \| I_+ \| + I_-(f_2) - 2\varepsilon \\ &\geq \| I_+ \| + \| I_- \| - 3\varepsilon, \end{aligned}$$

thus proving that $\| I \| = \| I_+ \| + \| I_- \|$. ∎

Theorem 7.11. To each bounded linear functional I on $C_c(X)$, where X is a locally compact Hausdorff space, there corresponds a *unique* finite signed weakly Borel measure μ on X such that

$$I(f) = \int f \, d\mu \qquad \text{for all } f \in C_c(X).$$

Also $\| I \| = | \mu | (X)$, where $| \mu |$ is the total variation of μ. ∎

Proof. By Lemma 7.8 and Theorem 7.10, $I = I_+ - I_-$, where I_+, I_- are positive linear functionals on $C_c(X)$ with $\| I \| = \| I_+ \| + \| I_- \|$;

and

$$I_+(f) = \int f\, d\mu_1$$

and

$$I_-(f) = \int f\, d\mu_2 \qquad \text{for all } f \in C_c(X).$$

where μ_1, μ_2 are unique *finite* weakly Borel measures on X. Letting $\mu = \mu_1 - \mu_2$,

$$I(f) = \int f\, d\mu \qquad \text{for } f \in C_c(X).$$

Clearly, $| I(f) | \leq \int | f | \, d | \mu | \leq \| f \| \cdot | \mu | (X)$ or $\| I \| \leq | \mu | (X)$. Conversely,

$$| \mu | (X) \leq \mu_1(X) + \mu_2(X) = \| I_+ \| + \| I_- \| = \| I \|.$$

Hence $\| I \| = | \mu | (X)$.

To prove uniqueness, let μ, ν be two finite signed measures such that for all $f \in C_c(X)$,

$$I(f) = \int f\, d\mu = \int f\, d\nu.$$

Then $\lambda = \mu - \nu$ is also a finite signed measure and for all $f \in C_c(X)$,

$$\int f\, d\lambda = 0.$$

This means that if $\lambda = \lambda^+ - \lambda^-$ is the Jordan decomposition of λ, then $\int f\, d\lambda^+ = \int f\, d\lambda^-$ for all $f \in C_c(X)$ so that λ^+ and λ^- induce the same positive linear functional on $C_c(X)$. By the uniqueness part of Theorem 7.9, $\lambda^+ = \lambda^-$ since λ^+ and λ^- are also regular. Therefore $\lambda = 0$ and $\mu = \nu$. ∎

Problems

7.4.1. (i) Let X be a metrizable noncompact space. For a given nonnegative extended real-valued number α let (α_n) be a sequence of nonnegative real numbers such that $\alpha = \sum_n \alpha_n$. Let (x_n) be a sequence in X with no cluster point. Define μ on $\text{Ba}(X)$ by

$$\mu(E) = \sum \{\alpha_n \colon x_n \in E\}.$$

Show μ is regular on $\text{Ba}(X)$ and $\mu(X) = \alpha$. (Use Theorem 7.3.)

(ii) If X is a σ-compact locally compact Hausdorff space that is not compact, let (K_n) be a sequence of compact G_δ sets such that $K_n{}^\circ$ is a Baire set and $K_n \subsetneqq K_{n+1}^\circ$, and $X = \bigcup_{n=1}^\infty K_n$. Let $C_n = K_{2n} - K_{2n-1}^\circ$ for $n = 1, 2, \ldots$. Let μ_n be any Baire measure on C_n such that $\mu_n(C_n) = \alpha_n$. [α and (α_n) are defined in (i).] Define μ on Ba(X) by

$$\mu(E) = \sum \mu_n(E \cap C_n).$$

Show that μ is a regular Baire measure such that $\mu(X) = \alpha$.

(iii) If X is a metrizable or σ-compact locally compact Hausdorff space and X is not compact, then show that there exists a discontinuous positive linear functional on $C_c(X)$.

7.4.2. Let $\bar{X}$ be the space of Example 7.11—the space of all ordinals less than or equal to Ω, the first uncountable ordinal.

(i) Prove that to each f in $C(\bar{X})$ there corresponds an $\alpha \neq \Omega$ such that f is constant on $\{\gamma \in \bar{X}: \alpha < \gamma\}$. (Compare Problem 7.3.13.)

(ii) If μ is the measure of Example 7.11, show

$$f(\Omega) = \int f\, d\mu$$

for each $f \in C(\bar{X})$.

(iii) What is the measure $\bar{\mu}$ of Theorem 7.8 corresponding to this functional?

7.4.3. *The Monotone Convergence Theorem for Nets.* Let μ be a regular Borel measure on a compact Hausdorff space X. Let $(f_\alpha)_{\alpha\in A}$ be an increasing net of continuous functions. Show that $f = \lim_\alpha f_\alpha \in L_1(\mu)$ if and only if $\sup_\alpha \| f_\alpha \|_1 < \infty$ and in this case $\lim_\alpha \| f - f_\alpha \|_1 = 0$.

• 7.5. Product Measures and Integration

Recall that if $\mathscr{A}$ and $\mathscr{B}$ are two σ-algebras on which σ-finite measures μ and ν, respectively, are defined, then the product $\mu\times\nu$ of μ and ν is the complete measure on a σ-algebra $\mathscr{M}$ containing $\mathscr{A}\times\mathscr{B}$ extending the function λ on the semialgebra of measurable rectangles given by $\lambda(A\times B) = \mu(A)\nu(B)$ (see Chapter 3, Section 3.4). *Throughout this section the product $\mu\times\nu$ will be assumed to be this product measure restricted to $\mathscr{A}\times\mathscr{B}$.* If μ and ν are σ-finite measures then $\mu\times\nu$ is the unique measure on $\mathscr{A}\times\mathscr{B}$ satisfying

$$\mu\times\nu(A\times B) = \mu(A)\nu(B)$$

for all A in $\mathscr{A}$ and B in $\mathscr{B}$. We will study, in particular, the product of measures μ and ν on σ-algebras of Borel sets. Furthermore, this section will continue the study of the Fubini–Tonelli Theorems on product integration introduced in Chapter 3, Section 3.4, for Borel-measurable functions on the product of two locally compact Hausdorff spaces. The main theorem in this section is Theorem 7.20.

Throughout this section the topological spaces X and Y under consideration will be assumed to be σ-compact locally compact Hausdorff spaces. In this case the σ-ring σ_r(compact sets in X) is identical to the σ-algebra of Borel sets $B(X)$. The results of this section could be proved without the assumption of σ-compactness for σ-rings of Borel sets as in [31], but the flavor is not lost and ease of notation is gained by assuming that all spaces are σ-compact.

Recall that any Borel measure on a σ-compact locally compact Hausdorff space is σ-finite.

If μ_0 and ν_0 are Baire measures on $B_a(X)$ and $B_a(Y)$, then $\mu_0 \times \nu_0$ is a Baire measure since

$$B_a(X) \times B_a(Y) = B_a(X \times Y). \tag{7.32}$$

In contrast, if μ and ν are Borel measures their product is not necessarily a Borel measure. The reason in simple: The domain of $\mu \times \nu$ is not necessarily a σ-algebra of Borel sets as sometimes

$$B(X) \times B(Y) \subsetneqq B(X \times Y).$$

(see Example 7.10). Our first goal in this section is to define a Borel measure on $B(X \times Y)$ which extends $\mu \times \nu$. The next result gives a criterion for when this can be done.

Theorem 7.12. Suppose μ and ν are nonzero Borel measures on $B(X)$ and $B(Y)$, respectively. There exists a unique regular Borel measure $\mu \otimes \nu$ (called the *tensor product* of μ and ν) on $B(X \times Y)$ which extends $\mu \times \nu$ if and only if μ and ν are regular Borel measures.† ∎

Proof. For the sufficiency proof, let μ_0 and ν_0 be the Baire measures which are the restrictions of μ and ν to $B_a(X)$ and $B_a(Y)$. We know from equation (7.32) that $\mu_0 \times \nu_0$ is a Baire measure on $B_a(X \times Y)$. Let $\mu \otimes \nu$ be the unique regular extension of $\mu_0 \times \nu_0$ to $B(X \times Y)$. (See Theorem 7.4.

† Recently D. H. Fremlin in *Canad. Math. Bull.* **19** (1976) has shown that a compact set in $X \times X$ need not be in the domain of the completion of $\mu \times \mu$. The reader may also consult for other related results Godfrey and Sion's paper in *Canad. Math. Bull.* **12** (1969).

Keep in mind that X and Y are σ-compact.) To show that $\mu \otimes \nu$ extends $\mu \times \nu$, it is sufficient to show (see Problem 7.5.2) that

$$\mu \otimes \nu(C \times D) = \mu \times \nu(C \times D)$$

for compact sets C and D. By the regularity of μ and ν, there exist compact G_δ sets (Proposition 7.11) G_C, G_D, and $G_{C\times D}$ such that

$$C \subset G_C \quad \text{and} \quad \mu(C) = \mu(G_C),$$
$$D \subset G_D \quad \text{and} \quad \nu(D) = \nu(G_D),$$

and

$$C \times D \subset G_{C\times D} \quad \text{and} \quad \mu \otimes \nu(C \times D) = \mu \otimes \nu(G_{C\times D}).$$

Since $\mu \otimes \nu$ and $\mu \times \nu$ agree on Baire sets in $X \times Y$, we have

$$\mu \times \nu(C \times D) \leq \mu \times \nu(G_{C\times D}) = \mu \otimes \nu(G_{C\times D}) = \mu \otimes \nu(C \times D),$$

and

$$\begin{aligned} \mu \otimes \nu(C \times D) &\leq \mu \otimes \nu(G_C \times G_D) = \mu \times \nu(G_C \times G_D) \\ &= \mu(G_C)\nu(G_D) = \mu(C)\nu(D) = \mu \times \nu(C \times D). \end{aligned}$$

Uniqueness is clear since any regular extension of $\mu \times \nu$ is a regular extension of $\mu_0 \times \nu_0$ on $B_a(X \times Y)$.

To prove the necessity, let ϱ be a regular extension of $\mu \times \nu$ to $B(X \times Y)$. It is then a regular extension of $\mu_0 \times \nu_0$ and hence by the sufficiency just proved it is the unique regular extension of $\mu' \times \nu'$, where μ' and ν' are the unique *regular* extensions of μ_0 and ν_0 to the σ-algebras $B(X)$ and $B(Y)$. Hence $\mu' \times \nu' = \mu \times \nu$ so that for all Borel sets A in X and B in Y we have

$$\mu'(A)\nu'(B) = \mu' \times \nu'(A \times B) = \mu \times \nu(A \times B) = \mu(A)\nu(B).$$

Since μ and ν are nonzero, we have $\mu = \mu'$ and $\nu = \nu'$. ∎

What happens in the case that μ and ν are arbitrary Borel measures? Can $\mu \times \nu$ be extended to a Borel measure? An answer is known in the case that one of μ or ν is regular. The Theorem 7.13 below shows that in this case $\mu \times \nu$ can be extended to two measures ϱ_1 and ϱ_2 on $B(X \times Y)$. First we prove some useful preliminary results.

If M is any subset of $X \times Y$, then M_x is the set

$$M_x = \{y\colon (x, y) \in M\} = \mathrm{Pr}_Y[M \cap (\{x\} \times Y)]$$

and M^y is the set

$$M^y = \{x\colon (x, y) \in M\} = \mathrm{Pr}_X[M \cap (X \times \{y\})],$$

where Pr_Y and Pr_X are the projection mappings. If $M \in B(X) \times B(Y)$, then $M_x \in B(Y)$ and $M^y \in B(X)$. Moreover, if $M \in B(X \times Y)$, then $M_x \in B(Y)$ and $M^y \in B(X)$ for all x in X and $y \in Y$. (See Problem 7.5.1.)

If μ and ν are Borel measures on $B(X)$ and $B(Y)$, respectively, and $M \in B(X \times Y)$, we define the extended real-valued functions Γ_M on X and Γ^M on Y, respectively, by

$$\Gamma_M(x) = \nu(M_x) \quad \text{and} \quad \Gamma^M(y) = \mu(M^y). \tag{7.33}$$

Proposition 7.15. Let C be compact in $X \times Y$.

(i) If ν is regular, $\Gamma_C\colon X \to R$ is Borel measurable.[†]

(ii) If μ is regular, $\Gamma^C\colon Y \to R$ is Borel measurable. ∎

Proof. (i) It must be shown that for each $a \in R$ the set $\{x\colon \Gamma_C(x) \geq a\}$ is in $B(X)$. Since Γ_C is a nonnegative function, it suffices to consider only positive numbers a. Denote $\{x\colon \Gamma_C(x) \geq a\}$ by A_a. Clearly, $A_a \subset \mathrm{Pr}_X(C)$. As Pr_X is continuous, $\mathrm{Pr}_X(C)$ is compact in X and A_a is contained in a compact set. To show that A_a is a Borel set, it suffices to show that A_a is closed, since then A_a will itself be compact as a subset of a compact set. To this end, let $z \in A_a{}^c$. This means $\nu(C_z) < a$, and by the regularity of ν there is an open Borel set V in Y such that $C_z \subset V$ and $\nu(V) < a$. Letting U be the open set

$$U = X - \mathrm{Pr}_X[C - (X \times V)],$$

clearly $x \in U$ if and only if $C_x \subset V$. Applying this criterion to z, this means U is a neighborhood of z disjoint from A_a.

(ii) The proof of (ii) is similar to that of (i). ∎

The impetus of Proposition 7.15 is that the hypotheses of the next two results are satisfied when μ and ν are regular.

Lemma 7.9. (i) If Γ_C is Borel measurable for each compact set C in $X \times Y$, then Γ_M is Borel measurable for each bounded Borel set M in $X \times Y$.

(ii) If Γ^C is Borel measurable for each compact set C in $X \times Y$, then Γ^M is Borel measurable for each bounded Borel set M in $X \times Y$. ∎

Proof. Again we prove (i) and note that the proof of (ii) is similar. Let $\mathscr{R}$ be the ring of all finite disjoint unions of proper differences of compact subsets of $X \times Y$ and let $\mathscr{R}_b$ be the ring of bounded Borel sets in

[†] Problem 7.6.13 gives an example in which Γ_C need not be Borel measurable for nonregular ν even with respect to the completion of μ.

$X \times Y$. Since

$$\mathscr{R} \subset \mathscr{R}_b \subset B(X \times Y)$$

and the σ-ring generated by $\mathscr{R}$ is $B(X \times Y)$, the σ-ring generated by $\mathscr{R}_b$ is also $B(X \times Y)$.

Let $M \in \mathscr{R}_b$. Choose compact sets C and K such that $M \subset C \times K$. Then M belongs to the σ-ring $B(X \times Y) \cap (C \times K)$ which is generated by the ring $\mathscr{R} \cap (C \times K)$. Let $\mathscr{M}$ be the class of all Borel sets N in $X \times Y$ with $N \subset C \times K$ and Γ_N Borel measurable. As Γ_N is measurable for each N in $\mathscr{R}$ (see Problem 7.5.3) and $\mathscr{R} \cap (C \times K) \subset \mathscr{R}$, we have

$$\mathscr{R} \cap (C \times K) \subset \mathscr{M}.$$

However, $\mathscr{M}$ is a monotone class. Indeed, if (N_n) is a nondecreasing sequence in $\mathscr{M}$, then with $N = \lim_n N_n$ we have

$$\begin{aligned}\{x \in X: \nu(N_x) > a\} &= \{x \in X: \lim_n \nu(N_{n_x}) > a\} \\ &= \bigcup_n \{x \in X: \nu(N_{n_x}) > a\}\end{aligned}$$

as $N_x = \lim_n N_{n_x}$ and (N_{n_x}) is a nondecreasing sequence for each x in X. This means Γ_N is measurable and $N \in \mathscr{M}$. Similarly, $\mathscr{M}$ can be shown to be closed with respect to nonincreasing sequences. Therefore $\mathscr{M}$ is a monotone class containing $\mathscr{R} \cap (C \times K)$, whereby $\mathscr{M}$ contains the σ-ring generated by $\mathscr{R} \cap (C \times K)$, namely, $B(X \times Y) \cap (C \times K)$. In particular $\mathscr{M}$ contains M, so Γ_M is Borel measurable. ∎

Theorem 7.13

(i) If Γ_C is Borel measurable for each compact set C in $X \times Y$, then there exists a unique Borel measure ϱ_1 on $B(X \times Y)$ such that

$$\varrho_1(M) = \int \Gamma_M \, d\mu \tag{7.34}$$

for all bounded Borel sets M in $X \times Y$. ϱ_1 is an extension of $\mu \times \nu$.

(ii) If Γ^C is Borel measurable for each compact set C in $X \times Y$, then there exists a unique Borel measure ϱ_2 on $B(X \times Y)$ such that

$$\varrho_2(M) = \int \Gamma^M \, d\nu \tag{7.35}$$

for all bounded Borel sets M in $X \times Y$. ϱ_2 is an extension of $\mu \times \nu$. ∎

Proof. (i) From Lemma 7.9 we know Γ_M is measurable for each bounded Borel set M. Inasmuch as $M \subset C \times K$ for compact sets C and K,

and

$$0 \leq \Gamma_M \leq \nu(K)\chi_C,$$

we can conclude that Γ_M is μ-integrable. Define ϱ_1' on the ring of bounded Borel sets $\mathscr{R}_b$ by

$$\varrho_1'(M) = \int \Gamma_M \, d\mu.$$

Clearly ϱ_1' is a nonnegative additive real-valued set function on $\mathscr{R}_b$. To show that ϱ_1' is a measure on $\mathscr{R}_b$ it suffices to show that if (M_n) is a nondecreasing sequence of sets in $\mathscr{R}_b$ converging to M in $\mathscr{R}_b$, then $\varrho_1'(M_n) \uparrow \varrho_1'(M)$. This is immediate from the Monotone Convergence Theorem.

Now let ϱ_1 be the unique extension of ϱ_1' to the σ-algebra $B(X \times Y)$ generated by $\mathscr{R}_b$. (See Theorem 2.4 in Chapter 2, giving special attention to the footnote there.) All that remains to be shown is that ϱ_1 is an extension of $\mu \times \nu$. As in Theorem 7.12, it suffices to show that ϱ_1 and $\mu \times \nu$ agree on sets $C \times K$, where C and K are compact. Note that $\Gamma_{C \times K} = \nu(K)\chi_C$. Hence

$$\varrho_1(C \times K) = \int \Gamma_{C \times K} \, d\mu = \nu(K)\mu(C) = \mu \times \nu(C \times K).$$

The proof of (ii) is similar. ∎

In light of Theorem 7.13 we have extensions ϱ_1 and ϱ_2 of $\mu \times \nu$ to $B(X \times Y)$. The next question is whether $\varrho_1 = \varrho_2$.

Example 7.19. (Where $\varrho_1 \neq \varrho_2$). Let $\bar{X}$ again be the set of ordinals less than or equal to Ω, the first uncountable ordinal. Let $X = \bar{X} - \{\Omega\}$. Let μ be the measure on $\bar{X}$ of Example 7.11 and let C be a compact set in $\bar{X} \times \bar{X}$. Is Γ_C Borel measurable? It must be shown that, for any positive real number a, the set A_a equal to $\{x \in \bar{X}\colon \Gamma_C(x) \geq a\}$ is a Borel set. It suffices to show that $A_a \cap X$ is closed in X. Since X is first countable it suffices to show that whenever (x_n) is a sequence in $A_a \cap X$ converging to x in X, then $x \in A_a \cap X$. By Problem 7.5.4, $\Gamma_C(x) \geq a$ so that $x \in A_a$.

Thus Γ_C is measurable and similarly Γ^C is measurable so that ϱ_1 and ϱ_2 are defined. To show $\varrho_1 \neq \varrho_2$ consider the set Z given by

$$Z = \{(x, y)\colon x < y < \Omega \text{ or } x = \Omega\}.$$

For any x,

$$\Gamma_Z(x) = \mu(Z_x) = 1$$

and for any y,

$$\Gamma^Z(y) = \mu(Z^y) = 0.$$

Hence

$$\varrho_1(Z) = \int \Gamma_Z \, d\mu = 1 \quad \text{and} \quad \varrho_2(Z) = \int \Gamma^Z \, d\mu = 0.$$

Theorem 7.14. If μ and ν are both regular, then so are ϱ_1 and ϱ_2; and indeed (by Theorem 7.12) $\varrho_1 = \varrho_2 = \mu \otimes \nu$. ▮

Proof. We show ϱ_1 is regular. It suffices to show by Proposition 7.11 that each bounded open measurable set U is inner regular. Given $\varepsilon > 0$, we must find a compact set C in $X \times Y$ with $C \subset U$ and $\varrho_1(U - C) < 2\varepsilon$. Now since U is bounded, $U \subset G \times H$, where G and H are compact sets in X and Y, respectively. If $\nu(H) = 0$, then $\varrho_1(U) = 0$ as

$$\varrho_1(U) \leq \varrho_1(G \times H) = \mu \times \nu(G \times H) = \mu(G)\nu(H) = 0.$$

However, if $\varrho_1(U) = 0$ we can simply take C to be the empty set and be finished. We assume therefore that $\nu(H) > 0$.

Since Γ_U is a measurable function with respect to $B(X)$, using Lusin's Theorem (Problem 3.1.13, in Chapter 3) and the regularity of μ, we may choose a compact set K in X so that $K \subset G$, $\mu(G - K) < \varepsilon/\nu(H)$ and Γ_U is continuous on K. Let $U' = U \cap (K \times Y)$. It is easy to see that

$$U - U' \subset (G - K) \times H$$

so that

$$\varrho_1(U - U') \leq \varrho_1(G - K) \times H) = \mu(G - K)\nu(H) < \varepsilon.$$

If we can find a compact set C in $X \times Y$ such that $C \subset U$ and $\varrho_1(U' - C) \leq \varepsilon$, then the relation

$$\varrho_1(U - C) \leq \varrho_1(U - U') + \varrho_1(U' - C) < 2\varepsilon$$

will complete the proof.

Now if $\mu(K) = 0$, then clearly from the definition of U', $\varrho_1(U') = 0$, in which case we can take C to be $\varnothing$ and the proof is complete. Let us therefore assume $\mu(K) > 0$.

Let $x \in K$. The x-section U_x of U is a Borel set and $U_x \subset H$ so that $\nu(U_x) < \infty$. By virtue of the regularity of ν, there exists a compact set $E(x)$ in Y with $E(x) \subset U_x$ and $\nu(U_x) < \nu\big(E(x)\big) + \varepsilon/\mu(K)$. This means $\Gamma_U(x) < \nu\big(E(x)\big) + \varepsilon/\mu(K)$. Since Γ_U is continuous on K, there exists an open neighborhood $V(x)$ of x, relative to the subspace K of X, such that

$$\Gamma_U(t) < \nu(E(x)) + \varepsilon/\mu(K) \qquad \text{for all } t \in V(x). \tag{7.36}$$

Since $[K \times E(x)] - U$ is compact, so is $\mathrm{Pr}_X([K \times E(x)] - U)$. Thus

$$V'(x) \equiv V(x) - \mathrm{Pr}_X([K \times E(x)] - U)$$

is open relative to K. If $t \in V(x)$, then $t \in V'(x)$ if and only if $E(x) \subset U_t$. Hence $V'(x)$ is an open neighborhood of x, relative to K, such that $V'(x) \times E(x) \subset U$. Let $D(x)$ be a compact neighborhood of x, relative to K, such that $D(x) \subset V'(x)$. Then $D(x) \times E(x) \subset U$ and $D(x) \subset V(x)$. By inequality (7.36)

$$\Gamma_U(t) < \nu(E(x)) + \varepsilon/\mu(K) \qquad \text{for all } t \in D(x).$$

It follows that

$$\Gamma_{U-[D(x)\times E(x)]}(t) < \varepsilon/\mu(K) \qquad \text{for all } t \in D(x). \tag{7.37}$$

We repeat the above argument for each x in K. As above we get compact sets $D(x)$ and $E(x)$ such that $D(x)$ is a neighborhood of x relative to K, $D(x) \times E(x) \subset U$, and inequality (7.37) is satisfied.

Since the interiors of the sets in the collection $[D(x)]_{x \in K}$ cover K, there exists by the compactness of K a finite subcovering, say

$$K \subset D(x_1) \cup \cdots \cup D(x_n).$$

Define C by

$$C \equiv \bigcup_{i=1}^{n} D(x_i) \times E(x_i).$$

Clearly C is compact and $C \subset U$. We assert that

$$\Gamma_{U'-C} \leq [\varepsilon/\mu(K)]\chi_K. \tag{7.38}$$

Since $\Gamma_{U'-C} \leq \Gamma_{U'}$ and $\Gamma_{U'}$ vanishes on $X - K$, inequality (7.38) is surely true at points of $X - K$. On the other hand, if $x \in K$ then $x \in D(x_i)$ for some x_i, and so

$$\Gamma_{U-D(x_i)\times E(x_i)}(x) < \varepsilon/\mu(K)$$

by inequality (7.37). Since $U' - C \subset U - D(x_i) \times E(x_i)$, we have $\Gamma_{U'-C}(x) < \varepsilon/\mu(K)$.

Using inequality (7.38) and the definition of ϱ_1 we have

$$\varrho_1(U' - C) = \int \Gamma_{U'-C}\, d\mu \leq [\varepsilon/\mu(K)]\mu(K) = \varepsilon. \qquad \blacksquare$$

Summarizing the results (Theorems 7.12–7.14), we may say ϱ_1 exists if ν is regular, ϱ_2 exists if μ is regular, $\mu \otimes \nu$ exists if μ and ν are regular, and $\varrho_1 = \varrho_2 = \mu \otimes \nu$ if μ and ν are regular.

Our final goal in this section is to give another criterion for ϱ_1 to equal ϱ_2 and to prove a Fubini–Tonelli theorem on iterated integration for the measure $\mu \otimes \nu$.

If $(X, \mathscr{A}, \mu)$ and $(Y, \mathscr{B}, \nu)$ are measure spaces and h is an extended real-valued measurable function on $X \times Y$, then we say the *iterated integral* $\int\int h\, d\nu\, d\mu$ exists if there exists a null set E in $\mathscr{A}$ and a μ-integrable function f on X such that h_x is ν-integrable for each x in $X - E$ and $\int h_x\, d\nu = f(x)$. By definition $\int\int h\, d\nu\, d\mu$ is then given to be $\int f(x)\, d\mu$. Similarly the iterated integral $\int\int h\, d\mu\, d\nu$ is defined.

Let us recall the Fubini and Tonelli Theorems of Chapter 3 (See Theorem 3.10).

Theorem 7.15. *Fubini's Theorem.* If $(X, \mathscr{A}, \mu)$ and $(Y, \mathscr{B}, \nu)$ are σ-finite measure spaces and if h is a $(\mu \times \nu)$-integrable function on $X \times Y$, then both iterated integrals of h exist and

$$\int\int h\, d\nu\, d\mu = \int h\, d(\mu \times \nu) = \int\int h\, d\mu\, d\nu. \qquad \blacksquare$$

Theorem 7.16. *Tonelli's Theorem.* If $(X, \mathscr{A}, \mu)$ and $(Y, \mathscr{B}, \nu)$ are σ-finite measure spaces and if h is a nonnegative measurable function on $X \times Y$ such that at least one of the iterated integrals of h exist, then h is $(\mu \times \nu)$-integrable. $\blacksquare$

Observe that in both theorems h is assumed to be measurable with respect to the measurable space $(X \times Y, \mathscr{A} \times \mathscr{B})$. In the case that X and Y are σ-compact locally compact Hausdorff spaces and we consider the measurable spaces $(X, B(X))$ and $(Y, B(Y))$, it may happen that a function h is measurable with respect to $B(X \times Y)$ but not with respect to $B(X) \times B(Y)$. It is natural therefore to ask whether the analogs of Theorems 7.15 and 7.16 hold for such functions h when μ and ν are Borel measures on $B(X)$ and $B(Y)$, respectively.

The measures ϱ_1 and ϱ_2 furnish an answer. Of course ϱ_1 exists if Γ_C is Borel measurable for each compact set in $X \times Y$. A similar statement applies to ϱ_2. For this reason we make the assumptions on Γ_C and Γ^C in the following analog of the Fubini Theorem.

Theorem 7.17. Suppose X and Y are σ-compact locally compact Hausdorff spaces.

(i) Suppose Γ_C is Borel measurable for each compact set C in $X \times Y$. If h is a ϱ_1-integrable function on $X \times Y$ [measurable with respect to $B(X \times Y)$], then $\iint h\,d\nu\,d\mu$ exists and equals $\int h\,d\varrho_1$.

(ii) Suppose Γ^C is Borel measurable for each compact set C in $X \times Y$. If h is a ϱ_2-integrable function on $X \times Y$, then the iterated integral $\iint h\,d\mu\,d\nu$ exists and is equal to $\int h\,d\varrho_2$. ∎

Proof. (i) Writing h as $h = h^+ - h^-$, where $h^+ = h \vee 0$ and $h^- = -(h \wedge 0)$, we may assume that $h \geq 0$.

First assume h is a simple Borel-measurable function with compact support. Then $h = \sum_{i=1}^n \alpha_i \chi_{E_i}$, where each E_i is a bounded Borel set. For each $i \in \{1, 2, \ldots, n\}$ according to equation (7.34),

$$\varrho_1(E_i) = \int \Gamma_{E_i}\,d\mu,$$

where again $\Gamma_{E_i}(x) = \nu(E_{i_x})$. For each $x \in X$,

$$\int \chi_{E_{i_x}} d\nu = \nu(E_{i_x}) < \infty,$$

and by linearity

$$\int h_x\,d\nu = \sum_{i=1}^n \alpha_i \nu(E_{i_x}) < \infty. \tag{7.39}$$

Hence

$$\int \left[\int h_x\,d\nu\right] d\mu = \int \left[\sum_{i=1}^n \alpha_i \nu(E_{i_x})\right] d\mu = \int \left[\sum_{i=1}^n \alpha_i \Gamma_{E_i}(x)\right] d\mu$$
$$= \sum_{i=1}^n \alpha_i \int \Gamma_{E_i}(x)\,d\mu = \sum_{i=1}^n \alpha_i \varrho_1(E_i) = \int h\,d\varrho_1. \tag{7.40}$$

Now assume h is any nonnegative ϱ_1-integrable function. There is a sequence of simple functions h_n such that $0 \leq h_n \uparrow h$. Since each Borel set in $B(X \times Y)$ (recall that $X \times Y$ is σ-compact) is the limit of an increasing sequence of bounded Borel sets, we may assume that each h_n has compact support. Hence using equations (7.39) and (7.40) we have

$$\int h\,d\varrho_1 = \lim_n \int h_n\,d\varrho_1 = \lim_n \int \left(\int h_{n_x}\,d\nu\right) d\mu < \infty. \tag{7.41}$$

Now for each $x \in X$, $(h_{n_x})_N$ is a nondecreasing sequence so that if we let

$$f_n(x) = \int h_{n_x}\, dv$$

then $(f_n(x))_N$ is a nondecreasing sequence of μ-integrable functions. For each $x \in X$, let $f(x) = \lim_n f_n(x)$. By the Monotone Convergence Theorem

$$\int f(x)\, d\mu = \lim_n \int f_n(x)\, d\mu = \lim_n \int \left[\int h_{n_x}\, dv \right] d\mu < \infty, \tag{7.42}$$

whence f is a μ-integrable function with

$$\int h_{n_x}\, dv = f_n(x) \uparrow f(x)$$

for each $x \in X$. Now for each $x \in X$, $h_{n_x} \uparrow h_x$, so that by the Monotone Convergence Theorem again

$$\int h_x\, dv = \lim_{n \to \infty} \int h_{n_x}\, dv = f(x). \tag{7.43}$$

Since f is μ-integrable there exists a set E with $\mu(E) = 0$ such that f is real-valued on $X - E$. Hence for $x \in X - E$, h_x is v-integrable by equation (7.43) and we can conclude thereby that the iterated integral $\iint h\, d\mu\, dv$ exists. Also combining equations (7.42) and (7.41) we have

$$\iint h\, d\mu\, dv = \int f(x)\, d\mu = \int h\, d\varrho_1.$$

The proof of statement (ii) is analogous. ∎

The next theorem is the analog of the Tonelli Theorem (Theorem 7.16) for the measurable space $(X \times Y, B(X \times Y))$.

Theorem 7.18. Suppose X and Y are σ-compact locally compact Hausdorff spaces.

(i) Suppose Γ_C is Borel measurable for each compact set C in $X \times Y$. If h is a nonnegative Borel function on $X \times Y$ [measurable with respect to $B(X \times Y)$] such that $\iint h\, dv\, d\mu$ exists, then h is ϱ_1-integrable.

(ii) Suppose Γ^C is Borel measurable for each compact set C in $X \times Y$, If h is a nonnegative Borel function on $X \times Y$ such that $\iint h\, d\mu\, dv$ exists, then h is ϱ_2-integrable. ∎

Proof. (i) Since $\int\int h\,d\nu\,d\mu$ exists, there exists a μ-integrable function f and a μ-null set E such that for $x \in X - E$, h_x is ν-integrable and $\int h_x\,d\nu = f(x)$.

There exists a sequence of simple functions h_n with compact support such that $0 \leq h_n \uparrow h$. As shown by equations (7.39) and (7.40), if $f_n(x)$ is defined to be $\int h_{n_x}\,d\nu$, then f_n is μ-integrable and

$$\int f_n\,d\mu = \int h_n\,d\varrho_1. \tag{7.44}$$

If $x \in X - E$, then as $h_{n_x} \leq h_x$ we have

$$f_n(x) = \int h_{n_x}\,d\nu \leq \int h_x\,d\nu = f(x).$$

Hence

$$\int f_n(x)\,d\mu \leq \int f(x)\,d\mu < \infty$$

or by equation (7.44)

$$\int h_n\,d\varrho_1 \leq \int f(x)\,d\mu.$$

By the Monotone Convergence Theorem

$$\int h\,d\varrho_1 = \lim_n \int h_n\,d\varrho_1 \leq \int f\,d\mu < \infty.$$

The proof of (ii) is similar. ∎

Using Theorem 7.17 and 7.18, we now have the following conditions for ϱ_1 to equal ϱ_2.

Theorem 7.19. Suppose X and Y are σ-compact locally compact Hausdorff spaces, and Γ_C and Γ^C are Borel measurable for each compact set in $X \times Y$. The following statements are equivalent:

(i) $\varrho_1 = \varrho_2$.

(ii) For any nonnegative Borel-measurable function h on $X \times Y$, whenever $\int\int h\,d\mu\,d\nu$ and $\int\int h\,d\nu\,d\mu$ exist, they are equal. ∎

Proof. (i) implies (ii) by using Theorems 7.18 and 7.17. Conversely, to show $\varrho_1 = \varrho_2$ when (ii) holds, it is sufficient to show $\varrho_1(C) = \varrho_2(C)$ for each compact set C in $X \times Y$. If C is compact in $X \times Y$, then χ_C is ϱ_1

and ϱ_2 integrable so that by Theorem 7.17, $\iint \chi_C \, d\mu \, d\nu$ and $\iint \chi_C \, d\nu \, d\mu$ exist and equal $\varrho_2(C)$ and $\varrho_1(C)$, respectively. By assumption, these iterated integrals are equal, so that $\varrho_1(C) = \varrho_2(C)$.

Combining Theorem 7.19 with Theorems 7.14, 7.17, and 7.18 we obtain a beautiful Fubini–Tonelli theorem for the measure $\mu \otimes \nu$ on $B(X \times Y)$.

Theorem 7.20. Assume μ and ν are regular Borel measures on the σ-compact locally compact Hausdorff spaces X and Y. Let $\mu \otimes \nu$ be the regular tensor product of μ and ν on $B(X \times Y)$. Let h be measurable with respect to $B(X \times Y)$.

(i) If h is $\mu \otimes \nu$-integrable, then both iterated integrals of h exist, and

$$\iint h \, d\nu \, d\mu = \iint h \, d\mu \, d\nu = \int h \, d(\mu \otimes \nu).$$

(ii) If h is nonnegative and one of the iterated integrals of h exists and is finite, then h is $\mu \otimes \nu$-integrable. ∎

Problems

7.5.1. Prove that the class of all sets $M \subset X \times Y$ such that $M_x \in B(Y)$ and $M^y \in B(X)$ is a σ-algebra containing all compact sets in $X \times Y$. Conclude that if $M \in B(X \times Y)$, then $M_x \in B(Y)$ and $M^y \in B(X)$.

7.5.2. In Theorem 7.12, prove that it is sufficient to show $\mu \otimes \nu(C \times D) = \mu \times \nu(C \times D)$ for compact sets C and D. [Hint: Show that $\mu \otimes \nu$ and $\mu \times \nu$ agree on the ring $\mathscr{R}$ generated by such rectangles. ($\mathscr{R}$ is the set of finite disjoint unions of rectangles of the form $(C_1 - C_2) \times (D_1 - D_2)$, where $C_2 \subset C_1$ and $D_2 \subset D_1$). Since the measures are σ-finite on $\mathscr{R}$, show they have a unique extension to the σ-ring generated by $\mathscr{R}$.]

7.5.3. Assume Γ_C is Borel measurable for each compact set C in $X \times Y$. Prove the following:

(i) Γ_M is measurable if $M = C - D$, where C and D are compact with $D \subset C$.

(ii) If M and N are disjoint Borel sets such that Γ_M and Γ_N are measurable, then $\Gamma_{M \cup N}$ is measurable.

7.5.4. Suppose C is a compact set in $X \times Y$ and (x_n) is a sequence in X converging to x in X. If $\Gamma_C(x_n) \geq a$ for all n, show that $\Gamma_C(x) \geq a$. [Hint: Let $E = \lim_n \sup C_{x_n}$. Show $\nu(E) \geq a$. Show $E \subset C_x$ and conclude $\nu(C_x) \geq a$.]

7.6. The Kakutani Fixed Point Theorem and the Haar Measure on a Compact Group

The theory of Haar measure is an important branch of measure theory and constitutes an extremely useful generalization of the theory of Lebesgue measure.

The Haar measure is a translation-invariant measure on a locally compact topological group [i.e., an algebraic group with locally compact topology where the mappings $(x, y) \to x \cdot y$ and $x \to x^{-1}$ are continuous]. The foundations of the theory of topological groups were laid around 1926–1927 by O. Schreier and F. Leja. To study the structure of certain topological groups, D. Hilbert in 1900 posed the following problem (now famous as Hilbert's fifth problem—fifth in the list of 23 problems he posed at the International Congress of Mathematics): Is every topological group that is locally Euclidean (i.e., every point has an open neighborhood homeomorphic to an open subset of R^n) necessarily a Lie group [i.e., a manifold that is a group and where the mappings $(x, y) \to xy$ and $x \to x^{-1}$ are analytic]? In 1933 A. Haar took a fundamental step towards the solution of this problem. He established the existence of a translation-invariant measure (now known as the Haar measure) on a second countable locally compact topological group. Soon after, in the same year, von Neumann utilized Haar's result and solved Hilbert's fifth problem in the affirmative for compact locally Euclidean groups. He also proved the uniqueness of the Haar measure in 1934, and later on in 1940 A. Weil extended Haar's result to all locally compact topological groups. Hilbert's fifth problem was solved completely (in 1952) in the affirmative by A. Gleason, D. Montgomery, and L. Zippin.

In this section we will present a fixed point theorem of S. Kakutani† and utilize it to prove the existence of the Haar measure on a compact topological group.

First we need a definition.

Definition 7.6. A family $\mathscr{F}$ of linear operators on a normed linear space X is called *equicontinuous* on a subset K of X if for every open set V with $0 \in V$ there is an open U with $0 \in U$ such that if $k_1, k_2 \in K$ and $k_1 - k_2 \in U$, then $\mathscr{F}(k_1 - k_2) \subset V$, i.e., $T(k_1 - k_2) \in V$ for every $T \in \mathscr{F}$. ▌

We will present here only a normed linear space version of the Ka-

† S. Kakutani, *Proc. Imp. Acad. Tokyo* **14**, 242–245 (1938).

kutani Theorem since this will suffice to serve our purpose (though more general versions can easily be obtained).

Theorem 7.21. *The Kakutani Fixed Point Theorem.* Let K be a nonempty compact convex subset of a normed linear space X, and let $\mathscr{F}$ be a group of linear operators on X. Suppose $\mathscr{F}$ is equicontinuous on K and $\mathscr{F}(K) \subset K$. Then there is $x \in X$ such that $T(x) = x$ for every $T \in \mathscr{F}$. ∎

Proof. By Zorn's Lemma there is a minimal nonempty compact convex set K_1 such that $\mathscr{F}(K_1) \subset K_1$. If K_1 is a singleton, there is nothing to prove. If K_1 is not a singleton, we will reach a contradiction to the minimality of K_1, and the theorem will follow.

Suppose there are $z_1 \neq z_2$, $z_1, z_2 \in K_1$. Let $y = z_1 - z_2$. Then there are positive r and r_1 such that

$$y \notin \bar{V} \text{ where } V \equiv \{x\colon \| x \| < r\} \tag{7.45}$$

and

$$\mathscr{F}(k_1 - k_2) \subset V \text{ if } k_1, k_2 \in K_1 \text{ and } k_1 - k_2 \in U \equiv \{x\colon \| x \| < r_1\}. \tag{7.46}$$

Let

$$U_0 = \bigcup_{n=1}^{\infty} \left\{ \sum_{i=1}^{n} a_i x_i\colon \sum_{i=1}^{n} a_i = 1, 0 \le a_i \le 1 \text{ and } x_i \in \mathscr{F}(U) \right\}.$$

Using the group property of $\mathscr{F}$, the reader can verify that

$$\mathscr{F}(U_0) = U_0.$$

Since $\mathscr{F}$ contains the identity operator, $U \subset U_0$ and $\inf\{t\colon t > 0 \text{ and } K_1 - K_1 \subset tU_0\} = t_0 < \infty$. Let $W = t_0 U_0$. Then it is clear that for each ε in $(0, 1)$,

$$K_1 - K_1 \not\subset (1 - \varepsilon)\overline{W} \tag{7.47}$$

and

$$K_1 - K_1 \subset (1 + \varepsilon)W. \tag{7.48}$$

Since $0 \in W$, $K_1 \subset \bigcup_{k \in K_1}(k + \frac{1}{2}W)$; and therefore by the compactness of K_1, there exist $k_i \in K_1$ with $1 \le i \le n$ such that

$$K_1 \subset \bigcup_{i=1}^{n} (k_i + \tfrac{1}{2}W). \tag{7.49}$$

Now we claim that

$$K_2 \equiv K_1 \cap \left\{ \bigcap_{k \in K_1} \left[k + \left(1 - \frac{1}{4n}\right) \overline{W} \right] \right\}$$

is a *nonempty* compact convex set satisfying

(i) $\mathscr{F}(K_2) \subset K_2$

and

(ii) $K_2 \neq K_1$.

This claim, once verified, will prove the theorem since K_2 will contradict the minimality of K_1. We prove the "nonempty" assertion last.

To prove (i), let $z \in K_2$ and $k \in K_1$. Since for $T \in \mathscr{F}$, $T^{-1}(K_1) \subset K_1$, then $k = T(k_0)$ for some $k_0 \in K_1$. Also $z \in k_0 + (1 - 1/4n)\overline{W}$ or $T(z) \in k + (1 - 1/4n) \cdot \overline{W}$, thus proving (i). Now to prove (ii), by (7.47) there are $x, y \in K_1$ such that $x - y \notin (1 - 1/4n) \cdot \overline{W}$. This means that $x \notin y + (1 - 1/4n)\overline{W}$ or $x \notin K_2$, thus proving (ii). To prove $K_2 \neq \varnothing$, we show that $p = (1/n)\sum_{i=1}^{n} k_i$ [k_i's as in (7.49)] belongs to K_2. Clearly $p \in K_1$, since K_1 is convex. Let $y \in K_1$. Then by (7.49)

$$y \in k_j + \tfrac{1}{2}W \tag{7.50}$$

for some j, where $1 \leq j \leq n$. Also, by (7.48), for each $i \neq j$ where $1 \leq i \leq n$,

$$y \in k_i + (1 + 1/4n)W. \tag{7.51}$$

By (7.50) and (7.51), we have

$$\begin{aligned} p &\in \frac{1}{n} \left\{ (y - \tfrac{1}{2}W) + (n-1)\left[y - \left(1 + \frac{1}{4n}\right) W \right] \right\} \\ &= y - \left(1 - \frac{1}{4n} - \frac{1}{4n^2}\right) W \end{aligned}$$

or $p \in y + (1 - 1/4n)W$ (since $W = -W$), proving that $p \in K_2$. The proof of the theorem is now complete. ∎

Before we go into the existence of the Haar measure, let us consider a few examples of topological groups. First a definition and a few basic remarks.

Definition 7.7. A *topological group* is a group G with a topology such that the mappings $(x, y) \to x \cdot y$ from $G \times G$ into G and $x \to x^{-1}$ from

G into G are continuous. Clearly, the continuity requirements in a topological group are equivalent to the requirement that $(x, y) \to x \cdot y^{-1}$ is continuous. ∎

Remarks

7.7. In a topological group the mappings $x \to z \cdot x$, $x \to x \cdot z$, and $x \to x^{-1}$ are all homeomorphisms.

7.8. Suppose a topological group G has the property (T_0): if $x \neq y$ and $x, y \in G$, then there is an open set V containing one of them but excluding the other. Note that $(T_1) \Rightarrow (T_0)$ (see Chapter 1, Section 1.4). But curiously enough, the following are equivalent in G:

(a) G is a T_0-space.
(b) G is a T_1-space.
(c) G is a T_2-space.
(d) $\cap \{U: U \text{ open and } e \in U\} = \{e\}$, where e is the identity of G.

Proof. (a) $\Rightarrow$ (b). Let $x \neq y$ and $x \in V$ open and $y \notin V$. Then $e \in x^{-1}V = W$ [open by (i)]. Now if $U = W \cap W^{-1}$, then $y \in yU$ (open). Now if $x \in yU$; then $x^{-1} \in Uy^{-1}$ (since $U = U^{-1}$), $x^{-1} \in x^{-1}Vy^{-1}$, or $y \in V$ (which is a contradiction). Hence $x \in yU$. Hence (b) is true.

(b) $\Rightarrow$ (c). Let $x \neq y$. Then by (b), $y \in V = G - \{x\}$ (open). Hence $e \in y^{-1}V$ (open). Since G is a topological group, there is open W with $e \in W$ and $WW^{-1} \subset y^{-1}V$. Let $U = G - \overline{yW}$. If $x \in \overline{yW}$, then $xW \cap yW \neq \varnothing$ and so $x \in yWW^{-1} \subset V$, which a contradiction. Hence $x \notin \overline{yW}$ and $x \in U$. Hence (c).

The proofs of (c) $\Rightarrow$ (d) $\Rightarrow$ (a) are easy and are left to the reader. ∎

7.9. A Hausdorff topological group is completely regular. A first countable topological group is metrizable. The proofs of these facts are somewhat involved and are omitted. The reader might consult Pontrjagin's text.[†]

7.10. Every real-valued continuous function f with *compact support* defined on a topological group G is *uniformly continuous* [i.e., given $\varepsilon > 0$, there exists an open set V with $e \in V$ such that $|f(x) - f(y)| < \varepsilon$ whenever $x^{-1}y \in V$].

Proof. Suppose K is the compact support of f. By using the compactness of K and the continuity of $(x, y) \to x \cdot y$, we can find for $\varepsilon > 0$ an open

[†] L. Pontrjagin, *Topological Groups*, Princeton University Press, Princeton, New Jersey (1939).

V_1 with $e \in V_1$ such that

$$|f(x) - f(x \cdot y)| < \varepsilon \tag{7.52}$$

whenever $x \in K$ and $y \in V_1$. Now for each $x \in A$, where

$$A \equiv \{y: |f(y)| \geq \varepsilon\} \subset \text{the interior of } K,$$

let V_x be an open set with $e \in V_x$ such that $x \cdot V_x \subset K$. Let $e \in W_x$ (open) such that $W_x \cdot W_x \subset V_x$. By the compactness of A there exist $x_1, x_2, \ldots, x_n \in A$ such that $A \subset \bigcup_{i=1}^n x_i W_{x_i}$ or $AW \subset K$, where $W = \bigcup_{i=1}^n W_{x_i}$. Therefore if $x \notin K$ and $y \in W^{-1}$, then

$$|f(x) - f(xy)| = |f(x \cdot y)| < \varepsilon. \tag{7.53}$$

From equations (7.52) and (7.53), taking $V = V_1 \cap W^{-1}$, we have for every $x \in G$ and $y \in V$,

$$|f(x) - f(xy)| < \varepsilon.$$

The proof of Remark (7.10) is now clear. ∎

7.11. Let G be a compact topological group and $C(G)$ be the real-valued continuous functions on G with "sup" norm. Let

$$\mathscr{F}_f = \left\{ \sum_{i=1}^n a_i f(s_i x): 0 \leq a_i \leq 1,\ \sum_{i=1}^n a_i = 1,\ s_i \in G, \text{ and } n \in N \right\}.$$

In other words, $\mathscr{F}_f$ is the convex hull of the set of all left translates of f [in $C(G)$]. Then $\overline{\mathscr{F}}_f$ is a compact subset of $C(G)$.

Proof. Suffice it to show that $\mathscr{F}_f$ is equicontinuous, since then the compactness of $\overline{\mathscr{F}}_f$ will follow by Arzela–Ascoli's Theorem (Chapter 1, Section 1.6). Since G is compact, by Remark 7.10, there is an open V with $e \in V$ and $x^{-1}y \in V$ such that $|f(sx) - f(sy)| < \varepsilon$ for every $s \in G$. The equicontinuity of $\mathscr{F}_f$ now follows. ∎

Examples

7.20. $R - \{0\}$ with the usual relative topology of R is a locally compact topological group under multiplication. R itself is also so under addition, and the Lebesgue measure is a translation-invariant measure on R.

7.21. The complex numbers of absolute value 1 is a compact topological group under multiplication and usual topology of R^2.

7.22. Let S be the set of all matrices of the form $\begin{pmatrix} x & y \\ 0 & 1 \end{pmatrix}$, where $x > 0$ and y is any real number. Then S is a locally compact topological group with usual matrix multiplication and with the topology induced by the usual topology of R^2 (in an obvious manner).

Theorem 7.22. There exists a unique regular probability measure μ [i.e., $\mu(G) = 1$] on every compact topological group G such that for every $s \in G$,

$$\text{(i)} \quad \int f_s \, d\mu = \int f \, d\mu,$$

$$\text{(ii)} \quad \int f^s \, d\mu = \int f \, d\mu,$$

and

$$\text{(iii)} \quad \int f_{-1} \, d\mu = \int f \, d\mu,$$

where $f \in C(G)$, $f_s(x) = f(sx)$, $f^s(x) = f(xs)$, and $f_{-1}(x) = f(x^{-1})$. This μ is called the *Haar measure* on G. ∎

Proof. If $L_s: f \to f_s$, then for $s \in G$, L_s is an isometry from $C(G)$ (with "sup" norm) into itself. Therefore, $\{L_s: s \in G\}$ is an *equicontinuous* group of linear operators on the Banach space $C(G)$. From Remark 7.11, $\bar{\mathscr{F}}_f$ is a compact convex subset of $C(G)$, and for $s \in G$, $L_s(\bar{\mathscr{F}}_f) \subset \bar{\mathscr{F}}_f$. By Theorem 7.21, there exists $f_0 \in \bar{\mathscr{F}}_f$ such that $f_0(xs) = f_0(x)$ for every x, $s \in G$; and therefore $f_0(x) = f_0(e)$ or f_0 is a constant c *which can be uniformly approximated by convex combinations of left translates of* f. If c' is a similar constant corresponding to the right translates of f, and c'' is a similar constant corresponding to the left translates of f_{-1}, we claim that $c = c' = c''$. To prove this, let $\varepsilon > 0$. Then there are $0 \leq a_i, b_j, c_k \leq 1$ with $\sum_{i=1}^n a_i = \sum_{j=1}^m b_j = \sum_{k=1}^p c_k = 1$ such that

$$\left| c - \sum_{i=1}^{n} a_i f(s_i x) \right| < \varepsilon, \tag{7.54}$$

$$\left| c' - \sum_{j=1}^{m} b_j f(x t_j) \right| < \varepsilon, \tag{7.55}$$

$$\left| c'' - \sum_{k=1}^{p} c_k f_{-1}(w_k x) \right| < \varepsilon \tag{7.56}$$

for all $x \in G$ and some (s_i), (t_j), and $(w_k) \in G$. In the inequality (7.54), writing t_j for x, multiplying by b_j, and then summing over j, we have

$$\left| c - \sum_{i\,j} a_i b_j f(s_i t_j) \right| < \varepsilon. \tag{7.57}$$

Similarly, inequalities (7.55) and (7.56) can be written as

$$\left| c' - \sum_{i,j} a_i b_j f(s_i t_j) \right| < \varepsilon \tag{7.58}$$

and

$$\left| c'' - \sum_{i,k} a_i c_k f(s_i w_k^{-1}) \right| < \varepsilon. \tag{7.59}$$

From inequalities (7.57) and (7.58), $c = c'$; also by writing (7.54) in a form similar to (7.59), we get $c = c''$. This argument also shows that there can be *only one* constant obtained as above corresponding to the translates of f or f_{-1}. If we call this constant $I(f)$, then we have

(A) $I(1) = 1$,
(B) $I(a \cdot f) = aI(f)$ for all reals a,
(C) $I(f) \geq 0$ for $f \geq 0$,
(D) $I(f_s) = I(f^s) = I(f_{-1})$, $s \in G$,

and

(E) $I(f + g) = I(f) + I(g)$.

We establish only (E). Let $\varepsilon > 0$. Then, as in (7.54), we get

$$\left| I(f) - \sum_{i=1}^{n} a_i f(s_i x) \right| < \varepsilon, \qquad x \in G, \tag{7.60}$$

where $\sum_{i=1}^{n} a_i = 1$, $0 \leq a_i$, and $s_i \in G$. Let $h(x) = \sum_{i=1}^{n} a_i g(s_i x)$. Then $h \in \mathscr{F}_g$ and $\mathscr{F}_h \subset \mathscr{F}_g$; therefore, since each of these sets contains a unique constant function, $I(g) = I(h)$. Hence there are $b_j > 0$, $\sum_{j=1}^{m} b_j = 1$, and $t_j \in G$ such that

$$\left| I(g) - \sum_{j=1}^{m} b_j h(t_j x) \right| < \varepsilon. \tag{7.61}$$

By replacing h with g, we have

$$\left| I(g) - \sum_{i,j} a_i b_j g(s_i t_j x) \right| < \varepsilon. \tag{7.62}$$

Writing (7.60) as

$$\left| I(f) - \sum_{i,j} a_i b_j f(s_i t_j x) \right| < \varepsilon, \tag{7.63}$$

we have by inequalities (7.62) and (7.63)

$$\left| I(f) + I(g) - \sum_{i,j} a_i b_j (f+g)(s_i t_j x) \right| < 2\varepsilon, \tag{7.64}$$

thus proving (E).

By the Riesz Representation Theorem (Theorem 7.10), there is a unique regular Borel measure μ satisfying $I(f) = \int f\, d\mu$ for all $f \in C(G)$. Since $I(1) = 1$, $\mu(G) = 1$. All the properties of μ in the theorem now follow from property (D) above. ∎

Problems

7.6.1. Suppose μ is the Haar measure in a compact group G. Show that
(i) $\mu(V) > 0$ for every open set $V(\neq \emptyset)$, and
(ii) $\mu(Bx) = \mu(xB) = \mu(B^{-1})$ for every Borel set B and $x \in G$.

7.6.2. Suppose μ is a weakly Borel measure (possibly infinite) on a Hausdorff topological group G such that
(i) $\mu(x \cdot K) = \mu(K)$ for all compact sets K and $x \in G$, and
(ii) $0 < \mu(V) < \infty$ for some open set V with compact closure.
Prove that (a) G is locally compact, and (b) G is compact if $\mu(G) < \infty$. [This means that it is impossible to have any meaningful translation-invariant measure on a non-locally-compact topological group.]

7.6.3. Let H be the component containing the identity of a topological group G. Prove that H is a subgroup of G such that $x^{-1}Hx = H$ for all $x \in G$.

7.6.4. Prove that every open subgroup of a topological group is closed.

7.6.5. Suppose G is a group with first countable topology such that
(i) $(x, y) \to x \cdot y$ is separately continuous, and
(ii) for any two compact sets A and B the set $AB^{-1} = \bigcup_{x \in B}\{y: yx \in A\}$ is compact.
Prove that G is a topological group.

7.6.6. Suppose that G is a group with a metric topology with property (i) of Problem 7.6.5 and the property $d(x, y) = d(xz, yz)$ for any x, y and $z \in G$. Then prove that G is a topological group. [Here it is relevant to mention a beautiful result of R. Ellis: Suppose G is a group with locally compact Hausdorff topology such that $(x, y) \to x \cdot y$ is separately continuous. Then G is a topological group. For a proof, see his paper.†]

† R. Ellis, *Duke Math. J.* **24**, 119–125 (1957).

7.6.7. Let G be a Hausdorff topological group and let μ be a weakly Borel measure such that

(i) $\mu(K \cdot x) = \mu(K)$ for every compact set K and $x \in G$,
(ii) $0 < \mu(V) < \infty$ for some open set V, and
(iii) $\mu(\{y\}) > 0$ for some $y \in G$.

Prove that G is discrete.

7.6.8. *Subsemigroups with Nonempty Interiors in a Compact Topological Group.* Let H be a subsemigroup of a compact topological group G such that H has a nonempty interior. Show that H is a compact subgroup of G. [Hint: Let S be the interior of H. Then $\bar{S}$ is a subsemigroup. If μ is the Haar measure of G, then for $x \in \bar{S}$ and any open set V, $\mu(x^{-1}V \cap \bar{S}) = \mu(V \cap \bar{S}) = \mu(Vx^{-1} \cap \bar{S})$. This means that $\bar{S} \cdot x = x \cdot \bar{S} = \bar{S}$, or $\bar{S}$ is a compact subgroup. Observe now that $S = \bar{S}$, since for $y \in \bar{S}$, $y \cdot S^{-1} \cap S \neq \emptyset$.]

7.6.9. Use Problem 7.6.8 to prove that every locally compact subsemigroup of positive Haar measure in a compact topological group is a compact subgroup. [Note: This result remains true without the requirement of "positive Haar measure." Actually, using an important result of K. Numakura (that a cancellative semigroup with compact topology and jointly continuous multiplication is a topological group), it is very easy to do Problems 7.6.8 and 7.6.9. The reader should try this.]

7.6.10. *A Fixed Point Theorem for Affine Maps.* Let T be a continuous map from a compact convex subset K of a normed linear space X into itself. Suppose that T is affine, i.e., for $0 \leq a \leq 1$ and x, $y \in K$, $T(ax + (1 - a)y) = aT(x) + (1 - a)T(y)$. Then T has a fixed point. [Hint: For $x \in K$ and $x_n = (1/n)\sum_{k=0}^{n-1} T^k(x)$, observe that $\| T(x_n) - x_n \| \to 0$ as $n \to \infty$.]

7.6.11. *The Markov–Kakutani Fixed Point Theorem.* Let $\mathscr{F}$ be a family of continuous maps from a compact convex subset K of a normed linear space X into itself such that for any T and S in this family, T is affine and $T(S(x)) = S(T(x))$ for all $x \in K$. Then $\mathscr{F}$ has a common fixed point. [Hint: For $T \in \mathscr{F}$, the set $K_T = \{x \in K\colon T(x) = x\}$ is nonempty by Problem 7.6.10. If $S \in \mathscr{F}$, then $S\colon K_T \to K_T$ and there exists $x \in K_T$ such that $S(x) = x$. An induction argument shows that every finite subfamily of $\mathscr{F}$ has a common fixed point. Now use the finite intersection property of the compact sets of common fixed points of the finite subfamilies of $\mathscr{F}$.] This theorem was first proven by A. Markov in 1936. In 1938, S. Kakutani gave an alternative proof of this theorem, along with an extension in the noncommutative case.

Apply the above fixed point theorem to show the existence of Banach limits on l_∞. For a discussion of Banach limits, see Problem 5.3.16 in Chapter 5.

7.6.12. *The Schauder Fixed Point Theorem.* It T is a continuous map from a compact convex subset of a Banach space into itself, then T has a fixed point.

The proof of this theorem is difficult and depends on the Brouwer theorem in Chapter 1. For a proof, the reader might consult [16]. Around 1930 J. Schauder first proved this theorem. In 1935, A. Tychonoff proved this theorem in the more general context of locally convex spaces.

This theorem is useful for various applications in differential and integral equations. Using this theorem, the existence part of the initial-value problem considered in Chapter 1, Theorem 1.24 can be proven easily assuming only the continuity of the function f and requiring no Lipschitz condition. Demonstrate this.

7.6.13. (*Johnson*). Let $Y = [0, \Omega]$ with the order topology and ν be the nonregular measure on Y as in Problem 7.3.13 or Example 7.11. Let A be a discrete group with two elements and $X = \times\{X_\lambda : \lambda \in [0, \Omega)\}$, where $X_\lambda \equiv A \; \forall \; \lambda$, with the product topology. Let μ be the Haar measure on the compact group X (with coordinatewise multiplication). Prove that $M = \{(x, y): x \in X, y \in Y \text{ and } x_\lambda = e \text{ if } \lambda \geq y\}$ is a compact set such that $\nu(M_x)$ is not measurable with respect to the completion of μ.

Appendix

A.1. Spectral Theory for Bounded Operators Revisited

In this section we utilize the knowledge of measure theory at our disposal to prove other versions of the Spectral Theorem for bounded self-adjoint operators—other than the resolution of the identity version given in Chapter 6. Although it is quite possible to generate a spectral measure from the resolution of the identity corresponding to a given self-adjoint operator, we prefer here to prove a more sophisticated spectral measure version of the Spectral Theorem making use of an elegant functional calculus version of the Spectral Theorem. Many of the results of this section will be utilized and duplicated in the next section, where we deal with unbounded operators.

Throughout this section, H denotes a complex Hilbert space.

Definition A.1. If X is a set and $\mathscr{R}$ is a ring of subsets of X, then a *positive-operator-valued measure* E is a function $E: \mathscr{R} \to L(H, H)$ such that

(i) $E(M) \geq 0$ for all M in $\mathscr{R}$, and
(ii) $E(\bigcup_{i=1}^{\infty} M_i) = \lim_n [\sum_{i=1}^{n} E(M_i)]$ (in the strong convergence sense) whenever (M_i) is a disjoint sequence of measurable sets whose union is also in $\mathscr{R}$.

If the values of E are projections, then E is called a *spectral measure*. If $X \in \mathscr{R}$ and $E(X) = I$, then spectral measure E is called *normalized*. ∎

Remark A.1. If E is a positive-operator-valued measure, then

(i) $E(\varnothing) = 0$;

(ii) E is finitely additive;

(iii) if $M \subset N$, then $E(M) \leq E(N)$;

(iv) $E(M \cup N) + E(M \cap N) = E(M) + E(N)$; and

(v) $E(M \cap N) = E(M)E(N)$ if and only if E is a spectral measure.

Theorem A.1. Let $E: \mathscr{R} \to L(H, H)$ be a function whose values are positive operators. Then E is a positive-operator-valued measure if and only if, for each h in H, the formula $\mu_h(M) = (E(M)h \mid h)$ defines a measure on $\mathscr{R}$. ∎

Proof. The necessity is clear. For the converse, let μ_h be a measure on $\mathscr{R}$ for each h and let (M_i) be a disjoint sequence of sets in $\mathscr{R}$ with $\bigcup_{i=1}^{\infty} M_i$ in $\mathscr{R}$. Now

$$\left(E\left(\bigcup_{i=1}^{\infty} M_i\right)h \mid h\right) = \mu_h\left(\bigcup_{i=1}^{\infty} M_i\right) = \lim_{n\to\infty} \mu_h\left(\bigcup_{i=1}^{n} M_i\right)$$
$$= \lim_{n\to\infty} \sum_{i=1}^{n} \mu_h(M_i) = \lim_{n\to\infty}\left(\sum_{i=1}^{n} E(M_i)h \mid h\right). \tag{A.1}$$

If

$$A_n = E\left(\bigcup_{i=1}^{\infty} M_i\right) - E\left(\bigcup_{i=1}^{n} M_i\right),$$

then $\| A_n \|$ is bounded† by $\| E(\bigcup_{i=1}^{\infty} M_i) \|$, and we have by Lemma 6.5 in Chapter 6 that

$$\| A_n h \|^4 = (A_n h \mid A_n h)^2 \leq (A_n h \mid h)(A_n{}^2 h \mid A_n h) \leq (A_n h \mid h) \| A_n \|^3 \| h \|^2,$$

so that by equation (A.1), $A_n h \to 0$. Hence $E(\bigcup_{i=1}^{\infty} M_i)$ is the strong limit of $\sum_{i=1}^{n} E(M_i)$ as $n \to \infty$. ∎

Proposition A.1. Let $\mathscr{R}$ be a ring of subsets of a set X, and suppose that for each vector h in H there is given a finite measure μ_h on $\mathscr{R}$. There exists a unique positive-operator-valued measure E on $\mathscr{R}$ such that $\mu_h(M) = (E(M)h \mid h)$ for all h in H and for all M in $\mathscr{R}$ if and only if for all vectors h and k and for each M in $\mathscr{R}$

(i) $[\mu_{h+k}(M)]^{1/2} \leq [\mu_h(M)]^{1/2} + [\mu_k(M)]^{1/2}$,

(ii) $\mu_{ch}(M) = | c |^2 \mu_h(M)$ for each scalar c,

† Since μ_h is a measure, A_n is a positive operator and $\| A_n \| = \sup_{\|h\|\leq 1}(A_n h \mid h)$.

(iii) $\mu_{h+k}(M) + \mu_{h-k}(M) = 2\mu_h(M) + 2\mu_k(M)$, and

(iv) for each M in $\mathscr{R}$ there exists a constant k_M such that

$$\mu_h(M) \leq k_M \| h \|^2. \qquad \blacksquare$$

Proof. Let M be arbitrary in $\mathscr{R}$. Define the real-valued function $\| \ \|_M$ on H by

$$\| h \|_M = [\mu_h(M)]^{1/2}.$$

As in the proof of Proposition 6.2 the conditions (i)–(iii) are equivalent to saying that $\| \ \|_M$ is a pseudonorm ($\| h \|_M = 0$ does not necessarily mean that $h = 0$) satisfying the Parallelogram Law, which in turn is equivalent to saying that

$$B_M(h, k) = \tfrac{1}{4}(\| h + k \|_M{}^2 - \| h - k \|_M{}^2 + i \| h + ik \|_M{}^2 - i \| h - ik \|_M{}^2)$$

defines a Hermitian sesquilinear form on H such that $B_M(h, h) = \| h \|_M{}^2$. The boundedness of B_M is equivalent to (iv). Hence by Theorem 6.6, conditions (i)–(iv) are equivalent to the existence for each M of a bounded self-adjoint operator $E(M)$ with $(E(M)h \mid h) = B_M(h, h) = \mu_h(M)$. By Theorem A.1, $E(M)$ defines a positive-operator-valued measure. $\blacksquare$

The next result is easily verified.

Proposition A.2. If E is a positive-operator-valued measure whose domain is a σ-ring, then

$$\sup\{\| E(M) \| : M \in \mathscr{R}\} < \infty. \qquad \blacksquare$$

From this point on through Theorem A.2, E will denote a positive-operator-valued measure on a σ-ring $\mathscr{R}$. If μ_h is the measure

$$\mu_h(M) = (E(M)h \mid h)$$

then by Proposition A.2 there exists a positive number K such that $\mu_h(M) \leq K \| h \|^2$ for all M and for all h.

If $f = g + ih$ is a complex-valued function on X, then f is *measurable* in case g and h are measurable. If μ is a measure on $\mathscr{R}$ and g and h are μ-integrable, then we say f is *μ-integrable* and define (as in Problem 3.2.24)

$$\int f \, d\mu \equiv \int g \, d\mu + i \int h \, d\mu.$$

f is μ-integrable if and only if $|f|$ is μ-integrable, and in this case

$$\left| \int f \, d\mu \right| \leq \int |f| \, d\mu.$$

Definition A.2. Suppose $\mathscr{F} \equiv \{\mu_h\}_{h\in H}$ is a family of measures on a σ-ring $\mathscr{R}$. [In particular $\mathscr{F}$ could be *generated by* E in the sense that $\mu_h(M) = (E(M)h \mid h)$ for each h and each M.] A measurable complex function on $(X, \mathscr{R})$ is *$\mathscr{F}$-integrable* (in particular *E-integrable* if $\mathscr{F}$ is generated by E) if it is μ_h-integrable for each h. If f is $\mathscr{F}$-integrable then for any ordered pair of vectors in H, we write

$$\int f\,d\mu_{h,k} = \tfrac{1}{4}\Big(\int f\,d\mu_{h+k} - \int f\,d\mu_{h-k} + i\int f\,d\mu_{h+ik} - i\int f\,d\mu_{h-ik}\Big). \blacksquare \quad \text{(A.2)}$$

Lemma A.1. For each pair of vectors h and k in H, the complex-valued function $L_{h,k}$ given by

$$L_{h,k}(f) = \int f\,d\mu_{h,k}$$

is a linear functional on the complex vector space of $\mathscr{F}$-integrable functions. Moreover if $\mathscr{F}$ is generated by E, then

$$\int \chi_M\,d\mu_{h,k} = (E(M)h \mid k) \quad \text{(A.3)}$$

for all M in $\mathscr{R}$. ∎

Proof. The fact that $L_{h,k}$ is linear follows from equation (A.2). To prove the latter part of the lemma, note that if $h \in H$, then

$$\int \chi_M\,d\mu_h = \mu_h(M) = (E(M)h \mid h).$$

Hence from equation (A.2) and the Polarization Identity (Proposition 6.1 in Chapter 6),

$$\begin{aligned}\int \chi_M\,d\mu_{h,k} &= \tfrac{1}{4}\big[(E(M)(h+k) \mid h+k) - (E(M)(h-k) \mid h-k) \\ &\quad + i(E(M)(h+ik) \mid h+ik) - i(E(M)(h-ik) \mid h-ik)\big] \\ &= (E(M)h \mid k).\end{aligned}$$ ∎

Lemma A.2. Let $\mathscr{F} = \{\mu_h\}_{h\in H}$ be any family of measures on a σ-ring $\mathscr{R}$. Let L be any linear functional on the vector space of $\mathscr{F}$-integrable functions. Then $L = L_{h,k}$ for some h and k if and only if

(i) whenever (f_n) is a sequence of $\mathscr{F}$-integrable functions such that $0 \le f_n \uparrow f$, where f is also $\mathscr{F}$-integrable, then $L(f_n) \to L(f)$, and

(ii) $L(\chi_M) = L_{h,k}(\chi_M)$ for all M in $\mathscr{R}$. ∎

Proof. To prove the necessity, if $0 \leq f_n \uparrow f$ as in (i), then

$$\int f_n d\mu_h \to \int f \, d\mu_h$$

for each h by the Monotone Convergence Theorem. Hence

$$\int f_n \, d\mu_{h,k} \to \int f \, d\mu_{h,k}$$

by equation (A.2).

To prove the sufficiency, by (ii), $L = L_{h,k}$ for all simple functions. If $f \geq 0$, choose a sequence of simple functions (f_n) such that $0 \leq f_n \uparrow f$. By (i)

$$Lf = \lim_{n \to \infty} L(f_n) = \lim_{n \to \infty} L_{h,k}(f_n) = L_{h,k}(f).$$

By linearity $Lf = L_{h,k}(f)$ for all $\mathscr{F}$-integrable functions f. ∎

Lemma A.3. If for each vector h and for each M in $\mathscr{R}$,

$$\mu_h(M) = \tfrac{1}{4}[\mu_{h+h}(M) - \mu_{h-h}(M) + i\mu_{h+ih}(M) - i\mu_{h-ih}(M)],$$

then

$$\int f \, d\mu_{h,h} = \int f \, d\mu_h \tag{A.4}$$

for each $\mathscr{F}$-integrable function f. In particular, if $\mu_h(M) = (E(M)h \mid h)$ for each M and h, then equation (A.4) is true for each E-integrable function f. ∎

Proof. Define Lf to be $\int f \, d\mu_h$ for each f. L is linear and satisfies (i) of Lemma A.2 by the Monotone Convergence Theorem. (ii) is satisfied since

$$L(\chi_M) = \mu_h(M) = \int \chi_M \mu_{h,h} = L_{h,h}(\chi_M).$$

Hence $L = L_{h,h}$. ∎

Proposition A.3. Suppose $\mathscr{F} = (\mu_h)_{h \in H}$ is a family of measures on a σ-ring $\mathscr{R}$ such that for each M in $\mathscr{R}$,

$$L_{\alpha h,k}(\chi_M) = \alpha L_{h,k}(\chi_M), \tag{A.5}$$

$$L_{h_1+h_2,k}(\chi_M) = L_{h_1,k}(\chi_M) + L_{h_2,k}(\chi_M), \tag{A.6}$$

$$L_{h,\alpha k}(\chi_M) = \bar{\alpha} L_{h,k}(\chi_M), \tag{A.7}$$

$$L_{h,k_1+k_2}(\chi_M) = L_{h,k_1}(\chi_M) + L_{h,k_2}(\chi_M) \tag{A.8}$$

for each scalar α and for all vectors in H. Then for each $\mathscr{F}$-integrable function f, the mapping

$$(h, k) \to \int f \, d\mu_{h,k} \equiv L_{h,k}(f)$$

is a sesquilinear form on H. Moreover if for each M in $\mathscr{R}$

$$L_{h,k}(\chi_M) = \overline{L_{k,h}(\chi_M)}, \tag{A.9}$$

then $L_{h,k}(f) = \overline{L_{k,h}(\bar{f})}$ for each $\mathscr{F}$-integrable function f. In addition, if equation (A.4) is true for some bounded measurable function f, and if $\mathscr{F}$ is a family of measures such that $\mu_h(X) \leq K \| h \|^2$ for some constant K independent of h, then the form above is also bounded. ∎

Proof. By use of Lemma A.1, statements (A.5)–(A.9) are readily verified to hold for each $\mathscr{F}$-integrable function f. For instance, to verify equation (A.9) define L on the class on $\mathscr{F}$-integrable functions by $L(f) = \overline{L_{k,h}(\bar{f})}$. Then L is linear and satisfies the conditions (i) and (ii) of Lemma A.1 for the pair (h, k) so that $L = L_{h,k}$. The last statement of the proposition follows from the inequality

$$\left| \int f \, d\mu_{h,h} \right| = \left| \int f \, d\mu_h \right| \leq \int |f| \, d\mu_h \leq \| f \|_\infty K \| h \|^2. \quad ∎$$

The last four lemmas have been preparatory to proving the next result. It will be most useful in the ensuing work of this section.

Theorem A.2. Let E be a positive-operator-valued measure on $(X, \mathscr{R})$. For each bounded measurable complex-valued function f on X, there exists a unique bounded operator T_f on H such that

$$(T_f h \mid h) = \int f \, d\mu_h \qquad \text{for all } h \in H. \tag{A.10}$$

Moreover,

$$(T_f h \mid k) = \int f \, d\mu_{h,k} \qquad \text{for all } h \text{ and } k \text{ in } H.$$

[Here μ_h is the measure $\mu_h(M) = (E(M)h \mid h)$.] ∎

Proof. Since for each h, $\mu_h(M) = (E(M)h \mid h)$, each of the equations (A.5), (A.6), (A.7), and (A.8) is true. This can be verified by equation

(A.3). Moreover, by Lemma A.2, equation (A.4) is true. Hence corresponding to the bounded sesquilinear form

$$(h, k) \to \int f \, d\mu_{h,k}$$

is a unique bounded operator T_f such that

$$(T_f h \mid k) = \int f \, d\mu_{h,k} \qquad \text{for all } h, k \text{ in } H.$$

The following identity shows that equation (A.10) is sufficient for uniqueness. If B is a bounded operator also satisfying equation (A.10), then for any h and k in H,

$$\begin{aligned} 4\int f \, d\mu_{h,k} &= \int f \, d\mu_{h+k} - \int f \, d\mu_{h-k} + i\int f \, d\mu_{h+ik} - i\int f \, d\mu_{h-ik} \\ &= (B(h+k) \mid h+k) - (B(h-k) \mid h-k) \\ &\quad + i(B(h+ik) \mid h+ik) - i(B(h-ik) \mid h-ik) \\ &= 4(Bh \mid k). \end{aligned}$$

∎

The operator corresponding to f in Theorem A.2 is denoted by $\int f \, dE$. Thus for all h and k in H we have

$$\left(\left(\int f \, dE\right) h \mid k\right) = \int f \, d\mu_{h,k}.$$

Remark A.2. Of interest are the following properties of the mapping $f \to \int f \, dE$.

(i) $\int (f+g) \, dE = \int f \, dE + \int g \, dE$.

(ii) $\int \alpha f \, dE = \alpha \int f \, dE$ for all scalars α.

(iii) $\int \bar{f} \, dE = (\int f \, dE)^*$.

(iv) $\int \chi_M \, dE = E(M)$ for each M in $\mathscr{R}$.

(v) $\int fg \, dE = \int f \, dE \int g \, dE$ if E is a spectral measure.

(vi) There exists a constant K such that $\| \int f \, dE \| \leq K \| f \|_\infty$ for all f.

Proof. Let us verify (iii), (vi), and (v), respectively. If we define L by $Lf = \overline{\int \bar{f} \, d\mu_{h,k}}$ for each E-integrable function, then the equation

$$L(\chi_M) = \overline{\int \chi_M \, d\mu_{h,k}} = (\overline{E(M)h \mid k}) = (E(M)k \mid h)$$

and Lemma A.2 show that $L = L_{h,k}$. Hence we have

$$\left(\left(\int f\,dE\right)^* h \mid k\right) = \overline{\left(\left(\int f\,dE\right)k \mid h\right)} = \overline{\int f\,d\mu_{k,h}} = \int \bar{f}\,d\mu_{h,k} = \left(\left(\int \bar{f}\,dE\right)h \mid k\right),$$

this means $(\int f\,dE)^* = \int \bar{f}\,dE$.

If f is a real-valued function,

$$\left(\left(\int f\,dE\right)h \mid h\right) = \left|\int f\,d\mu_h\right| \leq \| f \| \, \mu_h(X).$$

If f is a complex function $k + ig$, then

$$\left\|\int f\,dE\right\| = \left\|\int (k + ig)\,dE\right\| = \left\|\int k\,dE + i\int g\,dE\right\| \leq 2\,\| f \| \, \mu_h(X).$$

This equation verifies (vi).

Finally since $E(M \cap N) = E(M)E(N)$ when E is a spectral measure, (v) is valid if $f = \chi_M$ and $g = \chi_N$. If f and g are simple, (v) follows by linearity. If f and g are arbitrary bounded measurable functions, choose simple functions (f_n) and (g_n) such that $\| f_n - f \|_\infty \to 0$ and $\| g - g_n \|_\infty \to 0$. Then $\| f_n g_n - fg \|_\infty \to 0$ so that

$$\begin{aligned}\left\|\int fg\,dE - \int f\,dE\int g\,dE\right\| &\leq \left\|\int fg\,dE - \int f_n g_n\,dE\right\| \\ &\quad + \left\|\int f_n g_n\,dE - \int f_n g\,dE\right\| \\ &\quad + \left\|\int f_n g\,dE - \int fg\,dE\right\| \\ &\leq 2K\{\| fg - f_n g_n \|_\infty + \| f_n g_n - f_n g \|_\infty \\ &\quad + \| f_n g - fg \|_\infty\} \to 0.\end{aligned}$$

∎

Starting with a positive-operator-valued measure E on a σ-ring $\mathscr{R}$, we have constructed for each h in Hilbert space H a finite measure μ_h on $\mathscr{R}$ and then corresponded to E a mapping $f \to T_f$ on the class of all bounded complex-valued measurable functions on X such that equation (A.10) holds. Our next goal is to start with a fixed bounded self-adjoint operator T on H and associate with it a similar mapping from bounded measurable functions to operators in $L(H, H)$.

Associated with a spectral measure E defined on a measurable space $(X, \mathscr{R})$, where X is also a topological space, is the *spectrum* $\Sigma(E)$ of E.

It is defined as the complement in X of the union of all those open sets M in $\mathscr{R}$ for which $E(M) = 0$. Obviously $\Sigma(E)$ is a closed set in X. If $\Sigma(E)$ is compact, then E is called a *compact spectral measure*.

We need the information of the following lemma and proposition. As before, $B(R)$ denotes the σ-algebra of Borel sets on R.

Lemma A.4. If E is a spectral measure on $B(R)$, then

(i) for any M in $B(R)$, $E(M)$ is the "smallest" projection, denoted $\vee E(K)$, "greater than" all $E(K)$, where K is a compact subset of M (see page 355) [in symbols $E(M) = \vee E(K)$, where K is compact in M]; and

(ii) $E(R - \Sigma(E)) = 0$. ▮

Proof. (i) Since E is a spectral measure, $E(M) \geq E(K)$ for all compact subsets K of M; whereby $E(M) \geq \vee E(K)$. If $E(M) \neq \vee E(K)$, there exists a nonzero vector h in the range of $E(M)$ that is orthogonal to the range of each $E(K)$. Letting μ_h be the measure $\mu_h(M) = (E(M)h \mid h)$, which is regular by Theorem 7.2 in Chapter 7, we have $\mu_h(M) = \sup\{\mu_h(K): K \subset M,\ K \text{ compact}\}$. Since $\mu_h(K) = 0$ for each K, $\mu_h(M) = 0$ or $(E(M)h \mid h) = 0$. This means $h = E(M)h = 0$.

(ii) In view of (i), it suffices to show $E(K) = 0$ for K compact in $R - \Sigma(E)$. Since K can be covered by a finite collection of open sets, each of which has zero spectral measure, $E(K) = 0$. ▮

If E is a compact spectral measure on $B(R)$, let $f(\lambda) = \lambda \cdot \chi_{\Sigma(E)}$. Let $T = \int f\,dE$. The next proposition shows how $\sigma(T)$ is related to $\Sigma(E)$.

Proposition A.4. If E is a compact normalized spectral measure on $B(R)$ and $T = \int \lambda \cdot \chi_{\Sigma(E)}(\lambda)\,dE(\lambda)$, then $\Sigma(E) = \sigma(T)$. ▮

Proof. Since E is normalized, $\Sigma(E) \neq \varnothing$. If $\lambda_0 \in \Sigma(E)$, then $E(M) \neq 0$ for every open set M containing λ_0. Suppose $\lambda_0 \in \Sigma(E)$ but $\lambda_0 \notin \sigma(T)$. If $M = \{\lambda: |\lambda - \lambda_0| < 1/(2\,\|(T - \lambda_0)^{-1}\|)\}$, then there is a unit vector h in the range of $E(M)$ as $E(M) \neq 0$. Thus

$$\|Th - \lambda_0 h\|^2 = ((T - \lambda_0)^*(T - \lambda_0)h \mid h) = \int \overline{(\lambda - \lambda_0)}(\lambda - \lambda_0)\chi_{\Sigma(E)}(\lambda)\,d\mu_h(\lambda)$$

$$= \int |\lambda - \lambda_0|^2\,d\mu_h(\lambda) \leq \left(\frac{1}{2\,\|(T - \lambda_0)^{-1}\|}\right)^2.$$

However,

$$\| h \| = \| (T - \lambda_0)^{-1}(T - \lambda_0)h \| \leq \| (T - \lambda_0)^{-1} \| \frac{1}{2 \| (T - \lambda_0)^{-1} \|} = \frac{1}{2}.$$

This contradiction shows $\Sigma(E) \subset \sigma(T)$.

Conversely, if $\lambda_0 \notin \Sigma(E)$, we show $\lambda_0 \notin \sigma(T)$. If $\lambda_0 \notin \Sigma(E)$, there is an open set M containing λ_0 such that $E(M) = 0$. If $\delta = \inf\{| \lambda - \lambda_0 |: \lambda \in M^c\}$, then for every h we have as before that

$$\| Th - \lambda_0 h \|^2 = \int | \lambda - \lambda_0 |^2 \, d\mu_h(\lambda) = \int_{M^c} | \lambda - \lambda_0 |^2 \, d\mu_h(\lambda) \geq \delta^2 \| h \|^2.$$

Hence from Proposition 5.7, $(T - \lambda_0)^{-1}$ exists and $\lambda_0 \notin \sigma(T)$. ∎

Starting with a compact normalized spectral measure E on $B(R)$, we have obtained a bounded self-adjoint operator $T = \int \lambda \cdot \chi_{\Sigma(E)}(\lambda) \, dE(\lambda)$ on H. Moreover $\sigma(T) = \Sigma(E)$. We now wish to reverse the process in starting with a given bounded self-adjoint operator and obtaining a spectral (compact, normalized) measure on $B(R)$. By means of the next theorem—sometimes called the functional calculus form of the Spectral Theorem—we will have a neat way of obtaining this spectral measure. It is also a beautiful extension of the Continuous Functional Calculus Theorem (Theorem 6.10) to bounded measurable functions.

Let T be a bounded self-adjoint operator and let $\sigma(T)$ be its spectrum—a compact subset of the real axis. If $f \in C_1(\sigma(T))$ and f is real-valued, let $f(T)$ be the bounded linear operator in $L(H, H)$, defined as the image of f under φ in Theorem 6.10. It is easy to verify that if $h \in H$, the function on the real space of real-valued functions in $C(\sigma(T))$ given by $L_h(f) = (f(T)h \mid h)$ is a positive linear functional. Hence by Theorem 7.7 there is a unique Borel measure μ_h on $B(\sigma(T))$ such that

$$L_h(f) = (f(T)h \mid h) = \int f \, d\mu_h \tag{A.11}$$

for all $f \in C_1(\sigma(T))$ that are real-valued.

Let $\mathscr{F}$ now be the particular family of measures $\{\mu_h\}_{h \in H}$ satisfying equation (A.11). In particular if f is the function $\chi_{\sigma(T)}$ in $C_1(\sigma(T))$, then equation (A.11) gives for each μ_h

$$\mu_h(\sigma(T)) = \int \chi_{\sigma(T)} \, d\mu_h = (h \mid h) = \| h \|^2.$$

Each bounded measurable function on $\sigma(T)$ is thereby integrable for each μ_h, and for each such function g we can define $\int g \, d\mu_{h,k}$ as in equation

(A.2). Since for each continuous function f in $C_1(\sigma(T))$,

$$\begin{aligned}\int f\,d\mu_{h,h} &= \tfrac{1}{4}\Big[\int f\,d\mu_{h+h} - \int f\,d\mu_{h-h} + i\int f\,d\mu_{h+ih} - i\int f\,d\mu_{h-ih}\Big]\\ &= \tfrac{1}{4}[(f(T)(h+h)\mid h+h) - (f(T)(h-h)\mid h-h)\\ &\quad + i(f(T)(h+ih)\mid h+ih) - i(f(T)(h-ih)\mid h-ih)]\\ &= (f(T)h\mid h)\\ &= \int f\,d\mu_h;\end{aligned}$$

equation (A.4) is satisfied for continuous functions. Similarly a simple calculation shows that for each $f \in C_1(\sigma(T))$, equations (A.5)–(A.8) are valid (with f replacing χ_M). Since each bounded measurable function on $\sigma(T)$ can be approximated in L_1 (see Problem 7.3.12) by a continuous function on $\sigma(T)$, the equations (A.4)–(A.8) are seen to hold for all bounded measurable functions. This means (see Lemma A.3 and Proposition A.3) that for each bounded measurable function g on $\sigma(T)$ the mapping

$$(h, k) \to \int g d\mu_{h,k}$$

is a bounded sesquilinear form on H so that by Theorem 6.6, there exists a unique bounded operator, which we denote by $g(T)$, on H such that

$$(g(T)h \mid k) = \int g\,d\mu_{h,k} \qquad \text{for all } h,\ k \text{ in } H. \tag{A.12}$$

In particular for any h in H,

$$(g(T)h \mid h) = \int g\,d\mu_{h,h} = \int g\,d\mu_h, \tag{A.13}$$

and it is this equation for each h which determines $g(T)$ uniquely. This is shown by using the same equation employed in the proof of Theorem A.2.

If f is in $C_1(\sigma(T))$, $f(T)$ is $\varphi(f)$, where φ is the function of Theorem 6.10. By the properties of φ, $\varphi(\bar{f}) = (\varphi(f))^*$ or in other notation $\bar{f}(T) = (f(T))^*$. This relation is also true for any bounded measurable function g on $\sigma(T)$. Indeed for $f \in C_1(\sigma(T))$,

$$\overline{\int \bar{f}\,d\mu_{h,k}} = (\overline{f(T)^*h \mid k}) = (f(T)k \mid h) = \int f\,d\mu_{k,h}; \tag{A.14}$$

and since the class $C_1(\sigma(T))$ is dense in L_1, equation (A.14) is also valid for g. This means

$$(k \mid \bar{g}(T)h) = \overline{(\bar{g}(T)h \mid k)} = (g(T)k \mid h)$$

so that $(g(T))^* = \bar{g}(T)$. In particular if g is real-valued, $g(T)$ is self-adjoint.

To prove our next theorem, we have need of the following lemmas.

Lemma A.5. If g is a bounded nonnegative measurable function on $\sigma(T)$ and (g_n) is a sequence of measurable functions with $0 \leq g_n \leq g$ and $\int | g - g_n | \, d\mu_h \to 0$, then $g(T)h = \lim_{n\to\infty} g_n(T)h$. ∎

Proof. From equation (A.13) and the above hypothesis, $\lim(g_n(T)h \mid h) = (g(T)h \mid h)$. Also

$$\| g(T) - g_n(T) \| \leq \| g(T) \| + \| g_n(T) \| \leq 2 \| g(T) \|.^\dagger$$

Hence by Lemma 6.5 of Chapter 6,

$$\begin{aligned} \| [g(T) - g_n(T)]h \|^4 &= ([g(T) - g_n(T)]h \mid [g(T) - g_n(T)]h)^2 \\ &\leq ([g(T) - g_n(T)]h \mid h)([g(T) - g_n(T)]^2 h \mid [g(T) - g_n(T)]h) \\ &\leq ([g(T) - g_n(T)]h \mid h) \| g(T) - g_n(T) \|^3 \| h \|^2 \to 0. \quad ∎ \end{aligned} \tag{A.15}$$

Lemma A.6. If g is a nonnegative bounded measurable function and (f_n) is a sequence of continuous nonnegative functions such that $\| f_n \|_\infty \leq \| g \|_\infty$ and $\int | f_n - g | \, d\mu_h \to 0$ (such a sequence always exists), then $g(T)h = \lim f_n(T)h$. ∎

Proof. Let $g_n = f_n \wedge g$. [$g_n(x) = \min\{f_n(x), g(x)\}$.] Since the g_n satisfy the conditions of Lemma A.5, $g(T)h$ is the limit of $g_n(T)h$.

Also the sequence $(f_n - g_n)$ is a nonnegative sequence such that $\int (f_n - g_n) \, d\mu_h \to 0$. Moreover, the sequence $\| f_n(T) - g_n(T) \|$ is bounded since

$$\| f_n(T) - g_n(T) \| \leq \| f_n(T) \| + \| g_n(T) \| \leq 2 \| f_n(T) \|,$$

† $\| g_n(T) \| = \sup_{\|h\| \leq 1} (g_n(T)h \mid h) \leq \sup_{\|h\| \leq 1} (g(T)h \mid h) = \| g(T) \|$.

and $(\| f_n(T) \|)$ is a bounded sequence. [Since $f_n(T)$ is self-adjoint,

$$\begin{aligned}\| f_n(T) \| &= \sup_{\|h\|\leq 1} (f_n(T)h \mid h)\\ &= \sup_{\|h\|\leq 1} \int f_n \, d\mu_h \leq \sup_{\|h\|\leq 1} \int \| f_n \|_\infty \, d\mu_h \leq \sup_{\|h\|\leq 1} \int \| g \|_\infty \, d\mu_h\\ &= \| g \|_\infty \sup_{\|h\|\leq 1} \int d\mu_h = \| g \|_\infty .]\end{aligned}$$

By an inequality similar to (A.15) again, $[f_n(T) - g_n(T)]h$ converges to 0. We can conclude that $f_n(T)h$ converges to $g(T)h$. ∎

We have collected sufficient information to prove the following theorem.

Theorem A.3. *Functional Calculus Form of the Spectral Theorem.* Let T be a bounded self-adjoint operator on H. There is a unique mapping $\hat{\varphi}$ on the class of bounded measurable functions on $B(\sigma(T))$ in $L(H, H)$ such that

(i) $\hat{\varphi}(\alpha f + \beta g) = \alpha\hat{\varphi}(f) + \beta\hat{\varphi}(g)$,
$\hat{\varphi}(fg) = \hat{\varphi}(f)\hat{\varphi}(g)$,
$\hat{\varphi}(\bar{f}) = (\hat{\varphi}(f))^*$,
$\hat{\varphi}(1) = I$.

(ii) $\| \hat{\varphi}(f) \| \leq \| f \|_\infty$.

(iii) If $f(\lambda) = \lambda$, then $\hat{\varphi}(f) = T$.

(iv) If $f_n(\lambda) \to f(\lambda)$ for each λ and $(\| f_n \|_\infty)$ is bounded, then $\hat{\varphi}(f_n) \to \hat{\varphi}(f)$ strongly.

In addition $\hat{\varphi}$ satisfies

(v) If $Th = \lambda h$, then $\hat{\varphi}(f)h = f(\lambda)h$.

(vi) If $f \geq 0$, then $\hat{\varphi}(f) \geq 0$.

(vii) If $AT = TA$, then $A\hat{\varphi}(f) = \hat{\varphi}(f)A$. ∎

Proof. The mapping $\hat{\varphi}$ is given by $\hat{\varphi}(g) = g(T)$, where $g(T)$ is the unique bounded operator satisfying equation (A.12). $\hat{\varphi}$ is thus an extension of the mapping φ of Theorem 6.10 defined on $C_1(\sigma(T))$ and thereby satisfies (i), (ii), (iii), (v), and (vi) for continuous functions. The linearity of $\hat{\varphi}$ is readily verified using equation (A.13); the fact that $\hat{\varphi}(\bar{f}) = (\hat{\varphi}(f))^*$ and statement (vi) have been verified above.

To prove $\hat{\varphi}(fg) = \hat{\varphi}(f)\hat{\varphi}(g)$, we assume f and g are nonnegative and use linearity to treat the general case. Let (f_n) and (g_n) be sequences of continuous functions with $\| f_n \|_\infty \leq \| f \|_\infty$, $\| g_n \|_\infty \leq \| g \|_\infty$, $g_n \to g$ in

$L_1(\mu_h)$, and $f_n \to f$ in $L_1(\mu_{g(T)h} + \mu_h)$ for some h in H. By Lemma A.6,

$$f(T)[g(T)h] = \lim f_n(T)[g(T)h]$$

and

$$g(T)h = \lim g_n(T)h.$$

Hence for any positive integer m,

$$(f_m g)(T)h = \lim_{n\to\infty} (f_m g_n)(T)h = \lim_{n\to\infty} [f_m(T)g_n(T)]h = f_m(T)g(T)h,$$

whereby

$$(fg)(T)h = \lim_{n\to\infty} (f_n g)(T)h = \lim_{n\to\infty} [f_n(T)g(T)]h = [f(T)g(T)]h.$$

The proofs of the other statements—(ii), (v), and (vii)—are similar.

The uniqueness of $\hat{\varphi}$ and statement (iv) remain to be verified. Clearly, if (f_n) converges pointwise to f and $(\| f_n \|_\infty)$ is bounded, then by the Dominated Convergence Theorem (Theorem 3.3),

$$\int f_n \, d\mu_h \to f \, d\mu_h$$

for each h. Hence for each h, $f_n(T)h \to f(T)h$ as in equation (A.15), which means $f_n(T)$ converges strongly to $f(T)$.

Suppose ψ is any mapping satisfying (i)–(iv). By the uniqueness of φ of Theorem 6.10, $\hat{\varphi}$ and ψ agree on $C_1(\sigma(T))$. By linearity, to show ψ and φ are equal, it suffices to show they agree for real measurable functions. Let h be arbitrary but fixed in H and let VL denote the real vector lattice of real, bounded measurable functions on $\sigma(T)$. If we define I on VL by

$$I(f) = (\psi(f)h \mid h),$$

then I is a Daniell integral on VL [here we use (iv) to verify (D) of Definition 7.2]. By Proposition 7.4, there exists a unique (finite) measure ν_h on $\sigma(\text{VL}) = B(\sigma(T))$ such that

$$(\psi(f)h \mid h) = If = \int f \, d\nu_h, \qquad \text{for all } f \in \text{VL}.$$

Since μ_h is the unique measure on $B(\sigma(T))$ such that

$$(\hat{\varphi}(f)h \mid h) = \int f \, d\mu_h$$

for continuous real functions, and ψ and $\hat{\varphi}$ agree on continuous functions, it follows that $\nu_h = \mu_h$. As equation (A.13) determines $g(T)$ uniquely for nonnegative bounded measurable functions g, $\psi(g) = g(T) = \hat{\varphi}(g)$ for such functions. ∎

We now consider special bounded measurable functions on $\sigma(T)$ in order to generate a spectral measure on $B(R)$. For each Borel-measurable set M in R consider the characteristic function $\chi_{M \cap \sigma(T)}$. Define P: $B(R) \to L(H, H)$ by

$$P(M) = \hat{\varphi}(\chi_{M \cap \sigma(T)}).$$

In light of the properties of $\hat{\varphi}$, it is easy to convince oneself that P is a normalized spectral measure on $B(R)$.

By Theorem A.2, corresponding to each bounded measurable function g on $\sigma(T)$ is a unique bounded operator $\int g\, dP$ such that

$$\left(\left(\int g\, dP\right)h \mid h\right) = \int g\, d\mu_h \qquad \text{for all } h \text{ in } H, \tag{A.16}$$

where $\mu_h(M) = \big(P(M)h \mid h\big)$ for each h in H and for each M in $B\big(\sigma(T)\big)$. By equation (A.13), $g(T)$ is the unique operator such that

$$\big(g(T)h \mid h\big) = \int g\, d\tilde{\mu}_h \qquad \text{for each } h \text{ in } H, \tag{A.17}$$

where we use $\tilde{\mu}_h$ momentarily to denote the unique Borel measure on $B\big(\sigma(T)\big)$ such that

$$\big(f(T)h \mid h\big) = \int f\, d\tilde{\mu}_h$$

for all continuous functions f on $\sigma(T)$. Since for every M in $B\big(\sigma(T)\big)$

$$\tilde{\mu}_h(M) = \int \chi_M\, d\tilde{\mu}_h = \big(\hat{\varphi}(\chi_M)h \mid h\big) = \big(P(M)h \mid h\big) = \mu_h(M),$$

$\tilde{\mu}_h = \mu_h$. By the uniqueness of $g(T)$ in equation (A.17), we conclude that $g(T) = \int g\, dP$ for each bounded measurable function on $\sigma(T)$. In particular

$$T = \int \lambda\, dP(\lambda). \tag{A.18}$$

We are now ready to prove the following version of the Spectral Theorem.

Theorem A.4. *Spectral Measure Formulation of the Spectral Theorem.* There is a one-to-one correspondence between bounded self-adjoint operators T on H and normalized compact spectral measures E on $B(R)$. It is the following correspondence: $T = \int \lambda \cdot \chi_{\Sigma(E)}(\lambda)\, dE(\lambda)$. ∎

Proof. If E is a normalized compact spectral measure, let us consider $T = \int \lambda \cdot \chi_{\Sigma(E)}(\lambda)\, dE(\lambda)$ as given in Theorem A.2. It must be verified that this correspondence is a one-to-one correspondence.

To verify that the association of $E \to \int \lambda \cdot \chi_{\Sigma(E)}(\lambda)\, dE(\lambda)$ is onto, let S be any bounded self-adioint operator. Using Theorem A.3, let P be the normalized spectral measure on $B(R)$ given by $P(M) = \hat{\varphi}(\chi_{M \cap \sigma(S)})$. Clearly, P is compact and $\Sigma(P) = \sigma(S)$. Moreover, as we have established in equation (A.18)

$$S = \int \lambda \cdot \chi_{\sigma(S)}(\lambda)\, dP(\lambda),$$

so that P maps to S.

To verify the assertion $E \to \int \lambda \cdot \chi_{\Sigma(E)}(\lambda)\, dE(\lambda)$ is one-to-one, it suffices to show that when $T = \int \lambda \cdot \chi_{\Sigma(E)}(\lambda)\, dE(\lambda)$, then $E = P$, where P is the normalized compact spectral measure given by $P(M) = \hat{\varphi}(\chi_{M \cap \sigma(T)})$. We know from Proposition A.4 that $\Sigma(P) = \sigma(T) = \Sigma(E)$. To show that $E = P$, it suffices to show that $E(M) = P(M)$ for $M \subset \sigma(T)$ or that $E(M) = \hat{\varphi}(\chi_M)$. Now by Theorem A.2, $E(M)$ is the unique operator such that

$$(E(M)h \mid h) = \int \chi_M \, d\mu_h; \tag{A.19}$$

whereas by Theorem A.3, $\hat{\varphi}(\chi_M)$ is the unique operator such that

$$(\hat{\varphi}(\chi_M)h \mid h) = \int \chi_M \, d\tilde{\mu}_h, \tag{A.20}$$

where $\tilde{\mu}_h$ again denotes momentarily the unique Borel measure on $\sigma(T)$ such that

$$(f(T)h \mid h) = \int f \, d\tilde{\mu}_h$$

for all $f \in C(\sigma(T))$. By virtue of Remark A.2 and the fact that $T = \int \lambda \, dE$, the unique map of Theorem 6.10 in Chapter 6 is identical to the map of Theorem A.2 restricted to $C_1(\sigma(T))$. Hence for every continuous function f in $C_1(\sigma(T))$,

$$(f(T)h \mid h) = \int f \, d\mu_h$$

and $\mu_h = \tilde{\mu}_h$. This means, by equations (A.19) and (A.20), that $E(M) = \hat{\varphi}(\chi_M)$. ∎

Our final consideration in this section is the multiplication operator form of the Spectral Theorem. We will prove it for self-adjoint and normal operators. The proof is relatively easy and depends primarily on the Continuous Functional Calculus Theorem of Chapter 6. We begin with a definition and a preliminary proposition.

Definition A.3. A vector h in H is called a *cyclic vector* for T in $L(H, H)$ if the closed linear span of $\{T^n h: n = 0, 1, \ldots\}$ is H. ∎

Proposition A.5. Let T be a bounded self-adjoint operator with cyclic vector h_0. There then exists a unitary† operator $U: H \to L_2(\sigma(T), \mu_{h_0})$ such that

$$(UTU^{-1}f)(\lambda) = \lambda f(\lambda) \quad \text{a.e.}$$

[Here μ_{h_0} is the unique measure such that $(f(T)h_0 \mid h_0) = \int f \, d\mu_{h_0}$ for all f in $C(\sigma(T))$.] ∎

Proof. Define U on the dense subset

$$S = \{\varphi(f)h_0: f \in C_1(\sigma(T))\}$$

of H (φ is the unique map of Theorem 6.10 in Chapter 6) by $U(\varphi(f)h_0) = f$. To show U is well-defined on S suppose that $\varphi(f)h_0 = \varphi(g)h_0$. Then the equation

$$\begin{aligned} 0 = \| \varphi(f)h_0 - \varphi(g)h_0 \|^2 &= (\varphi(f-g)^*\varphi(f-g)h_0 \mid h_0) \\ &= (\varphi[\overline{(f-g)}(f-g)]h_0 \mid h_0) = \int |f-g|^2 \, d\mu_{h_0} \end{aligned} \qquad (A.21)$$

implies that $f = g$ a.e. with respect to μ_{h_0} or that $f = g$ in $L_2(\sigma(T), \mu_{h_0})$. An equation like (A.21) also shows that U is an isometric map from a dense subset of H onto a dense subset of $L_2(M, \mu_{h_0})$, namely, $C_1(\sigma(T))$. Hence U can be extended to an isometric map, also designated by U, from H onto $L_2(M, \mu_{h_0})$. If $f \in C_1(\sigma(T))$,

$$\begin{aligned} (UTU^{-1}f)(\lambda) &= [UT\varphi(f)h_0](\lambda) \\ &= [U\varphi(\lambda)\varphi(f)h_0](\lambda) \\ &= [U\varphi(\lambda f)h_0](\lambda) \\ &= \lambda f(\lambda). \end{aligned}$$

† By a unitary operator $U: H \to L_2$ we mean a linear isometry from H onto L_2.

By continuity, this equation is valid also for any f in $L_2(\sigma(T), \mu_{h_0})$ so that the proposition is proved. ∎

Inasmuch as not all bounded self-adjoint operators have cyclic vectors, the above proposition is not applicable in the general situation. Observe that if h is any vector in H, then the closed linear span M of $S = \{T^n h: n = 0, 1, 2, \ldots\}$ is *invariant* under T, that is, $TM \subset M$. Moreover h is a cyclic vector for M. Since $TM \subset M$ if and only if $T^*(M^\perp) \subset M^\perp$ by Proposition 6.14, it is also readily apparent that $T(M^\perp) \subset M^\perp$. To extend the above proposition to the arbitrary case, H must be decomposed into cyclic subspaces as in the next proposition.

If $(H_i)_{i \in I}$ is an arbitrary collection of Hilbert spaces each with inner product $(|)_i$, the *direct sum* $\bigoplus_{i \in I} H_i$ of $(H_i)_{i \in I}$ is the Hilbert space H of all families of element $(h_i)_{i \in I}$ with $h_i \in H_i$ such that the family $(\| h_i \|^2)_{i \in I}$ is summable. Addition, scalar multiplication, and the inner product in H are given by

$$x + y = (x_i + y_i)_{i \in I},$$

$$\alpha x = (\alpha x_i)_{i \in I},$$

and

$$(x \mid y) = \sum_{i \in I} (x_i \mid y_i)_i,$$

where $x = (x_i)_I$ and $y = (y_i)_I$.

Proposition A.6. Let T be a bounded self-adjoint operator on H. Then $H = \bigoplus_I H_i$, where $(H_i)_I$ is a family of mutually orthogonal closed subspaces of H such that

(i) T leaves H_i invariant, that is, $TH_i \subset H_i$ for each i.

(ii) For each i there exists a cyclic vector h_i for H_i, implying $H_i = \overline{\{f(T)h_i: f \in C_1(\sigma(T))\}}$ for some vector h_i. ∎

Proof. If $h \neq 0$ is any vector in H, the closed linear span H_1 of $S = \{T^n h: n = 0, 1, \ldots\}$ is invariant with cyclic vector h. If $H = H_1$, the theorem is proved.

If $H_1 \neq H$, there exists a vector h_2 orthogonal to H_1 and by the same process a subspace H_2 of H exists such that H_2 is orthogonal to H_1 and satisfies (i) and (ii) with h_2 the cyclic vector. If $H = H_1 \oplus H_2$, the proposition is proved. If not, we proceed using Zorn's Lemma as follows.

Let $\mathscr{F}$ be the family of all systems $(H_i)_{i \in J}$ consisting of mutually orthogonal subspaces of H satisfying (i) and (ii). By Zorn's Lemma, the family $\mathscr{F}$, ordered by the inclusion relation $\subset$, has a maximal element $(H_i)_{i \in I}$. Now $H = \bigoplus_I H_i$, for if not there exists a vector h_0 orthogonal to $\bigoplus_I H_i$ and a cyclic subspace H_0 satisfying (i) and (ii), contradicting the maximality of $(H_i)_I$. ▮

Theorem A.5. *Multiplication Operator Form of the Spectral Theorem.* Let T be a bounded self-adjoint operator on a Hilbert space H. Then there exists a family of finite measures $(\mu_i)_{i \in I}$ on $\sigma(T)$ and a unitary operator

$$U\colon H \to \bigoplus_{i \in I} L_2(\sigma(T), \mu_i)$$

such that

$$(UTU^{-1}f)_i(\lambda) = \lambda f_i(\lambda),$$

where $f = (f_i)_I$ is in $\bigoplus_{i \in I} L_2(\sigma(T), \mu_i)$. ▮

Proof. By Proposition A.6, $H = \bigoplus_{i \in I} H_i$ where each H_i is invariant and cyclic with respect to T. By Proposition A.5, for each i there exists a unitary operator $U_i\colon H_i \to L_2(\sigma(T), \mu_i)$ such that

$$(U_i T|_{H_i} U_i^{-1} f_i)(\lambda) = \lambda f_i(\lambda)$$

for $f_i \in L_2(\sigma(T), \mu_i)$. Define $U\colon \bigoplus_{i \in I} H_i \to \bigoplus_{i \in I} L_2(\sigma(T), \mu_i)$ by $U(h) = (U_i h_i)_{i \in I}$. It is easy to see that U satisfies the criteria of the theorem. ▮

If H is a separable Hilbert space, it is clear that the index set I in Proposition A.6 and hence in Theorem A.5 is the finite set $\{1, 2, \ldots, n\}$ for some $n \in N$ or the set N itself. Using this observation, we have the following important corollary of Theorem A.5.

Corollary A.1. Let T be a bounded self-adjoint operator on a Hilbert space H. Then there exists a measure space $(X, \mathscr{A}, \mu)$, a bounded function F on X, and a unitary map $U\colon H \to L_2(X, \mu)$ so that

$$(UTU^{-1}f)(x) = F(x)f(x) \text{ a.e.}$$

If H is separable, then μ can be chosen to be finite. ▮

Proof. Let X be the disjoint union of card(I) copies of $\sigma(T)$ and define μ on X by requiring that its restriction to the ith copy of $\sigma(T)$ be μ_i. If H is separable, in Proposition A.6 we can choose the cyclic vectors h_i so that $\| h_i \| = 2^{-i}$. In this case clearly $\mu(X) = \sum_{i=1}^{\infty} \mu_i(\sigma(T)) < \infty$ since $\mu_i(\sigma(T)) \leq 2^{-i}$ (see Proposition A.5). Letting U be the composition of the unitary operator from Theorem A.5 and the natural unitary operator from $\bigoplus_{i\in I} L_2(\sigma(T), \mu_i)$ onto $L_2(X, \mu)$, and letting F be the function whose restriction to the ith copy of $\sigma(T)$ is the identity function, the corollary follows. ∎

To extend the multiplication operator form of the Spectral Theorem to normal operators, we extend the continuous functional calculus to continuous functions of two variables in the following fashion (see also [42]).

Suppose T_1 and T_2 are two commuting self-adjoint operators. (Actually the argument below applies to any finite collection of commuting self-adjoint operators.) Let $S = \sigma(T_1)\times\sigma(T_2) \subset R\times R$. Let $\hat{\varphi}_1$ and $\hat{\varphi}_2$ be the unique mappings corresponding to T_1 and T_2, respectively, as in Theorem A.3 and let P_1 and P_2 be the respective compact spectral measures. Thus for $i = 1$ or 2, $P_i(M) = \hat{\varphi}_i(\chi_M)$ for all measurable M contained in $\sigma(T_i)$. Since by Lemma A.6, $\hat{\varphi}_i(\chi_M)$ is the strong limit of a sequence of polynomials in T_i, it is clear that for each measurable set A in $\sigma(T_1)$ and for each measurable set B in $\sigma(T_2)$, $P_1(A)$ and $P_2(B)$ commute.

Let f be any finite linear combination of functions of the form $\chi = \chi_{A\times B}$, where A and B are measurable in $\sigma(T_1)$ and $\sigma(T_2)$, respectively. Define $\chi(T_1, T_2)$ as $P_1(A)P_2(B)$ and define $f(T_1, T_2)$ by linearity. It is straightforward to check that f is well defined and also to verify that f can be written as $\sum \alpha_i\chi_{A_i\times B_i}$, where $(A_i\times B_i)\cap(A_j\times B_j) = \varnothing$ for $i\neq j$. Since f can be so written and

$$\| \chi(T_1, T_2) \| \leq \sup_{\lambda\in S} | \chi(\lambda) | ,$$

it is also true that for f

$$\| f(T_1, T_2) \| \leq \sup_{\lambda\in S} | f(\lambda) | .$$

The continuous mapping $f \to f(T_1, T_2)$ can thus be extended to uniform limits of functions such as f. By virtue of the fact that each function of the form $p(x, y) = x^i y^j$ on S can be approximated uniformly by functions like f and that polynomials in the variables x and y approximate uniformly continuous functions on S by Corollary 1.1 the extension to uniform limits includes in particular continuous functions on S.

We thereby have a mapping ψ from $C_1(\sigma(T_1)\times\sigma(T_2)) \to L(H, H)$. This mapping is easily seen to be linear, to preserve multiplication, and to satisfy $\psi(\bar{f}) = (\psi(f))^*$. Moreover the mapping $f(x, y) = x + iy$ is mapped by ψ to $T_1 + iT_2$, as is evident from (iv) of Theorem A.3 if we consider sequences of simple functions converging uniformly to $f_1(x, y) = x$ and $f_2(x, y) = y$, respectively.

Analogous to Definition A.3 we can consider for $h_0 \in H$ the closed linear span of the set $\{T_1{}^i T_2{}^j h_0 : i = 0, 1, 2, \ldots, j = 0, 1, 2, \ldots\}$. If this closed span is H, we can define as in Proposition A.5 a unitary operator U: $H \to L_2(\sigma(T_1) \times \sigma(T_2), \mu)$ where μ is the unique measure such that

$$(\psi(f)h_0 \mid h_0) = \int f\, d\mu$$

for all f in $C_1(\sigma(T_1) \times \sigma(T_2))$ and U is defined by the rule $U(\psi(f)h_0) = f$ for continuous functions.

Now if T is any normal operator, then T can be written as $T_1 + iT_2$, where T_1 and T_2 are the commuting self-adjoint operators given by

$$T_1 = \tfrac{1}{2}(T + T^*) \quad \text{and} \quad T_2 = \tfrac{1}{2}i(T - T^*).$$

Since $\psi(x + iy) = T_1 + iT_2 = T$, the unitary operator U satisfies for f in $C_1(\sigma(T_1) \times \sigma(T_2))$ the equation

$$\begin{aligned}(UTU^{-1}f)(x, y) &= [UT\psi(f)h_0](x, y)\\ &= [U\psi(x + iy)\psi(f)h_0](x, y)\\ &= [U\psi[(x + iy) \cdot f(x, y)]h_0](x, y)\\ &= (x + iy)f(x, y).\end{aligned}$$

By continuity, this equation is also valid for any $f \in L_2(\sigma(T_1) \times \sigma(T_2), \mu)$.

Thus for any normal T, if the closed linear span of $\{T_1{}^i T_2{}^j h_0 : i = 0, 1, 2, \ldots, j = 0, 1, 2, \ldots\}$ is H for some h_0 in H, then $T = T_1 + iT_2$ is unitarily equivalent to a multiplication operator just as for self-adjoint operators. By a similar application of Zorn's Lemma to that used in Proposition A.6 we arrive at the following generalization of Theorem A.5.

Theorem A.6. Let $T = T_1 + iT_2$ be a bounded normal operator on H. Then there exists a family of finite measures $(\mu_i)_{i \in I}$ on $\sigma(T_1) \times \sigma(T_2)$ and a unitary operator

$$U\colon H \to \bigoplus_{i \in I} L_2(\sigma(T_1) \times \sigma(T_2), \mu_i)$$

such that

$$(UTU^{-1}f)_i(x, y) = (x + iy) f_i(x, y) \quad \text{a.e.},$$

where $f = (f_i)_I$ is in $\bigoplus_{i \in I} L_2(\sigma(T), \mu_i)$. ∎

Immediately we obtain the following corollary of Theorem A.6, which is analogous to Corollary A.1.

Corollary A.2. Let T be a bounded normal operator on a Hilbert space H. Then there exists a measure space $(X, \mathscr{A}, \mu)$, a bounded complex function G on X, and a unitary map U: $H \to L_2(X, \mu)$ so that

$$(UTU^{-1}f)(\lambda) = G(\lambda) f(\lambda) \text{ a.e.}$$

If H is separable, μ can be taken to be a finite measure. ∎

A.2. Unbounded Operators and Spectral Theorems for Unbounded Self-Adjoint Operators

Most of the operators that occur in applications of the theory of Hilbert space to differential equations and quantum mechanics are unbounded. For this reason, we consider in this section basic definitions and theorems necessary to deal with unbounded operators and also we prove versions of the spectral theorem for unbounded operators. Let us first define precisely the type of operator we consider in this section.

Definition A.4. A linear operator T *in* (in contrast to "on") a Hilbert space H is a linear transformation on a linear subspace D_T of H into H. D_T is called the *domain* of T and $R_T \equiv \{Tf\colon f \in D_T\}$ is the *range* of T. ∎

If S and T are operators in H, we say T extends S; written $S \subset T$, if $D_S \subset D_T$ and $Sh = Th$ for all h in D_S. T is also called an *extension* of S.

T is *bounded* if T is bounded as a linear transformation from D_T into H.

Example A.1. Let $H = L_2(-\infty, \infty)$ and let T be defined in H on

$$D_T = \{f \in L_2(-\infty, \infty)\colon \lambda f(\lambda) \in L_2(-\infty, \infty)\}$$

by $Tf(\lambda) = \lambda f(\lambda)$. Since D_T contains all functions in $L_2(-\infty, \infty)$ which vanish outside a finite interval, D_T is a dense subset of H and contains moreover the characteristic functions $\chi_{[n,n+1)}$. Since $\| \chi_{[n,n+1)} \| = 1$ and

$$\| T\chi_{[n,n+1)} \|^2 = \int_n^{n+1} x^2\, dx \geq n^2 \qquad \text{for } n \geq 0,$$

T is clearly unbounded.

Analogous to bounded linear operators on H, we can also consider in many cases the adjoint of a linear operator in H whether it be bounded or unbounded by means of the following definition.

Definition A.5. Let $T: D_T \to H$ be an operator in H. Let D_{T^*} be the possibly empty set given by

$$D_{T^*} = \{h \in H\text{: corresponding to } h \text{ is a unique element } h^* \text{ in } H \text{ such that } (Tk \mid h) = (k \mid h^*) \text{ for all } k \text{ in } D_T\}.$$

If $D_{T^*} \neq \varnothing$, the mapping $T^*: D_{T^*} \to H$ given by $T^*h = h^*$ is called the *adjoint* of T. ∎

The questions that naturally arise are: When does T^* exist ($D_{T^*} \neq \varnothing$?) as an operator in H and when is $D_{T^*} = H$? Obviously, if T is a bounded linear operator on H, then $D_{T^*} = H$ and T^* is the "ordinary" adjoint defined earlier. The following proposition tells us in general when T^* has any meaning at all.

Proposition A.7. $D_{T^*} \neq \varnothing$ and T^* is a linear mapping in H with domain D_{T^*} if and only if D_T is dense in H. ∎

Proof. Suppose $\bar{D}_T = H$ and h and h^* are two vectors satisfying

$$(Tk \mid h) = (k \mid h^*) \qquad \text{for all } k \text{ in } D_T. \tag{A.22}$$

Then h^* is uniquely determined by h and equation (A.22), for if $h_1{}^*$ also satisfies equation (A.22), then

$$(k \mid h^* - h_1{}^*) = 0 \qquad \text{for all } k \text{ in } D_T$$

and as $\bar{D}_T = H$, $h^* = h_1{}^*$. This means $D_{T^*} \neq \varnothing$ since $0 \in D_{T^*}$. To show T^* is linear, let h_1 and h_2 be in D_{T^*}. Then

$$\begin{aligned}(k \mid \alpha_1 T^*h_1 + \alpha_2 T^*h_2) &= \bar{\alpha}_1(k \mid T^*h_1) + \bar{\alpha}_2(k \mid T^*h_2)\\ &= \bar{\alpha}_1(Tk \mid h_1) + \bar{\alpha}_2(Tk \mid h_2)\\ &= (Tk \mid \alpha_1 h_1 + \alpha_2 h_2),\end{aligned}$$

so that not only is $\alpha_1 h_1 + \alpha_2 h_2$ in D_{T^*} but $T^*(\alpha_1 h_1 + \alpha_2 h_2) = \alpha_1 T^*h_1 + \alpha_2 T^*h_2$.

Conversely, suppose D_T is not dense in H. Let h be in $H - \bar{D}_T$. By

Theorem 6.2, $h = h_1 + h_2$, where $h_1 \perp \bar{D}_T$ and $h_1 \neq 0$. Hence

$$(Tk \mid 0) = 0 = (k \mid h_1) \text{ for all } k \text{ in } D_T$$

and corresponding to 0 are two vectors 0 and h_1 that satisfy equation (A.22). Moreover this means that for every vector x in H for which there exists a vector x^* satisfying equation (A.22), there are two vectors x^* and $x^* + h_1$, that satisfy equation (A.22). This means $D_{T^*} = \varnothing$ and T^* cannot exist. ∎

The notion of the adjoint of an operator in H gives rise to the idea of self-adjointness of operators in H. This idea generalizes our previous concept of self-adjointness for bounded operators on H.

Definition A.6. An operator in H with $\bar{D}_T = H$ is *symmetric* if $T^* \supseteq T$. A symmetric operator is *self-adjoint* if $T^* = T$. ∎

The next result shows in particular that a self-adjoint operator with $D_T = H$ is always bounded.

Proposition A.8. The adjoint T^* of a linear operator on H ($D_T = H$) is a bounded operator in H. ∎

Proof. Assume T^* is not bounded. There then exists a sequence (h_n) in D_{T^*} such that $\| h_n \| = 1$ and $\| T^*h_n \| \to \infty$ as $n \to \infty$. Define the functionals φ_n on H by

$$\varphi_n(h) = (Th \mid h_n).$$

The φ_n are clearly bounded linear functionals on H. Moreover for each h, the sequence $(\varphi_n(h))$ is also bounded as shown by

$$| \varphi_n(h) | = | (Th \mid h_n) | \leq \| Th \| \; \| h_n \| = \| Th \|.$$

By the Principle of Uniform Boundedness (Theorem 5.13) we have for some constant C,

$$| \varphi_n(h) | \leq C \| h \|, \qquad n = 1, 2, \ldots. \tag{A.23}$$

Letting $h = T^*h_n$ in inequality (A.23) we have

$$| \varphi_n(T^*h_n) | = (T(T^*h_n) \mid h_n) = (T^*h_n \mid T^*h_n) = \| T^*h_n \|^2 \leq C \| T^*h_n \|,$$

which implies $\| T^*(h_n) \| \leq C$. This is a contradiction to the selection of (h_n). ∎

Corollary A.3. Any symmetric operator defined on H is bounded. ∎

Below are listed a few more important but trivially demonstrable facts.

Remarks

A.3. If S and T are linear operators in H, $S \subset T$ and $\bar{D}_S = H$, then $T^* \subset S^*$.

A.4. If T is a linear operator in H with $\bar{D}_T = \bar{D}_{T^*} = H$, then $T \subset T^{**}$.

A.5. Suppose T is a one-to-one operator in H so that T^{-1} is defined on $D_{T^{-1}} \equiv T(D_T)$. If $\bar{D}_T = \bar{D}_{T^{-1}} = H$, then T^* is one-to-one and $(T^*)^{-1} = (T^{-1})^*$.

Recall from Chapter 5 that an operator T in H is *closed* if whenever $h_1, h_2, \ldots$ in D_T converges to h in H and $Th_1, Th_2, \ldots$ converges to k in H, then $h \in D_T$ and $Th = k$. Equivalently T is closed if its graph $G_T = \{(h, Th): h \in D_T\}$ is closed in the direct sum $H \oplus H$. [$H \oplus H = \{(h, k): h \in H, k \in H\}$ with addition and scalar multiplication defined componentwise and the inner product given by $((h, k) \mid (h_1, k_1)) = (h \mid h_1) + (k \mid k_1)$.]

Proposition A.9. The adjoint T^* of T in H is closed. ∎

Proof. Let $g_1, g_2 \ldots$ be a sequence in D_{T^*} converging to g in H and suppose $T^*g_1, T^*g_2, \ldots$ converge to h in H. Then for any k in D_T,

$$(Tk \mid g) = \lim_n (Tk \mid g_n) = \lim_n (k \mid T^*g_n) = (k \mid h),$$

implying that $g \in D_{T^*}$ and $h = T^*g$. ∎

Definition A.7. An operator T in H is called *closable* if there is a closed operator S in H which extends T. The *closure* $\bar{T}$ of a closable operator T is the smallest closed operator extending T, that is, any closed operator extending T also extends $\bar{T}$. ∎

Obviously the closure exists for every closable operator.

Proposition A.10. Suppose T is an operator in H with $\bar{D}_T = H$. Then

(i) if T is closed, $\bar{D}_{T^*} = H$ and $T = T^{**}$,

(ii) $\bar{D}_{T^*} = H$ if and only if T is closable, in which case $\bar{T} = T^{**}$, and

(iii) if T is closable, $(\bar{T})^* = T^*$. ∎

Proof. Observe from the definition of T^* that the set $A = \{(T^*g, -g): g \in D_{T^*}\}$ is the set of all pairs $(g^*, -g)$ with $g \in D_{T^*}$ such that

$$(Tk \mid g) = (k \mid g^*) \qquad \text{for all } k \text{ in } D_T.$$

In other words A is the set of points $(g^*, -g)$ such that

$$\big((k, Tk) \mid (g^*, -g)\big)_{H\oplus H} = 0 \qquad \text{for all } k \text{ in } D_T,$$

where $(\mid)_{H\oplus H}$ is the inner product in $H \oplus H$. This means A is the orthogonal complement $G_T{}^{\perp}$ of the linear space G_T. If T is closed then $A^{\perp} = (G_T{}^{\perp})^{\perp} = G_T$.

Secondly, observe that if $\bar{D}_{T^*} = H$ so that T^{**} exists, then the graph of T^{**} consists of all the points (f, f^*) of $H \oplus H$ such that

$$(T^*h \mid f) = (h \mid f^*) \qquad \text{for all } h \in D_{T^*}$$

or for which

$$\big((T^*h, -h) \mid (f, f^*)\big)_{H\oplus H} = 0.$$

This implies $G_{T^{**}} = (G_T{}^{\perp})^{\perp}$, and in particular $G_{T^{**}} = G_T$ if T is closed.

The proofs of (i), (ii), and (iii) follow from these observations. For (i), assume $\bar{D}_{T^*} \subsetneqq H$ and let h be a nonzero element of H such that $(h \mid g) = 0$ for all $g \in D_{T^*}$. Then

$$\big((0, h) \mid (T^*g, -g)\big)_{H_1\oplus H_2} = 0 \qquad \text{for all } g \text{ in } D_{T^*}.$$

Since $A^{\perp} = G_T$, $(0, h) \in G_T$ so that $h = 0$. Hence $\bar{D}_{T^*} = H$. Observing that always $A \subset A^{**}$, the relation $G_{T^{**}} = G_T$ above shows that $A = A^{**}$.

In (ii), if T is closable then by (i), $\bar{D}_{(\bar{T})^*} = H$ and $\bar{T} = T^{**}$. However, $D_{(\bar{T})^*} \subset D_{T^*}$ so $\bar{D}_{T^*} = H$. Conversely, if $\bar{D}_{T^*} = H$ then T^{**} exists and $T \subset T^{**}$. If S is any closed extension of T, then G_S is closed and contains G_T. From the observations above, $G_{T^{**}} = (G_T{}^{\perp})^{\perp}$. Hence $G_{T^{**}}$ is the closure of G_T and $G_S \supset G_{T^{**}}$. Hence $\bar{T} = T^{**}$.

To prove (iii) notice that, if T is closable,

$$T^* = (\overline{T^*}) = T^{***} = (\bar{T})^*. \qquad ∎$$

A symmetric operator is always closable since $D_{T^*} \supset D_T$ and D_T is dense in H. If T is symmetric, T^* is a closed extension of T, so the closure T^{**} of T satisfies $T \subset T^{**} \subset T^*$. If T is self-adjoint, $T = T^{**} = T^*$. If T is closed and symmetric, $T = T^{**} \subset T^*$. Clearly, a closed symmetric operator is self-adjoint if and only if T^* is symmetric.

The following is an illustrative example of three closed operators in H.

Example A.2. As in Theorem 4.3, a complex-valued function f is absolutely continuous on $[\alpha, \beta]$, where $-\infty < \alpha < \beta < \infty$, if there exists an integrable function g on $[\alpha, \beta]$ such that

$$f(x) = \int_\alpha^x g(t)\, dt + f(\alpha).$$

Such a function f is continuous on $[\alpha, \beta]$ and differentiable a.e. with $f'(x) = g(x)$ a.e. on $[\alpha, \beta]$. Yet its derivative need not belong to $L_2(\alpha, \beta)$, as is shown if one considers the function $f(x) = x^{1/2}$ in $L_2[0, 1]$.

For the purpose of this example we consider three separate Hilbert spaces as follows:

$$H_1 = L_2[\alpha, \beta], \qquad \text{where } -\infty < \alpha < \beta < \infty,$$
$$H_2 = L_2[\alpha, \infty), \qquad \text{where } -\infty < \alpha < \infty,$$
$$H_3 = L_2(-\infty, \infty).$$

Also we consider three operators T_1 in H_1, T_2 in H_2, and T_3 in H_3 defined, respectively, on the following three domains:

$$D_1 = \{g \in H_1\colon g = f \text{ a.e. where } f \text{ is absolutely continuous on } [\alpha, \beta], \; f(\alpha) = 0 = f(\beta), \text{ and } f' \in L_2[\alpha, \beta]\},$$

$$D_2 = \{g \in H_2\colon g = f \text{ a.e. where } f \text{ is absolutely continuous on } [\alpha, \beta] \text{ for each } \beta > \alpha, \; f(\alpha) = 0, \text{ and } f' \in L_2[\alpha, \beta]\},$$

$$D_3 = \{g \in H_3\colon g = f \text{ a.e. where } f \text{ is absolutely continuous on } [\alpha, \beta] \text{ for each } -\infty < \alpha < \beta < \infty, \text{ and } f' \in H_3\}.$$

By definition, $T_1 g = if'$, $T_2 g = if'$, and $T_3 g = if'$ (g, f as above).

For each $i = 1, 2, 3$, $\bar{D}_i = H_i$. To show this for D_1 we recall that the linear subspace spanned by the set $\{x^n\colon n = 0, 1, 2, \ldots\}$ is dense in $L_2[\alpha, \beta]$ since the class of all complex polynomials is dense in $L_2[\alpha, \beta]$. However, each x^n is in $\bar{D}_1$ since each x^n can be approximated in $L_2[\alpha, \beta]$ by a function f in D_1 as illustrated in Figure 5. This means $\bar{D}_1 = H$. To

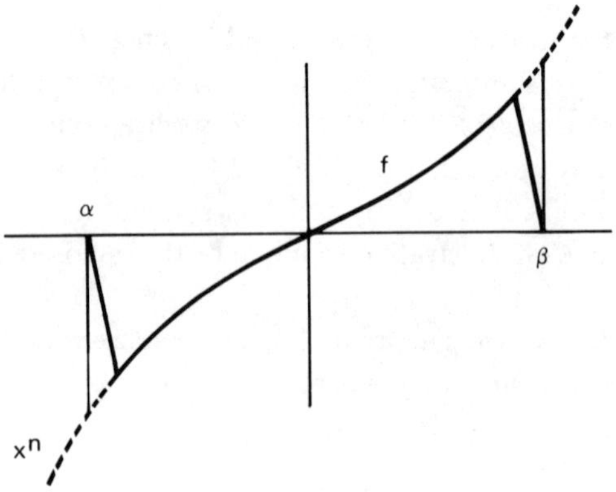

Fig. 5

prove $\bar{D}_2 = H$ and $\bar{D}_3 = H$, it is sufficient to observe analogously that the linear subspace spanned by the set $\{x^n e^{-x^2/2}: n = 0, 1, 2 \ldots\}$ is dense in $L_2(-\infty, \infty)$ {and hence their restrictions to $[\alpha, \infty)$ are dense in $L_2(\alpha, \infty)$} and then to approximate analogously each $x^n e^{-x^2/2}$ by a function f in D_1 or D_2.

The important observations to make here about T_1, T_2, and T_3 are the following: Each is an unbounded symmetric operator in H_i, respectively, and in particular T_3 is self-adjoint, whereas T_1 and T_2 are not self-adjoint. Moreover in each case $T_i = T_i^{**}$.

The verification that each T_i is unbounded is accomplished by considering functions in D_i of the following form. For $\alpha < \beta$ and $n \geq 2/(\beta - \alpha)$ define f_n by (see Figure 6)

$$f_n(x) = \begin{cases} n(x - \alpha) & \text{if } x \in [\alpha, \alpha + 1/n] \\ 2 - n(x - \alpha) & \text{if } x \in [\alpha + 1/n,\ \alpha + 2/n] \\ 0 & \text{if } x \in [\alpha + 2/n, \infty). \end{cases}$$

Clearly $f_n'(x) = n$, $-n$, or 0 on the respective intervals and

$$\| f_n \|^2 - \int_\alpha^{\alpha+2/n} | f_n(x) |^2 \leq 2/n$$

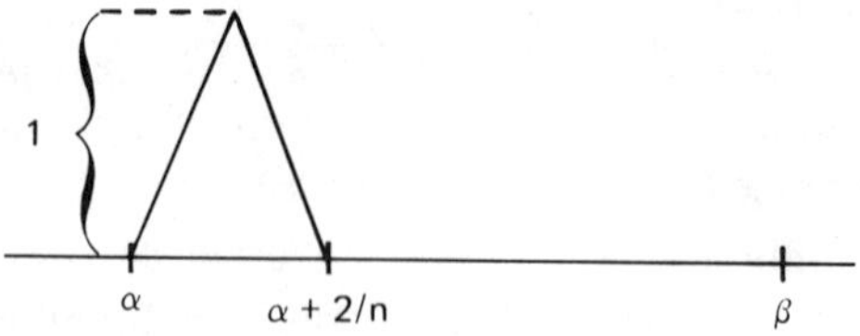

Fig. 6

while

$$\| if_n' \|^2 = \int_\alpha^{\alpha+2/n} n^2 \, dx = 2n.$$

These equations imply that

$$\frac{\| D_i f_n \|}{\| f_n \|} = D_i\left(\frac{f_n}{\| f_n \|}\right) \geq \frac{(2n)^{1/2}}{(2/n)^{1/2}} = n.$$

The verification that each T_i is symmetric is obtained in the following manner by integrating by parts. For f and g in the domain of T_1 and for $-\infty < \alpha < \beta < \infty$,

$$\begin{aligned}(if' \mid g) - (f \mid ig') &= i \int_\alpha^\beta f'(\xi)\overline{g(\xi)} \, d\xi + i \int_\alpha^\beta f(\xi)\overline{g'(\xi)} \, d\xi \\ &= if(\xi)\overline{g(\xi)} \Big|_\alpha^\beta = 0, \end{aligned} \tag{A.24}$$

whereby $(T_1 f \mid g) = (f \mid T_1 g)$. The verifications for T_2 and T_3 are similar. {Observe that if $f \in D_2$ then $\lim_{x\to\infty} f(x) = 0$ and likewise if $f \in D_3$ then $\lim_{x\to\pm\infty} f(x) = 0$. Indeed, for H_2 the equation

$$\begin{aligned}(f \mid f') + (f' \mid f) &= \lim_{x\to\infty} \int_\alpha^x [f(\xi)\overline{f'(\xi)} + f'(\xi)\overline{f(\xi)}] \, d\xi \\ &= \lim_{x\to\infty} [| f(x) |^2 - | f(\alpha) |^2]\end{aligned}$$

implies that $\lim_{x\to\infty} | f(x) |^2$ exists and since $f \in L_2(\alpha, \beta)$, $f(x) \to 0$ as $x \to \infty$.}

Let us next calculate the adjoint T_1^* of T_1. Let D_1^* be the set

$$D_1^* = \{g \in H_1\colon g = f \text{ a.e. where } f \text{ is absolutely continuous on } [\alpha, \beta], \; f' \in H_1\}.$$

Since equation A.24 still is valid for g in D_1^*, the domain of T_1^* contains D_1^* and $T_1^* g = if'$ for g in D_1^*, where g and f are as above. We wish to show that the domain of T_1^* is D_1^*, which will clearly show that $T_1 \subset T_1^*$ and $T_1 \neq T_1^*$. To this end, let f be in the domain of T_1^*. Let h be the absolutely continuous function given by

$$h(x) = \int_\alpha^x T_1^* f(\xi) \, d\xi + C,$$

where C is a constant chosen so that

$$\int_\alpha^\beta [f(\xi) + ih(\xi) \, d(\xi)] = 0.$$

For every g in D_1, an integration by parts gives that

$$\int_\alpha^\beta ig'(\xi)\,\overline{f(\xi)}\,d\xi = (T_1 g \mid f) = (g \mid T_1{}^*f) = \int_\alpha^\beta g(\xi)\overline{T_1{}^*f(\xi)}\,d\xi$$

$$= g(\xi)\overline{h(\xi)}\,\Big|_\alpha^\beta - \int_\alpha^\beta g'(\xi)\overline{h(\xi)}\,d\xi = i\int_\alpha^\beta ig'(\xi)\overline{h(\xi)}\,d\xi.$$

Hence

$$\int_\alpha^\beta g'(\xi)\overline{[f(\xi) + ih(\xi)]}\,d\xi = 0.$$

In particular, letting g be the function in D_1 given by

$$g(x) = \int_\alpha^x [f(\xi) + ih(\xi)]\,d\xi$$

we obtain that

$$\int_\alpha^\beta |f(\xi) + ih(\xi)|^2 = 0$$

or that a.e. we have

$$f(x) = -ih(x) = -i\int_\alpha^x T_1{}^*f(\xi)\,d\xi - iC$$

and h is absolutely continuous with $h'(x) = T_1{}^*f(x)$. Hence f is in $D_1{}^*$.

In an almost identical fashion the verification that $T_2{}^*g = if'$ on the domain

$D_2{}^* = \{f \in H_2$: f is absolutely continuous in each $[\alpha, \beta]$ with $\beta > \alpha$ and $f' \in H_2\}$

and that $T_3{}^*g = if'$ on the domain

$$D_3{}^* = D_3$$

can be carried out. Since $D_1 \subsetneqq D_1{}^*$, $D_2 \subsetneqq D_2{}^*$, and $D_3 = D_3{}^*$, clearly T_3 is self-adjoint and $T_1{}^{**}$, $T_2{}^{**}$, and $T_3{}^{**}$ are all defined.

It remains to show that $T_j = T_j{}^{**}$ for $j = 1$ and 2. In either case since $T_j \subset T_j{}^*$, we can say that $T_j \subset T_j{}^{**} \subset T_j{}^*$. It suffices to show therefore that $D_{T_j{}^{**}} \subset D_j$. Let $f \in D_{T_j{}^{**}}$. Then for all g in $D_j{}^*$ we have

$$(T_j{}^{**}f \mid g) = (f \mid T_j{}^*g),$$

and moreover since $T_j{}^{**}f = if'$ (because $T_j{}^{**} \subset T_j{}^*$) we have

$$0 = (if' \mid g) - (f \mid ig').$$

If $j = 1$, this means

$$0 = i \int_\alpha^\beta f'(\xi)\overline{g(\xi)}\, d(\xi) + i \int_\alpha^\beta f(\xi)\overline{g'(\xi)}\, d\xi$$
$$= if(\xi)\overline{g(\xi)} \Big|_\alpha^\beta = i[f(\beta)\overline{g(\beta)} - f(\alpha)\overline{g(\alpha)}].$$

By first letting $g(x) = (x - \alpha)/(\beta - \alpha)$ in $D_1{}^*$ and then letting $g = (\beta - x)/(\beta - \alpha)$ in $D_1{}^*$, we obtain $f(\alpha) = 0 = f(\beta)$ implying that $f \in D_1$. If $j = 2$, let $g(x) = e^{-(x-\alpha)}$ to obtain $f(\alpha) = 0$ so that f is again in D_2.

Interestingly enough, T_1 has uncountably many different self-adjoint extensions. Let $\gamma \in C$ with $|\gamma| = 1$ and define T_γ in H_1 on

$$D_{T_\gamma} = \{g \in L_2(\alpha, \beta)\colon g = f \text{ a.e. where } f \text{ is absolutely continuous on } [\alpha, \beta],\ f' \in H_1, \text{ and } f(\beta) = \gamma f(\alpha)\}$$

by $T_\gamma g = if'$. Each T_γ is self-adjoint and extends T_1. For each γ, we have $T_1 \subset T_\gamma \subset T_1{}^*$.

The next example presents a type of self-adjoint operator which will be shown in Theorem A.7 to be the "prototype" of all self-adjoint operators.

Example A.3. Let $(X, \mathscr{A}, \mu)$ be a measure space with μ a finite measure. Suppose that f is an extended real-valued measurable function on X which is finite a.e. Then the operator T_f in $L_2(X, \mu)$ defined by $T_f(g) = fg$ on

$$D_f = \{g \in L_2(X, \mu)\colon fg \in L_2(X, \mu)\}$$

is self-adjoint. Indeed, by considering the functions f_n defined as

$$f_n = \begin{cases} 1 & \text{on } N_n = \{x\colon |f(x)| < n\} \\ 0 & \text{otherwise,} \end{cases}$$

one can easily verify that $gf_n \in D_f$ for each $n = 1, 2, \ldots$ and for each g in $L_2(X, \mu)$. Moreover, since the functions gf_n are dense in $L_2(X, \mu)$, $\overline{D_f} = H$. T_f is thus clearly symmetric. In addition if h is an element in $D_{T_f{}^*}$, then by the Monotone Convergence Theorem,

$$\| T_f{}^*h \| = \lim_{n\to\infty} \| f_n \cdot T_f{}^*h \|$$
$$= \lim_{n\to\infty} \Big[\sup_{\|k\|=1} (k \mid f_n \cdot T_f{}^*(h))\Big]$$
$$= \lim_{n\to\infty} \Big[\sup_{\|k\|=1} (T_f(kf_n) \mid h)\Big]$$
$$= \lim_{n\to\infty} \Big[\sup_{\|k\|=1} (k \mid f_n fh)\Big] = \lim_{n\to\infty} \| f_n fh \| = \| fh \|.$$

Hence $fh \in L_2(X, \mu)$ and $h \in D_{T_f}$. Therefore, $D_{T_f} = D_{T_f^*}$ and T_f is self-adjoint.

As before for bounded linear operators on H, a scalar λ is an *eigenvalue* of an operator T in H if there exists a nonzero vector h in D_T such that $Th = \lambda h$. If T is a symmetric operator, the equation

$$\lambda(h \mid h) = (\lambda h \mid h) = (Th \mid h) = (h \mid Th) = (h \mid \lambda h) = \bar{\lambda}(h \mid h)$$

shows that eigenvalue λ is always real.

Again, the *resolvent set* $\varrho(T)$ of an operator in H with $\overline{D_T} = H$ is the the set of all scalars λ for which $R_{T-\lambda I}$ is dense in H and for which $(T - \lambda I)$ has a bounded inverse defined on $R_{T-\lambda I}$. The *spectrum* $\sigma(T)$ of T is the complement of $\varrho(T)$ and is clearly decomposed into three disjoint sets: the set of eigenvalues of T (sometimes called the *point spectrum*), the set of scalars λ for which $R_{T-\lambda I}$ is dense in H but for which $(T - \lambda I)^{-1}$ exists and is not bounded (called the *continuous spectrum*), and the set of scalars λ for which $(T - \lambda I)^{-1}$ exists but its domain $R_{T-\lambda I}$ is not dense (called the *residual spectrum*).

Proposition A.11. If T is a closed linear operator in H with $\overline{D_T} = H$ and $\lambda \in \varrho(T)$, then $(T - \lambda I)^{-1}$ is a bounded linear operator *on* (all of) H. ▌

Proof. $R_{T-\lambda I}$ is dense in H and since $(T - \lambda I)^{-1}$ is bounded on $R_{T-\lambda I}$, there exists a positive constant C such that

$$\| h \| \leq C \| (T - \lambda I)h \| \qquad \text{for all } h \text{ in } D_T. \tag{A.25}$$

If $k = \lim_n (T - \lambda I)h_n$, then by inequality (A.25) the $\lim_{n\to\infty} h_n$ exists, say h. Since T is closed, $(T - \lambda I)h = k$ and $k \in R_{T-\lambda I}$. Hence $R_{T-\lambda I} = H$. ▌

In light of Proposition A.11, we can see that the resolvent of a closed operator in H is the set of scalars λ for which $T - \lambda I$ is a bijection from D_T to H and for which $(T - \lambda I)^{-1}$ is bounded on H.

The next result establishes that the spectrum of a self-adjoint operator in H is contained in the real numbers.

Proposition A.12. Let T be a self-adjoint operator in H. Then $\varrho(T)$ contains all complex numbers with nonzero imaginary part. Moreover if $\operatorname{Im} \lambda \neq 0$ then

$$\| (T - \lambda I)^{-1} \| \leq \frac{1}{|\operatorname{Im} \lambda|} \tag{A.26}$$

and

$$\operatorname{Im}\big((T - \lambda I)h \mid h\big) = \operatorname{Im}(-\lambda) \, \| h \|^2 \qquad \text{for all } h \in D_T. \blacksquare \qquad \text{(A.27)}$$

Proof. Since T is self-adjoint, $(Th \mid h)$ is real for all h in D_T. Clearly equation (A.27) follows and by the Cauchy–Schwarz inequality,

$$| \operatorname{Im}(\lambda) | \; \| h \|^2 \leq | \big((T - \lambda I)h \mid h\big) | \leq \| (T - \lambda I)h \| \; \| h \|,$$

which implies that

$$| \operatorname{Im}(\lambda) | \; \| h \| \leq \| (T - \lambda I)h \|. \qquad \text{(A.28)}$$

If $\operatorname{Im}(\lambda) \neq 0$, this inequality implies that $(T - \lambda I)^{-1}$ exists as a linear operator on $R_{T-\lambda I}$ since $T - \lambda I$ is one-to-one by inequality (A.28). Now $R_{T-\lambda I}$ is also dense in H. Indeed, if not, there exists a nonzero vector k in D_T (which is dense in H) such that $\big((T - \lambda I)h \mid k\big) = 0$ for all h in D_T. However, then $\big(h \mid (T^* - \bar{\lambda} I)k\big) = \big(h \mid (T - \bar{\lambda} I)k\big) = 0$ for all h in D_T; and since $\bar{D}_T = H$, $(T - \bar{\lambda} I)k = 0$. This means $Tk = \bar{\lambda} k$ and $(Tk \mid k) = \bar{\lambda}(k \mid k)$, which contradicts the fact that $(Tk \mid k)$ is real. Therefore $R_{T-\lambda I}$ is dense in H. Moreover from inequality (A.28), $(T - \lambda I)^{-1}$ is bounded on $R_{T-\lambda I}$ with bound satisfying inequality (A.26). All this means $\lambda \in \varrho(T)$ when $\operatorname{Im} \lambda \neq 0$. ∎

The next proposition also shows that the resolvent set is open for a self-adjoint operator.

Proposition A.13. If T is a closed linear operator in H with $\bar{D}_T = H$, then the resolvent set is open and if λ and μ are in $\varrho(T)$,

$$R_\mu - R_\lambda = (\lambda - \mu) R_\lambda R_\mu,$$

where, as in Proposition 5.27 of Chapter 5, $R_\lambda = (T - \lambda I)^{-1}$. Moreover R_λ as a function on $\varrho(T)$ to $L(H, H)$ has derivatives of all orders. ∎

Proof. Observing that from Proposition A.11, R_λ is defined on H for $\lambda \in \varrho(T)$, the proof is identical to that of Proposition 5.27 of Chapter 5. ∎

Remarks. Here are some other easy to prove facts concerning the eigenvalues and the spectrum of an operator in H.

A.6. If $\bar{D}_T = \bar{D}_{T^*} = H$ and $T = T^{**}$, then $\lambda \in \sigma(T)$ if and only if $\bar{\lambda} \in \sigma(T^*)$.

A.7. If $\bar{D}_T = \bar{D}_{T^*} = H$ and $T = T^{**}$, then

$$\{h \in D_T \colon Th = \lambda h\} = [(T^* - \bar{\lambda} I) D_{T^*}]^\perp.$$

In particular if T is self-adjoint, λ is an eigenvalue of T if and only if $(T - \lambda I)D_T$ is not dense in H.

A.8. If T is a self-adjoint operator, $\lambda \in \varrho(T)$ if and only if $(T - \lambda I)D_T = H$.

Proof. (A.6) If $\lambda \in \varrho(T)$, then by Proposition A.11 $D_{(T-\lambda I)^{-1}} = H$. Since $(T - \lambda I)^* = T^* - \bar{\lambda} I$, by Remark A.5 we have $(T^* - \bar{\lambda} I)^{-1} = [(T - \lambda I)^{-1}]^*$. Since $[(T - \lambda I)^{-1}]^*$ is bounded, $\bar{\lambda} \in \varrho(T^*)$. Conversely, by reversing the argument, $\bar{\lambda} \in \varrho(T^*)$ implies $\lambda \in \varrho(T)$.

$$\begin{aligned}
\text{(A.7)}\quad Th - \lambda h = 0 &\Leftrightarrow (Th - \lambda h \mid k) = 0 \qquad \text{for all } k \text{ in } D_{T^*} \\
&\Leftrightarrow (h \mid (T^* - \bar{\lambda})k) = 0 \qquad \text{for all } k \text{ in } D_{T^*} \\
&\Leftrightarrow h \in [(T^* - \bar{\lambda} I)D_{T^*}]^{\perp}.
\end{aligned}$$

(A.8) If $\lambda \in \varrho(T)$, we know from Proposition A.11 that $(T - \lambda I)D_T = H$. Conversely, suppose $(T - \lambda I)D_T = H$. If $\lambda \notin R$, $\lambda \in \varrho(T)$ by Proposition A.12. If $\lambda \in R$, λ is not an eigenvalue by (A.7) and $T - \lambda I$ is self-adjoint. By Remark A.5, $(T - \lambda I)^{-1}$ is also self-adjoint and by Proposition A.8, $(T - \lambda I)^{-1}$ is bounded. Hence $\lambda \in \varrho(T)$. ∎

Example A.4. To illustrate some of the preceding theory let us consider three versions of the differentiation operator id/dt. Let T_1, T_2, T_3 be the operator id/dt on the respective domains†:

$$D_{T_1} = \{f \in L_2[0, 2\pi]: f \text{ is absolutely continuous on } [0, 2\pi] \text{ and } f(0) = 0\}$$

$$D_{T_2} = \{f \in L_2[0, 2\pi]: f \text{ is absolutely continuous on } [0, 2\pi] \text{ and } f(0) = f(2\pi)\}$$

$$D_{T_3} = \{f \in L_2[0, 2\pi]: f \text{ is absolutely continuous on } [0, 2\pi]\}.$$

From Example A.2 and the fact that D_1 of that example is contained in D_{T_1}, D_{T_2}, and D_{T_3}, each of these domains is dense in H. The spectrum of T_1 is empty, the spectrum of T_2 is the set of integers (which is also the set of eigenvalues of T_2), and the spectrum of T_3 is the whole complex plane. That the spectrum of T_1 is empty follows from the fact that for each λ the operator S_λ given by

$$(S_\lambda g)(t) = \int_0^t e^{-i\lambda(t-s)} g(s)\, ds, \qquad g \in L_2[0, 2\pi]$$

† Though not explicitly stated, these domains are here assumed to have the property: $g = f$ a.e. and $f \in D_{T_j} \Rightarrow g \in D_{T_j}$ and $T_j g = if'$.

is the inverse of $T_1 - \lambda I$. To calculate the spectrum of T_2, observe first that $\sigma(T_2)$ is a set of real numbers since T_2 is self-adjoint as in Example A.2. For each integer k, the function $f(t) = e^{-ikt}$ is a solution of $T_2 f = kf$ so that each integer is an eigenvalue. It is easy to verify that for each non-integer real number λ the equation

$$if'(t) - \lambda f(t) = g(t)$$

is solvable for each g in $L_2[0, 2\pi]$ so that by Remark A.8 each such λ is in $\varrho(T)$. Finally to calculate the spectrum $\sigma(T_3)$ observe that for each λ in C, the function $f(\lambda) = e^{-i\lambda t}$ is a solution of $if'(t) - \lambda f(t) = 0$ so that each λ in C is in fact an eigenvalue of T_3.

Let us now turn our attention to several versions of the Spectral Theorem for unbounded operators. In this case a multiplication operator form of the theorem leads nicely into a functional calculus form and then into a spectral measure version of the Spectral Theorem. The interested reader may also consult references [42] and [47].

The next theorem shows that all self-adjoint operators in H are unitarily equivalent in the sense of (ii) below to a self-adjoint operator of the type given in Example A.3.

Theorem A.7. *Multiplication Operator Form of the Spectral Theorem.* Let T be a self-adjoint operator in a Hilbert space H. There exists then a measure space $(X, \mathscr{A}, \mu)$, a unitary† operator U: $H \to L_2(X, \mu)$, and a measurable function F on X which is real a.e. such that

(i) $h \in D_T$ if and only if $F(\cdot)Uh(\cdot)$ is in $L_2(X, \mu)$

and

(ii) if $f \in U(D_T)$, then $(UTU^{-1}f)(\cdot) = F(\cdot)f(\cdot)$. ∎

Proof. To achieve the proof we utilize the multiplication operator form of the Spectral Theorem for bounded normal operators (Corollary A.2) by applying it to the operator $(T + i)^{-1}$. Let us first establish that this is a bounded normal operator.

By Propositions A.11 and A.12, $(T \pm i)^{-1}$ exist as bounded linear operators on H. In particular $R_{T\pm i} = H$ and $T \pm i$ are one-to-one operators. For any h and k in D_T, since T is self-adjoint,

$$\big((T - i)h \mid (T + i)^{-1}(T + i)k\big) = \big((T - i)^{-1}(T - i)h \mid (T + i)k\big).$$

† By a unitary operator U: $H \to L_2$ we mean a linear isometry from H onto L_2.

This implies that $((T+i)^{-1})^* = (T-i)^{-1}$. Since $(T+i)^{-1}$ and $(T-i)^{-1}$ commute by Proposition A.13, we have

$$(T+i)^{-1}((T+i)^{-1})^* = (T+i)^{-1}(T-i)^{-1} = ((T+i)^{-1})^*(T+i)^{-1}$$

and $(T+i)^{-1}$ is seen to be a normal operator.

By Corollary A.2, there is a measure space $(X, \mathscr{A}, \mu)$, a unitary operator $U\colon H \to L_2(X, \mu)$, and a bounded, measurable complex function G on X so that

$$\big(U(T+i)^{-1}U^{-1}f\big)(x) = G(x)f(x) \text{ a.e.} \tag{A.29}$$

for all f in $L_2(X, \mu)$.

Since Ker $(T+i)^{-1} = \{0\}$, $G(x) \neq 0$ a.e. Therefore if we define $F(x)$ as $G(x)^{-1} - i$ for each x in X, $|F(x)|$ is finite a.e. Now if $f \in U(D_T)$, then there exists a function g in $L_2(X, \mu)$ such that $f(\cdot) = G(\cdot)g(\cdot)$ in L_2. That this is so follows from the inclusions

$$U(D_T) \subset U(T+i)^{-1}(H) \subset U(T+i)^{-1}U^{-1}\big(L_2(X, \mu)\big). \tag{A.30}$$

Observing that $U(T+i)^{-1}U^{-1}$ is an injection, for any g in the range of $U(T+i)^{-1}U^{-1}$ we have from equation (A.29)

$$[U(T+i)^{-1}U^{-1}]^{-1}g(x) = [1/G(x)] \cdot g(x) \in L_2(X, \mu).$$

In particular for f in the set $U(D_T)$,

$$[U(T+i)^{-1}U^{-1}]^{-1}f(x) = [1/G(x)] \cdot f(x) \in L_2(X, \mu)$$

or

$$U(T+i)U^{-1}f(x) = [1/G(x)] \cdot f(x) \in L_2(X, \mu)$$

or

$$UTU^{-1}f(x) = [1/G(x)f(x) - if(x) = F(x)f(x) \in L_2(X, \mu).$$

This proves (ii) and the necessity of (i) provided F is real-valued, which we show below. For the converse of (i), if $F(x)Uh(x)$ is in $L_2(X, \mu)$, then there is a k in H so that $Uk = [F(x) + i]Uh(x)$. Thus $G(x)Uk(x) = G(x)[F(x) + i]Uh(x) = Uh(x)$, so $h = (T+i)^{-1}k$, whereby $h \in D_T$.

To finish the proof it must be established that F is real valued a.e. Observe that the operator in $L_2(X, \mu)$ defined by multiplication by F is self-adjoint since by (ii) it is "unitarily equivalent" to T. Hence for all

χ_M, M a measurable subset of X, $(\chi_M \mid F\chi_M)$ is real. However, if $\operatorname{Im} F > 0$ on a set of positive measure, then there exists a bounded set B in the plane so that $M = F^{-1}(B)$ has nonzero measure. Clearly $F\chi_M$ is in $L_2(X, \mu)$ since B is bounded and $\operatorname{Im}(\chi_M \mid F\chi_M) > 0$. This contradiction shows that $\operatorname{Im} F = 0$ a.e. ∎

Using the foregoing theorem we can prove the following result.

Theorem A.8. *Functional Calculus Form of the Spectral Theorem.* Let T be a self-adjoint operator in H. There is a unique map $\hat{\varphi}$ from the class of bounded Borel measurable functions on R into $L(H, H)$ so that

(i) $\hat{\varphi}(\alpha f + \beta g) = \alpha\hat{\varphi}(f) + \beta\hat{\varphi}(g)$,
$\hat{\varphi}(fg) = \hat{\varphi}(f)\hat{\varphi}(g)$,
$\hat{\varphi}(\bar{f}) = (\hat{\varphi}(f))^*$, $\hat{\varphi}(1) = I$.

(ii) $\| \hat{\varphi}(f) \| \leq K \| f \|_\infty$ for some $K > 0$.

(iii) If (f_n) is a sequence of bounded Borel functions converging pointwise to the identity function on R and $| f_n(x) | \leq | x |$ for all x and n, then for any $h \in D_T$, $\lim_n \hat{\varphi}(f_n)h = Th$.

(iv) If (f_n) converges pointwise to g and $(\| f_n \|_\infty)_N$ is bounded, then $\hat{\varphi}(f_n) \to \hat{\varphi}(g)$ strongly.

In addition

(v) If $Th = \lambda h$, $\hat{\varphi}(g)h = g(\lambda)h$.

(vi) If $h \geq 0$, then $\hat{\varphi}(h) \geq 0$. ∎

Proof. Define $\hat{\varphi}$ by

$$\hat{\varphi}(g) = U^{-1}\tau_{g(F)}U,$$

where F and U are as in the previous theorem and $\tau_{g(F)}: L_2(X, \mu) \to L_2(X, \mu)$ is given by $\tau_{g(F)}\psi = g(F(\cdot))\psi(\cdot)$. Using the previous theorem, the verification that $\hat{\varphi}$ satisfies the conditions of the theorem is routine but arduous. We illustrate by verifying $\hat{\varphi}(\bar{f}) = (\hat{\varphi}(f))^*$ and condition (iii).

First we show $\hat{\varphi}(\bar{f}) = (\hat{\varphi}(f))^*$. For any h and k in H we have

$$\begin{aligned}(\hat{\varphi}(f)h \mid k) &= (U^{-1}f[F(\cdot)]U(h)(\cdot) \mid k)\\ &= (U^{-1}f[F(\cdot)]U(h)(\cdot) \mid U^{-1}[U(k)(\cdot)])\\ &= (f[F(\cdot)]U(h)(\cdot) \mid U(k)(\cdot))\\ &= \int f[F(\cdot)]U(h)(\cdot)\overline{U(k)(\cdot)}\, d\mu(\cdot).\end{aligned}$$

Similarly

$$\begin{aligned}(h \mid \hat{\varphi}(\bar{f})k) &= (U^{-1}[U(h)(\cdot)] \mid U^{-1}\bar{f}[F(\cdot)]Uk(\cdot)) \\ &= \int U(h)(\cdot)\overline{\bar{f}[F(\cdot)]U(k)(\cdot)}\, d\mu(\cdot) \\ &= \int f[F(\cdot)U(h)(\cdot)\overline{U(k)(\cdot)}\, d\mu(\cdot),\end{aligned}$$

whereby $(\hat{\varphi}(f))^* = \hat{\varphi}(\bar{f})$.

Next we verify condition (iii). First observe that since $|f_n(x)| \leq |x|$ and $\lim_{n\to\infty} f_n(x) = x$, for any h in D_T

$$|f_n(F(\cdot))U(h)(\cdot) - F(\cdot)U(h)(\cdot)|^2 \to 0 \tag{A.31}$$

and

$$|f_n(F(\cdot))U(h)(\cdot) - F(\cdot)U(h)(\cdot)|^2 \leq 4\,|F(\cdot)U(h)(\cdot)|^2 \tag{A.32}$$

where the right side of equation (A.32) is integrable by (i) of the previous theorem. Hence for any h in D_T

$$\int |f_n(F(\cdot))U(h)(\cdot) - F(\cdot)U(h)(\cdot)|^2\, d\mu \to 0 \tag{A.33}$$

by the Dominated Convergence Theorem. However, for any h in D_T

$$\begin{aligned}&\| U^{-1}[f_n(F(\cdot))]U(h)(\cdot) - Th \|_H \\ &\quad = \| U^{-1}[f_n(F(\cdot))]U(h)(\cdot) - U^{-1}UTU^{-1}U(h) \|_H \\ &\quad \overset{(a)}{=} \| U^{-1}[f_n(F(\cdot))]U(h)(\cdot) - U^{-1}F(\cdot)U(h)(\cdot) \|_H \\ &\quad \overset{(b)}{=} \| f_n(F(\cdot))U(h)(\cdot) - F(\cdot)U(h)(\cdot) \|_{L_2(X,\mu)},\end{aligned}$$

where we have used (ii) of the previous theorem at (a) and the fact that U is unitary at (b). Since the final expression equals the expression in (A.33), $\hat{\varphi}(f_n)h \to Th$.

The uniqueness of $\hat{\varphi}$ must yet be established. The proof is not trivial and we need to do some preliminary work first. This we proceed to do.

Observe that corresponding to each mapping ψ from the class of bounded Borel functions on R into $L(H, H)$ that satisfies (i)–(iv) of Theorem A.8 there is a normalized spectral measure E on $B(R)$ given by $E(M) = \psi(\chi_M)$. Particularly E satisfies

(1) $E(M)$ is a projection for each M in $B(R)$,

(2) $E(\varnothing) = 0$ and $E(R) = I$,

(3) If $M = \bigcup_{i=1}^{\infty} M_i$ where $M_i \cap M_j = \varnothing$ for $i \neq j$, then

$$E(M) = \lim_{n\to\infty} \sum_{i=1}^{n} E(M_i) \text{ (strong sense),}$$

and

(4) $E(M \cap N) = E(M)E(N)$.

The spectral measure which corresponds to $\hat{\varphi}$ is denoted by P.

By Theorem A.2, if E is *any* normalized spectral measure on $B(R)$ (in particular E could correspond to ψ or be P), then for each bounded complex-valued measurable function f on R, there exists a unique bounded operator T_f such that

$$(T_f h \mid h) = \int_{-\infty}^{\infty} f(\lambda)\, d\mu_h(\lambda) \qquad \text{for all } h \text{ in } H, \tag{A.34}$$

where μ_h is the finite measure $\mu_h(M) = \big(E(M)h \mid h\big)$. T_f is denoted by $\int f\,dE$. Observe that if E is P, then for any bounded Borel function f on R, $\int f\,dP = \hat{\varphi}(f)$. This is readily verified for characteristic functions χ_M for M in $B(R)$ since

$$\big(\hat{\varphi}(\chi_M)h \mid h\big) = \big(P(M)h \mid h\big) = \mu_h(M) = \int \chi_M\, d\mu_h.$$

From this it follows for simple functions, and from (iv) of the theorem, it follows for nonnegative bounded functions. By additivity it is true for all bounded measurable functions.

If E is any normalized spectral measure on $B(R)$, we can make equation (A.34) be valid for arbitrary measurable functions by the following procedure. If g is any measurable function ($|g|$ finite), then the set

$$D_g = \left\{h \,\middle|\, \int_{-\infty}^{\infty} |g|^2\, d\mu_h < \infty\right\}$$

is dense in H. This follows from the fact that for each h in H,

$$h = \lim_{n\to\infty} \sum_{i=1}^{n} E(S_i)h,$$

where $S_i = \{\lambda \in R: i - 1 \leq |g(\lambda)|^2 < i\}$ for $i = 1, 2, 3, \ldots$ and the fact that for each n, $E\big(\bigcup_{i=1}^{n} S_i\big)h$ is in D_g. Therefore if g is a nonnegative measurable function, defining $T_g h$ for each h in D_g by

$$T_g h = \lim_{n\to\infty} T_{g\wedge n\cdot\chi_{[-n,n]}} h,$$

where $T_{g\wedge n\cdot\chi_{[-n,n]}}$ is given by equation (A.34), we define a linear operator T_g on a dense linear subspace of H. By the Monotone Convergence Theorem T_g satisfies for each h in D_g the equation

$$(T_g h \mid h) = \int g \, d\mu_h. \tag{A.35}$$

Now if g is measurable and $| g |$ finite we write g as $g_1 - g_2 + i(g_3 - g_4)$, where each g_i is nonnegative, measurable, and integrable with respect to μ_h for each h in D_g. On D_g we define T_g as the linear operator

$$T_g = T_{g_1} - T_{g_2} + i(T_{g_3} - T_{g_4}).$$

By virtue of the fact that each T_{g_i} satisfies equation (A.35), T_g is easily seen to be well-defined and to satisfy equation (A.35).

Equation (A.35) for each h in D_g actually determines T_g uniquely. To see this, we define for each h and k in D_g the sum $\int g \, d\mu_{h,k}$ of the integrals as in Definition A.2. By a straightforward calculation, it is easy to check that

$$4\int g \, d\mu_{h,k} \equiv \int g \, d\mu_{h+k} - \int g \, d\mu_{h-k} + i\int g \, d\mu_{h+ik} - i\int g \, d\mu_{h-ik} = 4(Sh \mid k) \tag{A.36}$$

for any operator S satisfying equation (A.35) for all h in D_g. Hence for any h and k in D_g, $(T_g h \mid k) = (Sh \mid k)$ and since $\overline{D_g} = H$, $T_g h = Sh$ for any h in H.

As before, we write $T_g = \int g \, dE$ if T_g is the unique linear operator on D_g satisfying equation (A.35).

Observe that if g is a real-valued function, then T_g is a symmetric operator on D_g. Indeed, writing g as $g_1 - g_2$, where g_1 and g_2 are nonnegative functions, for all h and k in D_g,

$$\begin{aligned}
(T_g h \mid k) &= (T_{g_1} h \mid k) - (T_{g_2} h \mid k) \\
&= \lim_{n\to\infty} (T_{g_1\wedge n\cdot\chi_{[-n,n]}} h \mid k) - \lim_{n\to\infty} (T_{g_2\wedge n\cdot\chi_{[-n,n]}} h \mid k) \\
&= \lim_{n\to\infty} (h \mid T_{g_1\wedge n\cdot\chi_{[-n,n]}} k) - \lim_{n\to\infty} (h \mid T_{g_2\wedge n\cdot\chi_{[-n,n]}} k) \\
&= (h \mid T_{g_1} k) - (h \mid T_{g_2} k) \\
&= (h \mid T_g k).
\end{aligned}$$

If T is a given self-adjoint operator in H and P is the normalized spectral measure corresponding to T via $P(M) = \hat{\varphi}(\chi_M)$, the question arises whether $T = \int \lambda \, dP$. An affirmative answer is easily obtained from (iii) of Theorem A.8. To show this we must first establish that D_T the domain of T is actually $\{h \mid \int_{-\infty}^{\infty} |\lambda|^2 \, d\mu_h < \infty\}$, which we denote by D_λ.

First we show $D_T \subset D_\lambda$. Recall that if f_n is the function $\lambda \cdot \chi_{[-n,n]}$ then

$$\hat{\varphi}(f_n) = \int f_n \, dP,$$

and

$$[\hat{\varphi}(f_n)]^2 = \hat{\varphi}(f_n^2) = \int f_n^2 \, dP.$$

Hence if $h \in D_T$, then

$$\begin{aligned}(Th \mid Th) &= \lim_n \, (\hat{\varphi}(f_n)h \mid \hat{\varphi}(f_n)h) \\ &= \lim_n \, (\hat{\varphi}(f_n)^*\hat{\varphi}(f_n)h \mid h) \\ &= \lim \, ((\hat{\varphi}(f_n))^2 h \mid h) \\ &= \lim_n \int \lambda^2 \chi_{[-n,n]} \, d\mu_h \\ &= \int \lambda^2 \, d\mu_h,\end{aligned}$$

which implies $\int \lambda^2 \, d\mu_h < \infty$ and $h \in D_\lambda$.

Now if $T_1 = \int \lambda \, dP$, then T_1 is symmetric with domain the set D_λ. Also for any h in D_T we have by (iii) of Theorem A.8,

$$\begin{aligned}(Th \mid h) &= \lim_n \, (\hat{\varphi}(f_n)h \mid h) = \lim_n \left[\left(\int f_n \, dP\right)h \mid h\right] \\ &= \lim_n \int \lambda \cdot \chi_{[-n,n]} \, d\mu_h = \int \lambda \, d\mu_h = (T_1 h \mid h).\end{aligned}$$

Hence $T \subset T_1$. This means $T_1^* \subset T^* = T$ and since $T_1 \subset T_1^*$, $T_1 \subset T$. Consequently $T = T_1$ and $D_T = D_\lambda$. More importantly, $T = \int \lambda \, dP$.

In view of the rather lengthy discussion above, it should be clear that to prove $\hat{\varphi}$ is unique it suffices to show that when E is any normalized spectral measure such that

$$T = \int \lambda \, dP = \int \lambda \, dE \tag{A.37}$$

then $P = E$. Indeed, if P corresponds to $\hat{\varphi}$ and E corresponds to ψ, then equation (A.37) is true and if $E = P$ then $\hat{\varphi}$ and ψ agree on characteristic functions, simple functions, and then all bounded measurable functions. Proof of the uniqueness of $\hat{\varphi}$ in this manner will also enable us to quickly prove the final formulation of the spectral theorem—Theorem A.9 below.

We seek to show then that if E is any normalized spectral measure such that

$$T = \int \lambda \, dP = \int \lambda \, dE$$

then $E = P$. First we need some lemmas.

Lemma A.7. If E is a normalized spectral measure and A is a bounded self-adjoint operator such that

$$A = \int \lambda \, dE,$$

then E is compact. ∎

Proof. Suppose $\sigma(A) \subset (m, M)$. Let h be any vector in H and let $k = E(-\infty, m)h$. Then

$$E(-\infty, \lambda)k = \begin{cases} E(-\infty, \lambda)k & \text{if } \lambda < m \\ k & \text{if } \lambda \geq m. \end{cases}$$

Hence

$$(Ak \mid k) = \int_{-\infty}^{\infty} \lambda \, d\mu_k, \qquad \text{where } \mu_k(M) = (E(M)k \mid k)$$

$$= \int_{-\infty}^{m} \lambda \, d\mu_k \leq m \, \| E(-\infty, m)k \|^2 = m \, \| k \|^2.$$

However if $k \neq 0$,

$$(Ak \mid k) \geq \inf_{\|h\|=1} (Ah \mid h)(k \mid k) > m(k \mid k)$$

so that $k = 0$. Hence $E(K) = 0$ for all $K \subset (-\infty, m)$. Similarly $E(K) = 0$ if $K \subset (M, \infty)$. Hence E is compact. ∎

Lemma A.8. Let T be an operator in H and Q be a projection. If $QT \subset TQ$ (meaning $D_T = D_{QT} \subset D_{TQ} \equiv \{h \in H\colon\ Qh \in D_T\}$ and $QTh = TQh$ for each h in D_T), then $QT = TQ$ (meaning $D_{QT} = D_{TQ}$). ∎

Proof. Since $QT \subset TQ$, then $QTQ \subset TQ^2 = TQ$. However, since $D_{QTQ} = D_{TQ}$, $QTQ = TQ$. Inasmuch as $(I - Q)T \subset T(I - Q)$ also, it follows in the same way that $(I - Q)T(I - Q) = T(I - Q)$. Moreover

$$T = QT + (I - Q)T \subset TQ + T(I - Q) = T$$

since if $h \in D_{TQ}$ and $h \in D_{T(I-Q)}$, then $h \in D_T$ as $h = Qh + (I - Q)h$. Since the extremes of this inequality are equal,

$$QT + (I - Q)T = TQ + T(I - Q).$$

Applying Q to both sides, one obtains

$$QT + Q(I - Q)T = QTQ + QT(I - Q)$$

or

$$QT = TQ + Q(I - Q)T(I - Q) = TQ. \qquad \blacksquare$$

Now for any normalized spectral measure E such that

$$T = \int \lambda \, dE \tag{A.38}$$

we have $ET = TE$, that is $E(M)T = TE(M)$ for all M in $B(R)$. To see this, first observe that for any M, $E(M)$ maps D_T into D_T. For if $k \in D_T$, then

$$\int_{-\infty}^{\infty} | \lambda |^2 \, d\mu_{E(M)k} \leq \int | \lambda |^2 \, d\mu_k < \infty,$$

where $\mu_{E(M)k}(N) = \big(E(N)E(M)k \mid k\big) \leq \big(E(N)k \mid k\big) = \mu_k(N)$ for all N in $B(R)$. Hence $E(M)k \in D_T$ by definition of D_T. Secondly, for any k in D_T, we have

$$\begin{aligned} (Tk \mid k) &= \int \lambda \, d\mu_k \\ &= \sum_{n \in Z} \int \lambda \, \chi_{[n-1,n)} \, d\mu_k \\ &= \lim_{n \to \infty} (T_0 k + T_1 k + T_{-1} k + \cdots + T_n k + T_{-n} k \mid k), \end{aligned} \tag{A.39}$$

where $T_n = \int \lambda \chi_{[n-1,n]} \, dE$, a bounded self-adjoint operator on H. Now

$$\| T_n k \|^2 = (T_n{}^2 k \mid k) = \int | \lambda |^2 \chi_{[n-1,n)} \, d\mu_k$$

so

$$\sum_{n \in Z} \| T_n k \|^2 = \int | \lambda |^2 \, d\mu_k < \infty$$

since $k \in D_T$. Hence $\sum T_n k$ converges in H and from equation (A.39), $Tk = \sum T_n k$. Now $E(M)T_n = T_n E(M)$ since

$$E(M)T_n = \int \chi_E \cdot \lambda \cdot \chi_{[n-1,n)}\, dE = T_n E(M).$$

Since $E(M)$ is continuous, for $k \in D_T$,

$$TE(M)k = \sum_{n \in Z} T_n E(M)k = \sum_{n \in Z} E(M)T_n k = E(M)\bigl(\sum_n T_n k\bigr) = E(M)Tk.$$

Now apply Lemma A.8.

Lemma A.9. Let P be the spectral measure corresponding to the self-adjoint operator T in H via $\hat{\varphi}$. Let A be any operator in $L(H, H)$ such that $AT = TA$. Then for any integer n, $AH_n \subset H_n$, where $H_n = P_n(H)$ and $P_n = \int \chi_{[n-1,n]}\, dP$. ∎

Proof. By definition, for any M in $B(R)$ and $f \in L_2(X, \mu)$,

$$[UP(M)U^{-1}]f(\cdot) = (UU^{-1}\tau_{\chi_M(F)}UU^{-1})f(\cdot) = \chi_M(F(\cdot))f(\cdot),$$

where F, U, and μ are given in Theorem A.7. Let R_M be the range of $UP(M)U^{-1}$ in $L_2(X, \mu)$. First let $M = [-1, 1] \subset R$. Then

$$F^{-1}(M) = \{x \in X\colon -1 \leq F(x) \leq 1\}.$$

Clearly $R_M = \{f \in L_2(X, \mu)\colon f(x) = 0 \text{ a.e. if } x \notin F^{-1}(M)\}$. It is easy to see that also

$$R_M = \{f \in L_2(\mu)\colon \| [F(\cdot)]^n f(\cdot) \|_2 \text{ is bounded for } 1 \leq n < \infty\}.$$

[If $f(x) = 0$ for $x \notin F^{-1}(M)$, then

$$\int | [F(x)]^n f(x) |^2\, d\mu \leq \int | f(x) |^2\, d\mu.$$

Conversely, suppose $g \in L_2$, $\| F^n g \|_2$ is bounded, and $g(x) \neq 0$ for $x \in B$ for some set B of positive measure on which $| F(x) | > 1$. Then

$$\int | F(x)^n g(x) |^2\, d\mu \to \infty \text{ as } n \to \infty.]$$

Now suppose $f \in R_M$. Then $\| F(\cdot)^n f(\cdot) \|_2$ is bounded and for

$f \in U(D_T)$,

$$\| F(\cdot)^n(UAU^{-1})[f(\cdot)] \|_2 = \| (UT^nU^{-1})(UAU^{-1})f(\cdot) \|_2$$
$$= \| UAU^{-1}UT^nU^{-1}f(\cdot) \|_2 \leq \| A \| \| F(\cdot)^n f(\cdot) \|_2,$$

which is bounded. Hence $UAU^{-1}(R_M) \subset R_M$ when $M = [-1, 1]$

If $M = [a, b] = \{\lambda \in R: |\lambda - \lambda_0| \leq r\}$, then

$$F^{-1}(M) = \left\{x \in X: \left|\frac{F - \lambda_0}{r}(x)\right| \leq 1\right\}.$$

Since A commutes with T, UAU^{-1} will also commute with multiplication by $(F - \lambda_0)/r$; again in this case R_M will be invariant under UAU^{-1}. If now $M = [n - 1, n)$, then $M = \bigcup_{k=1}^{\infty} M_k$ when (M_k) is an increasing sequence of closed intervals. Since $R_M = \{f \in L_2(X, \mu): f(x) = 0 \text{ a.e. if } x \notin F^{-1}(M)\}$ clearly $R_{M_1} \subset R_{M_2} \subset \cdots \subset R_M$ and by the Dominated Convergence Theorem, $\overline{[\bigcup_{k=1}^{\infty} R_{M_k}]} = R_M$. For any k and f in R_{M_k}, $UAU^{-1}f \in R_{M_k} \subset R_M$. Since R_M is closed, $UAU^{-1}(R_M) \subset R_M$ and R_M is invariant under UAU^{-1}. This means

$$UAU^{-1}[UP_nU^{-1}(L_2)] \subset UP_nU^{-1}(L_2)$$

or

$$AP_n(H) \subset P_n(H). \qquad \blacksquare$$

For each integer n, let us continue to let $P_n = \int \chi_{[n-1,n]}\, dP$, that is $P_n = P([n - 1, n))$, and let $H_n = P_n(H)$. Since

$$I = P(R) = \lim_{n\to\infty} [P_0 + P_1 + P_{-1} + \cdots + P_n + P_{-n}]$$

in the strong sense, for each h in H, $h = \sum_{n\in z} h_n$, where $h_n = P_n(h)$. This means $H = \oplus H_n$, the direct sum of the orthogonal family of subspaces H_n.

As shown above, if E is a normalized spectral measure satisfying equation (A.38), then $TE = ET$. From Lemma A.9, $E(M)(H_n) \subset H_n$ for all $M \in B(R)$ and each n. Hence each H_n is invariant under $P(M)$ and $E(M)$, $M \in B(R)$. Thereby we can define $\hat{E}_n$ and $\hat{P}_n$ by

$$\hat{E}_n(M) = E(M)|_{H_n} \quad \text{and} \quad \hat{P}_n(M) = P(M)|_{H_n}$$

{Note that for each M in $B(R)$, $\hat{P}_n(M) = P([n - 1, n) \cap M)$.} Since

$$T = \int \lambda\, dE,$$

we have for $h \in H_n \cap D_T$

$$(T|_{H_n} h \mid h) = \int \lambda \, dv_h(\lambda),$$

where $v_h(M) = (E(M)h \mid h) = (\hat{E}_n(M)h \mid h)$, so that $T|_{H_n} = \int \lambda \, d\hat{E}_n$. Similarly $T|_{H_n} = \int \lambda \, d\hat{P}_n$.

Since $\hat{P}_n$ is a compact normalized spectral measure, the equality $T|_{H_n} = \int \lambda \, d\hat{P}_n$ also tells us that $T|_{H_n}$ is a bounded self-adjoint operator defined on H_n. This follows from the one-to-one correspondence established between compact spectral measures and bounded self-adjoint operators established in Theorem A.4. Since

$$T|_{H_n} = \int \lambda \, d\hat{P}_n = \int \lambda \, d\hat{E}_n,$$

$\hat{P}_n$ and $\hat{E}_n$ are both compact normalized spectral measures (Lemma A.7) corresponding to $T|_{H_n}$. As there can be only one such measure,

$$E|_{H_n} = \hat{E}_n = \hat{P}_n = P|_{H_n}.$$

Since $H = \bigoplus H_n$ and for each M, $E(M)$ and $P(M)$ are bounded operators, if $h = \sum_{i \in Z} h_i$,

$$\begin{aligned} E(M)h &= \lim_{n\to\infty} E(M)\left(\sum_{i=-n}^{n} h_i\right) \\ &= \lim_{n\to\infty} \sum_{i=-n}^{n} E(M)h_i = \lim_{n\to\infty} \sum_{i=-n}^{n} \hat{E}_i(M)h_i \\ &= \lim_{n\to\infty} \sum_{i=-n}^{n} \hat{P}_i(M)h_i = P(M)h. \end{aligned}$$

Hence $E = P$.

We have at last completed the proof of Theorem A.8. ∎

Our next result summarizes our preceding discussion and proof of the uniqueness of $\hat{\varphi}$ in a nutshell.

Theorem A.9. *Spectral Measure Version of the Spectral Theorem.* There is a one-to-one correspondence between self-adjoint operators T in H and normalized spectral measures P on $B(R)$. The correspondence is given by $T = \int \lambda \, dP$. Moreover for each real-valued measurable function f [with respect to $B(R)$] there is a unique self-adjoint operator $f(T)$ given by

$$f(T) = \int f \, dP,$$

where $T = \int \lambda \, dP$. If f is bounded, $f(T) = \hat{\varphi}(T)$, where $\hat{\varphi}$ is given by Theorem A.8. ∎

The remaining portion of this chapter is devoted to giving some applications of the Spectral Theorem.

Knowledge of the spectral measure corresponding to an operator T can be most valuable in obtaining complete knowledge of the operator T in regard to determining its domain, the value of inner product $(Th \mid k)$ for h and k in H, the spectrum and eigenspaces of T, and operator functions $f(T)$ of T for f measurable. In Theorem A.10 below we show how the spectrum of T is related to the spectral measure corresponding to T.

A few crucial observations should be made regarding the Spectral Theorem. First, for any measurable (real or complex valued) function f on R, the unique operator $f(T)$ satisfying equation (A.35) is defined. Its domain is the set $\{h \mid \int_{-\infty}^{\infty} |f|^2 \, d\mu_h < \infty\}$, dense in H. Secondly, for any h in $D_{f(T)}$

$$\| f(T)h \|^2 = \int_{-\infty}^{\infty} | f(\lambda) |^2 \, d\mu_h(\lambda). \tag{A.40}$$

Equation (A.40) is easily verified for the case when f is a nonnegative measurable function. From the discussion preceding equation (A.35), we have

$$\begin{aligned} \| f(T)h \|^2 &= \lim_{n\to\infty} \| f \wedge n \cdot \chi_{[-n,n]}(T)h \|^2 \\ &= \lim_{n\to\infty} ([f \wedge n \cdot \chi_{[-n,n]}(T)]^2 h \mid h) \\ &= \lim_{n\to\infty} \int_{-\infty}^{\infty} [f \wedge n \cdot \chi_{[-n,n]}(\lambda)]^2 \, d\mu_h(\lambda) \\ &= \int_{-\infty}^{\infty} f^2 \, d\mu_h . \end{aligned}$$

In case f is any measurable function, $f = f_1 - f_2 + i(g_3 - g_4)$ and $|f|^2 = f_1^2 + f_2^2 + g_3^2 + g_4^2$. In this situation equation (A.40) can easily be seen to hold.

Recall from Problem 6.5.8 in Chapter 6 that there is a one-to-one correspondence between spectral functions on R and normalized spectral measures on $B(R)$. It is analogous to the correspondence between Borel measures on R and distribution functions. Given a spectral measure P on $B(R)$, the spectral function E on R is given by $E(x) = P(-\infty, x]$. Theorem A.9 thus implies a one-to-one correspondence between self-adjoint operators T in H and spectral functions E on R. If f is a measurable function, then for any h in H

$$\int f(\lambda)\, d\mu_h(\lambda) = \int f(\lambda)\, d(E(\lambda)h \mid h),$$

whichs accords with Definition 3.10 in Chapter 3.

The applications given below are more easily stated in terms of spectral functions than spectral measures.

Using equation (A.40), the following theorem relating the spectrum of a self-adjoint operator T to properties of its spectral function is obtained.

Theorem A.10. Let T be a self-adjoint operator in H and let E be the spectral function on R corresponding to T. Then

(i) the spectrum $\sigma(T)$ is a subset of the real numbers;

(ii) λ_0 is an eigenvalue of T if and only if

$$E(\lambda_0) \neq E(\lambda_0-) \equiv \lim_{\substack{\lambda<\lambda_0 \\ \lambda\to\lambda_0}} E(\lambda);$$

moreover the eigenspace of H corresponding to λ_0 is $R_{E(\lambda_0)-E(\lambda_0-)}$;

(iii) λ_0 is in the continuous spectrum if and only if $E(\lambda_0) = E(\lambda_0-)$, but $E(\lambda_1) < E(\lambda_2)$ whenever $\lambda_1 < \lambda_0 < \lambda_2$; and

(iv) the residual spectrum of T is empty. ∎

Proof. (i) has been proved in Proposition A.12. Clearly

$$T - \lambda_0 I = \int (\lambda - \lambda_0)\, dE(\lambda)$$

and by (A.40) for h in D_T

$$\| (T - \lambda_0 I)h \|^2 = \int | \lambda - \lambda_0 |^2\, d(E(\lambda)h \mid h). \tag{A.41}$$

Hence $Th = \lambda_0 h$ if and only if $E(\lambda_0+) = E(\lambda_0) = E(\lambda)$ for all $\lambda \geq \lambda_0$ and $E(\lambda_0-) = E(\lambda)$ for all $\lambda < \lambda_0$. [Recall $E(\lambda)h$ is right continuous in λ for each h, $E(\lambda)h \to 0$ as $\lambda \to -\infty$, and $E(\lambda)h \to h$ as $\lambda \to \infty$.] In other words, $Th = \lambda_0 h$ if and only if $h = [E(\lambda_0) - E(\lambda_0-)]h$. This verifies statement (ii).

We next verify (iv). If the residual spectrum is not empty, then there is a real number λ_0 for which $(T - \lambda_0 I)^{-1}$ exists but its domain $R_{T-\lambda_0 I}$ is not dense. This means that there is a nonzero h_0 in H that is orthogonal to $R_{T-\lambda_0 I}$; that is, $((T-\lambda_0 I)h \mid h_0) = 0$ for all h in D_T. Hence

$$(Th \mid h_0) = (\lambda_0 h \mid h_0) = (h \mid \lambda_0 h_0)$$

so that $\lambda_0 h_0$ is in D_{T^*} and $T^* h_0 = \lambda_0 h_0$. Since T is self-adjoint, $Th_0 = \lambda_0 h_0$ and λ_0 is an eigenvalue of T. Since the point spectrum and residual spectrum are disjoint, this is a contradiction.

It remains to show (iii). From (ii) and (iv), if $E(\lambda_0) = E(\lambda_0-)$ then λ_0 is either in the resolvent of T or in the continuous spectrum. Now λ_0 is in the resolvent if and only if there exists a positive constant k such that

$$\| (T-\lambda_0 I)h \| \geq k \| h \| \qquad \text{for all } h \text{ in } D_T.$$

In other words from equation (A.41) it is necessary and sufficient that

$$\int_{-\infty}^{\infty} (\lambda-\lambda_0)^2 \, d(E(\lambda)h \mid h) \geq k^2 \| h \|^2. \tag{A.42}$$

Now if there exist λ_1 and λ_2 with $\lambda_1 < \lambda_0 < \lambda_2$ such that $\lambda_0 - \lambda_1 = \lambda_2 - \lambda_0 < k$ and $E(\lambda_1) \neq E(\lambda_2)$, then

$$\int_{-\infty}^{\infty} (\lambda-\lambda_0)^2 \, d(E(\lambda)h \mid h) < k^2 \int_{-\infty}^{\infty} d(E(\lambda)h \mid h) = k^2 \| h \|^2$$

with $h = [E(\lambda_2) - E(\lambda_1)]x$ for x in H. Since this contradicts inequality (A.42), if $E(\lambda_1) \neq E(\lambda_2)$, λ_0 is in the continuous spectrum of T. Conversely, if λ_0 is in the continuous spectrum, $E(\lambda_0-) = E(\lambda_0)$ by (ii). Moreover if there exists λ_1 and λ_2 with $\lambda_1 < \lambda_0 < \lambda_2$ and $E(\lambda_1) = E(\lambda_2)$ {implying $E(\lambda)$ is constant on $[\lambda_1, \lambda_2]$}, then the function $f(\lambda) = 1/(\lambda - \lambda_0)$ is bounded almost everywhere and $f(T) = (T - \lambda_0 I)^{-1}$ by Theorem A.8 is the bounded inverse defined on H of $T - \lambda_0 I$. This means λ_0 is in the resolvent of T, a contradiction. Hence for all λ_1 and λ_2 with $\lambda_1 < \lambda < \lambda_2$, $E(\lambda_1) \neq E(\lambda_2)$. ∎

The following example illustrates the use of the preceding theorem.

Example A.5. Let T be the multiplication operator in $L_2(-\infty, \infty)$ considered in Example A.1. It is routine to show T is symmetric. To show T is self-adjoint it must be shown that $D_{T^*} \subset D_T$. Suppose $g \in D_{T^*}$. Then for every $f \in D_T$

$$\int_{-\infty}^{\infty} x f(x)\overline{g(x)} \, dx = (Tf \mid g) = (f \mid T^* g) = \int_{-\infty}^{\infty} f(x)\overline{T^* g(x)} \, dx$$

whence

$$\int_{-\infty}^{\infty} f(x)[\overline{g(x)}x - \overline{T^* g(x)}] \, dx = 0.$$

Let $[a, b]$ be a finite interval and define h by $h(x) = [xg(x) - T^*g(x)]\chi_{[a,b]}$. Then $\int_{-\infty}^{\infty}[h(x)]^2\,dx = 0$ so that $h(x) = 0$ almost everywhere. Since $[a, b]$ is an arbitrary interval $xg(x) = T^*g(x)$ almost everywhere or $xg(x) = T^*g(x) \in L_2(-\infty, \infty)$. Hence $g \in D_T$.

Let E be the function on R into $L(L_2(-\infty, \infty), L_2(-\infty, \infty))$ given by

$$E(\lambda)g = g \cdot \chi_{(-\infty,\lambda]}.$$

E is easily checked to be a projection-valued, nondecreasing, right-continuous (in the strong sense) function with $E(\lambda) \to 0$ as $\lambda \to -\infty$ and $E(\lambda) \to I$ as $\lambda \to \infty$. Moreover, for any f in D_T and g in $L_2(-\infty, \infty)$,

$$\begin{aligned}(Tf \mid g) &= \int_{-\infty}^{\infty} xf(x)\overline{g(x)}\,dx \\ &= \int_{-\infty}^{\infty} xd\int_{-\infty}^{x} f(t)\overline{g(t)}\,dt \\ &= \int_{-\infty}^{\infty} xd\int_{-\infty}^{\infty} \overline{g(t)}E(x)[f(t)]\,dt \\ &= \int_{-\infty}^{\infty} x\,d(E(x)f \mid g).\end{aligned}$$

This means E is the unique spectral function such that $T = \int \lambda\,dE$.

What is the spectrum of T? Note that for all λ, $E(\lambda) = E(\lambda-)$ and $E(\lambda_1) < E(\lambda_2)$ if $\lambda_1 < \lambda_2$. Hence the continuous spectrum makes up the entire set of real numbers.

It is known that the position operator T of the preceding example is unitarily equivalent to the so-called momentum operator, the operator T_3 of Example A.2. This means there exists a unitary operator F on $L_2(-\infty, \infty)$ such that

$$F(D_{T_3}) = D_T \quad \text{and} \quad T_3 = F^{-1}TF.$$

The operator F is known as the Fourier–Plancherel operator. The proof of this unitary equivalence is not trivial and not given here.† However, the spectrum of T_3 can be easily analyzed to be exactly that of T by means of the following theorem.

Theorem A.11. If S and T are unitarily equivalent operators in a Hilbert space H, then the point spectrum, continuous spectrum, and residual spectrum of S are the same as that of T. ∎

† The proof is given in [27], page 135.

The proof is trivial and is omitted.

Example A.6. *Application of the Spectral Theorem in Solving the Schrödinger Equation.* An equation that occurs in quantum mechanics is the time-dependent Schrödinger equation given by

$$i\frac{du}{dt} = Au(t),$$

where $u(t)$ is an element of a Hilbert space H, A is a self-adjoint operator in H, and t is a time variable with $u(t) \in D_A$. An initial condition is $u(0) = u_0 \in D_A$. The derivative of u is given as the

$$\lim_{\Delta t \to 0} \frac{u(t+\Delta t) - u(t)}{\Delta t}$$

in the strong topology of H.

The Spectral Theorem enables us to solve the Schrödinger equation. Let e^{-itA} be the bounded operator on H given by

$$e^{-itA} = \int_{-\infty}^{\infty} e^{-it\lambda}\, dP(\lambda),$$

where $A = \int \lambda\, dP$. We wish to show that

$$\frac{d}{dt}(e^{-itA}h) = -iA(e^{-itA}h) \tag{A.43}$$

for every h in D_A.

To prove this, compute the following limit:

$$\begin{aligned}
&\lim_{\Delta t \to 0} \left\| \left[\frac{e^{-i(t+\Delta t)A} - e^{-itA}}{\Delta t} + ie^{-itA}A \right] h \right\|^2 \\
&\quad = \lim_{\Delta t \to 0} \int_{-\infty}^{\infty} \left| \frac{e^{-i(t+\Delta t)\lambda} - e^{-it\lambda}}{\Delta t} + ie^{-it\lambda}\lambda \right|^2 d\big(E(\lambda)h \mid h\big) \\
&\quad = \lim_{\Delta t \to 0} \int_{-\infty}^{\infty} \left| \frac{e^{-i(\Delta t)\lambda} - 1}{\Delta t} + i\lambda \right|^2 d\big(E(\lambda)h \mid h\big).
\end{aligned}$$

Letting $M =$ max of $| [e^{-i(\Delta t)} - 1]/\Delta t + i |^2$ for $\Delta t \in R$, the integrand above is bounded by $M\lambda^2$, which is integrable since $h \in D_A$. Using the Lebesgue Dominated Convergence Theorem, the limit can be taken inside to the integrand. Since the limit of the integrand is zero, the above limit is zero.

Hence for every h in D_A,

$$\frac{d}{dt}(e^{-itA}h) = -ie^{-itA}Ah. \tag{A.44}$$

Equation (A.43) follows from (A.44) since for h in D_A

$$e^{-itA}Ah = Ae^{-itA}h. \tag{A.45}$$

This follows from the fact that if h is in D_A, then $e^{-itA}h$ is in D_A since by equation (A.40)

$$\begin{aligned}\| E(M)e^{-itA}h \|^2 &= \int \chi_M \mid e^{-it\lambda} \mid^2 dE_h(\lambda) \\ &= \int \chi_M \, dE_h(\lambda) = \| E(M)h \|^2.\end{aligned}$$

Equation (A.45) then follows by an argument similar to that after Lemma A.4.

The solution $u(t) = e^{-itA}u_0$ of the Schrödinger equation is unique. To prove this, suppose $v(t)$ in D_A is a solution. Then for any k in H

$$\begin{aligned}\frac{d}{ds}\big(e^{-i(t-s)A}v(s) \mid k\big) &= \lim_{\Delta s \to 0} \frac{\big(e^{-i[t-(s+\Delta s)]A}v(s+\Delta s) \mid k\big) - \big(e^{-i(t-s)A}v(s) \mid k\big)}{\Delta s} \\ &= \lim_{\Delta s \to 0}\left(\frac{e^{-i[t-(s+\Delta s)]A} - e^{-i(t-s)A}}{\Delta s} v(s+\Delta s) \,\middle|\, k\right) \\ &\quad + \lim_{\Delta s \to 0}\left(e^{-i(t-s)A}\,\frac{v(s+\Delta s) - v(s)}{\Delta s} \,\middle|\, k\right) \\ &= \left(-\frac{d}{dt} e^{-i(t-s)A}v(s) \,\middle|\, k\right) + \left(e^{-i(t-s)A}\frac{dv}{ds} \,\middle|\, k\right) \\ &= \big(ie^{-i(t-s)A}Av(s) \mid k\big) + \big(e^{-i(t-s)A}[-iAv(s)] \mid k\big) = 0.\end{aligned}$$

Hence for all k in H

$$0 = \int_0^t \frac{d}{ds}\big(e^{-i(t-s)A}v(s) \mid k\big)\,ds = \big(e^{-i0A}v(t) \mid k\big) - \big(e^{-itA}v(0) \mid k\big),$$

and since $v(0) = u_0$ and $e^{-i0A} = I$ we have

$$v(t) = e^{-itA}u_0.$$

Uniqueness is thus proved.

Bibliography

1. Akhiezer, N. I., and Glazman, I. M., *Theory of Linear Operators in Hilbert Space*, Frederick Ungar Publishing Co., New York (1961, 1963).
2. Ash, R. B., *Measure, Integration, and Functional Analysis*, Academic Press, New York (1972).
3. Asplund, E., and Bubgart, L., *A First Course in Integration*, Holt, Rinehart, and Winston, New York (1966).
4. Bachman, G., *Elements of Abstract Harmonic Analysis*, Academic Press, New York (1964).
5. Bachman, G., and Narici, L., *Functional Analysis*, Academic Press, New York (1966).
6. Banach, S., *Théorie des Opérations Linéaires*, Monografje Matematyczne, Warsaw (1932).
7. Bauer, H., *Probability Theory and Elements of Measure Theory*, Holt, Rinehart, and Winston, New York (1972).
8. Berberian, S. K., *Introduction to Hilbert Space*, Oxford University Press, New York (1961).
9. Berberian, S. K., *Measure and Integration*, MacMillan, New York (1965).
10. Berberian, S. K., *Notes on Spectral Theory*, Van Nostrand, Princeton, New Jersey (1966).
11. Berberian, S. K., On the extension of Borel measures, *Prov. Amer. Math. Soc.* **16**, 415–418 (1965).
12. Bourbaki, N., *Topologie Générale*, Herman, Paris (1951).
13. Brown, A. L., and Page, A., *Elements of Functional Analysis*, Van Nostrand Reinhold Co., London (1970).
14. Carleson, L., On the convergence and growth of partial sums of Fourier series, *Acta Math.* **116**, 135–157 (1966).
15. Cohen, P. J., *Set Theory and the Continuum Hypothesis*. W. A. Benjamin, Inc., New York (1966).
16. Dunford, N., and Schwartz, J., *Linear Operators. Part I: General Theory*, Wiley (Interscience), New York (1958).
17. Dunford, N., and Schwartz, J., *Linear Operators. Part II*, Wiley (Interscience), New York (1964).

18. Epstein, B., *Linear Functional Analysis*, W. B. Saunders Co., Philadelphia (1970).
19. Fano, G., *Mathematical Methods of Quantum Mechanics*, McGraw-Hill, New York (1971).
20. Graves, L. M., *The Theory of Functions of Real Variables* (2nd ed.), McGraw-Hill Company, New York (1956).
21. Halmos, P. R., *Lectures on Ergodic Theory*, Chelsea, New York (1956).
22. Halmos, P. R., *A Hilbert Space Problem Book*, Van Nostrand Co., Princeton, New Jersey (1967).
23. Halmos, P. R., *Introduction to Hilbert Space and the Theory of Spectral Multiplicity*, Chelsea Publishing Co., New York (1951).
24. Halmos, P. R., *Measure Theory*, D. Van Nostrand, Princeton, New Jersey (1950).
25. Halmos, P. R., *Naive Set Theory*, Van Nostrand, Princeton, New Jersey (1960).
26. Hardy, G. H., Weierstrass's non-differentiable function, *Trans. Amer. Math. Soc.* **17**, 301–325 (1916).
27. Helmberg, G., *Introduction to Spectral Theory in Hilbert Space*, North Holland, Amsterdam (1969).
28. Hewitt, E., and Stromberg, K., *Real and Abstract Analysis*, Springer-Verlag, Berlin (1965).
29. Hunt, R. A., On the convergence of Fourier series, pp. 235–255 in *Orthogonal Expansions and Their Continuous Analogs*, Southern Illinois University Press, Carbondale, Illinois (1968).
30. Johnson, R. A., A compact non-metrizable space such that every closed subset is a G-delta, *Amer. Math. Monthly* **77**, 172–176 (1970).
31. Johnson, R. A., On product measures and Fubini's theorem in locally compact spaces, *Trans. Amer. Math. Soc.* **123**, 112–129 (1966).
32. Johnson, R. A., On the Lebesgue decomposition theorem, *Proc. Amer. Math. Soc.* **18**, 628–632 (1967).
33. Johnson, R. A., Some types of Borel measures, *Proc. Amer. Math. Soc.* **22**, 94–99 (1969).
34. Kelley, J. L., *General Topology*, Van Nostrand, Princeton, New Jersey (1955).
35. Levin, M., and Stiles, W., On the regularity of measures on locally compact spaces, *Proc. Amer. Math. Soc.* **16**, 201–206 (1972).
36. Liusternik, L. A., and Sobolev, V. J., *Elements of Functional Analysis*, Fredrick Ungar Publishing Company, New York (1961).
37. Lorch, E. R., On certain implications which characterize Hilbert space, *Ann. of Math.* **49**, 523–532 (1948).
38. Luther, N. Y., Lebesgue decomposition and weakly Borel measures, *Duke Math. J.*, **35**, 601–615 (1968).
39. McShane, E. J., *Integration*, Princeton University Press, Princeton (1944).
40. Monroe, M. E., *Measure and Integration* (2nd ed.), Addison-Wesley, Reading, Massachusetts (1971).
41. Naimark, M. A., *Normed Rings* (translated from Russian), P. Noordhoof, Ltd., Groningen, The Netherlands (1964).
42. Nelson, E., *Topics in Dynamics I: Flows*, Princeton University Press, Princeton, New Jersey (1969).
43. Naylor, A. W., and Sell, G. R., *Linear Operator Theory in Engineering and Science*, Holt, Rinehart and Winston, New York (1971).
44. Oxtoby, J. C., *Measure and Category*, Springer-Verlag, New York (1971).

45. Phillips, E., *An Introduction to Analysis and Integration Theory*, Intext Educational Pub., Scranton, Toronto, London (1971).
46. Prugovečki, E., *Quantum Mechanics in Hilbert Space*, Academic Press, New York (1971).
47. Reed, M., and Simon, B., *Functional Analysis*, Academic Press, New York (1972).
48. Riesz, F., and Sz-Nagy, B., *Functional Analysis*, Fredrick Ungar Publishing Co., New York (1955).
49. Roman, P., *Some Modern Mathematics for Physicists and Other Outsiders*, Vol. 2, Pergamon Press Inc., New York (1975).
50. Ross, K. A., and Stromberg, K., Baire sets and Baire measures, *Ark. Mat.* **6**, 151–160 (1965).
51. Royden, H. L., *Real Analysis* (2nd ed.), The MacMillan Co., New York (1968).
52. Rudin, W., *Real and Complex Analysis*, McGraw-Hill, New York (1966).
53. Sabharwal, C. L., and Alexiades, V., On the extension of Lebesgue measure, unpublished.
54. Saks, S., *Theory of the Integral*, Warszawa-Lwów (1937).
55. Segal, I. E., and Kunze, R. A., *Integrals and Operators*, McGraw-Hill, New York (1968).
56. Sehgal, V. M., A fixed point theorem for mappings with a contractive iterate, *Proc. Amer. Math. Soc.* **23**(3), 631–634 (1969).
57. Sehgal, V. M., On fixed and periodic points for a class of mappings, *J. London Math. Soc.* **5**(2) 571–576 (1972).
58. Serrin, J., and Varberg, D., A general chain rule for derivatives and the change of variables formula for the Lebesgue integral, *Amer. Math. Monthly* **76**, 514–520 (1969).
59. Stone, M. H., *Linear Transformations in Hilbert Space*, American Mathematical Society, Providence, Rhode Island (1964).
60. Sucheston, L., Banach Limits, *Amer. Math. Monthly*, **74** (3), 308–311 (1967).
61. Suppes, P., *Axiomatic Set Theory*, Van Nostrand, Princeton, New Jersey (1960).
62. Taylor, A. E., *Introduction to Functional Analysis*, John Wiley and Sons, Inc., New York (1958).
63. Titchmarsh, E. C., *The Theory of Functions* (2nd ed.), Oxford University Press, London and New York (1939).
64. Vaidyanathaswamy, R., *Set Topology*, Chelsea, New York (1960).
65. Varberg, D. E., On absolutely continuous functions, *Amer. Math. Monthly* **72**, 831–841 (1965).
66. Vulikh, B. Z., *Introduction to Functional Analysis for Scientists and Technologists*, Pergamon Press Ltd., Oxford (1963).
67. Whitley, R., An elementary proof of the Eberlein–Šmulian theorem, *Math. Ann.* **172**, 116–118 (1967).
68. Willard, S., *General Topology*, Addison-Wesley, Reading, Massachusetts (1970).
69. Yosida, K., *Functional Analysis* (4th ed.), Springer-Verlag, New York (1974).
70. Zaanen, A. C., *Linear Analysis*, North Holland, Amsterdam (1953).

Symbol and Notation Index

Set and element notations

Notation used in set operations and relations

Function notations

Measure theoretic notations

Topological notations

Function space notations

Normed space notations

Norm notations

Operator notations

Miscellaneous notation

Subject Index